nuclei in the cosmos

Carlos A Bertulani

Texas A&M University, USA

NEW JERSEY • LONDON • SINGAPORE • BEIJING • SHANGHAI • HONG KONG • TAIPEI • CHENNAI

Published by

World Scientific Publishing Co. Pte. Ltd.
5 Toh Tuck Link, Singapore 596224
USA office: 27 Warren Street, Suite 401-402, Hackensack, NJ 07601
UK office: 57 Shelton Street, Covent Garden, London WC2H 9HE

Library of Congress Cataloging-in-Publication Data
Bertulani, Carlos A.
Nuclei in the Cosmos / Carlos Bertulani.
p. c
Includes bibliographical references.
ISBN 978-9814417662 (hardcover : alk. paper)
1. Nuclear astrophysics. 2. Nucleosynthesis. I. Titl
QB464.B458 2013
523.01'97--dc23
2012049837

British Library Cataloguing-in-Publication Data
A catalogue record for this book is available from the British Library.

Typeset by Stallion Press
Email: enquiries@stallionpress.com

Printed in Singapore

Preface

This book has its origin in lecture notes of several nuclear astrophysics courses that the author has taught at universities worldwide, such as the Federal University of Rio de Janeiro, Michigan State University, University of Arizona, and the Texas A & M University-Commerce. The original notes were enlarged substantially and now contain an amount of material somewhat larger than usually taught in a one semester course on cosmology or nuclear astrophysics. The original text has been altered to bring it up to date. However, astrophysics is a rapidly changing field, in which new astronomical findings suddenly change the interest and emphasis of mainstream research and public curiosity.

The general idea of the book is to present the basic information to guide the student through the vast amount of information needed to understand nuclear/particle astrophysics and cosmology. Although there is some reference to observations, experiments, or measurements when necessary, there is no attempt to describe the equipment and methods of astronomy and experimental physics in a systematic and consistent way. The description of historic events in astronomy, nuclear, and particle physics are mentioned sporadically, but there is no commitment to giving a detailed account of these facts. Astrophysics is a very vast field, requiring knowledge of numerous branches of physics. The goal of this book is to gather this information in a coherent and concise way, a difficult task by any measure.

The adequate level for the complete understanding of this book corresponds to a student at the end of undergraduate study in physics, including, besides the basic physics, a course of modern physics and a first course of quantum mechanics. But, students of exact sciences, and of technology in general, can still benefit from a good part of the subjects presented in this book.

This book is dedicated to the memory of my dear son Daniel Bertulani, a proud member of the United States Air Force, who passed away on

September 11, 2012. Daniel will be remembered for being a caring person, who would volunteer his services and help people around him without asking anything in return. His companionship and charismatic smile were the best gifts that my wife Eliete, his brother Henrique, and I had for 28 years of our lives. He will be (very) missed by all who have had the pleasure of knowing him. We love you, Dani.

Carlos Bertulani
Commerce, Texas, January 2013

Contents

Chapter 1

Introduction

1.1. Nuclear Physics and Astrophysics

The evolution of the Universe is the object of study of cosmology and astrophysics. Nuclear astrophysics studies the synthesis of heavy nuclei starting from lighter ones under the temperature and pressure conditions existing in the stars. Nuclear physics studies the behavior of nuclei under normal conditions or in excited states, as well as the reactions among them. Chemistry studies the structure of atomic molecules and the reactions among them. Finally, biology studies the formation and development of large molecular agglomerates that compose living beings. In any of these sciences the objective is to understand complex structures starting from simpler structures and from the interactions among them.

Nuclear Astrophysics is the field concerning "the synthesis and evolution of atomic nuclei, by thermonuclear reactions, from the *Big Bang* to the present [Boy08,Ili07,RR88]. What is the origin of the matter of which we are made?". Our high entropy Universe, presumably resulting from the Big Bang, contains many more photons per particle of matter with mass, e.g., electrons, protons and neutrons. Because of the high entropy and the low density of matter (on stellar scales) at any given temperature as the Universe expanded, there was time to manufacture elements only up to helium. The major products of cosmic nucleosynthesis remained hydrogen and helium. Stars formed from this primordial matter and they used these elements as fuel to generate energy like a giant nuclear reactor. In the process, the stars could shine and manufacture higher atomic number elements like carbon, oxygen, calcium and iron of which we and our world are made. The heavy elements are either dredged up from the core of the star to the surface from which they are dispersed by stellar wind or directly ejected into the interstellar medium when a (massive) star explodes.

This stardust is the source of heavy elements for a new generation of stars and Sun-like systems.

The Sun is slowly burning hydrogen into helium. It is not exactly the same now as when it first started burning hydrogen in its core and will start to look noticeably different once it exhausts all the hydrogen that it can burn. In other words, nuclear reactions in the interiors of stars determine the evolution or the life-cycle of the stars, apart from providing them with internal power for heat and light and manufacturing all the heavier elements that the early Universe could not.

There is a correspondence between the evolutionary state of a star, its external appearance and internal core conditions and the nuclear fuel it burns — a sort of a mapping between the astronomer's Hertzsprung–Russell diagram and the nuclear physicist's chart of the nuclides [Ber07].

1.2. The Milky Way

The stars we can see with the naked eye all belong to our galaxy, the *Milky Way*, which is a spiral galaxy, about 30 kpc across and about 1 kpc thick. The kpc, *kiloparsec*, is the common astronomical unit of distance; 1 parsec is 3.1×10^{16} m, or 3.26 light years, *ly*. Thus light travels across our galaxy in about 100,000 y. The distance between the Earth and the Sun is called the *Astronomical Unit*, 1 AU $= 1.5 \times 10^{13}$ cm. The parsec is also defined as the distance away that an object must have in order to produce a parallax shift of 1 second of arc, which is 4.85×10^{-6} radians, i.e.,

$$1 \text{ pc} = \frac{1 \text{ AU}}{4.85 \times 10^{-6} \text{ rad}} = 3.09 \times 10^{18} \text{ cm} = 3.26 \text{ ly}. \tag{1.1}$$

Thus, the distance in parsecs is just one over the parallax shift in arcseconds.

The Milky Way contains some 200×10^9 stars, and interstellar dust and gas (~200 pc thick) which spreads out to a diameter of about 50 kpc (hot gas atoms, the *halo*). Our Sun is located at the outer edge of one of the spiral arms, about 8.5 kpc from the galactic center. The dust limits the sight towards the center to only a few kpc; without this dust the galaxy center would shine equally bright as our Sun. The stars in our galaxy move tangentially around its center with angular velocities increasing closer to the center, indicating the existence of a heavy central object.

The Milky Way (see Fig. 1.1) belongs to the *Local Group*, a cluster of some 20 galaxies which include the *Large Magellanic Cloud*, our nearest galaxy, 50 kpc away, and the *Andromeda galaxy*, 650 kpc away. The Local Group is part of the larger Virgo supercluster. The visible Universe contains

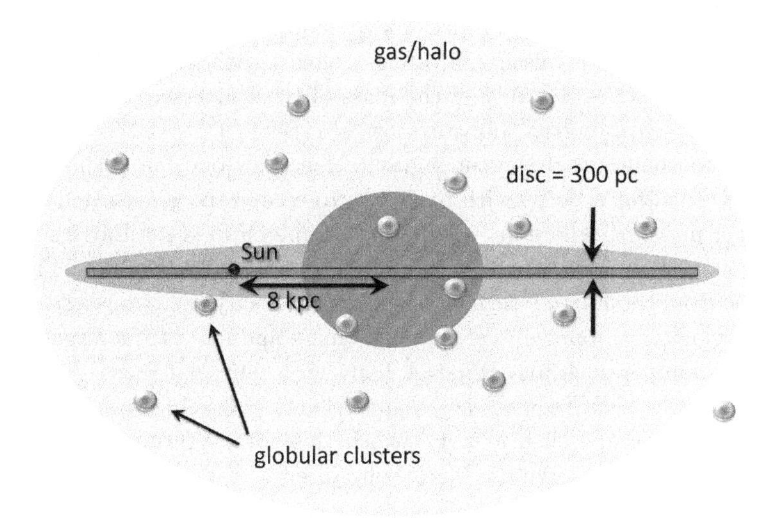

Figure 1.1. Structure of the Milky Way (not to scale).

some 10^{10} galaxies. The galaxies only fill a fraction of space, the rest appears void of matter.

In the 1930s Hubble discovered that galaxies on the whole are equally distributed in all directions of space as observed from the Earth [Hu29]. Thus space on a large scale seems to be isotropic. This idea of uniformity of the Universe is called the *cosmological principle*. This information has been deduced from celestial mechanics (motion of bodies according to Newton's fundamental laws) and from spectroscopic analysis of light and other kinds of radiation. It has been found that the mass of our Sun is 1.99×10^{30} kg $= 1$ solar mass, $M_{\odot}$. The mass of the Milky Way is larger than 2×10^{11} $M_{\odot}$. About 10% of the mass is interstellar gas, and 0.1% is dust (typically particles with diameter 0.01–0.1 μm). The interstellar gas density varies considerably in our galaxy; in our part of space it varies from about 10^9 (in dark clouds) to 10^5 atoms m^{-3} (on the average ~1 atom cm^{-3}). Though it contains mainly H and He, large and rather complex molecules containing H, C (up to C_{15} molecules), N and O (including amino acids) have also been discovered (for an extended discussion, see [Cla84,Cho01,TG04]).

1.3. Dark Matter

Astronomical models of the Universe indicate that it will expand forever if the observed galaxies alone account for the total mass of the Universe.

Almost 90% of the mass needed for a slowing down and ultimately contracting Universe is missing. Most cosmologists believe that the mass of the observed galaxies is <10% of the mass of the Universe, the main part consisting of "*dark matter*" (for a review, see [Ber10]).

From mechanics and Newton's gravitational law one can calculate the velocity needed for a body, with mass m_x, to escape the gravitational pull of a larger mass, m, where $m_x \ll m$. For example, if m is the Earth's mass (5.94×10^{24} kg), a rocket (mass m_x) must have a velocity of about 11 km/s to escape from the Earth's surface (the escape velocity, v_e). Conversely, for a given velocity, v_e, one can calculate the mass and size of the large body needed to hamper such an escape. A body with our solar mass, $M_\odot$, but a radius of only 3 km, requires an escape velocity $> 3 \times 10^8$ m/s. Thus not even light will escape such a body, which therefore is termed *black hole*. Though we cannot observe the black holes directly, some secondary effects can be observed.

Astronomical observations of star movements support the existence of black holes. For example, from movements of stars close to our galactic center, it is believed that a black hole is located at *Sagittarius* SgrA* in the center of the Milky Way, with a mass larger than 3×10^6 $M_\odot$. The radius of such a hole would be of the same size as that of our Sun. The density of matter in the hole would be several million times the density of our Sun (average value for the Sun is about 1400 kg/m^{-3}). Obviously matter cannot be in the same atomic state (i.e., nuclei surrounded by electrons) as we know on Earth. Instead we must assume that the electron shells are partly crushed; we refer to this as *degenerate matter*. For completely crushed atoms, matter will mainly consist of compact nuclei. For example, for calcium the nuclear density is $\sim 2.5 \times 10^{17}$ kg/m^3. For completely crushed nuclei, the matter will consist of individual nucleons, with even larger densities.

Even if black holes are a considerable portion of the missing mass, this will not be enough. A detailed analysis of the variation in luminosity (a factor of about 2.5) for some 10 million double stars in the Large Magellanic Cloud gives support for the existence of nearby "*gravitational micro-lenses*", which are believed to be unborn stars (so-called *brown dwarfs*) of sizes $\sim 10^{-1}$ $M_\odot$. When such a dark object passes the line of sight to a distant star it acts as a focusing lens for the light, thereby temporarily increasing that star's apparent luminosity. As these gravitational micro-lenses seem to be especially abundant in the halo of our galaxy (and presumably in halos of other galaxies), they — together with neutron stars and black holes — account for some of the missing matter. But they are not responsible for the bulk of the dark matter. Indeed, dark matter poses one of the major scientific challenges of our time.

1.4. Hertzsprung–Russell Diagram

Spectral analysis of the light received from astronomical objects have provided us with information of their (surface) temperature (from their continuous spectrum) and outer chemical composition (from identification of spectral line frequencies), while bolometric measurements have given their luminosity (energy flux density). In 1911 Hertzsprung and Russell discovered that if the luminosity and color (or temperature) of stars in different galaxies were compared with similar type of stars in the Milky Way, the stars become distributed according to a certain pattern [Rus14], the so-called *Zero Age Main Sequence* (ZAMS) of stars. The diagram shows a mass sequence of stars in the process of burning hydrogen into helium following the diagram beginning at the lower right side, going up along the main sequence into the red giants, then to the left and down, decreasing in size and luminosity to end as blue or white dwarfs.

1.4.1. *Luminosity*

Hertzsprung–Russell (HR) *diagrams*, like the one in Fig. 1.2, are valid for stars of about 0.7–70 $M_\odot$: from such diagrams conclusion can be drawn about the size (or mass) and relative age of the star, as it is assumed that stars of a given mass follow the same evolution as they age. In general, only the surface (photosphere) of a star is visible. The photosphere is a region of only a few 100 km deep for the Sun but many AU for a red giant. In general the luminosity of a *black body* is

$$L = 4\pi R^2 \sigma T^4, \tag{1.2}$$

where R is the star radius and σ is the Stefan–Boltzmann constant (see Chapter 4). This means that you can increase luminosity by either upping the radius or the temperature. Stars do both. On the other hand with the radius constant, the luminosity versus temperature in a log–log diagram is a straight line. The first attempts to plot the luminosity of stars against temperature were done by Hertzsprung and Russell, after whom such diagrams are named. Such diagrams are one of the major working tools in stellar astronomy. Both the luminosity and the temperature are not

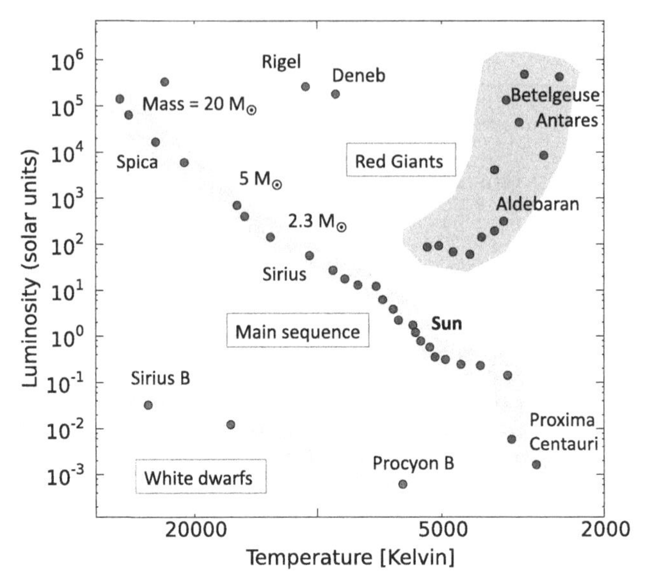

Figure 1.2. HR diagram showing surface temperature, for stars born on the main sequence as a function of their absolute luminosity relative to the solar luminosity $L_{\odot}$ and their temperature in Kelvin.

straightforward numbers. Of course, to obtain the luminosity of a star an absolute distance calibration is necessary.

From the *Stefan–Bolzmann law*, Eq. (1.2), we conclude that stars above the main sequence on the HR diagram (higher luminosity), with the same temperature as cooler main sequence stars, have greater surface areas (larger radii). Also, stars that have the same luminosity as dimmer main sequence stars, but are to the left of them (hotter) on the HR diagram, have smaller surface areas (smaller radii). Bright, cool stars are therefore very large. These enormous stars are called *red giants* and lie above the main sequence line. Similarly, stars that are very hot and yet still dim must have small surface areas. These small, hot stars are called *white dwarfs* and lie below the main sequence. They can have radii as small as the Earth, having temperatures around 10,000 K. From the Stefan–Boltzmann law we also notice that by holding R constant and plotting the luminosity as the temperature varies, we can generate a set of diagonal lines in the HR diagram. These constant radius lines would go from the upper left to the lower right (negative slope) of the diagram and pass by the three shaded regions in the graph of Fig. 1.2. The categories of stars included in these shaded areas are main sequence stars, red giants, and white dwarfs.

1.4.2. *Magnitudes*

The apparent luminosity, F, which we observe with our telescopes, is related to the *absolute luminosity*, L^*, i.e., the total energy flux in all directions from a star, by the relation

$$F = \frac{L^*}{4\pi d^2}, \tag{1.3}$$

where d is the distance from the star. The historical classification of stars into brightness classes is now usually replaced by their *relative (apparent) magnitude*, m^*. The *absolute magnitude*, $\mathcal{M}$, is a measure of the star's luminosity (or *brightness*). It is also defined so that a difference of 5 magnitudes corresponds to a factor of exactly 100 times in intensity. In this scale, the Sun has $\mathcal{M} = +5$. The apparent magnitude is defined so that if the star is at 10 parsecs (or 32.6 ly) distance from us, then its apparent magnitude is equal to its absolute magnitude. This leads to the relation between the magnitudes m_1 and m_2 for two stars with corresponding apparent luminosities F_1 and F_2,

$$m_2 - m_1 = 2.5 \log \left(\frac{F_1}{F_2} \right). \tag{1.4}$$

For example, the apparent visual magnitudes of some stars are: (a) $m = -1.5$ for Sirius (brightest star), (b) $m = -26.8$ for the Sun, and (c) $m = 25$ for the faintest object visible from Earth telescopes.

The apparent magnitude is easy to determine because one only needs to measure the apparent luminosity, with no consideration given to how distance is influencing the observation. Hence, it is not so useful because it mixes up the intrinsic luminosity of the star (which is related to its internal energy production) and the effect of distance (which has nothing to do with the intrinsic structure of the star). Thus, the absolute magnitude is more useful: if one knows the true distance to the star, then one can use the inverse square law to determine how its apparent brightness would change if it moved from its true position to a standard distance of 10 parsecs.

1.4.3. *Color index*

Colors of stars are even more complex to define. Star color indices were defined by using the response of photographic plates with band widths spanning the Ultraviolet, Blue and Visual (UBV) spectra. The color index is the Blue magnitude minus the Visual magnitude, where the magnitude is given by Eq. (1.4). Hence, hot stars are characterized by small, in fact negative, color index while cold stars have large color index. Astronomers also correlate the color index with the effective surface temperature of a star. The HR diagram is in fact a plot of the luminosity of a star or the *bolometric magnitude* (total energy emitted by a star) versus its surface temperature, or its color index.

In general, a spectral classification

$$\mathrm{O, B, A, F, G, K, M}$$

with 1–10 subgroups is used (Sun: G2), which is actually pretty well correlated to the temperature [JM53]. O stars are the hottest and the letter sequence indicates successively cooler stars up to the coolest M class. A useful mnemonic for remembering the spectral type letters is "*Oh Be A Fine Girl/Guy Kiss Me*". Informally, O stars are called "blue", B "blue–white", A stars "white", F stars "yellow–white", G stars "yellow", K stars "orange", and M stars "red", even though the actual star colors perceived by an observer may deviate from these colors depending on visual conditions and individual stars observed.

For bulk spectroscopy, often a so called *color index* (typically B–V) is used which comes from comparing intensities through different color filters in particular for surveying stellar clusters which are all at the same distance. Hot stars give off more blue light than red; cool stars give off more red light than blue. Colored filters are used to measure different wavelengths of light from stars. The magnitude of the star is measured first through a standardized B-band ("blue") filter. Then the star's magnitude is measured through a V-band ("visible", peaking in green) filter. The value of V is subtracted from B to get the *B–V color index*. Theorists in general prefer real temperatures. Normally temperatures are plotted in descending order (left blue–right red).

One way to classify the luminosity of a star is through its spectrum. But, instead of looking at which lines (wavelengths) are present, we look at the width of the lines. It turns out that more luminous stars have narrower spectral lines than less luminous ones. The spectral classification scheme of stars has been dilligently carried out by numerous astronomers (see, for example, Refs. [Gin01,Way02]).

1.4.4. *Mass–luminosity relation*

There is a correlation between a main sequence star's mass and its luminosity. This is because stars shine due to nuclear fusion reactions in their core. This will be discussed in detail in the following chapters. The more luminous the stars are, the more reactions are taking place in their cores. Massive stars live shorter lives than the common small stars because even though they have a larger amount of hydrogen for nuclear reactions, their rate of consuming fuel is very much greater. It is rather simple to estimate how long stars can continue consuming fuel, as the lifetime is equal to the amount of fuel/consumption rate. For the main sequence stars we find the brightest star to have the highest surface temperature and be of blue color. The main sequence stars spend most of their life burning hydrogen and have mass that is related to their luminosity: $L = const \times M^{\nu}$, with $\nu = 3.5$ to 4.0. In fact, for non-radiative pressure dominated stars, we obtain the star luminosity–mass relation

$$L \sim M^{3.5-4}. \tag{1.5}$$

This equation only applies to main sequence stars with masses $2M_{\odot} < M < 20M_{\odot}$ and does not apply to red giants or white dwarfs. For stars bigger than $20M_{\odot}$, one finds

$$L \sim M. \tag{1.6}$$

The equations above are also obtained empirically by determining the mass of stars in binary systems to which the distance is known. After enough stars are plotted, stars will form a line on a logarithmic plot and the slope of the line gives the proper value of the power of M.

If one assumes that a considerable mass fraction of each star is consumed in stellar evolution, then the lifetime of a star is given by $\tau \sim M/L$. From Eqs. (1.5) and (1.6), we get

$$\begin{aligned} \tau &\sim M^{-3} \quad \text{for } M < 20M_{\odot}, \\ \tau &\sim const \quad \text{for } M \gg 20M_{\odot}. \end{aligned} \tag{1.7}$$

Typically for the Sun 10^{10} years are expected, while B stars only live to about 10^7 years. Estimated lifetimes for other stars are listed in Table 1.1. Stars that have fewer elements heavier than helium in them compared to other stars will have slightly shorter lifetimes than those given in Table 1.1.

Some stars have times in their evolutionary track where they are not in equilibrium. These stars pulsate, expanding and contracting due to a competition between the thermal pressure in the star and its gravitational force. Stars called *Cepheid Variables* are an important type of star in this

Table 1.1. Representative lifetimes of stars as a function of their masses.

Mass (solar masses)	Time (years)	Spectral type
60	3 million	O3
30	11 million	O7
10	32 million	B4
3	370 million	A5
1.5	3 billion	F5
1	10 billion	G2 (Sun)
0.1	1000s billions	M7

state. They are used as distance indicators (*standard candles*) because the period of their pulsation varies in proportion with their luminosity. It was observed that the longer the period of variation of a Cepheid variable, the greater its luminosity [Lea08] and then one was able to correlate the brightnesses of Cepheids with those of known types of ordinary stars [Sh14], tying the relative distance scale to an absolute one. Thus, we can observe a Cepheid, note how long it takes for its brightness to vary and plot that information on an already established graph to find out its intrinsic luminosity. Comparing this true brightness (its "absolute magnitude") with its apparent brightness as seen in the sky allows us to calculate how far away it is, using the inverse-square law of brightness.

1.4.5. *Stellar evolution*

Stellar evolution is most adequately described on a HR diagram, and for example the Sun after consuming most of its hydrogen fuel will contract its core while expanding its outer layers to a radius that will include the Earth. The contraction at first raises the luminosity and then the Sun will expand and redden, or move up and then to the right in a HR diagram. At a later stage the helium fuel will ignite in the contracted core and the Sun will move to the left on an *asymptotic branch* on the HR diagram. At the end of helium burning the Sun will further contract to a white dwarf and reside forever at the lower bottom left of the HR diagram. For main sequence stars the luminosity is given by the Stefan–Boltzmann law, Eq. (1.2), i.e., $L = 4\pi R^2 \sigma T_{\mathit{eff}}^4$, where T_{eff} is the effective temperature, because stars do not have a well defined surface and are not perfect black body radiators.

Above and to the right of the main sequence we find the *red giant* stars, characterized by large luminosity and therefore easily seen in the sky. This class includes only a small number of stars, a few percent of the known stars.

The redness of these stars arises from their large radii and they represent a star in its later stages of evolution, after it has consumed its hydrogen fuel in the core and consist mainly of helium. The *subgiants* are believed to be stars that expand their outer envelope while contracting their helium cores, leading to the burning of helium. The *horizontal branch* stars, on the other hand, are believed to be at various stages of helium burning. The *supergiant* stars are believed to be stars at the advanced stages of their stellar evolution and perhaps approaching the end of their energy-generating lifetime.

In the lower left corner of the HR diagram we find the white dwarfs representing approximately 10% of known stars, which are very dense stars of mass comparable to a solar mass, with considerably smaller radii, comparable to the Earth's radius. Due to the small surface area these stars have large surface temperature (blue color) in order to allow them to radiate their luminosity. These group composes of the Universe's cemetery of stars that are inactive and simply radiate their pressure energy. The white dwarfs are so dense that the electron degeneracy keeps them from collapsing, hence cannot have a mass larger then approximately $1.4\,M_{\odot}$, the *Chandrasekhar limit*, beyond which the electron degeneracy cannot overcome the gravitational collapse. Such massive stars, or cores of massive stars, collapse to a neutron star or a black hole under their own gravitational pressure.

1.5. Hubble's Law

In the 1920s Edwin Hubble used the period–luminosity relation for variable stars to establish the distances to various galaxies and proved that they lie far outside our Milky Way. In the course of his work, he discovered that all galaxies, except for those in the Virgo cluster, show a *spectral redshift* (i.e., an increased distance between known frequency lines) [Hu29]. This is assumed to be due to the *Doppler effect*, which modifies the wavelength of the radiation emitted by stars in the galaxies due to their relative velocity to an observer on Earth. When a star (source of the light waves) is moving toward the Earth, each successive wave crest in the light wave is emitted from a position closer to Earth than the previous wave crest. Therefore each wave crest takes less time to reach the Earth than the previous wave crest. The time between the arrival of successive wave crests at the Earth is reduced, causing an increase in the frequency of light. The opposite effect happens when a star is moving away from the Earth. Then each wave crest is emitted from a position farther from the Earth than the previous wave crest, and the arrival time between successive waves crests is increased, reducing the frequency (compare with the lowering of the pitch from the horn of a train moving away from us). Therefore, Hubble's observation of the redshift, or increased wavelength of the emitted radiation, implies that the galaxies are moving away from us.

Early measurements of the distance of stars from the Earth used the parallax method [Cla84]. It was found that the nearest star, Alpha-Centauri visible in the southern hemisphere (a triple star system composed of Alpha-Centauri Proxima, A and B) produced 1.52 sec of arc of angular displacement, or a parallax of 0.76 arc sec. The Earth average orbit radius 1 AU = 149.6 Mkm, where AU is the *Astronomical Unit*, or approximately 8 light minutes. One then finds by a simple triangulation that the star distance is hopelessly far from us, at a distance of approximately 4.2 ly. Modern day optical telescopes have an accuracy of the order of 0.01 sec of an arc and with the use of interferometry one can improve the resolution to 0.001 sec of an arc. Hence, the parallax method has a limited use, for stars closer than 1 kpc.

The *redshift* z for a radiation with wavelength λ_0 is defined as

$$z = \frac{\lambda - \lambda_0}{\lambda_0} = \frac{Hd}{c}, \tag{1.8}$$

where λ is the radiation wavelength as measured on Earth and d is the distance of the galaxy from the Earth. The right-hand-side of this Equation is based on Hubble's observation that the redshift increases with the distance of the galaxy from the Earth, where H is the *Hubble constant.* For velocities $v \ll c$, Eq. (1.8) becomes $z = v/c$ (see derivation of Eq. (B.78) in Appendix B). Hence,

$$v_r = Hd, \quad \text{or} \quad \frac{\dot{R}}{R} = H, \tag{1.9}$$

where v_r is the radial velocity, and we also use the notation $R = d$ and $\dot{R} = v_r$. This is the common expression of *Hubble's law.* After measuring the redshift, which we can do by passing a galaxy's light through a spectrogram, we can deduce the distance using Hubble's law. This technique is the astronomer's basic tool for finding the distances to the farthest things in the Universe. If the redshift is plotted against the apparent magnitude of the brightest star in a large number of galaxies it is seen to increase with decreasing luminosity, which is interpreted as more distant (fainter) galaxies move away faster from us than the closer ones [Hu29]. Except for the galaxies in the Local Group, all galaxies recede from us with velocities up to 20,000 km/s. Hence it is concluded that the *Universe expands.*

If the Universe is expanding, the galaxies were once much closer to each other. If the rate of expansion has been unchanged, the inverse of the Hubble constant, H^{-1}, would represent the age of the Universe. In Eq. (1.9) v_r is the radial velocity of a galaxy at distance d from us. But velocity is just distance divided by time, i.e., $v_r = d/t_0$, where t_0 is the time the expansion has been going on, assuming a constant speed. Thus, $d/t_0 = Hd$, or $t_0 = 1/H$. t_0 is only an upper limit of the *age of the Universe.* According to present estimates a H value of 0.05–0.1 ms^{-1} pc^{-1} corresponds to an age of 10–20 Gy (Gy = Gigayears, or Aeon, 10^9 years). Cosmologists also give the age in the "scale factor" $(1+z)$-values, i.e., redshift values; e.g., we would observe a z-value of 10 for an object about one billion years old from the formation of the Universe. Figure 1.3 shows that experimental data are in good agreement with Hubble's law.

Although it seems natural to believe that the expansion of the Universe is slowing down due to gravitational pull, in fact the recent expansion is accelerating. This observation is one of the major discoveries of modern cosmology [Rie98,Per99].

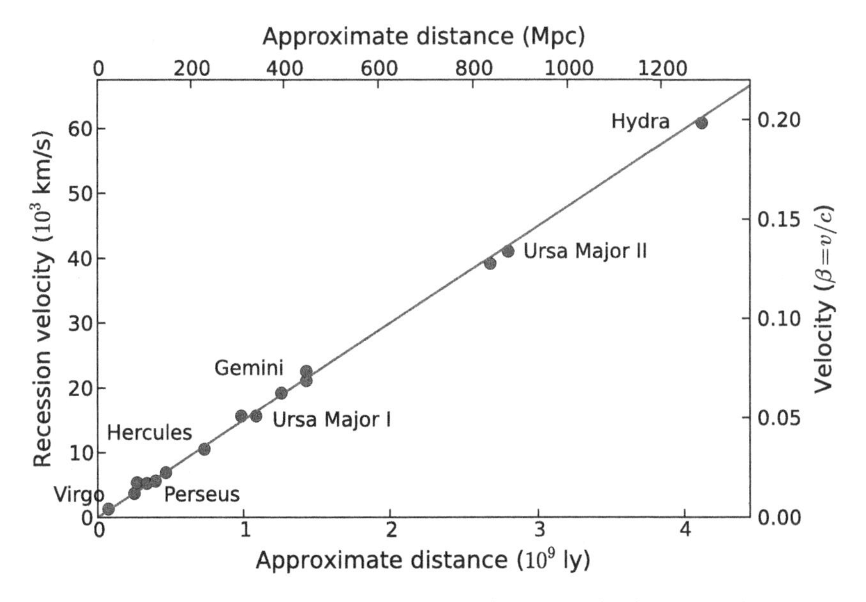

Figure 1.3. Observed relation between the distance and velocity of galaxies.

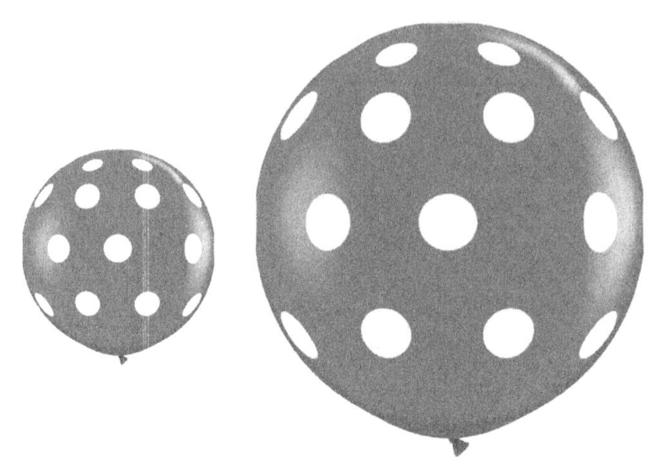

Figure 1.4. Distances between cluster of galaxies are increasing uniformly, as represented here by circles on the surface of an expanding balloon.

1.6. Stars and Galaxies

First generation stars are stars that coalesced from the primordial dust that included approximately 24% helium and 76% hydrogen with traces of lithium. Some of these stars are small enough that they have not evolved and are still burning hydrogen, while others have already converted to dwarfs. The Sun is not a first generation star and has burned its hydrogen fuel for the last 4.6 billion years and will do so for approximately 5 more billion years. First generation stars are expected to have very small amounts of elements heavier then carbon (sometimes generically referred to as *metals*). One defines the *metallicity* of a star (also called Z) as the proportion of its matter made up of chemical elements other than hydrogen and helium. The metallicity of the Sun is approximately 1.8 percent by mass. For other stars, the metallicity is often expressed as "[Fe/H]", which represents the logarithm of the ratio of a star's iron abundance compared to that of the Sun. Iron is not the most abundant heavy element, but it is among the easiest to measure with spectral data in the visible spectrum. Therefore,

$$[\mathrm{Fe/H}] = \log_{10}\left(\frac{N_{\mathrm{Fe}}}{N_{\mathrm{H}}}\right)_{star} - \log_{10}\left(\frac{N_{\mathrm{Fe}}}{N_{\mathrm{H}}}\right)_{Sun}, \tag{1.10}$$

where N_{Fe} and N_{H} are the number of iron and hydrogen atoms per unit of volume, respectively. Stars with metallicity of -3 to -4 are believed to be primordial with ages in the range of 10 to 14 billion years. It should be emphasized that while the metallicity of a star is measured on its surface, one needs to know the core metallicity and hence one needs to introduce stellar atmospheric models, and thus these data in some cases are model dependent.

Star clusters reside far away from us, and may contain as many as $10^5 - 10^7$ stars. A set of these, called *globular clusters*, contain stars in spherical distribution with a radius in the range of approximately 10 parsecs. Some clusters may include only a few stars. Globular clusters are scattered over the entire sky, but are strongly concentrated toward Sagittarius (Galactic centre); usually they contain $\sim 10^{5-6}$ mostly old stars tightly bound in a quasi-spherical distribution ~ 75 ly in diameter. It is believed that the age of stars in the globular cluster is of the order of 14 ± 3 Gy, or as old as the Universe itself. Within this cluster we find a relatively young class with blue giants as the most luminous, called population I, and an older class with red giants as the most luminous members, called population II. *Open clusters* lie along the Galactic plane in all directions; they are loose groupings of $\sim 10^{2-3}$ mostly young stars, with a much more diffuse and less symmetric distribution than globular clusters. The HR diagram indicates that globular clusters are $\sim$9–10 billion years old.

Population I stars include the Sun and tend to be luminous, hot and young, concentrated in the disks of spiral galaxies. They are particularly found in the spiral arms. With the model of heavy element formation in supernovae, this suggests that the gas from which they formed had been seeded with the heavy elements formed from previous giant stars. *Population II stars* tend to be found in globular clusters and the nucleus of a galaxy. They tend to be older, less luminous and cooler than population I stars. They have fewer heavy elements, either by being older or being in regions where no heavy-element producing predecessors would be found. Astronomers often describe this condition by saying that they are "*metal poor*", and the "*metallicity*" is used as an indication of age. Populations I and II stars differ in their age and their content of metals. Besides populations I and II there are also *population III stars.* These were stars with very low metallicity stars, other than a little Li, formed within $10^6 - 10^7$ years of the Big Bang (independent of metallicity). They no longer exist, but affected the environment of the early Universe.

The initial composition of the galaxy was probably uniform. Thus, if there exists no mechanism capable of concentrating the metals in the disk of the galaxy, an overwhelming fraction of its metals must have been synthesized within the galaxy. This implies that nucleosynthesis is a natural process to describe the evolution of stars. The metal content of the galaxy increases with time since the matter out of which stars form is being cycled through an increasing number of stellar generations. And the differences in metallicity between the two stellar populations suggest that population I stars formed later during the history of the galaxy when the interstellar medium became much more metal rich.

About 10% of galaxies reside in big clusters. However, most of the cluster mass is hot gas ($10^6 - 10^8\,$K) as revealed by X-ray observation. Our galaxy is part of a local group (including Andromeda) which seems to be an appendix of the Virgo cluster. All the stars we know reside in galaxies. Galaxies come in three basic groups: Elliptic, spiral and irregular. The masses range from 10^5 to $10^{12} M_\odot$, though that may be open to the lower end. Generally speaking, irregular galaxies (Magellanic Clouds) are the smallest ones. Elliptical galaxies come in all sizes, but some are the very biggest galaxies we know (giant elliptic). This may be a result of spiral galaxy mergers. Elliptic galaxies seem to be pretty void of gas and star formation is at least at a very low level.

The mass range of spiral galaxies is more limited: $10^{10} - 10^{11}\,M_\odot$. The arms of a spiral galaxy are an entirely dynamic phenomenon: The blue stars just live and die there. Spiral galaxies are divided into spirals and bar spirals with a bar going out from the center. Due to tightness and looseness of the arms they are classified with the letters a (tight), b, c (loose). Any spiral galaxy (disk) is surrounded by an elliptical halo. There is evidence that black holes play an important role in galaxy formation as the existence of many early quasars indicates. There is also clearly dark matter associated with galaxies.

1.7. Principles and Observations

The simplicity of the Universe is based on the following *cosmological principle*. The Universe, on average, looks the same from any point. If the Universe is locally isotropic, as viewed from any point, it should also be uniform. So the cosmological principle states: Our Universe is approximately isotropic and homogeneous, as viewed by any observer at rest. More accurately, the matter in the Universe is homogeneous and isotropic when averaged over very large scales.

This assumption is being tested continuously as we actually observe the large scale distribution of galaxies. Figure 1.5 shows the distribution of

Figure 1.5. Redshift survey containing some 15,000 galaxies, covering over 83% of the sky up to redshifts of $z \leq 0.05$. We show here the projection of the galaxy distribution in galactic coordinates. (Collected by the Wilkinson Microwave Anisotropy Probe (WMAP) — Image courtesy of NASA.)

measured galaxies over a 30° swath of the sky. In cosmology the simplicity of the Universe appears on sufficiently large scales. A few decades ago it was assumed that homogeneity applies on scales above 10 Mpc. However, the recent discovery of giant filaments and voids, as well as large-scale streaming motions, suggests that one may need to go to scales of order 100 Mpc. Since the scale of the observable Universe is around 6000 Mpc, this is only a factor of 60 larger, but it still gives about $60^3 \sim 2 \times 10^5$ cells, which is a large enough number to apply statistics to.

A long time ago cosmologists believed in a much stronger principle, called the perfect cosmological principle, which says that the Universe appears the same from all points and at all times. In other words, there can have been no evolution. This contradicts the observations which show that the real Universe does not satisfy this principle because we now have evidence that the Universe evolves (expansion of the Universe). Another principle, the status of which remains ambiguous, is the so-called *anthropic principle*. This principle claims that certain features of the Universe — such as the values of the physical constants — are determined by the requirement that life should arise, because otherwise we could not be here asking questions about it.

That is what we know at the present moment about the Universe as a whole:

1. The Universe is expanding from a hot and dense initial state called the Big Bang.

2. In the Big Bang the light elements were synthesized.
3. We believe that there was a period of *inflation* which led to many observable properties of the Universe [Gut97,Gut98].
4. Any observable large scale structure was seeded by some small perturbations which are the relics of quantum fluctuations.
5. This structure is dominated by cold dark matter.
6. At the present moment the Universe seems to be expanding with an acceleration rather than a deceleration as cosmologists used to think in the 20th century.

These issues will be discussed in details in the next chapters. For the moment, it is just enough to remember these observations and conclusions.

1.8. Cosmic Microwave Background

For a long time after its discovery the *Cosmic Microwave Background* (CMB) radiation did seem to be uniform over sky [PW65]. The CMB originated about 400,000 years after the Big Bang, known as the period of *recombination* or *decoupling*, when the temperature of the Universe was about 3000 K (this is explained in later chapters, i.e., Chapter 9). This temperature corresponds to an energy of about 0.25 eV (see Appendix A), which is much less than the 13.6 eV ionization energy of hydrogen. Since then, the CMB temperature has decreased by a factor of 1,000 due to the expansion of the Universe. As the Universe expands, the CMB photons are redshifted, decreasing the radiation's temperature inversely proportional to a parameter called the Universe's scale length a (Chapter 2). The temperature T of the CMB as a function of redshift, z, can be shown to be proportional to the temperature of the CMB as observed in the present day (2.725 K or 0.235 meV),

$$T = 2.725(1 + z). \tag{1.11}$$

When observational technology improved, an anisotropy in the CMB was detected. Only very sensitive instruments can detect tiny fluctuations in the CMB temperature. These fluctuations bring direct information about the physical conditions in the very early Universe, about the nucleosynthesis and the origin of galaxies and formation of the large scale structure of the Universe. In other words, with the help of the CMB fluctuations one can determine the basic parameters of the Big Bang theory. It is possible to say that any map of the CMB temperature is a direct image of the remote past of our Universe. These fluctuations (ΔT is the deviation of T from the average temperature) are found to be

$$\frac{\Delta T}{T} \simeq 10^{-5}, \tag{1.12}$$

and contain extremely valuable information about the origin, evolution, and content of the Universe. A more detailed picture of the early Universe is obtained if we measure the polarization of the CMB. This polarization is unavoidably generated due to scattering of the CMB photons on free electrons during the epoch of highly ionized plasma. The information obtained with the help of polarization measurements and some theoretical models provides new clues about events which took place when the age of the Universe was only tiny fraction of second (based on "reasonable" physics assumptions).

We now describe the usual techniques for characterizing the temperature field. By measuring the CMB temperature in progressively smaller and smaller patches of the sky, variations are measured in terms of an "angular fluctuation spectrum", or the amplitude or strength of temperature fluctuations as a function of angular size. One defines the normalized temperature in direction $\hat{\mathbf{n}}$ on the celestial sphere by the deviation from the average: $\Delta T(\hat{\mathbf{n}})/T$. Then one calculates the multipole decomposition of this temperature field in terms of spherical harmonics $Y_{\ell m}$,

$$\alpha_{\ell m} = \int d\cos\theta \; d\phi \; Y_{\ell m}(\theta, \phi) \frac{\Delta T}{T}(\hat{\mathbf{x}}). \tag{1.13}$$

If the sky temperature field arises from random fluctuations, then the field is fully characterized by its power spectrum, or *temperature–temperature correlation* spectrum

$$\left\langle \frac{\Delta T}{T}(\hat{\mathbf{x}}) \frac{\Delta T}{T}(\hat{\mathbf{y}}) \right\rangle. \tag{1.14}$$

The order m describes the angular orientation of a fluctuation mode, but the degree (or multipole) ℓ describes its characteristic angular size. Thus, in a Universe with no preferred direction, we expect the power spectrum to be independent of m. Finally, we define the angular power spectrum C_ℓ by

$$\langle \alpha_{\ell m} \alpha_{\ell' m'} \rangle = \delta_{\ell \ell'} \delta_{m m'} C_\ell, \tag{1.15}$$

where the brackets denote an ensemble average over the sky. The best estimate of C_ℓ is then from the average over m,

$$\left\langle \frac{\Delta T}{T}(\hat{\mathbf{x}})\frac{\Delta T}{T}(\hat{\mathbf{y}}) \right\rangle = C_\ell. \tag{1.16}$$

The temperature–temperature correlation spectrum is often plotted in the form

$$(\Delta T)^2 = \frac{\ell(\ell+1)}{2\pi} C_\ell T_{CMB}^2, \tag{1.17}$$

where T_{CMB} is the blackbody temperature of the CMB. This is the variance (or power) per logarithmic interval in ℓ and is expected to be (nearly) uniform over much of the spectrum. This normalization is useful for calculating the contributions to the fluctuations in the temperature in a given pixel from a range of ℓ values,

$$(\Delta T)^2 = \int_{l_{min}}^{l_{max}} \frac{\ell(\ell+1)}{2\pi} C_\ell T_{CMB}^2. \tag{1.18}$$

Figure 1.6 shows the current understanding of the temperature power spectrum. The region below $\ell \sim 20$ indicates the initial conditions. These modes correspond to Fourier modes at the time of decoupling (to be discussed in Chapter 9), with wavelengths longer than the horizon scale. Note that were the sky describable by random white noise, the C_ℓ spectrum would be flat and the temperature power spectrum, defined by Eq. (1.17), would have risen in this region like ℓ^2. At high ℓ values, there are *acoustic oscillations*, which are damped at even higher ℓ values. The positions and heights of the acoustic oscillation peaks reveal fundamental properties about the geometry and composition of the Universe.

In the early Universe before decoupling, rapid scattering couples photons and baryons into a plasma which behaves as a perfect fluid. Initial quantum overdensities create potential (gravitational) wells — inflationary seeds of the Universe's structure. Infall of the fluid into the potential wells is resisted by its pressure, thus forming acoustic oscillations: periodic compression (overdensities in the fluid; hot spots) and rarefactions (underdensities; cold spots). These acoustic oscillations of the early Universe are frozen at recombination and give the CMB spectrum a unique signature. They give the microwave background its characteristic peak structure. The peaks correspond, roughly, to resonances in which the photons decouple when a

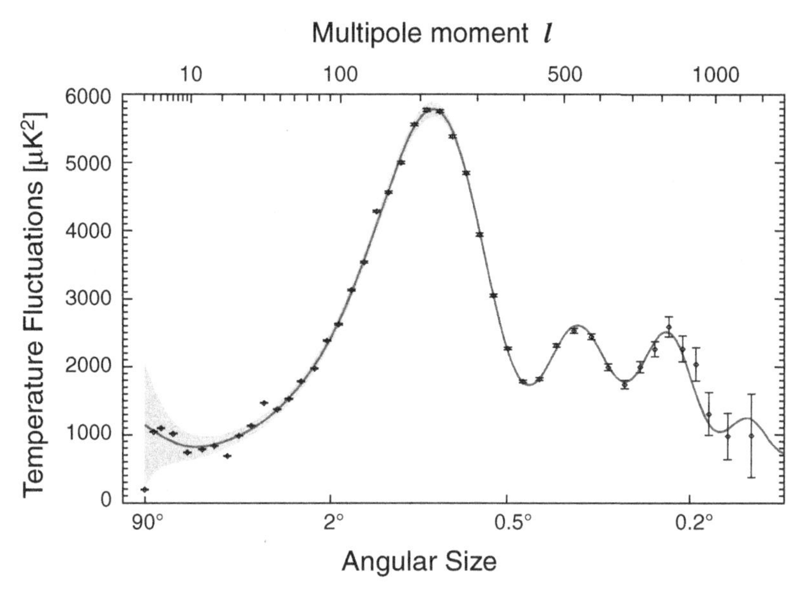

Figure 1.6. The temperature–temperature correlation power spectrum. Data from the Wilkinson Microwave Anisotropy Probe (WMAP) [Jar11], and high-ℓ data from other experiments are shown, in addition to the best cosmological model to the WMAP data alone. Note the multipole scale on the top and the angular scale on the bottom. (Figure courtesy of the WMAP science team — NASA.)

particular mode is at its peak amplitude. The CMB data reveals that the initial inhomogeneities in the Universe were small.

1.9. Exercises

1. Two radar pulses sent out from the Earth at 6:00 AM and 8:00 AM one day bounce off an alien spaceship and are detected on Earth at 3:00 PM and 4:00 PM. You are not sure, however, which reflected pulse corresponds to which emitted pulse. Is the spaceship moving toward Earth or away? If its speed is constant (but less than c), when will it (or did it) pass by the Earth?
2. The scale of the Universe observable now is about $L_{obs} \simeq 10^4$ Mpc. Using the speed of light, transform this number in a time scale. What does this tell you? (see Appendix B.)
3. (a) Explain briefly what is meant by the cosmological principle and (b) what observational data support it. (c) What is the approximate scale at which homogeneity is reached?

4. The average radial velocity of galaxies in the Hercules cluster is 10,800 km/s.
   (a) Using $H = 70\,\mathrm{km/s/Mpc}$, find the distance to this cluster. Give your answer in megaparsecs and in light-years.
   (b) How would your answer to (a) differ if the Hubble constant had a smaller value? A larger value? Explain your answers.
5. Assume (wrongly) that the matter density in the Universe can be explained by some hypothetical dark objects of mass $M = 10^{-3} M_\odot$, where $M_\odot$ is solar mass (these objects are as massive as Jupiter). The contribution of these objects to the density of the Universe is $\rho_o = 10^{-34}\,\mathrm{kg\ m^{-3}}$. Estimate the average distance between these objects.
6. Suppose that in Sherwood Forest, the average radius of a tree is $R = 1\,\mathrm{m}$ and the average number of trees per unit area is $a = 0.005\,\mathrm{m^{-2}}$. If Robin Hood shoots an arrow in a random direction, how far, on average, will it travel before it strikes a tree?
7. The *Olber's paradox.* (a) Show that in an infinite static Universe with a uniform density of sources the night sky should be as bright as the Sun. (b) Explain qualitatively why the evolution and expansion of the Universe can resolve this paradox.
8. Consider three galaxies in an expanding Universe located at points a, b and c. Prove that if the Hubble law is valid for an observer at a, then it is also valid for observers at b and at c.
9. We can define the Hubble constant, H by the relationship $cz = Hd$ where c is the speed of light and d is the distance to the object. (a) Show that $H = 70\,\mathrm{km/s/Mpc}$ makes dimensional sense. (b) Simplify this value of H so that it has SI units.
10. In the spectrum of the galaxy NGC 4839, the K line of singly ionized calcium has a wavelength 403.2 nm. (The wavelength of this line for an atom at rest is 393.3 nm).
    (a) What is the redshift of this galaxy?
    (b) Determine the distance to this galaxy using the Hubble law with $H = 70\,\mathrm{km/s/Mpc}$.
11. Since we are made mostly of water, we are very efficient at absorbing microwave photons. If we were in intergalactic space, approximately how many CMB photons would we absorb per second? (Assume we are spherical.) What is the approximate rate, in watts, at which we would absorb radiative energy from the CMB? Ignoring other energy inputs and outputs, how long would it take the CMB to raise our temperature

by one nanoKelvin (10^{-9} K)? (Assume our heat capacity is the same as pure water, $C = 4200\,\mathrm{J\ kg^{-1}\ K^{-1}}$.)

12. A hypothesis once used to explain the Hubble relation is the "tired light hypothesis". The tired light hypothesis states that the Universe is not expanding, but that photons simply lose energy as they move through space (by some unexplained means), with the energy loss per unit distance being given by the law

$$\frac{dE}{dr} = -\kappa E,$$

where κ is a constant. Show that this hypothesis gives a distance redshift relation which is linear in the limit $z \ll 1$. What must be the value of κ in order to yield a Hubble constant of $H = 70\,\mathrm{km\ s^{-1}\ Mpc^{-1}}$?

13. Suppose that the Milky Way galaxy is a typical size, containing $\sim 10^{11}$ stars, and that galaxies are typically separated by a distance of one Mpc. Estimate the density of the Universe in SI units. How does this compare with the density of the Earth? ($1\,M_\odot \simeq 2 \times 10^{30}$ kg, 1 pc $\simeq 3 \times 10^{16}$ m.)

14. What evidence can you think of to support the assertion that the Universe is charge neutral, and hence contains an equal number of protons and electrons?

15. In the real Universe the expansion is not completely uniform. Rather, galaxies exhibit some random motion relative to the overall Hubble expansion, known as their *peculiar velocity* and caused by the gravitational pull of their near neighbors. Supposing that a typical (e.g., root mean square) galaxy peculiar velocity is $600\,\mathrm{km\ s^{-1}}$, how far away would a galaxy have to be before it could be used to determine the Hubble constant to ten per cent accuracy, supposing

    (a) The true value of the Hubble constant is $100\,\mathrm{km\ s^{-1}\ Mpc^{-1}}$?
    (b) The true value of the Hubble constant is $50\,\mathrm{km\ s^{-1}\ Mpc^{-1}}$?

    Assume in your calculation that the galaxy distance and redshift can be measured exactly. Unfortunately, that is not true of real observations.

16. Astronomers often state the distance to a remote galaxy in terms of its *distance modulus*, which is the difference between the apparent magnitude m and the absolute magnitude $\mathcal{M}$.

    (a) By measuring the brightness of supernova 1994I in the galaxy M51, the distance modulus for this galaxy was determined to be $m - \mathcal{M} = 29.2$. Find the distance to M51 in megaparsecs (Mpc).

(b) A separate distance determination, which involved measuring the brightnesses of planetary nebulae in M51, found $m - \mathcal{M} = 29.6$. What is the distance to M51 that you calculate from this information?

(c) What is the difference between your answers to parts (a) and (b)? Compare this difference with the 750 kpc distance from Earth to M31, the Andromeda galaxy.

17. The peak of the energy density distribution of a black-body at $f_{peak} \simeq 2.8k_BT/h$ implies that f_{peak}/T is a constant. Evaluate this constant in SI units (see appendix A for useful numbers). The Sun radiates approximately as a black-body with $T_\odot \simeq 5800\,\mathrm{K}$. Compute f_{peak} for solar radiation. Where in the electromagnetic spectrum does the peak emission lie?

18. The cosmic microwave background has a black-body spectrum at a temperature of 2.725 K. Repeat the previous problem to find the peak frequency of its emission, and also find the corresponding wavelength. Confirm that the peak emission lies in the microwave part of the electromagnetic spectrum. Finally, compute the total energy density of the microwave background.

Chapter 2

Space and Matter

2.1. Introduction

We want to learn about the Universe, its size, its shape and its age. Also, we want to understand the distribution of matter and how this distribution arose. Even more ambitiously, we want to know how the Universe started and how it will end. These are all bold questions to ask, involving tremendous theoretical and observational effort invested over the past century. It is not totally inaccurate to say that modern cosmology started with Einstein's theory of *general relativity* (GR). GR is the overarching framework for modern cosmology, and we cannot avoid discussing this subject with at least a brief account of some of the most important features of this theory [Ein16].

The basic postulate of the general relativity states that a uniform gravitational field is equivalent to (which means is not distinguishable from) a uniform acceleration. A person cannot see locally the difference between standing on the surface of some gravitating body (for example the Earth) and moving in a rocket with corresponding acceleration. All bodies in a given gravitational field move in the same manner, if initial conditions are the same. In other words, in a given gravitational field all bodies move with the same acceleration. In the absence of a gravitational field, all bodies move also with the same acceleration relative to the non-inertial frame. Thus we can formulate the *principle of equivalence* which says: locally, any non-inertial frame of reference is equivalent to a certain gravitational field. The important consequence of the principle of equivalence is that locally gravitational fields can be eliminated by proper choice of the frame of reference. Such frames of reference are called locally inertial or Galilean frames of reference. There is no experiment to distinguish between being weightless far out from gravitating bodies in space and being in free-fall in a gravitational field. Globally (not locally), "actual" gravitational fields can be distinguished from corresponding non-inertial frame of reference by its behavior at infinity: gravitational fields generated by gravitating bodies fall with distance.

2.2. Friedmann Equation

2.2.1. *Energy and pressure*

The first law of thermodynamics states that, for an isolated system (e.g., the Universe)

$$dU + dW = dQ \tag{2.1}$$

where U stands for the internal energy, W is the work done by the system and Q is the heat transfer. Ignoring any heat transfer, $dQ = 0$, and writing $dW = Fdr = pdV$ where F is the force, r is the distance characterizing the size of the system, p is the pressure and V is the volume, then

$$dU = -pdV. \tag{2.2}$$

On the other hand, denoting by ρ the energy density, the energy is expressed as

$$U = \rho V, \tag{2.3}$$

from which it follows that

$$\frac{dU}{dt} = \frac{d\rho}{dt}V + \rho\frac{dV}{dt} = -p\frac{dV}{dt}, \tag{2.4}$$

where the last equality results from Eq. (2.2). Since $V \propto r^3$, then $(dV/dt)/V = 3(dr/dt)/r$. Thus,

$$\frac{d\rho}{dt} = -3(\rho + p)\frac{1}{r}\frac{dr}{dt}. \tag{2.5}$$

Assuming all energy to be in the form of matter, the relation between matter, M, density, ρ, and radius, r, means that

$$\rho = \frac{M}{4\pi r^3/3}, \tag{2.6}$$

so that

$$\frac{d\rho}{dt} = \frac{d\rho}{dr}\frac{dr}{dt} = -3\rho\frac{1}{r}\frac{dr}{dt}, \tag{2.7}$$

which yields, together with Eq. (2.5), the *matter pressure*,

$$p = 0. \tag{2.8}$$

That is, if no kinetic energy is taken into account, pressure is zero for a system with mass M. This is the same pressure as the ideal gas law for zero temperature.

The above result changes if radiation is included in the total energy U. Let us consider radiation modes in a cavity based on analogy with a string held fixed at two points separated by a distance L. The possible wavelengths, λ, of a standing wave on the string obey the relation

$$L = \frac{n\lambda}{2}, \tag{2.9}$$

$n = 1, 2, 3, \ldots$. Radiation travels at the velocity of light, so that

$$c = f\lambda = f\frac{2L}{n}, \tag{2.10}$$

where f is the frequency. Planck's formula for the energy of a quantum of radiation with frequency $f = \omega/2\pi$ is $U = \hbar\omega = hf$, where h is Planck's constant. Thus,

$$U = \frac{1}{L}\frac{nhc}{2} \propto V^{-1/3}, \tag{2.11}$$

where $V = L^3$ is the volume of a cube of length L. Using Eq. (2.2) the pressure becomes

$$p = -\frac{dU}{dV} = \frac{1}{3}\frac{U}{V}. \tag{2.12}$$

This, together with $\rho = U/V$, yields the *radiation pressure*,

$$p = \frac{\rho}{3} = \frac{\gamma}{3}\rho. \tag{2.13}$$

In summary,

$$p = \frac{\gamma}{3}\rho, \tag{2.14}$$

where $\gamma = 1$ for *radiation* and $\gamma = 0$ for *matter*.

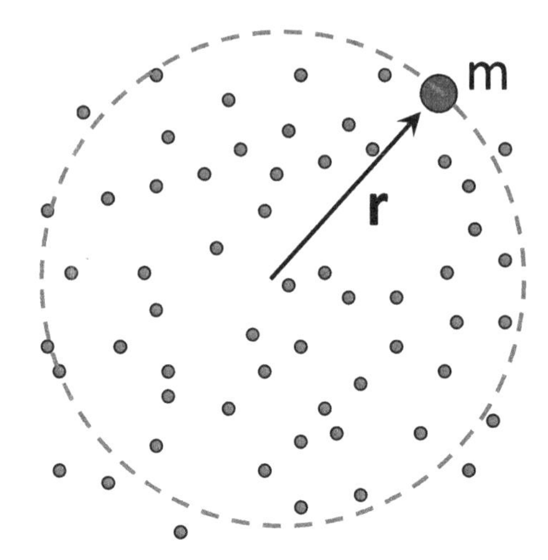

Figure 2.1. Particle of mass m interacting with a dust gas of mass M within a sphere with a radius r.

2.2.2. *Friedmann equation*

Let us consider now a particle with mass m in the gravitational potential energy field of a uniform dust gas of mass M, as shown in Fig. 2.1. The Newtonian result that the gravitational field inside a uniform shell is zero also holds in general relativity, and is known as *Birkhoff's theorem* [Bir23]. Thus, the gravitational force on the particle is the same as if the enclosed mass within radius r is localized entirely at the origin, $r = 0$. If it is located a distance r from the center of the dust, the total energy E of the particle is then given by

$$E = T + V = \frac{1}{2}m\dot{r}^2 - G\frac{Mm}{r} = \frac{1}{2}mr^2\left(H^2 - \frac{8\pi G}{3}\rho\right), \tag{2.15}$$

where $\dot{r} = dr/dt$, $H = \dot{r}/r$ is the Hubble constant, G is Newton's constant, and we used $M = \rho 4\pi r^3/3$ in the last passage.

The escape velocity of the test particle is the velocity needed to overcome the gravitational pull of the dust gas, never turning back. This is just $v_{esc} = \sqrt{2GM/r} = \sqrt{(8\pi G/3)\rho r^2}$, and the Eq. (2.15) can also be written as

$$\dot{r}^2 = v_{esc}^2 - k', \tag{2.16}$$

with $k' = -2E/m$. The constant k' can either be negative, zero or positive, corresponding to the total energy E being positive, zero or negative.

For a particle in motion near the Earth, this would correspond to the particle escaping (unbound), orbiting (critical case) or returning (bound) to Earth because the speed $\dot{r}$ is greater, equal to or smaller than the escape speed v_{esc}.

Equation (2.15) is re-arranged as

$$H^2 = \frac{8\pi G}{3}\rho + \frac{2E}{mr^2}. \tag{2.17}$$

Writing the distance in terms of a *scale factor* a and a constant length s as $r(t) = a(t)s$, and defining $k = k'/s^2 = -2E/ms^2$, it follows that

$$H^2 = \left(\frac{\dot{a}}{a}\right)^2 = \frac{8\pi G}{3}\rho - \frac{k}{a^2}, \tag{2.18}$$

since $\dot{r}/r = \dot{a}/a$ and $\ddot{r}/r = \ddot{a}/a$. This is the *Friedmann equation* [Fri23]. It specifies the speed of recession.

Even though Friedmann equation was derived for matter, it is also true for radiation. It is also true for vacuum, with $\Lambda = 8\pi G\rho_{vac}$, where Λ is the cosmological constant and ρ_{vac} is the vacuum energy density (next section). Exactly the same equation is obtained from the general relativistic Einstein field equations. The factor k can be rescaled so that instead of being negative, zero or positive it takes on the values -1, 0 or $+1$. In Newtonian mechanics this corresponds to unbound, critical or bound trajectories. From a geometric point of view, this corresponds to an *open, flat* or *closed Universe* [KT90].

The acceleration for the Universe is obtained from Newton's second equation, i.e.

$$-G\frac{mM}{r^2} = m\ddot{r}. \tag{2.19}$$

In terms of the density and the scale a, we get from Eq. (2.18) (with $k = 0$),

$$\frac{F}{mr} = \frac{\ddot{r}}{r} = \frac{\ddot{a}}{a} = -\frac{4\pi G}{3}\rho. \tag{2.20}$$

Another way to get this equation is to take the time derivative of Eq. (2.18) (valid for matter and radiation)

$$\frac{d}{dt}\dot{a}^2 = 2\dot{a}\ddot{a} = \frac{8\pi G}{3}\frac{d}{dt}(\rho a^2). \tag{2.21}$$

Upon using Eq. (2.5) the acceleration equation is obtained as

$$\frac{\ddot{a}}{a} = -\frac{4\pi G}{3}(\rho + 3p) = -\frac{4\pi G}{3}(1+\gamma)\rho, \tag{2.22}$$

which reduces to Eq. (2.20) for the matter equation of state ($\gamma = 0$). Exactly the same equation is obtained from the Einstein field equations.

2.2.3. *The cosmological constant*

In both Newtonian and relativistic cosmology the Universe is unstable to gravitational collapse. Both Newton and Einstein believed that the Universe is static. In order to obtain this Einstein introduced a repulsive gravitational force, called the *cosmological constant*, and Newton could have done exactly the same thing, had he believed the Universe to be finite. In order to obtain a possibly zero acceleration, a positive term (conventionally taken as $\Lambda/3$) is added to the acceleration Eq. (2.22) as

$$\frac{\ddot{a}}{a} = -\frac{4\pi G}{3}(\rho + 3p) + \frac{\Lambda}{3}, \tag{2.23}$$

which, with the proper choice of Λ, can give the required zero acceleration for a static Universe. Again exactly the same equation is obtained from the Einstein field equations. Our derivation is entirely equivalent to just adding a repulsive gravitational force in Newton's Law. The question now is how this repulsive force enters the energy Eq. (2.18).

Identifying the force from

$$\frac{\ddot{r}}{r} = \frac{\ddot{a}}{a} = \frac{F_{rep}}{mr} = \frac{\Lambda}{3} \tag{2.24}$$

and using

$$F_{rep} = \frac{\Lambda}{3}mr = -\frac{dV_{rep}}{dr}, \tag{2.25}$$

gives the potential energy as

$$V_{rep} = -\frac{1}{2}\frac{\Lambda}{3}mr^2. \tag{2.26}$$

which is just a simple repulsive harmonic oscillator. Replacing this into the conservation of energy equation,

$$E = T + V = \frac{1}{2}m\dot{r}^2 - G\frac{Mm}{r} - \frac{1}{2}\frac{\Lambda}{3}mr^2 = \frac{1}{2}mr^2\left(H^2 - \frac{8\pi G}{3}\rho - \frac{\Lambda}{3}\right), \tag{2.27}$$

gives

$$H^2 = \left(\frac{\dot{a}}{a}\right)^2 = \frac{8\pi G}{3}\rho - \frac{k}{a^2} + \frac{\Lambda}{3}. \tag{2.28}$$

Equations (2.23) and (2.28) constitute the fundamental equations of motion that are used in all discussions of Friedmann models of the Universe.

Imagine that a hole has been drilled from one side of the Earth, through the center and to the other side. If a ball is dropped into the hole, it will execute harmonic motion. We show this by noting that gravity is an inverse square law for point masses M and m separated by a distance r as given by $F = GMm/r^2$. Yet, if one of the masses is a mass distribution represented by a constant density ρ, then the force on a mass m at a distance r from the center of the distribution is $F = (4/3)\pi Gm\rho r$. The force rises linearly with the distance because the amount of matter enclosed also increases. Thus, the gravitational force for a constant mass distribution increases like Hooke's law and thus oscillatory solutions are encountered. This sheds light on our repulsive oscillator found above. In this case we want the gravity to be repulsive, but the cosmological constant acts just like the uniform matter distribution.

One often writes the cosmological constant in terms of a vacuum energy density as $\Lambda = 8\pi G\rho_{vac}$ so that the velocity and acceleration equations become

$$H^2 = \left(\frac{\dot{a}}{a}\right)^2 = \frac{8\pi G}{3}\rho - \frac{k}{a^2} + \frac{\Lambda}{3} = \frac{8\pi G}{3}(\rho + \rho_{vac}) - \frac{k}{a^2}, \tag{2.29}$$

and

$$\frac{\ddot{a}}{a} = -\frac{4\pi G}{3}(1+\gamma)\rho + \frac{\Lambda}{3} = -\frac{4\pi G}{3}(1+\gamma)\rho + \frac{8\pi G}{3}\rho_{vac}. \tag{2.30}$$

2.2.4. *Energy density and dark energy*

The density, ρ, of matter in the Universe determines its shape. Dividing both sides of Eq. (2.28) by H^2 yields

$$1 = \frac{8\pi G\rho}{3H^2} - \frac{k}{H^2a^2} + \frac{\Lambda}{3H^2}. \tag{2.31}$$

Each of the terms in this equation has special significance. The *mass density* is

$$\Omega_m = \frac{8\pi G\rho}{3H^2}, \tag{2.32}$$

the *curvature density* is

$$\Omega_k = -\frac{k}{H^2 a^2}, \tag{2.33}$$

and the *vacuum energy density*, or *dark energy*, is

$$\Omega_\Lambda = \frac{\Lambda}{3H^2}. \tag{2.34}$$

Another quantity of interest is the *critical density*, given by

$$\rho_{crit} = \frac{3H^2}{8\pi G}, \tag{2.35}$$

in terms of which the mass density can be written as $\Omega_m = \rho/\rho_{crit}$. In terms of the present value of the Hubble parameter the critical density is,

$$\rho_{crit} = 1.88 \times 10^{-29} h^2 \text{ g cm}^{-3}, \tag{2.36}$$

where

$$h = \frac{H}{100\,\text{km Mpc}^{-1}\ \text{s}^{-1}} \simeq 0.70. \tag{2.37}$$

The constant h is known as the *Hubble unit* with value 1 if H is 100 km Mpc^{-1}.

Defining

$$\Omega = \Omega_m + \Omega_\Lambda, \tag{2.38}$$

the Friedmann equation can be rewritten as (with $\Lambda = 0$)

$$(\Omega - 1)H^2 = \frac{k}{a^2} \tag{2.39}$$

so that $k = 0, +1, -1$ corresponds to $\Omega = 1$, $\Omega > 1$ and $\Omega < 1$.

The value of Ω, at least on relatively small scales, seems to depend on scale. Indeed, the contribution to Ω from visible matter associated with stars and hot gas is quite small, $\Omega \approx 0.003 - 0.01$. On somewhat larger scales, that of galactic halos or small groups of galaxies, $\Omega \approx 0.02 - 0.1$. On galaxy cluster scales, it appears that Ω may be as large as 0.3. And while there is some evidence, the observations are far from conclusive in indicating a value of Ω as large as 1, meaning a *flat Universe.*

The matter density decreases with the radius of the Universe as $r(t = 0)/r(t) = a_0^3/a^3$. Thus, we can write a mixture of matter and dark energy by[1]

$$\rho = \rho_m + \rho_\Lambda = \rho_{m0}\left(\frac{a_0}{a}\right)^3 + \rho_\Lambda, \tag{2.40}$$

and the Friedmann equation becomes

$$\left(\frac{\dot{a}}{a}\right)^2 - H^2\Omega_{m0}\left(\frac{a_0}{a}\right)^3 - H^2\Omega_{\Lambda 0} = -\frac{k}{a^2}. \tag{2.41}$$

Using $k = 0$ (flat Universe), $\Omega_{m0} = 1 - \Omega_{\Lambda 0}$ and, for simplicity $a_0 = 1$ (in appropriate units), we get

$$\left(\frac{\dot{a}}{a}\right)^2 = H^2\left[(1 - \Omega_{\Lambda 0})\frac{1}{a^3} + \Omega_{\Lambda 0}\right]. \tag{2.42}$$

Integrating over time, with t_0 denoting the age of the Universe, we get

$$\begin{aligned} H_0 t_0 &= \int_0^1 da \frac{\sqrt{a}}{\sqrt{1 - \Omega_{\Lambda 0} + \Omega_{\Lambda 0} a^2}} \\ &= \frac{2}{3\sqrt{\Omega_{\Lambda 0}}} \ln\left(\frac{1 + \sqrt{\Omega_{\Lambda 0}}}{\sqrt{1 - \Omega_{\Lambda 0}}}\right), \end{aligned} \tag{2.43}$$

where H_0 is the present value of the Hubble constant.[2] We thus see that, as $\Omega_{\Lambda 0} \to 1$, then $t_0 \to \infty$. It is thus necessary to have some matter to keep the age of the Universe finite.

We can turn this argument around. Assuming the age of the Universe to be $t_0 \simeq 13.7\,\mathrm{Gy}$ we get $\Omega_{\Lambda 0} \simeq 0.72$, or $\Omega_{m0} = 0.28$, i.e. only 28% of the Universe is matter and 78% is dark energy. Observations also indicates that only 4% of the Universe is baryonic (normal) matter, and that the remaining 24% is in some other still unknown form, a *dark matter*. Dark matter and dark energy thus compose about 95% of the Universe.

---

[1] Here, the index "0" means the present value of the variables.

[2] We willl often drop the index "0" for this quantity, whenever there is no ambiguity.

2.2.5. *Static Universe*

The static Universe requires $a = a_0 =$ constant and thus $\ddot{a} = \dot{a} = 0$. From Eq. (2.23), $\ddot{a} = 0$ requires that

$$\Lambda = 4\pi G(\rho + 3p) = 4\pi G(1+\gamma)\rho. \tag{2.44}$$

If there is no cosmological constant ($\Lambda = 0$) then either $\rho = 0$ which is an empty Universe, or $p = -\rho/3$ which requires negative pressure. Both of these alternatives were unacceptable to Einstein and therefore he concluded that a cosmological constant was present, i.e. $\Lambda \neq 0$. From Equation (2.44) this implies

$$\rho = \frac{\Lambda}{4\pi G(1+\gamma)}, \tag{2.45}$$

and because ρ is positive this requires a positive Λ. Inserting Eq. (2.45) into Eq. (2.28), it follows that

$$\Lambda = \frac{3(1+\gamma)}{3+\gamma}\left[\left(\frac{\dot{a}}{a_0}\right)^2 + \frac{k}{a_0^2}\right]. \tag{2.46}$$

Now imposing $\dot{a} = 0$ and assuming a matter equation of state ($\gamma = 0$) implies $\Lambda = k/a_0^2$. However the requirement that Λ be positive forces $k = +1$, giving

$$\Lambda = \frac{1}{a_0^2} = \text{constant}. \tag{2.47}$$

Thus the cosmological constant for the static Universe is not any value but rather simply the inverse of the scale factor squared, where the scale factor has a fixed value in this static model.

Using Eq. (2.44), we obtain that the static Universe is closed with the scale factor (which in this case gives the radius of curvature) given by (*Einstein's radius*)

$$a_0 = \frac{1}{\sqrt{4\pi G\rho_0}}. \tag{2.48}$$

Using $\rho_0 = \rho_{crit}$, the numerical value of Einstein's radius is of order of 10^{10} light years.

It is worth noting that even though the model is static, it is unstable: if perturbed away from the equilibrium radius, the Universe will either expand to infinity or collapse. If we increase a from a_0, then the Λ term will dominate the equations, causing a runaway expansion, whereas if we decrease a from a_0, the dust term will dominate, causing collapse. Therefore, this model is also physically unsound, and this is a far worse problem than the (to Einstein) unattractive presence of Λ.

2.2.6. *Matter and radiation profiles*

Equation (2.5) can be rewritten as

$$\dot{\rho} + 3(\rho + p)\frac{\dot{a}}{a} = 0, \tag{2.49}$$

or

$$\frac{d}{dt}(\rho a^3) + p\frac{da^3}{dt} = 0. \tag{2.50}$$

From Eq. (2.13), $p = \gamma\rho/3$, from which follows that

$$\frac{d}{dt}(\rho a^{3+\gamma}) = 0. \tag{2.51}$$

Integrating this we obtain

$$\rho = \frac{c}{a^{3+\gamma}}, \tag{2.52}$$

where c is a constant. This shows that the density falls as a^{-3} for matter-dominated and a^{-4} for radiation-dominated Universes.

2.2.7. *The age of the Universe*

Writing Eq. (2.28) as

$$\dot{a}^2 = \frac{8\pi G}{3}(\rho + \rho_{vac})a^2 - k, \tag{2.53}$$

we infer that the present day value of k is

$$k = \frac{8\pi G}{3}(\rho_0 + \rho_{0vac})a_0^2 - H_0^2 a_0^2, \tag{2.54}$$

with $H_0^2 = (\dot{a}/a_0)^2$. Present day values of quantities have been denoted with a subscript 0. Substituting Eq. (2.54) into Eq. (2.53) yields

$$\dot{a}^2 = \frac{8\pi G}{3}(\rho a^2 - \rho_0 a_0^2 + \rho_{vac} a^2 - \rho_{0vac} a_0^2) - H_0^2 a_0^2. \tag{2.55}$$

Integrating the equation above gives the expansion age of the Universe

$$t_0 = \int_0^{a_0} \frac{da}{\dot{a}} = \int_0^{a_0} \frac{da}{\sqrt{\frac{8\pi G}{3}(\rho a^2 - \rho_0 a_0^2 + \rho_{vac} a^2 - \rho_{0vac} a_0^2) - H_0^2 a_0^2}}. \tag{2.56}$$

For a non-varying cosmological constant $\rho_{vac} = \rho_{0vac}$ and because $a^2 < a_0^2$, then a non-zero cosmological constant will give an age for the Universe larger than it would have been obtained were it not present.

2.3. Solutions of the Friedmann Equation

2.3.1. *Deceleration parameter*

We consider here a flat, $k = 0$, Universe. Results for an open or closed Universe with $k \neq 0$ can easily be obtained and are shown schematically in Fig. 2.2. Currently the Universe is in a matter dominated phase whereby the dominant contribution to the energy density is due to matter. However the early Universe was radiation dominated and the very early Universe was vacuum dominated. With $k = 0$, there will only be one term on the right hand side of Eq. (2.29) depending on what is dominating the Universe. For a matter ($\gamma = 0$) or radiation ($\gamma = 1$) dominated Universe the right hand side of Eq. (2.29) will be of the form $1/a^{3+\gamma}$ (ignoring the vacuum energy), whereas for a vacuum dominated Universe the right hand side will be a constant. The solution to the Friedmann equation for a radiation dominated Universe will thus be (from $a\,da \propto dt$)

$$a \propto t^{1/2} \quad \text{(radiation dominated Universe)}, \tag{2.57}$$

while for the matter dominated case it will be (from $a^{1/2} da \propto dt$)

$$a \propto t^{2/3} \quad \text{(matter dominated Universe)}. \tag{2.58}$$

One can see from $\ddot{a}$ that these results give negative acceleration, corresponding to a decelerating expanding Universe.

From the Hubble rate in Eq. (1.8), we get

$$\dot{H} = -H^2 + \frac{\ddot{a}}{a} = -H^2 \left(1 - \frac{\ddot{a}}{H^2 a}\right) = -H^2(1+q), \tag{2.59}$$

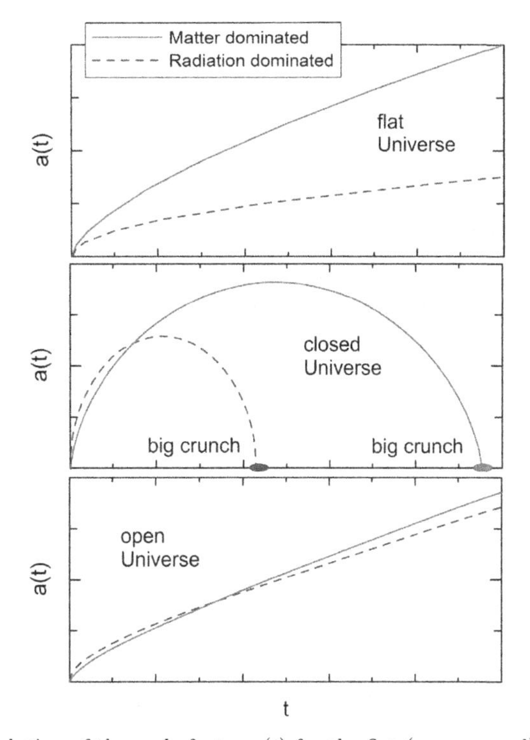

Figure 2.2. Evolution of the scale factor $a(t)$ for the flat (upper panel), closed (middle panel) and open (lower panel) Friedmann Universe.

where we defined the *deceleration parameter*,

$$q = -\frac{\ddot{a}}{H^2 a}. \tag{2.60}$$

In this approximation, the Universe is modeled by the dust approximation, i.e., $p = 0$, so that Eq. (2.30) (with $\rho_{vac} = 0$) yields

$$\rho = \frac{3H^2}{4\pi G} q. \tag{2.61}$$

Plugging this result into the Friedmann Equation (2.18), one gets

$$-k = a^2 H^2 (1 - 2q). \tag{2.62}$$

Since both $a \neq 0$ and $H \neq 0$, for flat Universe ($k = 0$) we get $q = 1/2$. We also get $q > 1/2$ if $k = 1$ and $q < 1/2$ if $k = -1$. When combined with Eq. (2.61), this yields the *critical density*, Eq. (2.35), the density needed to yield the flat Universe.

It is important to note that the quantity q provides the relationship between the density of the Universe and the critical density,

$$q = \frac{\rho}{2\rho_{crit}}. \tag{2.63}$$

2.3.2. *Flat Universe*

In this case, $k = 0$ and $q = 1/2$. We consider the matter-dominated and the radiation-dominated Universes separately.

Matter-dominated (dust approximation)

We have $p = 0$ and $a^3\rho = const$ and, using Eq. (2.53) (with $\Lambda = 0$), we get

$$\frac{\dot{a}^2}{a^2} = \frac{8\pi G}{3}\rho_0\left(\frac{a_0}{a}\right)^3, \tag{2.64}$$

which leads to

$$\int a^{1/2}da = \frac{2}{3}a^{3/2} + C = \sqrt{\frac{8\pi G\rho_0 a_0^3}{3}}t. \tag{2.65}$$

At the Big Bang, $t = 0$, $a = 0$, so $C = 0$. Since the Universe is assumed flat, $k = 0$, $\rho_0 = \rho_{crit}$, we have

$$\frac{a}{a_0} = (6\pi G\rho_0)^{1/3}t^{2/3} = (6\pi G\rho_{crit})^{1/3}t^{2/3} = \left(6\pi G\frac{3H_0^2}{8\pi G}\right)^{1/3} t^{2/3}$$
$$= \left(\frac{3H_0}{2}\right)^{2/3} t^{2/3}. \tag{2.66}$$

From this we compute the age of the Universe t_0, which corresponds to the Hubble rate H_0 and the scale factor $a = a_0$, to be

$$t_0 = \frac{2}{3H_0}. \tag{2.67}$$

Using the present value of H_0 given by Eq. (2.37), we get

$$t_0 \simeq 9.1 \times 10^9 \text{ years} \equiv 9.1\ \mathcal{A}\ \text{(aeon)}. \tag{2.68}$$

The model above in known as the *Einstein–de Sitter model.*

Radiation-dominated

We have $p = \rho/3$ and $a^4\rho = const$ and, using Eq. (2.53) (with $\Lambda = 0$), we get

$$\int a\ da = 2a^2 + C = \sqrt{\frac{8\pi G\rho_0 a_0^4}{3}}t. \tag{2.69}$$

Again, at the Big Bang, $t = 0$, $a = 0$, thus $C = 0$, and ρ_0 is equal to the critical density ρ_{crit}. Therefore,

$$\frac{a}{a_0} = \left(\frac{2\pi}{3}G\rho_{crit}\right)^{1/4} t^{1/2} = \left(\frac{H_0}{2}\right)^{1/2} t^{1/2}. \tag{2.70}$$

For a radiation-dominated Universe, the age of the Universe would be much longer ($t_0 \sim 27\,\mathcal{A}$) than for a matter-dominated Universe.

2.3.3. *Closed Universe*

In this case, $k = 1$ and $q > 1/2$. Here we only present the results, leaving the derivations for the exercises.

Matter-dominated

We have $p = 0$ and $a^3\rho = const$. The solutions are obtained in parametric form:

$$\frac{a}{a_0} = \frac{q}{2q-1}(1 - \cos\eta), \quad \text{with } \frac{t}{t_0} = \frac{q}{2q-1}(\eta - \sin\eta). \tag{2.71}$$

Radiation-dominated

We have $p = \rho/3$ and $a^4\rho = const$. The solutions are also obtained in parametric form:

$$\frac{a}{a_0} = \sqrt{\frac{2q}{2q-1}}\sin\eta, \quad \text{with } \frac{t}{t_0} = \sqrt{\frac{2q}{2q-1}}(1 - \cos\eta). \tag{2.72}$$

In both matter- and radiation-dominated closed Universes, the evolution is cycloidal — the scale factor grows at an ever-decreasing rate until it reaches a point at which the expansion is halted and reversed. The Universe then starts to compress and it finally collapses in the *Big Crunch* (see Figs. 2.2 and 2.3).

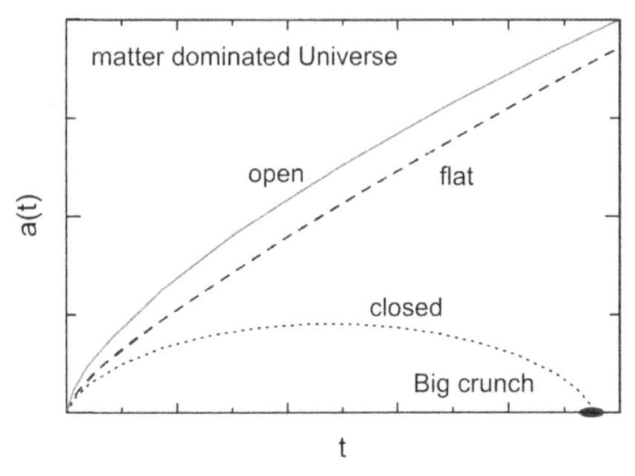

Figure 2.3. Evolution of the scale factor $a(t)$ for the flat, closed and open matter-dominated Friedmann Universe.

2.3.4. *Open Universe*

In this case, we have $k = -1$ and $q < 1/2$ and the solutions for matter-dominated and radiation-dominated Universe are discussed next.

Matter-dominated

We have $p = 0$ and $a^3\rho = const.$ The solutions are obtained in parametric form,

$$\frac{a}{a_0} = \frac{q}{2q-1}(\cosh\eta - 1), \quad \text{with } \frac{t}{t_0} = \frac{q}{2q-1}(\sinh\eta - \eta). \tag{2.73}$$

Radiation-dominated

We have $p = \rho/3$ and $a^4\rho = const.$ The solutions are also obtained in parametric form,

$$\frac{a}{a_0} = \sqrt{\frac{2q}{1-2q}}\sinh\eta, \quad \text{with } \frac{t}{t_0} = \sqrt{\frac{2q}{1-2q}}(\cosh\eta - 1). \tag{2.74}$$

Figure 2.3 summarizes the evolution of the scale factor $a(t)$ for open, flat and closed matter-dominated Universes. In earlier times, one can expand the trigonometric and hyperbolic functions to leading terms in powers of η, and the a and t dependence on η for the different curvatures are given in the Table 2.1. This shows that at early times the curvature of the Universe does not matter. The singular behavior at early times is essentially independent

Table 2.1. Solutions of Friedmann equation for matter-dominated Universe.

Curvature	All η		Small η		
k	a/a_0	t	a	t	$a(t)$
0	$(6\pi G\rho_0)^{1/3}t^{2/3}$	—	$\propto t^{2/3}$	—	$\propto t^{2/3}$
1	$\frac{q}{2q-1}(1-\cos\eta)$	$\frac{q}{2q-1}(\eta-\sin\eta)$	$\propto \eta^2$	$\propto \eta^3$	$\propto t^{2/3}$
-1	$\frac{q}{1-2q}(\cosh\,\eta-1)$	$\frac{q}{1-2q}(\sinh\,\eta-\eta)$	$\propto \eta^2$	$\propto \eta^3$	$\propto t^{2/3}$

of the curvature of the Universe, or k. The Big Bang is a matter dominated singularity.

2.3.5. *The flatness problem*

According to Eq. (2.39), the density parameter satisfies an equation of the form

$$|\Omega - 1| = \frac{|\rho(t) - \rho_c|}{\rho_c} = \frac{1}{\dot{a}^2(t)}. \tag{2.75}$$

The present day value of Ω is known only roughly, $0.1 \leq \Omega \leq 2$. On the other hand $1/\dot{a}^2(t) \sim 1/t^2$ in the early stages of the evolution of the Universe, so the quantity $|\Omega - 1|$ was extremely small. One can show that in order for Ω to lie in the range $0.1 \leq \Omega \leq 2$ now, the early Universe must have had

$$|\Omega - 1| = \frac{|\rho(t) - \rho_c|}{\rho_c} \leq 10^{-59}. \tag{2.76}$$

This means that if the density of the Universe were initially greater than ρ_c, say by $10^{-55}\rho_c$, it would be closed and the Universe would have collapsed long time ago. If on the other hand the initial density were $10^{-55}\rho_c$ less than ρ_c, the present energy density in the Universe would be vanishingly low and the life could not exist. The question of why the energy density in the early Universe was so fantastically close to the critical density is usually known as the flatness problem.

2.3.6. *The horizon problem*

The redshift, given by Eq. (1.8), is a natural consequence of the Doppler effect. As the Universe expands at a rate r, the wavelength of a particle scales as $\lambda = \lambda_0/r$ which, combined with Eq. (1.8) yields $z = (1-r)/r$, or

$$r = \frac{1}{1+z}. \tag{2.77}$$

Gravitational redshift is observed when a receiver is located at a higher gravitational potential than the source. The physical explanation is that the particle loses a fraction of the energy (and hence increases its wavelength) by overcoming the difference in the potential (climbing out of the potential well). *Comoving coordinates* are those in which an observer is comoving with the Hubble expansion. Only for these observers in the comoving coordinates, is the Universe isotropic. Otherwise, portions of the Universe will exhibit a systematic bias: portions of the sky will appear systematically blue- or red-shifted.

Comoving horizon is defined as the total portion of the Universe visible to the observer. It represents the sphere with radius equal to the distance the light could have traveled (in the absence of interactions) since the Big Bang ($t=0$). In time dt, light travels a comoving scaled distance $d\eta = dx/a = cdt/a$, where dx is a physical distance. In units of $c = 1$, this becomes

$$\eta = \int_0^t \frac{dt'}{a(t')}, \tag{2.78}$$

where η is called the *conformal time*. Because it is a monotonically increasing variable of time t, it can be used as an independent variable when discussing the evolution of the Universe (just like the time t, temperature T, redshift z and the scale factor a). In some approximations, Eq. (2.78) above can be analytically solved. For instance, in a matter-dominated Universe $\eta \propto \sqrt{a}$ and in a radiation-dominated Universe $\eta \propto a$.

Cosmological time is identical to locally measured time for an observer at a fixed comoving spatial position, that is, in the local comoving frame. *Proper distance* is equal to the locally measured distance in the comoving frame for nearby objects. To measure the proper distance between two distant objects, one imagines that one has many comoving observers in a straight line between the two objects, so that all of the observers are close to each other, and form a chain between the two distant objects. All of these observers must have the same cosmological time. Each observer measures their distance to the nearest observer in the chain, and the length of the chain, the sum of distances between nearby observers, is the total proper distance.

The portions of the sky on our comoving horizon which are separated by more than η are not causally connected, because there has not been an "exchange of information" between these regions. This means that, in the absence of interaction, these parts should have evolved differently and reached different temperatures. In a physical system homogeneity can be established only in regions which were in communication. In the case of the Universe the scale of these regions is set by the horizon radius $r \sim ct$, where t is the age of the Universe. When the age of the Universe was 10^{-37} s and its temperature $T = 10^{28}$ K (GUTs epoch) then $r \sim 3 \times 10^{-27}$ cm. So if we consider a region of this size there will be homogeneity. The question arises that as the Universe expands what will be the extent of homogeneity now? As the temperature has come down from $T = 10^{28}$ K to $T = 3$ K, the scale factor has increased by 3×10^{27}. So the size of the homogeneous region at present turns out to be $\sim 3 \times 10^{-27} \times 3 \times 10^{27} = 9$ cm. But from observations of CMB we find homogeneity in a region whose radius is of the order of 10^{28} cm. So how did two disconnected regions "come to know" of each others conditions? This is known as the *horizon problem.*

2.4. The Curvature of Space-Time

We have derived the Friedmann equation without use of *General Relativity* (GR). But the derivation of this equation from general relativity brings new insights in our understanding of the Universe. Although a detailed account of the theory general relativity (and of cosmology, in general) is beyond the scope of this book, we discuss in Appendix C a summary of the theory. In this section we discuss the main consequences of GR for cosmology and stellar evolution without emphasis on the derivation of the equations. We will return to many of these topics in the following chapters and we will need some of this knowledge to understand standard phenomena.

2.4.1. *Einstein field equations*

In Appendix C we show the main ingredients leading to *Einstein's field equations.* Here we summarize some of their properties and common metrics. The GR equations are

$$G_{\mu\nu} = 8\pi G T_{\mu\nu} - \Lambda g_{\mu\nu}, \tag{2.79}$$

where the *Einstein tensor* is defined as

$$G_{\mu\nu} = R_{\mu\nu} - \frac{1}{2} R g_{\mu\nu}. \tag{2.80}$$

$R_{\mu\nu}$ is the *Ricci tensor* and R is known as the *Riemann scalar.* Einstein equations are a set of 16 coupled equations which will give the *spacetime metric* $g_{\mu\nu}$ (buried inside $G_{\mu\nu}$) for a given *energy–momentum tensor,* $T_{\mu\nu}$. Actually there are only 10 independent equations because of the symmetry $g_{\mu\nu} = g_{\nu\mu}$. Alternatively, we can write Einstein equations as

$$R_\mu^\nu - \frac{1}{2}R = 8\pi G T_\mu^\nu - \Lambda = 8\pi G \tilde{T}_\mu^\nu, \tag{2.81}$$

where

$$\tilde{T}_\mu^\nu = \begin{pmatrix} \rho - \dfrac{\Lambda}{8\pi G} & 0 & 0 & 0 \\ 0 & -p + \dfrac{\Lambda}{8\pi G} & 0 & 0 \\ 0 & 0 & -p + \dfrac{\Lambda}{8\pi G} & 0 \\ 0 & 0 & 0 & -p + \dfrac{\Lambda}{8\pi G} \end{pmatrix}. \tag{2.82}$$

The new energy–momentum tensor $\tilde{T}_\mu^\nu = T_\mu^\nu - \Lambda/8\pi G$ reveals the nature of the cosmological constant — it is a source of energy density and the inverse pressure (opposing the pressure of matter). Indeed, this is what led to the coining of the name *dark energy.* Einstein introduced the cosmological constant because he was dissatisfied that otherwise his field equations did not allow, apparently, for a static Universe: gravity would cause a Universe which was initially at dynamic equilibrium to contract [Ga70].

The solutions of the Einstein field equations are *metrics* of spacetime. These metrics describe the structure of the spacetime including the inertial motion of objects in the space. As the field equations are non-linear, they cannot always be completely solved (i.e., without making approximations). For example, there is no known complete solution for a spacetime with two massive bodies in it (which is a theoretical model of a binary star system). Approximations are usually made in these cases. But there are numerous cases where the field equations have been solved completely. These solutions lead to the prediction of black holes and to different models of evolution of the Universe.

After obtaining $g_{\mu\nu}$ from Einstein's field equations, we can calculate the paths of light rays, the orbits of planets, etc. In practice the solution of the Einstein field equations are exceedingly difficult and only a few exact solutions are known. In practice, the way one usually solves Einstein's equations is by specifying a metric in general terms which contains unknown coefficients. This metric is substituted into the Einstein equations and one solves for the unknown coefficients.

The *Friedmann–Robertson–Walker* (FRW) (or Friedmann–Lemaître–Robertson–Walker) metric is the metric appropriate to a homogeneous and isotropic Universe but where size can change with time [Fri22,Fri24, Lem27,Lem31,Lem33,Rob35,Rob36]. The *Schwarzschild* metric describes spacetime in the vicinity of a non-rotating massive spherically-symmetric object [Sch16]. This will be relevant to discuss the structure of compact stars. We give a brief description of both.

2.4.2. *The FRW metric*

The metric of special relativity is $ds^2 = c^2dt^2 - (dx^2 + dy^2 + dz^2)$. Clearly the spatial part is a 3-d Euclidean flat space. In Appendix D we show that the spatial metric for a homogeneous, isotropic curved space with size $a(t)$ that can change in time is

$$ds^2 = a^2(t)\left[\frac{dr^2}{1-kr^2} + r^2(d\theta^2 + \sin^2\theta d\phi^2)\right], \tag{2.83}$$

where $k = 0,\ +1, -1$ for flat, closed and open *hyperspheres*, respectively.

Replacing the spatial part of the special relativity metric with (2.83) we have the *Friedmann–Robertson–Walker* (FRW) metric

$$ds^2 = c^2dt^2 - a^2(t)\left[\frac{dr^2}{1-kr^2} + r^2(d\theta^2 + \sin^2\theta d\phi^2)\right]. \tag{2.84}$$

where $a(t)$ is called the scale factor and the constant k can be 0 or ± 1, depending on the curvature. The Friedmann–Robertson–Walker metric is an exact solution of Einstein's field equations of general relativity; it describes a simply connected, homogeneous, isotropic expanding or contracting Universe (see Appendix D).

The coordinates r, θ, ϕ are such that the circumference of a circle corresponding to t, r, θ all being constant is given by $2\pi a(t)r$, the area of a sphere corresponding to t and r constant is given by $4\pi a^2(t)r^2$, but the physical radius of the circle and sphere is given by

$$R_U = a(t)\int_0^r \frac{dr'}{\sqrt{1-kr'^2}}. \tag{2.85}$$

For $k = +1$ the Universe is closed (but without boundaries), and $a(t)$ may be interpreted as the "radius" of the Universe at time t. If $k = 0$, or $+1$, the Universe is (flat, or open) and infinite in extent.

The time coordinate t appearing in the FRW metric is the so-called *cosmic time*. It is the time measured on the clock of an observer moving along with the expansion of the Universe. The isotropy of the Universe makes it possible to introduce such a global time coordinate. Observers at different points can exchange light signals and agree to set their clocks to a common time t when, e.g., their local matter density reaches a certain value. Because of the isotropy of the Universe, this density will evolve in the same way in the different locations, and thus once the clocks are synchronized they will stay so.

2.4.3. *The Schwarzschild metric*

The Schwarzschild solution is another simple and useful solution of the Einstein field equations. It describes spacetime in the vicinity of a non-rotating massive spherically-symmetric object. Working with coordinates (t, r, θ, ϕ) labeled 0 to 3, the metric in its most general form has 10 independent components, each of which is an arbitrary function of 4 variables. The solution is assumed to be spherically symmetric, static and in the vacuum. As explained in Appendix D, these assumptions lead to the Schwarzschild metric

$$ds^2 = c^2\left(1 - \frac{2Gm}{c^2 r}\right)dt^2 - \left(1 - \frac{2Gm}{c^2 r}\right)^{-1} dr^2 - r^2(d\theta^2 - \sin^2\theta d\phi^2). \tag{2.86}$$

In deriving the Schwarzschild metric the static assumption is stronger than required, as the Birkhoff's theorem states that any spherically symmetric vacuum solution of Einstein's field equations is stationary. *Birkhoff's theorem* has the consequence that any pulsating star which remains spherically symmetric cannot generate *gravitational waves* (as the region exterior to the star must remain static).

Consider a static, spherically symmetric star residing alone in an otherwise empty Universe. Since the star's interior is spherical, it is reasonable to assume that the exterior will be spherical; and since the exterior is also vacuum ($T_{\mu\nu} = 0$), its spacetime geometry must be that of Schwarzschild. If the circumference of the star's surface is $2\pi R$ and its surface area is $4\pi R^2$, then that surface must reside at the location $r = R$ in the Schwarzschild coordinates of the exterior. In other words, the spacetime geometry will be described by the Schwarzschild metric, Eq. (2.86), at radii $r > R$, but by something else inside the star, at $r < R$.

Particles with finite rest masses reside on the star's surface. They move along *timelike* world lines. From the Schwarzschild metric the proper time is $d\tau^2 = -ds^2 = c^2(1 - 2GM/c^2R)dt^2$ along those world lines. Thus $d\tau^2$ is positive (timelike world line) if and only if $R > 2GM/c^2 = 3.0\,\text{km} \times M/M_\odot$. Thus, a static star with total mass-energy M can never have a circumference smaller than $2\pi R = 4\pi GM/c^2$. The Sun satisfies this constraint by a huge margin: $R = 7 \times 10^5$ km. A one-solar-mass white dwarf star satisfies it by a smaller margin: $R \simeq 6 \times 10^3$ km. And a one-solar-mass neutron star satisfies it by only a modest margin: $R \simeq 10$ km.

2.4.4. *Gravitational waves*

Some binary pulsars exhibit orbital decay, consistent with energy losses due to the emission of gravitational waves. By analogy with electromagnetic waves, produced when a charged particle is accelerated in an electromagnetic field, *gravitational waves* are produced when mass is accelerated in a gravitational field. Gravity waves are a direct prediction of GR, as the field equations take the form of wave equations. The effect is predicted to be very small and several sensitive experiments are being prepared to measure gravity waves from a number of different astrophysical sources (e.g. binary pulsars, coalescing black holes).

To show how gravitational waves arise from Einstein's theory of general relativity, consider a situation where the gravitational field is weak but not static. In the absence of gravity, space-time is flat and is characterized by the Minkowski metric, $\eta_{\mu\nu} = (0, -1, -1, -1)\delta_{\mu\nu}$. A weak gravitational field can be considered as a small "perturbation" on the flat *Minkowski metric*, Eq. (B.43),

$$g_{\mu\nu} = \eta_{\mu\nu} + h_{\mu\nu}, \quad |h_{\mu\nu}| \ll 1. \tag{2.87}$$

Indices of any tensor can be raised or lowered using $\eta^{\mu\nu}$ or $\eta_{\mu\nu}$ respectively as the corrections would be of higher order in the perturbation, $h_{\mu\nu}$. We can therefore write,

$$g^{\mu\nu} = \eta^{\mu\nu} - h^{\mu\nu}. \tag{2.88}$$

Under a Lorentz transformation the perturbation transforms as a second-rank tensor,

$$h_{\alpha\beta} = a^{\mu}_{\alpha} a^{\nu}_{\beta} h_{\mu\nu}. \tag{2.89}$$

The equations obeyed by the perturbation, $h_{\mu\nu}$, are obtained by writing the Einstein's equations to first order. To first order, the *Christoffel symbol*, Eq. (C.25), is,

$$\Gamma^{\lambda}_{\mu\nu} = \frac{1}{2}\eta^{\lambda\rho}[\partial_\mu h_{\rho\nu} + \partial_\nu h_{\mu\rho} - \partial_\rho h_{\mu\nu}] + \mathcal{O}(h^2), \tag{2.90}$$

and the *Riemann curvature tensor* reduces to

$$R_{\mu\nu\rho\sigma} = \eta_{\mu\lambda}\partial_\rho \Gamma^{\lambda}{}_{\nu\sigma} - \eta_{\mu\lambda}\partial_\sigma \Gamma^{\lambda}{}_{\nu\rho}. \tag{2.91}$$

The Ricci tensor is obtained to first order as

$$R_{\mu\nu} \approx R^{(1)}_{\mu\nu} = \frac{1}{2}\left[\partial_\lambda\partial_\nu h^{\lambda}{}_{\mu} + \partial_\lambda\partial_\mu h^{\lambda}{}_{nu} - \partial_\mu\partial_\nu h - \Box h_{\mu\nu}\right], \tag{2.92}$$

where, $h = \eta^{\mu\nu}h_{\mu\nu}$ and $\Box \equiv \eta^{\lambda\rho}\partial_\lambda\partial_\rho$ is the D'Alembertian in flat space-time. Contracting again with $\eta^{\mu\nu}$, the Ricci scalar is obtained as

$$R = \partial_\lambda\partial_\mu h^{\lambda\mu} - \Box h. \tag{2.93}$$

The Einstein tensor, $G_{\mu\nu}$, in the limit of weak gravitational field is

$$\begin{aligned} G_{\mu\nu} = R_{\mu\nu} - \frac{1}{2}\eta_{\mu\nu}R \simeq \frac{1}{2}[&\partial_\lambda\partial_\nu h^{\lambda}_{\mu} + \partial_\lambda\partial_\mu h^{\lambda}{}_{\nu} - \eta_{\mu\nu}\partial_\mu\partial_\nu h^{\mu\nu} \\ &+ \eta_{\mu\nu}\Box h - \Box h_{\mu\nu}]. \end{aligned} \tag{2.94}$$

Einstein's field equations (2.79) do not have unique solutions as any solution to these equations will not remain invariant under a "gauge" transformation. As a result, equations (2.79) will have infinitely many solutions. In other words, the decomposition (2.88) of $g_{\mu\nu}$ in the weak gravitational field approximation does not completely specify the coordinate system in space-time. When we have a system that is invariant under a gauge transformation, we *fix* the gauge and work in a selected coordinate system. One such coordinate system that satisfies the gauge condition is

$$g^{\mu\nu}\Gamma^{\lambda}{}_{\mu\nu} = 0. \tag{2.95}$$

In the *weak field limit*, this condition reduces to

$$\partial_\lambda h^{\lambda}{}_{\mu} = \frac{1}{2}\partial_\mu h. \tag{2.96}$$

This condition is often called the *Lorentz gauge*. In this selected gauge, the linearized Einstein equations simplify to,

$$\Box h_{\mu\nu} - \frac{1}{2}\eta_{\mu\nu}\Box h = -16\pi G T^{\mu\nu}. \tag{2.97}$$

Defining $\bar{h}_{\mu\nu}$, as

$$\bar{h}_{\mu\nu} = h_{\mu\nu} - \frac{1}{2}\eta_{\mu\nu}h, \tag{2.98}$$

the Lorentz gauge condition further reduces to

$$\partial_\mu \bar{h}^\mu{}_\lambda = 0. \tag{2.99}$$

The Einstein equations are then

$$\Box \bar{h}_{\mu\nu} = -16\pi G T^{\mu\nu}. \tag{2.100}$$

Outside the source, where $T_{\mu\nu} = 0$, Eq. (2.100) reduces to

$$\Box \bar{h}_{\mu\nu} = 0, \tag{2.101}$$

or

$$\left(-\frac{\partial^2}{c^2 \partial t^2} + \Delta\right) \bar{h}_{\mu\nu} = 0, \tag{2.102}$$

where $\Delta = \partial_i \partial_i$, with $i = 1, 2, 3$. This is just a wave equation, for waves traveling at the speed of light. Its solutions can be written as superpositions of plane waves with frequencies ω and wave vectors $\mathbf{k}$, in the form $A_{\mu\nu}\cos(\omega t - \mathbf{k}\cdot\mathbf{x})$, where $\omega = c|\mathbf{k}|$ and $A_{\mu\nu}$ has constant components. Note that solutions of this kind will exist also in the complete absence of sources (i.e., vacuum spacetime with $T_{\mu\nu}$ everywhere). Although the latter solutions are by themselves unphysical, they illustrate that the gravitational field has dynamics of its own, independent of matter.

As gravitational waves pass through space-time, they cause small ripples. The stretching and shrinking is on the order of 1 part in 10^{21} even with a strong gravitational wave source. Due to their small magnitude, gravitational waves are difficult to detect. Large astronomical events that could create measurable space-time waves include the collapse of a neutron star, a black hole or the Big Bang.

2.4.5. *Black holes*

According to Birkhoff's theorem, the spacetime geometry outside an imploding, spherical star must be that of Schwarzschild. This means, in particular, that an imploding, spherical star cannot produce any gravitational waves, as such waves would break the spherical symmetry. As it applies to a static star, so also for an imploding one, because real particles live on its surface, the world line of that surface, with $r = R(t)$,

$\theta = $ constant, $\phi = $ constant, must be timelike. Consequently, at each point along the world line it must lie within the local light cones.

The radial edges of the light cones are generated by the world lines of radially traveling photons, i.e., photons with world lines of constant θ, ϕ, and varying t, r.[3] Setting to zero the ds^2 of Eq. (2.86), we see that along these null, radial world lines,

$$ds^2 = 0 = c^2 \left(1 - \frac{2GM}{c^2 r}\right) dt^2 - \left(1 - \frac{2GM}{c^2 r}\right)^{-1} dr^2, \tag{2.103}$$

or

$$\frac{dt}{dr} = \pm \frac{1/c}{1 - 2GM/c^2 r}. \tag{2.104}$$

Integrating this differential equation we obtain

$$r + \left(\frac{2GM}{c^2}\right) \ln \left| \left(\frac{c^2 r}{2GM}\right) - 1 \right| = \pm ct + \text{constant}. \tag{2.105}$$

Since the world line of the star's surface is confined to the interiors of the local light cones, the fact that $dt/dr \to \pm\infty$ in Eq. (2.104) prevents the star's world line $r = R(t)$ from ever, in any finite time t, reaching the gravitational radius, or *Schwarzschild radius*

$$R_s = \frac{2GM}{c^2}. \tag{2.106}$$

Only after a lapse of infinite coordinate time t will the star's surface reach the gravitational radius $r = 2GM/c^2$. Although the implosion to $R_s = 2GM/c^2$ requires infinite Schwarzschild time t, it requires only a finite proper time τ as measured by an observer who rides inward on the star's surface. In fact, one can show that the proper time is

$$\tau \simeq \left(\frac{R_0^3}{2GM}\right)^{1/2} = 15 \text{ microseconds } \times \left(\frac{R_0 c^2}{2MG}\right)^{3/2} \frac{M}{M_\odot}, \tag{2.107}$$

where R_0 is the star's initial radius when it first begins to implode freely. This implosion time is equal to $1/(4\sqrt{2})$ times the orbital period of a test particle at the radius of the star's initial surface. For a star with mass and initial radius equal to that of the Sun, it is about 30 minutes; for a neutron

[3]Spherical symmetry dictates that if a photon starts out traveling radially, it will always continue to travel radially.

star that has been pushed over the maximum mass limit by accretion of matter from its surroundings, is about 0.1 milliseconds.

The region of strong, vacuum gravity left behind by the implosion of the star is called a *black hole*. The *horizon*, $r = 2GM/c^2$, is the surface of the hole, and the region $r < 2GM/c^2$ is its interior. The spacetime geometry of the black hole, outside and at the surface of the star which creates it by implosion, is that of Schwarzschild. The horizon is defined as the boundary between spacetime regions that can and cannot communicate with the external Universe. It forms initially at the star's center, and then expands to encompass the surface at the precise moment when the surface penetrates the gravitational radius.

The singularity can possibly be seen by means of light rays that emerge from its vicinity. However, because the future light cones are all directed into it, no light-speed or sub-light-speed signals can ever emerge from it. In fact, because the outer edge of the light cone is tilted inward at every event inside the gravitational radius, no signal can emerge from inside the gravitational radius to tell external observers what is going on there. In effect, the gravitational radius is an absolute event horizon for our Universe, a horizon beyond which we cannot see, except by plunging through it.

Due to quantum fluctuations, particle–antiparticle pairs are created near the event horizon. One particle falls into the singularity as the other escapes. Antiparticles that escape radiate as they annihilate with matter. Hawking [Haw75] calculated the blackbody temperature of the black hole to be

$$T = \frac{\hbar c^3}{8\pi kGM}, \tag{2.108}$$

where k is the Boltzmann constant. The power radiated is

$$W(T) = 4\pi\sigma R^2 \left(\frac{\hbar c^3}{8\pi kGM}\right)^4, \tag{2.109}$$

where σ is the Stefan–Boltzmann constant from blackbody theory (discussed in Chapter 4). Small primordial black holes ($< 10^{-19}\, M_\odot$) can be detected by their *Hawking radiation*, but it is negligible for other black holes. The energy expended to pair production at the event horizon decreases the total mass-energy of the black hole, causing the black hole to slowly evaporate with a lifetime

$$\tau = \frac{M^3}{3v\hbar} = 8.3 \times 10^{-26}\, M^3\, \mathrm{s/g^3}, \tag{2.110}$$

where v is a constant, $v = 1/(15,360\pi)$. Thus, the smaller the mass the shorter the lifetime. Solar-mass black holes would live much longer than the age of the Universe, but small black holes ($< 10^{14}$ g, about the size of a mountain) would explode in a certain amount of time.

2.4.6. *Inflation*

Inflation [Gut81] occurs when the vacuum energy contribution dominates the ordinary density and curvature terms in Equation (2.29). Assuming these are negligible and substituting $\Lambda = \text{constant}$, one gets $\dot{a}^2/a^2 = \Lambda/3$, resulting in the solution

$$a \propto \exp(\sqrt{\Lambda/3}t) \quad \text{(inflationary Universe).} \tag{2.111}$$

This corresponds to a so-called *de Sitter Universe*, characterized by a metric

$$ds^2 = -dt^2 + e^{2Ht}[d\chi^2 + \chi^2(d\theta^2 + \sin\theta^2 d\theta^2)]. \tag{2.112}$$

The acceleration is positive, corresponding to an accelerating expanding Universe called an *inflationary Universe.*

We are heading toward de Sitter Universe, because the density of dark energy remains constant, while the matter density scales as a^{-3} and radiation density as a^{-4}, which makes the dark energy an ever-increasing part of the cosmic inventory. The exponential expansion of the scale factor means that the physical distance between any two observers will eventually be growing faster than the speed of light. At that point those two observers will, of course, not be able to have any contact anymore. Eventually, we will not be able to observe any galaxies other than the Milky Way and a handful of others in the gravitationally-bound Local Group cluster of galaxies.

Before $t = 10^{-3}$ s, the Universe was dominated by quantum mechanics. A measure for this point in time is the (reduced) Compton wavelength of the Universe

$$R_c = \frac{\hbar}{Mc}, \tag{2.113}$$

where M is all the mass in the Universe. Equating the Compton wavelength and the Schwartzschild radius (Eq. (2.106)) leads to the so called

Planck mass (within a factor of $\sqrt{2}$)

$$M_{Pl} = \left(\frac{\hbar c}{G}\right)^{1/2} = 1.2 \times 10^{19}\ \text{GeV}/c^2, \tag{2.114}$$

which is equivalent to a temperature $T = 10^{32}$ K, using $M_{Pl}c^2 \sim kT$. Via the uncertainty relation from quantum mechanics, $\Delta E \Delta t \sim \hbar/2$, or $\Delta x \Delta p \sim \hbar/2$, one can get a Planck time and length, i.e. $t_{Pl} = 5 \times 10^{-44}$ s and $l_{Pl} = 3 \times 10^{-33}$ cm. At these lengths and times, time itself becomes undefined.

In typical models of inflation the phase transition takes place at temperatures around the Grand Unification Temperature (GUT) when the Universe was about 10^{-34} s old. Suppose that the Universe stayed in the inflationary state for 10^{-32} s. This may appear to be a short time, but in fact inflation lasted for 100 hundred times the age of the Universe at the time inflation started. Consider a small region with radius around, say, 10^{-23} cm before inflation. After inflation the volume of the region has increased by a factor of $(e^{100})^3 = 1.9 \times 10^{130}$. This huge increase solves some of the problems previously mentioned:

(a) *The flatness problem* — In the relativistic cosmological models the k/a^2 term ($k = \pm 1$) could have been important. However, with the sudden inflation the scale factor has increased by $\sim 10^{29}$ and the value of k/a^2 has been reduced by 10^{58}. So, even if we started from a large curvature term there is no other mechanism necessary to arrive at the nearly flat Universe. During inflation the energy density of the Universe is constant, whereas the scale factor increases exponentially. This means that Ω must have been exponentially close to unity, $\Omega = 1$ to an accuracy of many decimal places.

(b) *The horizon problem* — Due to inflation the small size particle horizon is increased by a factor of 10^{29}, which solves the problem.

After the end of inflation the vacuum energy of the field driving inflation was transferred to ordinary particles, so a reheating of the Universe took place. During inflation the Universe itself supercooled with $T \sim e^{-Ht}$. The period of inflation and reheating is strongly non-adiabatic, since there is an enormous generation of entropy at reheating. After the end of inflation, the Universe restarts in an adiabatic phase with the standard conservation of the energy. In fact the Universe restarts from very special initial conditions that the horizon, flatness and monopole problems are avoided.

2.5. Exercises

1. The Copernican principle asserts that no point in the Universe is special. That means that an observer will find the Universe to have the same large-scale properties regardless of where she or he carries out the observations. Prove that if we assume that the Universe is isotropic and that the Copernican principle is valid, then the Universe must be homogeneous.
2. To illustrate that Newtonian theory is not applicable to an infinite Universe, show that gravitational forces from two halves of the homogeneous Universe are infinite and should destroy any body, a contradiction with everyday experience. There is also a flaw in this reasoning. What is it?
3. Use the value of H to calculate the critical density in SI units and also in hydrogen atoms per cubic centimeter.
4. In the solar system frame, two events are measured to occur 3.0 hr apart in time and 1.5 hr apart in space. Observers in an alien spaceship measure the two events to be separated by only 0.5 hr in space. What is the time separation between the events in the alien's frame? (Read Appendix B.)
5. Suppose you are planning a trip in which a spacecraft is to travel at a constant velocity for exactly six months, as measured by a clock on board the spacecraft, and then return home at the same speed. Upon return, the people on Earth will have advanced exactly one hundred years into the future. According to special relativity, how fast must you travel? (Read Appendix B.)
6. The four-velocity u^α of a particle with nonzero rest mass is defined as $u^\alpha = dx^\alpha/d\tau$, where τ is the proper time. The proper time is the time measured by an observer riding along with the particle. Such an observer therefore sees no motion in the spatial coordinates: $ds^2 = -c^2dt^2 + dx^2 + dy^2 + dz^2 = -c^2dt^2$. Remember that the time interval dt depends on the observer; for this special observer (who has $dx = dy = dz = 0$), we call the time coordinate τ. The proper time interval $d\tau$ is therefore related to the invariant interval ds by $ds^2 = -c^2d\tau^2$. With this in mind, what is the squared magnitude of the four-velocity?
7. Suppose that the Universe were full of regulation baseballs, each of mass $m = 0.145$ kg and radius $r = 0.0369$ m. If the baseballs were distributed uniformly throughout the Universe, what number density of baseballs would be required to make the density equal to the critical density? (Assume non-relativistic baseballs.) Given this density of baseballs,

how far would you be able to see, on average, before your line of sight intersected a baseball? In fact, we can see galaxies at a distance $\gg c/H \gg 4000\,\text{Mpc}$; does the transparency of the Universe on this length scale place useful limits on the number density of intergalactic baseballs?

8. Show that the first law of thermodynamics

$$dE = TdS - PdV$$

leads to

$$S = \frac{4aT^3V}{3}$$

for thermal radiation.

9. Estimate the mass density of galaxies in the Universe in SI units, given a typical separation of 1 Mpc and a typical mass of 10^{11} $M_\odot$. Compare this with the critical density of the Universe.
10. Suppose you are in an infinitely large, infinitely old Universe in which the average density of stars is $n_s = 10^9\,\text{Mpc}^{-3}$ and the average stellar radius is equal to the Sun's radius: $R_s = R_\odot = 7 \times 10^8$ m. How far, on average, could you see in any direction before your line of sight struck a star? (Assume standard Euclidean geometry holds true in this Universe.) If the stars are clumped into galaxies with a density $n_G = 1\,\text{Mpc}^{-3}$ and average radius $R_G = 2000\,\text{pc}$, how far, on average, could you see in any direction before your line of sight hit a galaxy?
11. If $\Omega_{matter} = 0.3$, $\Omega_{vac} = 0.7$ and $T_{CMB} = 2.73\,\text{K}$, at what redshift was the radiation energy density (really, the density of relativistic particles, which includes both photons and neutrinos) comparable to the vacuum energy density?
12. In this chapter we examined solutions for the expansion when the Universe contained either dust ($p = 0$) or radiation $p = \rho/3$. Suppose we have a more general equation of state $p = w\rho$ where w is a constant in the range $-1 \leq w \leq 1$. Find solutions for $\rho(a)$, $a(t)$ and hence $\rho(t)$ for Universes containing such matter. Assume $k = 0$ in the Friedmann equation. What is the solution if $p = -\rho$?
13. Prove Eqs. (2.71) and (2.72) for the closed Universe (*Hint*: introduce the variable change $d\eta = dt/a$, where η is the *conformal time*).
14. Prove Eqs. (2.73) and (2.74) for the open Universe.

15. If $\rho = 3 \times 10^{-27}$ kg m^{-3}, what is the radius of curvature a_0 of Einstein's static Universe? How long would it take a photon to circumnavigate such a Universe?
16. Show that the age of the matter-dominated Universe is given by

$$t_0 = \frac{1}{H_0} \begin{cases} \dfrac{1}{1-2q} - \dfrac{q}{(1-2q)^{2/3}} \cosh^{-1}\left(\dfrac{1-q}{q}\right), & \text{for } q < \dfrac{1}{2} \\ \dfrac{1}{1-2q} + \dfrac{q}{(2q-1)^{2/3}} \cos^{-1}\left(\dfrac{1-q}{q}\right), & \text{for } q \geq \dfrac{1}{2}. \end{cases} \tag{2.115}$$

Plot it and discuss.
17. Metric for an open Universe: for a $k = -1$ Friedmann cosmology ($\Lambda = 0$), with $\rho = p = 0$, show that the FRW metric line element becomes

$$c^2 d\tau^2 = c^2 dt^2 - c^2 t^2 [dr^2 + \sinh^2 r (d\theta^2 + \sin^2 \theta d\phi^2)].$$

18. (a) Show that the general relativistic relation between recession velocity and cosmological redshift is

$$v_{rec}(t, z) = \frac{c}{a_0} \dot{a}(t) \int_0^z \frac{dz'}{H(z')},$$

where $H(z')$ is the Hubble constant at redshift z'. (b) Show that the special relativistic relation between velocity of an object and redshift is

$$v(z) = c \frac{(1+z)^2 - 1}{(1+z)^2 + 1}.$$

(c) Show that both relativistic relations are approximately $v \sim cz$ at small distance.
19. Suppose the energy density of the cosmological constant is equal to the present critical density $\rho_\Lambda = \rho_0 = 5200\,\text{MeV}\,\text{m}^{-3}$. What is the total energy of the cosmological constant within a sphere 1 AU in radius? What is the rest energy of the Sun ($E = Mc^2$)? Comparing these two numbers, do you expect the cosmological constant to have a significant effect on the motion of planets within the Solar System?
20. Consider Einstein's static Universe, in which the attractive force of the matter density ρ is exactly balanced by the repulsive force of the cosmological constant, $\Lambda = 4\pi G \rho$. Suppose that some of the matter is converted into radiation (by stars, for instance). Will the Universe start to expand or contract? Explain your answer.

Chapter 3

Particles

3.1. Introduction

In the previous chapter we learned how the solutions of the Friedmann equation have profound implications for our understanding of the Universe. We have also mentioned matter and radiation, but we did not specify, for example, what matter is composed of. In this chapter we will discuss the basic properties of elementary particles. Some of them have a more recursive role on most of the important properties of stellar and cosmic evolution. They are the electron, neutrino, photon and nucleon. We will discuss the characteristics of heavier particles first.

The proton, with symbol p, is the nucleus of the hydrogen atom, has charge $+e$ of the same absolute value as that of the electron, and mass

$$m_p = 938.271998(38)\,\mathrm{MeV/c^2}, \tag{3.1}$$

where the values in the parenthesis are the errors in the last two digits. The proton has spin $\frac{1}{2}$.

The *neutron*, with symbol n, has charge zero, spin $\frac{1}{2}$, and mass

$$m_n = 939.565330(38)\,\mathrm{MeV/c^2}. \tag{3.2}$$

The neutrons and protons obey the Pauli principle and thus they are fermions, having spin $\frac{1}{2}$. We recall that particles with fractional spin $(2n + 1)/2$ are *fermions*, and that particles with integer spin are *bosons*. The proton and the neutron have similar properties in several aspects, and it is convenient to utilize the generic name of *nucleon* for both.

The force that keeps the nucleus bound is the *nuclear force*. It acts between two nucleons of any type and, contrary to the Coulomb force, it is of short range. To a good extent, the nuclear force has its origin in the exchange of virtual particles with finite rest mass between the nucleons.

These particles are called *mesons.* The π-meson (or simply, *pion*) interacts strongly with nucleons but has a very short lifetime and decays into a *muon*, which has a longer lifetime and does not interact strongly with other particles. The muon does not enter into the description of the nuclear force and is classified among the *leptons*, the family of light particles to which the electron belongs.

The pion exists in three charge states, π^+, π^0 and π^-. The π^+ and π^- have the same mass, 139.56995(35) MeV, the same mean lifetime, $\tau = 2.6 \times 10^{-8}$ s, and decay almost exclusively by the process

$$\pi^+ \to \mu^+ + \nu_\mu, \qquad \pi^- \to \mu^- + \bar{\nu}_\mu, \tag{3.3}$$

where μ^+, μ^- are the positive and negative muons, ν_μ is the *muonic neutrino* and $\bar{\nu}_\mu$, the corresponding antineutrino. Only a small fraction, 1.2×10^{-4}, of pions decays by

$$\pi^+ \to \mathrm{e}^+ + \nu_e, \qquad \pi^- \to \mathrm{e}^- + \bar{\nu}_e, \tag{3.4}$$

yielding a positron (electron) and an *electron neutrino* (*antineutrino*). The neutrinos are particles with zero charge and very small mass. Electron neutrinos have a significant role in β-decay theory.

Analysis of pion–nucleon and pion–deuteron reactions leads to the conclusion that the pion spin is zero. The pions are bosons and obey *Bose statistics.*

3.2. Properties of Elementary Particles

3.2.1. *Antiparticles*

For each particle in nature there is a corresponding antiparticle, with the same mass, as well as electric charge with the same magnitude but opposite sign. Antiprotons and antineutrons are called *antinucleons.* The magnitude of every quantity associated to a particle is identical to that of the corresponding antiparticle but, as we shall see soon, there are other quantities besides charge for which the values for particles and antiparticles have opposite signs.

The mesons π^+ and π^- are antiparticles of each other. In this case it is not important to define which is the particle and which is the antiparticle, since mesons are not normal constituents of matter. In the case of the π^0, particle and antiparticle coincide, since charge and magnetic moment are zero.

The term *antimatter* refers to the collection of antiparticles. The Universe is almost exclusively filled with matter while existing antimatter is of secondary origin resulting from relativistic collisions of matter. This is true not only of the antiparticles produced in accelerators but also of the antiparticles encountered in cosmic rays.

3.2.2. *Spin, isospin and baryonic number*

The elementary particles exist in groups of approximately the same mass, but with different charges. The neutron mass, for example, is about the same as that of the proton, and the mass of the neutral pion, π^0, is approximately equal to that of the charged pions, π^+ and π^-. The proton and the neutron can be seen as two *charge states* of a single particle, using the name nucleon to identify this particle.

In the theory of atomic spectra a state that has *multiplicity* $(2s + 1)$, has spin s.[1] It is common, however, to refer to the *spin quantum number* simply as *spin* s. One example is the Zeeman effect which is the energy splitting among the $(2s + 1)$ states of an atom in a magnetic field.

The nucleon has, relative to its charge, a multiplicity $2 \times \frac{1}{2} + 1 = 2$. By analogy with the theory of the atomic spectra, we create a quantity called *isospin*, $t = \frac{1}{2}$, to obtain the multiplicity $2t + 1 = 2$. The isospin cannot be identified with an angular momentum, and does not have any connection with the spatial properties of the nucleon. However, we can deal with the isospin in the same way we deal with the angular momentum. The addition of the isospin of several particles can be treated with the vector model for addition, in analogy to that used in the atomic spectra theory. For example, when we add two spins $\frac{1}{2}$, the total spin can be 0 or 1. That is also the case of the sum of isospins of two nucleons (each one has spin $\frac{1}{2}$).

The $(2s+1)$ states of a system with spin s are denoted by the $(2s+1)$ distinct values of the z component of $\mathbf{s}$ (in units of $\hbar$):

$$s_z = -s, -s + 1, \ldots, s - 1, s. \tag{3.5}$$

By analogy, the $(2t + 1)$ states of a system with isospin t are denoted by the $(2t + 1)$ distinct values of the t_z component,

$$t_z = -t, -t + 1, \ldots, t - 1, t. \tag{3.6}$$

[1]The spin is a vector with modulus $\hbar\sqrt{s(s + 1)}$.

The direction of the third axis in charge space is chosen in such a way that $t_z = +\frac{1}{2}$ for the proton and $t_z = -\frac{1}{2}$ for the neutron.

The pion has three charge states; thus it has isospin $t = 1$. The three pions form a charge *multiplet*, or isospin multiplet, with multiplicity $2t + 1 = 3$. The state $t_z = +1$ is attributed to the π^+, $t_z = 0$ to the π^0 and $t_z = -1$ to the π^-. This is connected to the convention that was adopted for the nucleons and is necessary for the validity of Eq. (3.7) that follows.

The isospin magnitude is an invariant quantity in a system governed by the strong interaction. In the electromagnetic interactions this quantity is not necessarily conserved and we shall verify ahead that this is the only conservation law that has different behavior in relation to these two forces.

The number of nucleons (plus some related particles — see below) before and after a reaction is always the same. One often uses the quantity, B, called *baryonic number*, that is always conserved in reactions. We attribute to the proton and to the neutron the baryonic number $B = 1$, and to the antiproton and antineutron $B = -1$. To the pions we ascribe $B = 0$ (also for electrons, neutrinos, muons and photons). In this way the conservation of baryonic number is extended to all reactions. This principle is extended to the leptons, defining a *leptonic number*, which is also conserved in reactions.

From the isospin and baryonic number definition we can write the charge q, in units of e, as

$$q = t_z + \frac{B}{2}. \tag{3.7}$$

Since the antiparticle of a particle of charge q and baryonic number B has charge $-q$ and baryonic number $-B$, it must have also a third isospin component $-t_z$, where t_z is the isospin z-component of the corresponding particle.

3.2.3. *Strangeness and hypercharge*

Strange particles form two distinct groups. One of them consists of particles heavier than the nucleons and decaying into them, and are called *hyperons*. The symbols Λ, Σ, Ξ and Ω are utilized for the several hyperons. Because it decays into a nucleon, a hyperon is a baryon and has baryonic number 1. They also have spin $\frac{1}{2}$, so they are fermions. The other group of strange particles are bosons with spin 0 and are called *mesons-K*, or *kaons*.

Typical reactions involving strange particles are:

$$\begin{aligned} \Lambda^0 &\rightarrow p + \pi^-, & \Lambda^0 &\rightarrow n + \pi^0, \\ \Xi^- &\rightarrow \Lambda^0 + \pi^-, & \Xi^0 &\rightarrow \Lambda^0 + \pi^0, \\ K^0 &\rightarrow \pi^+ + \pi^-, & \Lambda^0 &\rightarrow \pi^- + p, \\ \pi^- + p &\rightarrow K^0 + \Lambda^0. \end{aligned} \tag{3.8}$$

The *interaction time* for reactions involving nucleons and pions is obtained approximately by the time in which a pion, with velocity near that of light, travels a distance equal to the nuclear force range. This time is about $r/c \simeq 10^{-23}$ s, which is much less than the mean lifetime of the Λ^0 ($\tau = 2.5 \times 10^{-10}$ s), or of other strange particles. Experimentally the rate at which lambdas, or other strange particles, are produced is consistent with an interaction time of the order of 10^{-23} s. The reason why strange particles are produced so fast but decay so slowly is that the strong interactions (those acting between nucleons, or between pions and nucleons) are responsible for the production of strange particles. The reactions in which only one strange particle takes part, as in its decay, proceed through the *weak interaction*, similar to β-decay or the decay of muons and charged pions.

The production of strange particles can be explained by the introduction of a new quantum number, the *strangeness*, and postulating that the strangeness is conserved in the strong interactions. For example, two strange particles, but with opposite strangeness, could be produced by means of the strong interaction in a collision between a pion and a nucleon. The strangeness, however, is not conserved in the decay of a strange particle, and this decay is attributed to the weak interaction.

3.2.4. *Quarks and gluons*

The great number of hadrons, as shown in Table 3.1, and their apparently complex distribution leads to the question of whether these particles are complex structures composed of simpler entities. The inspiration for that came from the symmetries observed when one put mesons and baryons in plots of strangeness versus the t_z-component of isospin. The type of observed symmetry is a characteristic of the group called SU_3, where three basic elements can generate singlets (the mesons η' and ϕ), octets (the other eight mesons and the eight spin 1/2 baryons) and decuplets (the spin 3/2 baryons). These three basic elements, initially conceived only as mathematical entities able to generate the necessary symmetries, ended acquiring the status of real elementary particles, called *quarks*. To obtaining

Table 3.1. Attributes of particles that interact strongly. The baryons are the particles with baryonic number $B \neq 0$; the mesons have $B = 0$. S is the strangeness, t the isotopic spin and t_z its projection; s is the particle spin and m its mass. Baryons have positive intrinsic parity, mesons have negative ones.

	B	S	t	t_z	s	m (MeV/c^2)
p	+1	0	1/2	+1/2	1/2	938.272
n	+1	0	1/2	−1/2	1/2	939.565
$\overline{\text{p}}$	−1	0	1/2	−1/2	1/2	938.272
$\overline{\text{n}}$	−1	0	1/2	+1/2	1/2	939.565
Λ	+1	−1	0	0	1/2	1115.68
Σ^+	+1	−1	1	+1	1/2	1189.4
Σ^0	+1	−1	1	0	1/2	1192.6
Σ^-	+1	−1	1	−1	1/2	1197.4
$\overline{\Lambda}$	−1	+1	0	0	1/2	1115.68
$\overline{\Sigma^+}$	−1	+1	1	−1	1/2	1189.4
$\overline{\Sigma^0}$	−1	+1	1	0	1/2	1192.6
$\overline{\Sigma^-}$	−1	+1	1	+1	1/2	1197.4
Ξ^0	+1	−2	1/2	+1/2	1/2	1315
Ξ^-	+1	−2	1/2	−1/2	1/2	1321
$\overline{\Xi^0}$	−1	+2	1/2	−1/2	1/2	1315
$\overline{\Xi^-}$	−1	+2	1/2	+1/2	1/2	1321
Ω^-	+1	−3	0	0	3/2	1672
π^0	0	0	1	0	0	134.976
π^+	0	0	1	+1	0	139.567
π^-	0	0	1	−1	0	139.567
K$^+$	0	+1	1/2	+1/2	0	493.7
K$^-$	0	−1	1/2	−1/2	0	493.7
K^0	0	+1	1/2	−1/2	0	497.7
$\overline{K^0}$	0	−1	1/2	+1/2	0	497.7

hadronic properties, these three quarks, presented in *flavors up (u), down (d)* and *strange (s)* must have the characteristic values shown in Table 3.2.

The most striking fact is the existence of particles with fractional charge (a fraction of the electron charge). We can construct a nucleon by combining three quarks (neutron = udd) and is natural to attribute to quarks a baryonic number $B = 1/3$. The pions, by their turn, are obtained by the junction of a quark and an antiquark: ($\pi^+ = \text{u}\bar{\text{d}}$), ($\pi^0 = \text{d}\bar{\text{d}}$), ($\pi^- = \text{d}\bar{\text{u}}$),

Table 3.2. Characteristic quantum numbers of quarks.

Flavor	Charge	Spin	Strangeness
up	+2/3	1/2	0
down	−1/3	1/2	0
strange	−1/3	1/2	−1

where the properties of an antiparticle for the quarks are obtained in the conventional way.

To reproduce the properties of the other baryons and mesons, the strange quarks have to play a role. A hyperon like the Σ^0, for example, has the constitution ($\Sigma^0 = \mathrm{uds}$) while a meson has ($\mathrm{K}^+ = \mathrm{u\bar{s}}$). It is convenient to say at this point that a certain combination of quarks does not necessarily has lead to only one particle. In the case of the combination above, we also have the possibility to build the hyperon $\Sigma^{*0} = \mathrm{uds}$. The reason for this is that, besides other quantum numbers that will be discussed ahead, a combination of three fermions can give rise to particles with different spin. If we consider the quark's orbital angular momentum as zero, which is true for all particles that we discussed, the total spin of the three quarks can be 1/2 or 3/2. The hyperon Σ^0 corresponds to the first case and the hyperon Σ^{*0} to the second.

A first difficulty in the theory appears when we examine the particles ($\Delta^{++} = \mathrm{uuu}$), ($\Delta^- = \mathrm{ddd}$), and ($\Omega^- = \mathrm{sss}$). Since the three quarks in each case are fermions with $l = 0$, it is clear that at least two of them would be in the same quantum state, which violates the Pauli principle. To overcome this difficulty, one has to introduce a new quantum number, the *color*: the quarks, besides the flavors up, down or strange, would also have a color, red (R), green (G) or blue (B), or an *anticolor*, $\bar{\mathrm{R}}, \bar{\mathrm{G}}$ or $\bar{\mathrm{B}}$. It is clear that, in the same way as the flavor, the color has nothing to do with the usual notion that we have of that property. The introduction of this new quantum number solves the above difficulty, since a baryon like the Δ^{++} is now written as $\Delta^{++} = \mathrm{u}_R\mathrm{u}_G\mathrm{u}_B$, the problems with the Pauli principle are eliminated. The addition of three new quantum numbers expands the number of possible hadrons but a new rule comes to play, limiting the possibilities of color combination: *all the possible states of hadrons are colorless*, where colorless in this context means absence of color or white color. The white is obtained when, in a baryon, one adds three quarks, one of each color. In this sense the analogy with the common colors works, since the addition of red, green and blue gives white. In a

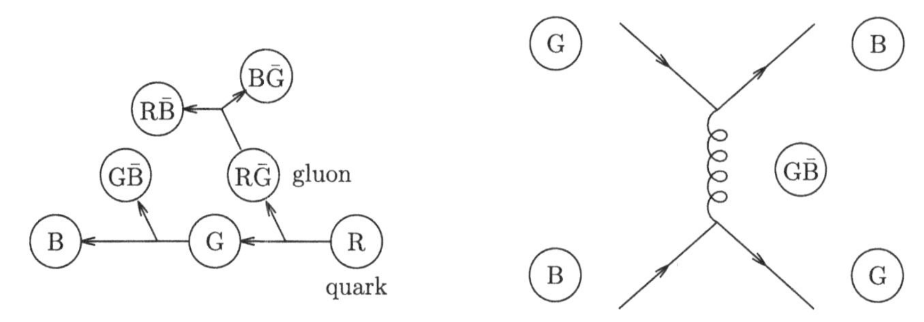

Figure 3.1. (a) Forces between quarks mediated by gluon exchange. (b) Diagram showing how a quark B changes to a quark G, and vice-versa, by the exchange of a gluon GB.

meson, the absence of color results from the combination of a color and the respective anticolor. Another way to present this property is to understand the anticolor as the complementary color. In this case, the analogy with the common colors also works and the pair color–anticolor also results in white.

The concept of color is not only useful to solve the problem with the Pauli principle. It has a fundamental role in quark interaction processes. The accepted theory for this interaction establishes that the force between quarks works by the exchange of massless particles with spin 1, called *gluons*. These gluons always carry one color and one different anticolor and in the mediation process they interchange the respective colors; one example is seen in Fig. 3.1. One can also see in the Fig. 3.1b that the gluons themselves can emit gluons.

The fields around hadrons where exchange forces act by means of colors are denominated *color fields* while the gluons, which are the exchanged particles, turn to be the field particles of the strong interaction. In this task they replace the pions that, in the new scheme, are composite particles. The fact that the gluons have colors and can interact mutually makes the study of color fields (quantum chromodynamics) particularly complex.

A fourth quark flavor has received the designation of *charm* (c). This c quark has a charge of $+2/3$; it has strangeness zero but has a new quantum number, the charm C, with an attributed value $C = 1$. The J/Ψ particle is interpreted as a $c\bar{c}$ state, called *charmonium*, by analogy with the positronium $e\bar{e}$.

Two more quarks exist. The quark b (from *bottom* or *beauty*), has charge $-1/3$ and a new quantum number, the *beauty* B^*. The quark b has

Table 3.3. Properties of the elementary particles — in the upper Table each quark can appear in three colors, R, G and B. Only one member of the particle–antiparticle pair appears in the Table.

Quarks	Charge	Spin	Strangeness	Charm	Beauty	Truth
u	+2/3	1/2	0	0	0	0
d	−1/3	1/2	0	0	0	0
s	−1/3	1/2	−1	0	0	0
c	+2/3	1/2	0	1	0	0
b	−1/3	1/2	0	0	−1	0
t	+2/3	1/2	0	0	0	1

Leptons	Mass (MeV/c²)	Charge	Spin	Half-life (s)
e^-	0.511	−1	1/2	∞
ν_e	0	0	1/2	∞
μ^-	105.66	−1	1/2	2.2×10^{-6}
ν_μ	0	0	1/2	∞
τ^-	1784	−1	1/2	3.4×10^{-13}
ν_τ	0	0	1/2	∞

Field particles	Mass (GeV/c²)	Charge	Spin
Photon	0	0	1
$W^\pm$	81	1	1
Z^0	93	0	1
Gluons	0	0	1
Graviton	0	0	2

$B^* = -1$. The last quark is the t (from *top* or *true*), with charge +2/3. This quark was identified in experiments conducted at Fermilab in 1993 [Ab94].

The theory of quarks, with its colors and flavors, have created a scheme in which a great number of experimental facts can be explained. High-energy electron beams have indeed detected an internal structure in nucleons with all the features of quarks. However, one can never pull out a quark from a hadron and study its properties separately. To explain this impossibility, a theory of *Asymptotic Freedom* was developed *confining* quarks permanently to the hadrons. One consequence is that their masses cannot be directly determined, since it depends on the binding energies, which are also unknown. The quark model enjoys a high prestige in the theory of elementary particles and there is a substantial reduction in the number of elementary particles, that is, the point particles without an internal structure. These are the quarks, the leptons and the field bosons. A sketch of the properties of these particles is shown in Table 3.3.

3.3. Nuclear Properties

3.3.1. *Sizes*

The radius of protons and neutrons that compose the nucleus is of the order of 1 fm. Suppose that a nucleus has A nucleons and that nucleons are distributed inside a sphere of radius R. If the nucleons can be considered as small hard spheres of radius r_0 in contact with each other, we can write $A = (4\pi R^3/3)/(4\pi r_0^3/3)$ or,

$$R \cong r_0 A^{1/3}. \tag{3.9}$$

Empirically one finds $r_0 \simeq 1.2$ fm.

3.3.2. *Binding energies*

For a nucleus ${}^A_Z\text{X}$, with proton number Z and neutron number $N = A - Z$, the binding energy is given by

$$B(Z, N) = \{Zm_\text{p} + Nm_\text{n} - m(Z, N)\}c^2, \tag{3.10}$$

where m_p is the proton mass, m_n the neutron mass and $m(Z, N)$ the mass of the nucleus.

The binding energy defined by Eq. (3.10) is always positive. In Fig. 3.2 we show the binding energy per nucleon, B/A, as a function of A. The average value of B/A increases fast with A for light nuclei, decreasing slowly from 8.5 MeV to 7.5 MeV beginning with $A \cong 60$, where it has a maximum. We can say that for $A > 30$ nuclei the binding energy B is approximately proportional to A.

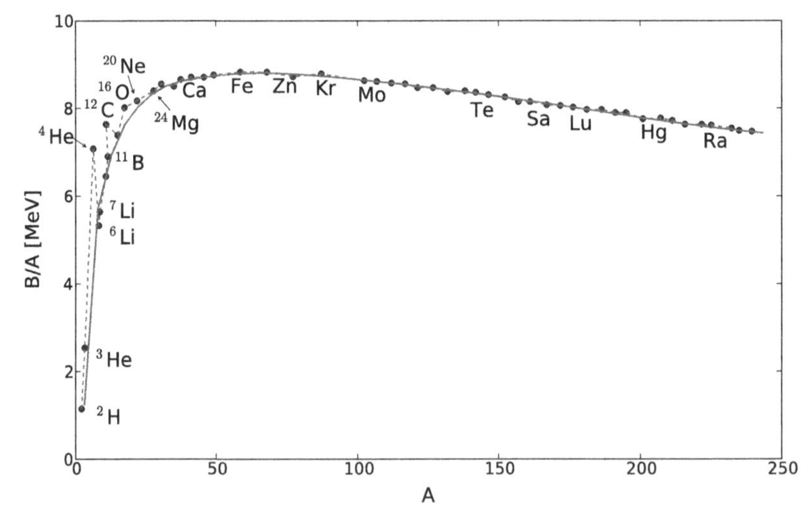

Figure 3.2. Average experimental values of B/A for nuclei, and the corresponding curve calculated by equations (3.16) and (3.21).

In the light nuclei region four points are observed whose binding energy per nucleon are greater than the local average: ${}^{4}_{2}\text{He}$, ${}^{8}_{4}\text{Be}$, ${}^{12}_{6}\text{C}$ and ${}^{16}_{8}\text{O}$. The nuclei ${}^{20}_{10}\text{Ne}$ and ${}^{24}_{12}\text{Mg}$ also lie in the upper part of the plot. Notice that these nuclei have equal and even proton and neutron number.

The initial rise of the B/A curve indicates that the fusion of two light nuclei produces a nucleus with greater binding energy per nucleon, releasing energy. This is the origin of energy production in the stars. The initial stage in the evolution of a star is the production of helium by means of hydrogen fusion; in later stages the production of heavier elements occurs by fusion of lighter nuclei. It is not difficult to conclude from Fig. 3.2 that if a star follows the normal course of its evolution without big accidents it will end as a cold cluster of $A \cong 60$ nuclei, since from that time on nuclear fusion is no longer energetically advantageous.

On the other side, for heavy nuclei the division into approximately equal parts (*nuclear fission*) releases energy. Figure 3.2 shows that in this case, the energy gain is nearly 1 MeV per nucleon and thus about 200 MeV is gained in each event. The nuclear fission process is the basis of nuclear reactors operation, where neutrons strike heavy elements (normally uranium or plutonium) causing them to fission and to produce more neutrons, forming chain reactions. It is also the basis of war artifacts.

The fact that the "saturation of nuclear forces" binding energy per nucleon is approximately constant for $A > 30$ is due to the *saturation of the nuclear forces*. Each nucleon is bound to $(A-1)$ other nucleons, in such a way that there are in total $A(A-1)/2$ nucleon–nucleon bindings in a nucleus with mass number A. Thus, if the range of nucleon–nucleon forces were greater than the nuclear dimension, the binding energy B should be proportional to the number of bindings between them, that is, B should be proportional to A^2. Since this is not the case, one concludes that the nucleon–nucleon forces have a range much smaller than the nuclear radius.

The binding energy B is the energy necessary to separate all the protons and all the neutrons of a nucleus. Another quantity of interest is the *separation energy* of a nucleon from the nucleus. The separation energy of a neutron from a nucleus (Z, N) is given by

$$S_n(Z,N) = \{m(Z,N-1) + m_n - m(Z,N)\}c^2 = B(Z,N) - B(Z,N-1). \tag{3.11}$$

In the same way we can define the separation energy of a proton or an α-particle. The separation energy can vary from a few MeV to about 20 MeV. It depends very much on the structure of the nucleus. One observes that S_n is greater for nuclei with an even number of neutrons. We can define a *pairing energy* as the difference between the separation energy of a nucleus with an even number of neutrons and one of a neighboring nucleus, that is,

$$\delta_n(Z, N) = S_n(Z, N) - S_n(Z, N - 1), \tag{3.12}$$

where N is even. One observes experimentally that both δ_n and δ_p are about 2 MeV.

When one plots the separation energy versus Z or N one sees that at the values $2, 8, 20, 50, 82, 126, 184$[2] the separation energy changes abruptly. These values are known as *magic numbers* and nuclei with magic Z (or N) have the last proton (or neutron) shell complete, similarly to what occurs in atomic physics with the closed shells of electrons in noble gases.

3.3.3. *Angular momentum*

The nucleus is a quantum system composed of A nucleons. The nucleons are fermions (half-spin particle): the laws of quantum mechanics for the addition of angular momenta establish that the total angular momentum, or the "spin" of the nucleus, is equal to $n\hbar$ (n integer) if A is even and $(n + \frac{1}{2})\hbar$ if A is odd. This is a direct consequence of the nucleon spin value, $\frac{1}{2}\hbar$, since orbital angular momenta only contribute with integer values of $\hbar$.

We can summarize the experimental observations by:

$I = n\hbar$ for odd–odd nuclei
$I = (n + 1/2)\hbar$ for odd nuclei (even–odd or odd–even)
$I = 0$ for even–even nuclei,

where I is the total angular momentum quantum number and n is an integer number greater than or equal to zero.

The correct determination of the nuclear spin values is an important element in the greater or smaller acceptance of models that try to describe the nuclear properties.

3.3.4. *Excited states*

The set of A nucleons that form the nucleus possesses, as all quantum bound systems, a sequence of excited energy states, above the ground state (the most bound) to which one ascribes energy zero. The values of

[2]The last two values refer only to neutrons.

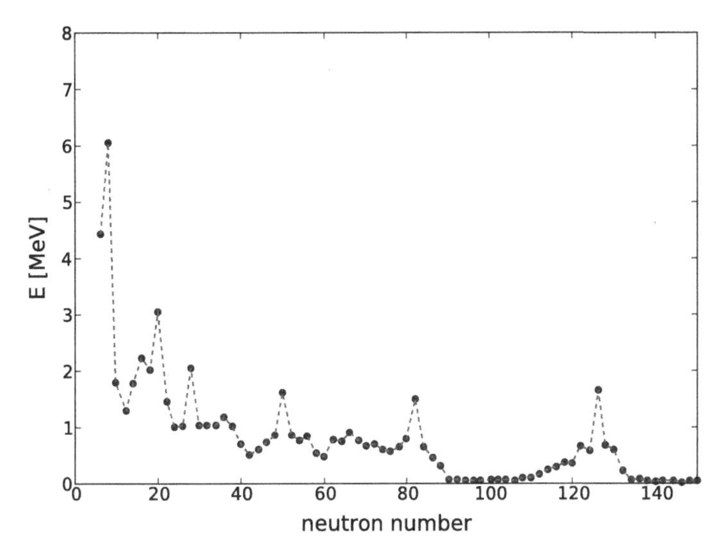

Figure 3.3. Energy of the first excited state, averaged over the stable even–even nuclei, as a function of the neutron number.

the energy of these states are normally presented in diagrams that also give, when known, the values of spin and parity corresponding to each state.

The distribution of states can vary enormously from nucleus to nucleus. ${}^{4}_{2}$He, for example, has its first excited state around 19 MeV while nuclei like ${}^{182}_{73}$Ta, ${}^{198}_{79}$Au, ${}^{223}_{88}$Ra and ${}^{223}_{90}$Th have more than 50 states below 1 MeV. But it is a general rule that the density of states increases rapidly with the energy, forming practically a continuum for high energies.

The energies of the first excited states are also affected by the presence of a magic number of protons or neutrons in the nucleus. Figure 3.3 shows the variation of the average energy of the first excited state of stable even–even nuclei as a function of the number of neutrons. The maxima at the magic numbers $8, 20, 28, 50, 82$ and 126 are evident.

3.3.5. *Nuclear stability*

In common situations an excited state decays to a lower energy state of the same nucleus with the emission of a γ-ray. The sequence of these transitions leads normally to the ground state of that nucleus. However, the ground state itself may not be stable for many nuclides, which can decay into other nuclides by the spontaneous emission of one or more particles or fragments. The several options for the transformation of an unstable nucleus will be discussed in detail in later Chapters and are described briefly in what follows:

a) *β^- and β^+ decays* — Stable light nuclei have a proton number Z similar to the neutron number N. In heavy nuclei a greater number of neutrons is necessary to compensate the Coulomb force between the protons. In both cases, when a nucleus has a value N greater than necessary for equilibrium it can decay by the emission of an electron and an antineutrino (*β^--decay*) in the form

$$^A_Z\mathrm{X}_N \to {}^A_{Z+1}\mathrm{Y}_{N-1} + \mathrm{e}^- + \bar{\nu_e}, \tag{3.13}$$

to come to a situation of greater equilibrium. If, on the other side, N is less than necessary, the *β^+-decay* can occur

$$^A_Z\mathrm{X}_N \to {}^A_{Z-1}\mathrm{Y}_{N+1} + \mathrm{e}^+ + \nu_e, \tag{3.14}$$

the emitted particles being now one positron and one neutrino. In both cases the product nucleus Y is not necessarily stable, and can also decay by the same way or by another form of disintegration.

b) *Electron capture* — This process consists in the capture of an atomic electron by the nucleus, giving rise to a decrease of the proton number and an increase of the neutron number by 1. The effect is the same as in the β^+-decay, and electron capture can compete strongly with it in heavy nuclei.

c) *α-decay* — In this disintegration mode an α-particle (^{4}He nucleus) is emitted, being the process energetically allowed for heavy nuclei. One example is the decay

$$^{238}_{92}\mathrm{U} \to {}^{234}_{90}\mathrm{Th} + \alpha, \tag{3.15}$$

with the emission of a 4.2 MeV α-particle. α-disintegration is responsible for the non-existence of stable elements with $Z > 83$.

d) *Light fragment emission* — nuclei heavier than ^{4}He can also be emitted in the few cases where the process is energetically allowed. The emission of ^{14}C, ^{24}Ne, ^{28}Mg and ^{32}Si by heavy nuclei has been experimentally observed [Pr89] but they are all rare and difficult to measure.

e) *Fission* — The energetics of fission was discussed in Section 3.3. It is necessary to remark that the occurrence of this process can take place spontaneously for very heavy nuclei. It is responsible, together with α-emission, for the extremely short half-lives of $Z > 100$ nuclides.

The processes described above take place in unstable nuclei at their ground state. They can also proceed from an excited state, but only in special situations. The basic option for an excited state is to decay to other states of lower energy, emitting γ-radiation or ejecting an electron from the atomic shells (internal conversion). We can also mention that for states of high excitation energy, the emission of a nucleon is a process that can be energetically possible and can compete with γ-emission.

Figure 3.4 shows the distribution in Z and N of the stable nuclides and of the known unstable ones. Among the latter, the great majority is

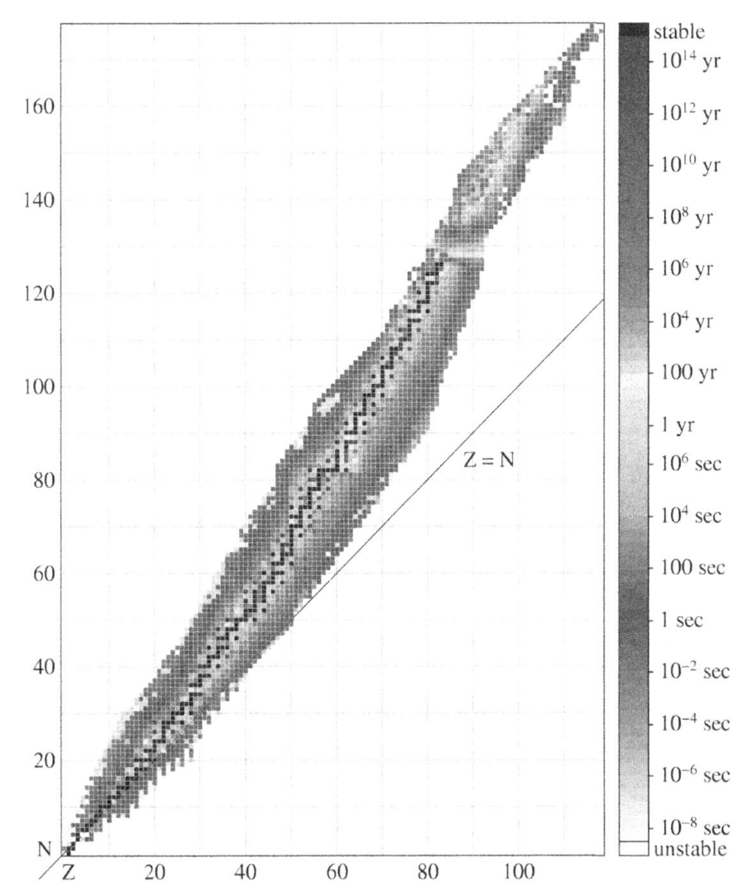

Figure 3.4. Distribution of stable nuclides (black squares) and of the known unstable nuclides (empty squares) as a function of the atomic number Z and of the neutron number N. The scale on right relates the isotope lifetime to its color on the chart. (Data from the National Nuclear Data Center, Brookhaven National Laboratory, USA.)

artificially produced in the laboratory and only a few of them exist in nature in significant amounts. From this last group, $^{235}_{92}$U, $^{238}_{92}$U and $^{232}_{90}$Th are of large importance in nuclear engineering.

The stable nuclei define a band in Fig. 3.4 called *β-stability line*. The nuclides to the right (or below) of that line have an excess of neutrons and are unstable by β^- emission while the nuclides to the left (or below) the line have an excess of protons and tend to decay by β^+-emission. The more distant from the line of β-stability the nuclide is, the more unstable it will be. Recently, nuclei with very unequal balance between proton and neutron numbers (called *exotic nuclei*), such as ^{11}Li and ^{8}He, have been studied.

It is worth establishing at this point the nomenclature for nuclides that have some common characteristics. Thus we define:

Isotopes: nuclides that have the same Z, that is, the same number of protons. It is common to use the name isotope as a synonym for nuclide and in this case the term is also used in the singular.
Isotones: nuclides that have the same number N of neutrons.
Isobars: nuclides that have the same mass number $A = Z + N$.

The set of 284 stable nuclei, distributed in 83 elements, shows some striking features:

a) Light nuclei have approximately the same number of protons and neutrons, that is, $Z \cong N$. Heavy nuclei have $N > Z$.
b) Even-Z nuclei are much more numerous than odd-Z nuclei and even-N nuclei are much more numerous than odd-N nuclei.
c) As a consequence of (b), even A nuclides are much more numerous. Among them the even–even are most common. In fact, there are only a few examples of stable odd–odd nuclides: ^{2}H, ^{6}Li, ^{10}B and ^{14}N.
d) From the 20 elements that have only one isotope, only ^{9}Be has Z even.
e) The element that has the greatest number of stable isotopes is Sn, with 10.

3.4. Nuclear Models

3.4.1. *The liquid drop model*

The liquid drop model is based on the hypothesis that the nucleus has a similar behavior to a liquid, because of the saturation of the forces between its constituents. The binding energy is given by

$$B(Z,A) = a_V A - a_S A^{2/3} - a_C Z^2 A^{-1/3} - a_A \frac{(Z-A/2)^2}{A} + \frac{(-1)^Z + (-1)^N}{2} a_P A^{-1/2}. \quad (3.16)$$

Substituting into Eq. (3.10) we obtain an equation for the mass of a nucleus

$$m(Z,A) = Zm_p + (A-Z)m_n - a_V A + a_S A^{2/3} + a_C Z^2 A^{-1/3} + a_A \frac{(Z-A/2)^2}{A} - \frac{(-1)^Z + (-1)^N}{2} a_P A^{-1/2}. \quad (3.17)$$

This is also known as the von Weizsäcker mass formula [Wei35].

The first term of Eq. (3.16) is known as *volume energy.* It is based on the experimental fact that the binding energy per nucleon is approximately constant, thus the total binding energy is proportional to A. The surface nucleons contribute less to the binding energy since they only feel the nuclear force from the inner side of the nucleus. The number of nucleons in the surface should be proportional to the surface area, $4\pi R^2 = 4\pi r_0^2 A^{2/3}$ fm. This is the reason for the second term in Eq. (3.16).

The binding energy should also be smaller due to the Coulomb repulsion between the protons. The Coulomb energy of a charged sphere with homogeneous distribution, and total charge Ze, is given by $(3/5)(Ze)^2/R = (3/5)(e^2/r_0^2)(Z^2/A^{1/3})$. Thus, the *Coulomb energy* contributes negatively to the binding energy with a part given by the third term of Eq. (3.16).

If the nucleus has a different number of protons and neutrons, its binding energy is smaller than for a symmetric nucleus. This *asymmetry term* also contributes negatively and it is given by the fourth term in Eq. (3.16).

The binding energy is larger when the proton and neutron numbers are even (even–even nuclei) and it is smaller when one of the numbers are odd (odd nuclei) and, furthermore, when both are odd (odd–odd nucleus). Thus, we introduce a *pairing term*

$$B_{\text{pairing}} = \begin{cases} +\delta & \text{for even–even nuclei} \\ 0 & \text{for odd nuclei} \\ +\delta & \text{for odd–odd nuclei.} \end{cases} \quad (3.18)$$

Empirically we find that

$$\delta \cong a_P A^{-1/2}. \quad (3.19)$$

This is the last term in Eq. (3.16).

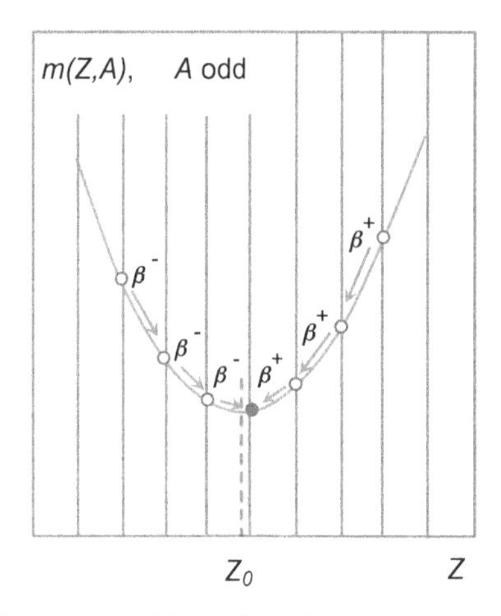

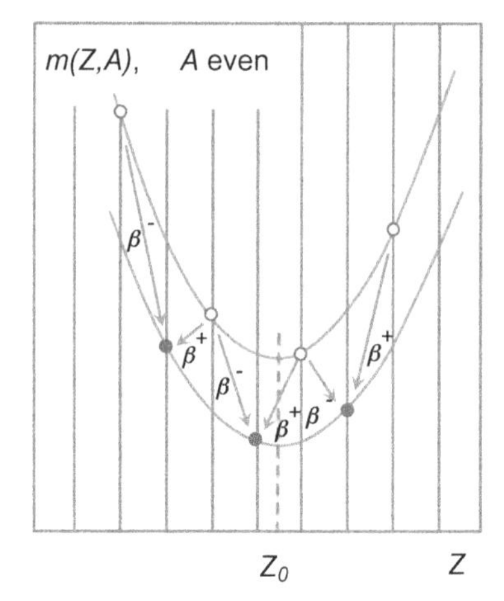

Figure 3.5. Mass of nuclei with a fixed A. The stable nuclei are represented by full circles.

Figure 3.2 shows a comparison of Eq. (3.16) to experimental data for odd nuclei. For $A < 20$ the agreement with experiment is not good. This is expected, since light nuclei are not so similar to a liquid drop.

Equation (3.17) allows us to deduce important properties. Observe that this equation is quadratic in Z. For odd nuclei one has a parabola, as seen in Fig. 3.5.

For A even, we obtain two parabolas due to the pairing energy $\pm\delta$. Nuclei with a given Z can decay into neighbors by β^+ (positron) or β^-(electron) particle emission. In Fig. 3.5 we see that for a nucleus with odd A there is only one stable isobar, while for even A countless stable isobars are possible.

Fixing the value of A, the number of protons Z_0 for which $m(Z, A)$ is minimum is obtained by

$$\left|\frac{\partial m(Z,A)}{\partial Z}\right|_{A=\text{const}} = 0. \tag{3.20}$$

From (3.17) we get

$$Z_0 = \frac{A}{2}\left(\frac{m_n - m_p + a_A}{a_C A^{2/3} + a_A}\right) = \frac{A}{1.98 + 0.015A^{2/3}}. \tag{3.21}$$

We see from Eq. (3.21) that stability is obtained with $Z_0 < A/2$, that is, with a number of neutrons larger than that of protons. We know that this

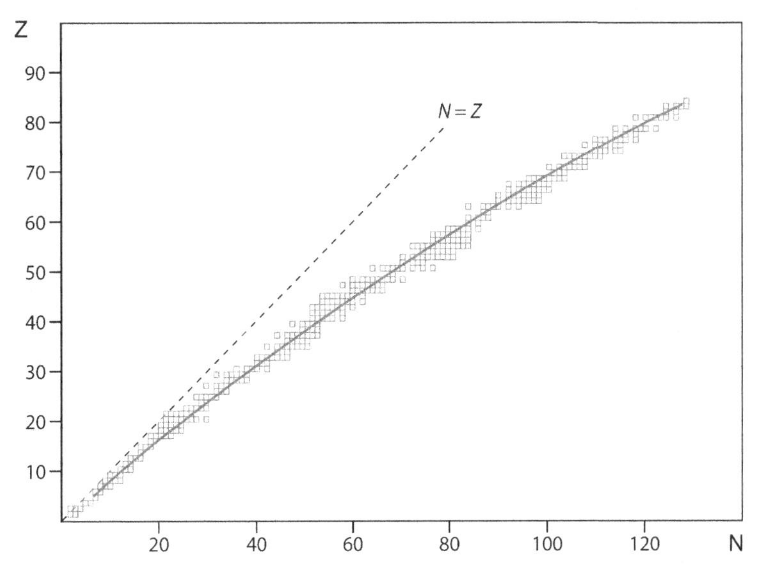

Figure 3.6. Location of the stable nuclei in the plane $N - Z$. The full line is the curve of Z_0 against $N = A - Z_0$, obtained from Eq. (3.21).

in fact happens and Fig. 3.6 shows that the stability line obtained with Eq. (3.21) accompanies perfectly the valley of stable nuclei.

From Eq. (3.17) we can also ask if a given nucleus is or is not unstable against the emission of an α-particle. For this it is necessary that

$$E_\alpha = [m(Z, A) - m(Z - 2, A - 4) - m_\alpha]c^2 > 0, \tag{3.22}$$

a condition satisfied for $A \gtrsim 150$. We can also verify the possibility of a heavy nucleus to fission, that is, of breaking into two pieces of approximately the same size. This will be possible if

$$E_f = [m(Z, A) - 2m(Z/2, A/2)]c^2 > 0, \tag{3.23}$$

a relation valid for $A \gtrsim 90$.

3.4.2. *Non-relativistic Fermi gas*

At very high temperatures, typical within stellar environments, the atoms might lose part or all of their electrons due to collisions and interactions with other particles. The material within a star is thus formed by a hot mixture of nuclei, atoms and electrons, the so-called *plasma*. Electrons are the lightest particles which can move freely within a stellar plasma and are responsible for a major portion of the pressure within some stars. But their motion is restricted by the Pauli principle which has profound consequences for the calculation of pressure, especially at low temperatures. For simplicity, let us consider a *Fermi gas*, i.e., a gas composed of fermions (particles obeying the Pauli principle) such as the electrons, at zero temperature.

For non-relativistic motion, the electrons in the Fermi gas model obey the Schrödinger equation for a free particle,

$$-\frac{\hbar^2}{2m_e}\nabla^2\Psi = -\frac{\hbar^2}{2m_e}\left[\frac{\partial^2}{\partial x^2}+\frac{\partial^2}{\partial y^2}+\frac{\partial^2}{\partial z^2}\right]\Psi = E\Psi, \tag{3.24}$$

where m is the fermion mass and E its energy. Let us assume that the region to which the electrons are limited to is the interior of a cube. The final results will be independent of this hypothesis. In this way, Ψ will have to satisfy the boundary conditions $\Psi(x,y,z) = 0$ for $x = 0, y = 0, z = 0$ and $x = a, y = a, z = a$, where a is the side of the cube. The solution is given by

$$\Psi(x,y,z) = A\sin(k_x x)\sin(k_y y)\sin(k_z z), \tag{3.25}$$

with

$$k_x a = n_x\pi, \quad k_y a = n_y\pi \quad \text{and} \quad k_z a = n_z\pi, \tag{3.26}$$

where n_x, n_y and n_z are positive integers and A is a normalization constant.

For each group (n_x, n_y, n_z) we have an energy

$$E(n_x,n_y,n_z) = \frac{\hbar^2 k^2}{2m_e} = \frac{\hbar^2}{2m_e}(k_x^2+k_y^2+k_z^2) = \frac{\hbar^2\pi^2}{2m_e a^2}n^2, \tag{3.27}$$

where $n^2 = n_x^2 + n_y^2 + n_z^2$ (not to be confused with particle density!). Equations (3.26) and (3.27) represent the quantization of a particle in a box, where $\mathbf{k} \equiv (k_x, k_y, k_z)$ is the momentum (divided by $\hbar$) of the particle in the box. Due to the Pauli principle, a given momentum can only be occupied by at most two electrons with opposite spins.

Consider the space of vectors $\mathbf{k}$: for each cube of side length π/a there exists, in this space, only one point that represents a possible solution in the form (3.25). The possible number of solutions (see Fig. 3.7) $n(k)$ with the magnitude of $\mathbf{k}$ between k and $k+dk$ is given by the ratio between the volume of the region shown in the figure and the volume $(\pi/a)^3$ for each

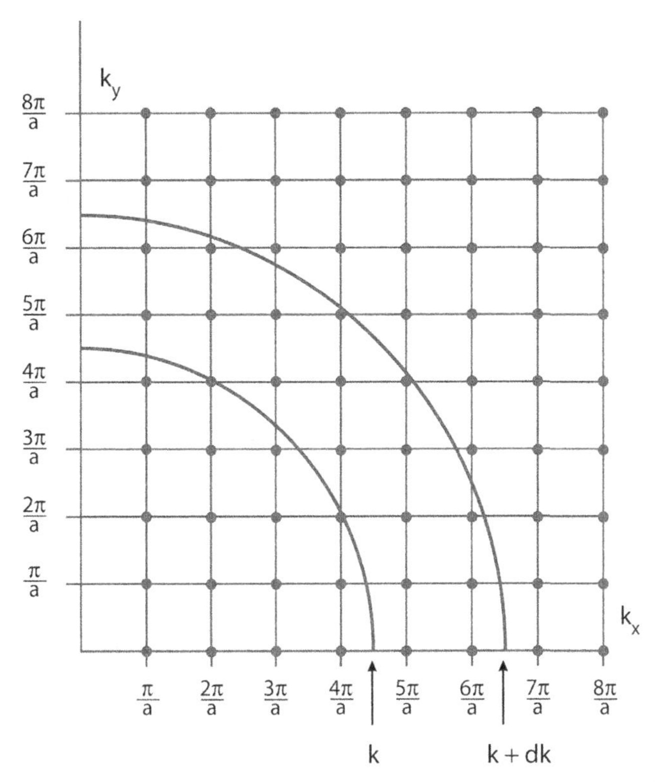

Figure 3.7. Allowed states in part of momentum space contained in the plane $k_x k_y$. Each state is represented by a point in the lattice.

allowed solution in the k-space. One obtains,

$$dn(k) = \frac{1}{8} 4\pi k^2 dk \frac{1}{(\pi/a)^3}, \tag{3.28}$$

where $4\pi k^2 dk$ is the volume of a spherical shell in the **k**-space with radius between k and $k + dk$. Only 1/8 of the shell is considered, since only positive values of k_x, k_y and k_z are necessary for counting all the states with eigenfunctions defined by Eq. (3.25). With the aid of Eq. (3.27) we can make the energy appear explicitly in Eq. (3.28):

$$dn(E) = \frac{\sqrt{2}\, m_e^{3/2} a^3}{2\pi^2 \hbar^3} E^{1/2} dE. \tag{3.29}$$

The total number of possible states of the system is obtained by integrating (3.29) from 0 to the minimum value of the energy needed to include all the N electrons. This value, E_F, is called the *Fermi energy*. Thus, we obtain

$$n(E_F) = \frac{\sqrt{2}\, m_e^{3/2} a^3}{3\pi^2 \hbar^3} E_F^{3/2} = \frac{N}{2}, \tag{3.30}$$

where the last equality is due to the mentioned fact that a given state can be occupied by two electrons with opposite spin. Inverting Eq. (3.30) we obtain

$$E_F = \frac{\hbar^2}{2m_e} (3\pi^2 n_e)^{2/3}, \tag{3.31}$$

where $n_e = N/a^3$ is the electron number density.

The number of fermions with energy between E and $E + dE$, given by (3.29), is plotted in Fig. 3.8 as a function of E. This distribution of particles is referred to as the *Fermi distribution* for $T = 0$, which is characterized by the absence of any particle with $E > E_F$. This corresponds to the ground state of the electron gas. An excited state ($T > 0$) can be obtained by the passage of a electron to a state above the Fermi level, leaving a vacancy (hole) in its energy state previously occupied.

Note that the Fermi energy only depends on the number density of the confined electrons. The mean energy of the electrons is given by

$$\bar{E} = \frac{2}{N} \int_0^{E_F} E \, dn(E) = \frac{3E_F}{5}, \tag{3.32}$$

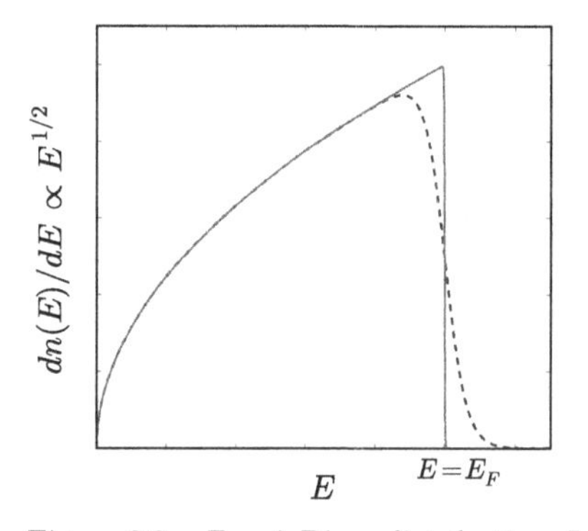

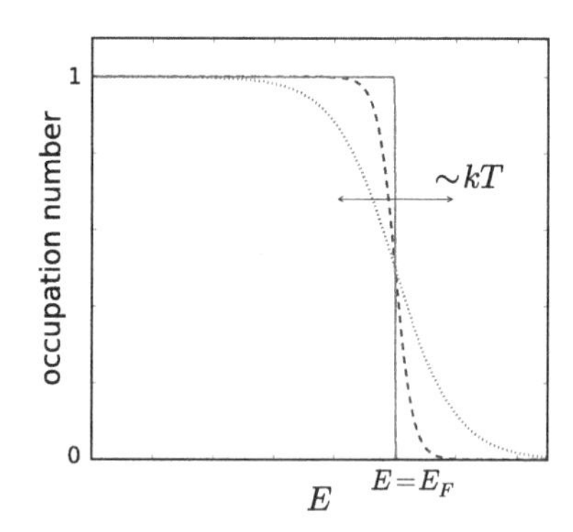

Figure 3.8. Fermi–Dirac distribution. Left: For $T = 0$ the distribution abruptly drops to zero at $E = E_F$, whereas for $T > 0$ it drops to zero within an energy interval of $\Delta E \approx kT$. Right: Occupation number distribution. The three curves are for $T = 0$ (solid), T_1 (dashed) and $T_2 > T_1$ (dotted).

Now, according to classical physics, the mean thermal energy of the electrons is $3kT/2$, where T is the electron gas temperature. Thus, if $kT \ll E_F$ then our original assumption that the electrons are cold is valid. Note that, in this case, the electron energy is much larger than that predicted by classical physics. Electrons in this state are termed *degenerate*. On the other hand, if $kT \gg E_F$ then the electrons are hot, and are essentially governed by classical physics. Electrons in this state are termed *non-degenerate.*

The total energy of a degenerate electron gas is

$$E_{total} = N\bar{E} = \frac{3}{5}NE_F. \tag{3.33}$$

3.4.3. *Fermi gas at $T \neq 0$*

In the previous chapter we have discussed the probability, or occupancy distribution, of a Fermi gas, appropriate for electrons, nucleons, and other fermions. We have obtained the level occupancy for $T = 0$. Now it is appropriate to discuss what happens when $T > 0$. Let us check what happens as we raise T, but keep $kT \ll E_F$. By heating a Fermi gas, we excite some electrons. Only electrons with energies close to E_F can be excited: empty states are available only above the Fermi energy. Thus, some states above E_F (within a narrow energy range $\sim kT$ around E_F) will be populated and some states below E_F depleted. This is shown in Fig. 3.8.

For $T \neq 0$ instead of a unit occupancy up to the Fermi level, the occupancy distribution is given by

$$n(E) = \frac{1}{\exp[(E - E_F)/kT] + 1}, \tag{3.34}$$

to account for the "smearing" of the distribution around the Fermi energy as T increases. This is known as the *Fermi–Dirac distribution* function (occupancy) giving the mean number of fermions in a particular quantum state. Though the occupancy $n(E)$ is less than unity for fermions, it is not a probability because the sum over all states (including the spin degeneracy g_i) is the total number of particles, i.e., $\sum_i g_i n(E_i) = N$.

One of the greatest successes of the free electron model and Fermi–Dirac statistics is the explanation of the T dependence of the heat capacity of a Fermi gas. The *heat capacity* is given by

$$C_V = \frac{dE_t(T)}{dT}, \tag{3.35}$$

where E_t is the total internal energy of the Fermi gas. To calculate the heat capacity, we need to know how E_t depends on temperature. By heating a Fermi gas, we populate some states above the Fermi energy E_F and deplete some states below E_F. This modification is significant within a narrow energy range $\sim kT$ around E_F (we assume that the system is cold — strong degeneracy). The fraction of electrons "transferred" to higher energies is $\sim kT/E_F$, the energy increase for the electrons "jumping" the Fermi energy being $\sim kT$. Thus, the increase of the internal energy with temperature is proportional to $N \times (kT/E_F) \times (kT) \sim N(kT)^2/E_F$. Thus, $C_V = dE_t/dT \propto Nk^2T/E_F$. An exact calculation yields

$$C_V = \frac{\pi^2}{2} Nk \frac{kT}{E_F}. \tag{3.36}$$

For a classical ideal gas, following the Boltzmann distribution, a similar calculation yields $C_V = 3NkT/2$. The Fermi gas heat capacity is much smaller (by $kT/EF \ll 1$) than that of a classical ideal gas with the same energy and pressure. The small heat capacity is a direct consequence of the Pauli principle: most of the electrons cannot change their energy, only a small fraction kT/E_F of the electrons are excited out of the ground state.

The calculation of number of electrons at a given state with energy E_i requires the knowledge of the *density of states* g_i for electrons. This is in fact a continuous function of the energy E and has been calculated previously, Eq. (3.29). That is, $g(E) = \sqrt{2}\, m^{3/2}V/(2\pi^2\hbar^3)E^{1/2}$. So, for electrons, the number of particles with energy E is given by

$$n_{FD}(E) = \frac{\pi}{2}\left(\frac{8m}{h^2}\right)^{3/2} \frac{E^{1/2}}{\exp\left[(E - E_F)/kT\right] + 1}. \tag{3.37}$$

It is easy to see that Eq. (3.37) reproduces the Fermi distribution at $T = 0$ while for high temperatures such that $kT \gg E$, one recovers the Boltzmann distribution, $n_B(E) \sim e^{-E/kT}$. At $T = 0$, all the states with $E < E_F$ have the occupancy $= 1$, all the states with $E > E_F$ have the occupancy $= 0$ (i.e., they are unoccupied). With increasing T, the step-like function is smeared over the energy range $\sim kT$.

3.4.4. *The Fermi gas model of the nucleus*

As with the case of electrons, the nucleons within a nucleus obey the Pauli principle. Thus, we can also devise a Fermi gas model for the nucleons as we did for the electrons. The protons compose one Fermi gas and the neutrons compose the other, with their corresponding Fermi energies. Assuming the proton and neutron Fermi energies to be the same (i.e., $N = Z$), instead of Eq. (3.30), we get

$$n(E_F) = \frac{\sqrt{2}\, m^{3/2} a^3}{3\pi^2 \hbar^3} E_F^{3/2} = \frac{A}{4}, \tag{3.38}$$

where the last equality is because a given state can be occupied by four nucleons. Inverting Eq. (3.38) we obtain

$$E_F = \frac{\hbar^2}{2m} \left(\frac{3\pi^2 n_N}{2} \right)^{2/3}, \tag{3.39}$$

where n_N is now the nucleon density. We assume that the maximum energy is the same for both nucleons, which means equal number densities for protons and neutrons. If that case is not true, the Fermi energy for protons and neutrons will be different. If n_p and n_n are the respective proton and neutron densities, we will have

$$E_F(p) = \frac{\hbar^2}{2m} (3\pi^2 n_p)^{2/3}, \tag{3.40}$$

and

$$E_F(n) = \frac{\hbar^2}{2m} (3\pi^2 n_n)^{2/3}, \tag{3.41}$$

for the corresponding Fermi energies.

If we use $n_N = 1.72 \times 10^{38}$ nucleons/cm^3 = 0.172 nucleons/fm^3, which is the approximate density of all nuclei with $A \gtrsim 12$, we obtain

$$k_F = \frac{\sqrt{2mE_F}}{\hbar} = 1.36\ \text{fm}^{-1}, \tag{3.42}$$

which corresponds to

$$E_F = 37\ \text{MeV}. \tag{3.43}$$

We know that the separation energy of a nucleon is of the order of 8 MeV. Thus, the nucleons are not inside a well with infinite walls as one assumes in the Fermi gas model, but in a well with depth $V_0 \cong (37{+}8)$ MeV = 45 MeV (Fig. 3.9).

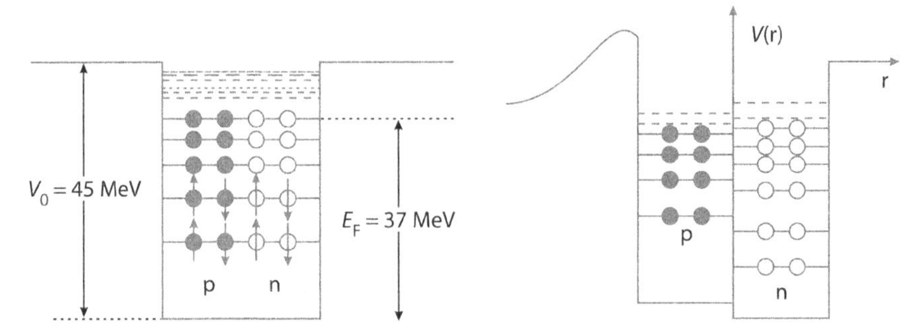

Figure 3.9. Left: Potential well and states of a Fermi gas for $T = 0$. In two of the levels we show the spins associated to each nucleon. Right: When one takes into account the Coulomb force the potentials for protons and neutrons are different and we can imagine a well for each nucleon type.

3.4.5. *The shell model*

The exact Hamiltonian for a problem of A bodies can be written has

$$H = \sum_{i}^{A} T_i(\mathbf{r}_i) + V(\mathbf{r}_1, \ldots, \mathbf{r}_A), \tag{3.44}$$

where T is the kinetic energy operator and V the potential function.

If we restrict ourselves to two body interactions (i.e., nucleon–nucleon interaction), Eq. (3.44) takes the form

$$H = \sum_{i}^{A} T_i(\mathbf{r}_i) + \frac{1}{2} \sum_{ji} V_{ij}(\mathbf{r}_i, \mathbf{r}_j). \tag{3.45}$$

In the shell model, the nucleon i does not feel the potential $\sum_j V_{ij}$, but a central potential $U(r_i)$, that only depends on the coordinates of nucleon i. This potential can be introduced in Eq. (3.45), with the result

$$H = \sum_{i}^{A} T_i(\mathbf{r}_i) + \sum_{i}^{A} U(r_i) + H_{res}, \tag{3.46}$$

$$H_{res} = \frac{1}{2} \sum_{ji} V_{ij}(\mathbf{r}_i, \mathbf{r}_j) - \sum_{i} U(r_i). \tag{3.47}$$

H_{res} refers to the *residual interactions*, that is, to the part of potential V not embraced by the central potential U. The hope of the shell model is that the contribution of H_{res} is small or, in another way, that the *shell model Hamiltonian,*

$$H_0 = \sum_{i=1} H_0^{(i)} = \sum_{i=1}^{A} [T_i(r_i) + U(r_i)], \tag{3.48}$$

represents a good approximation for the exact expression of H.

The solutions $\Psi_1(\mathbf{r}_1), \Psi_2(\mathbf{r}_2), \ldots$, of equation

$$H_0^{(i)} \Psi_i = E \Psi_i, \tag{3.49}$$

with respective eigenvalues $E_1, E_2, \ldots$, are called *orbits* or *orbitals*. In the shell model prescription the A nucleons fill the orbitals of lower energy in a way compatible with the Pauli principle. Thus, if the sub-index 1 of Ψ_1, which represents the group of quantum numbers of the orbital 1, includes spin and isospin, we can say that a first nucleon is described by $\Psi_1(\mathbf{r}_1), \ldots,$ the Ath by $\Psi_A(\mathbf{r}_A)$. The wavefunction

$$\Psi = \Psi_1(\mathbf{r}_1)\Psi_2(\mathbf{r}_2)\cdots\Psi_A(\mathbf{r}_A) \tag{3.50}$$

is a solution of Equation (3.49) with eigenvalues

$$E = E_1 + E_2 + \cdots E_A \tag{3.51}$$

and it would be, in principle, the wavefunction of the nucleus, with energy E, given by the shell model. We should have in mind, however, that we are treating a fermion system and that the total wavefunction should be antisymmetric for an exchange of coordinates of two nucleons. Such a wavefunction is obtained from (3.50) for the construction of the *Slater determinant*

$$\Psi = \frac{1}{\sqrt{A!}} \begin{vmatrix} \Psi_1(\mathbf{r}_1) & \Psi_1(\mathbf{r}_2) & \cdots & \Psi_1(\mathbf{r}_A) \\ \Psi_2(\mathbf{r}_1) & \Psi_2(\mathbf{r}_2) & \cdots & \Psi_2(\mathbf{r}_A) \\ \vdots & \vdots & \ddots & \vdots \\ \Psi_A(\mathbf{r}_1) & \Psi_A(\mathbf{r}_2) & \cdots & \Psi_A(\mathbf{r}_A) \end{vmatrix}, \tag{3.52}$$

where the change of coordinates (or of the quantum numbers) of two nucleons changes the sign of the determinant.

The nuclear potential is usually assumed to have a form of a finite "well" with width R and diffuseness a, given by a *Woods–Saxon* form

$$U = \frac{U_0}{1 + \exp[(r - R)/a]},$$

where $U_0 < 0$, R and a are adjustable parameters. To it is added a *spin–orbit* interaction in the form

$$\alpha f(r) \mathbf{l} \cdot \mathbf{s}, \tag{3.53}$$

where α is a strength parameter, and $f(r)$ is a radial function that should be obtained by comparison with experiments. A spin–orbit term already appears in atomic physics as a result of the interaction between the magnetic moment of the electrons and the magnetic field created by its orbital motion. In nuclear physics this term has a different nature and it is related to the quantum field properties of an assembly of nucleons.

The addition of the spin–orbit term, Eq. (3.53), changes the energy values $E = \int \Psi^* H \Psi$. If we suppose now that the spin–orbit term is small and that it can be treated as a small perturbation, the wavefunctions in (4.55) are basically the ones of a central potential. Since $\mathbf{j} = \mathbf{l} + \mathbf{s}$, we have $\mathbf{l} \cdot \mathbf{s} = (j^2 - l^2 - s^2)/2$ and

$$\begin{aligned} \int \Psi^* \, \mathbf{l} \cdot \mathbf{s} \, \Psi &= \frac{l}{2}, & \text{for } j = l + \frac{1}{2}, \\ \int \Psi^* \, \mathbf{l} \cdot \mathbf{s} \, \Psi &= -\frac{1}{2}(l+1), & \text{for } j = l - \frac{1}{2}. \end{aligned} \tag{3.54}$$

Thus, the spin–orbit interaction removes the degeneracy in j and, anticipating that the best experimental result will be obtained if the orbitals for larger j have the energy lowered, we admit a negative value for α, what allows us to write for the energy increment:

$$\Delta E|_{j=l+1/2} = -|\alpha| \langle f(r) \rangle \frac{l}{2}, \tag{3.55}$$

$$\Delta E|_{j=l-1/2} = +|\alpha| \langle f(r) \rangle \frac{1}{2}(l+1), \tag{3.56}$$

where $\langle f(r) \rangle$ is the average value of the spin–orbit strength.

Figure 3.10 exhibits the level scheme of a central potential with the introduction of the spin–orbit interaction. It is easy to see the effect of (3.55) and (3.56) in the energy distribution of the levels. Denominating now a shell as a group of levels of closed energy, not necessarily associated

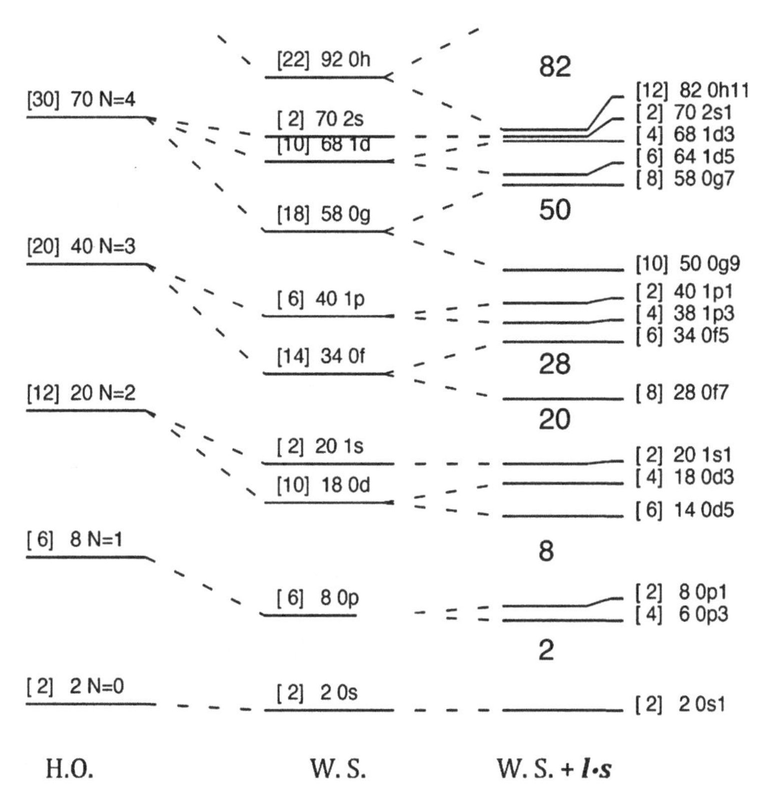

Figure 3.10. Level scheme showing the break of the degeneracy in j caused by the spin–orbit interaction term and the emergence of the magic numbers in the shell closing. The values inside brackets indicate the number of nucleons of each type that the level admits and the values outside the total number of nucleons of each type up to that level. The ordering of the levels is not rigid, and one could obtain level inversions when the form of the potential changes [MJ55]. Also shown are the number of nodes, N of the radial wavefunction. The notation 0p3 means $N = 0$, $l = 1$ and $j = 3/2$, and similarly for the other cases.

to only one principal quantum number of the oscillator, we obtain a perfect description of all the magic numbers.

We will use this level scheme[3] to establish what one denominates by *single particle model* or *extreme shell model*. In this version, the model admits that an odd nucleus is composed of an inert even–even core plus an unpaired nucleon, and that this last nucleon determines the properties of

[3]Each level of Fig. 3.10, characterized by the quantum numbers n, l, j, contains $2j + 1$ nucleons of a same type and is also referred to as a sub-shell.

the nucleus. We can determine, starting from the Fig. 3.10, which is the state of the unpaired nucleon.

Let us take ^{17}O as example. This nucleus has a shell closed with 8 protons but has a remaining neutron above the closed core of 8 neutrons. A fast examination of Fig. 3.10 indicates that this neutron finds itself in the level $1d_{5/2}$. We can, otherwise, say that the neutron *configuration* of this nucleus is

$$(1s_{1/2})^2(1p_{3/2})^4(1p_{1/2})^2(2s_{1/2})^2(1d_{5/2})^1$$

being evident that in this definition we list the filled levels for the neutrons, with the upper indices, equal to $2j+1$, indicating the number of particles in each one of them. It is also common to restrict the configuration to the levels partially filled, the sub-shells being completely ignored. Thus, the configuration of neutrons in ^{17}O would be $(1d_{5/2})^1$. The prediction of the model is, in this case, that the total angular momentum of the ground state of ^{17}O is $\frac{5}{2}$ and the parity is positive ($l = 2$). This prediction is in agreement with experiments. For similar reasons, the model predicts that the ground state of ^{17}F is a $\frac{5}{2}^+$ state and this is indeed the measured value.

The magic numbers manifest themselves in many observed nuclear properties, such as masses, shown in Fig. 3.11 in terms of the difference of

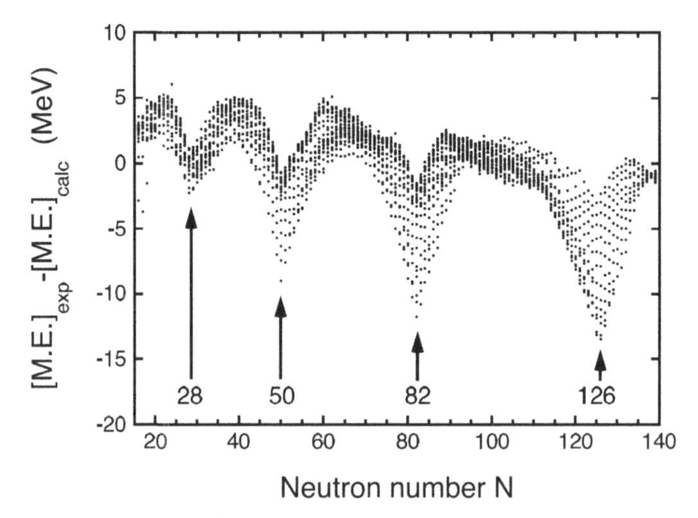

Figure 3.11. Difference between experimental ground-state atomic mass excess and the mass excess predicted by the spherical macroscopic part of the Finite-Range Droplet Mass (FRDM) formula [Mo95] versus neutron number.

the measured ground-state atomic mass excess from its mean value that is calculated by using a smooth semiempirical mass formula. At the locations of magic neutron numbers, the atomic mass excess is smaller, resulting in a smaller atomic mass and a larger binding energy according to Eq. (3.10). This provides unambiguous evidence for the shell structure of nuclei. The synthesis of the heavy elements is strongly influenced by the magic neutron numbers of N = 50, 82, and 126.

3.5. Exercises

1. The four-momentum of a particle with nonzero rest mass m_0 is $p^\mu = m_0 u^\mu$.[4] Note that "rest mass" means the mass measured by an observer riding along with the particle. What is the squared magnitude of the four-momentum?
2. Which of the following processes are absolutely forbidden and why?

$$\begin{array}{ll} \text{(a)} & \pi^0 + \mathrm{n} \rightarrow \pi^- + \mathrm{p} \\ \text{(b)} & \mathrm{p} + \mathrm{e}^- \rightarrow \gamma + \gamma \\ \text{(c)} & \mathrm{n} \rightarrow \mathrm{p} + \mathrm{e}^- + \bar{\nu}_e \\ \text{(d)} & \mathrm{n} \rightarrow \mathrm{p} + \mathrm{e}^+ + \nu_e \\ \text{(e)} & \gamma + \mathrm{p} \rightarrow \bar{\mathrm{n}} + \pi^+ \end{array}$$

3. Verify the expression (3.7) for p, n, $\bar{\mathrm{p}}$, $\bar{\mathrm{n}}$, π^+, π^- and π^0.
4. Using relation (3.7), show that the quarks up and down are members of an isospin doublet $t_z = \pm 1/2$.
5. Find the quark composition of particles in Table 3.1.
6. In a scale where a water droplet (radius = 1 mm) is increased to the size of the Earth (radius = 6400 km), what would be the radius of a ^{238}U nucleus?
7. (a) Using the relativistic expression for the momentum–energy relation (see Appendix B), find the de Broglie wavelength $\lambda = h/p$ for protons with kinetic energy 500 keV and 900 MeV. (b) Repeat the calculation using a non-relativistic expression for the momentum. (c) Repeat (a) and (b) for electrons with the same energies.
8. Find the approximate density of nuclear matter in g/cm^3.

---

[4]For a review on the theory of Special Relativity, see Appendix B.

9. Radium-226 is a radioactive element whose nucleus emits an α particle (a helium nucleus) with a kinetic energy of about 7.8×10^{-13} J (4.9 MeV). To what amount of mass in kg is the energy equivalent to this kinetic energy? What is: (a) the ratio of this equivalent mass and the rest mass of the Radium nucleus, (b) $\beta = v/c$ for the alpha particle, and (c) the recoil kinetic energy of the daughter nucleus (the nucleus resulting from the emission of the alpha particle)? Consult the literature for the masses of Radium-226 and the daughter nucleus (Radon-222).
10. Find the binding energy of the last neutron in ^{4}He and of the last proton in ^{16}O. How do these energies compare with B/A for these nuclei? What this says about the ^{4}He stability relative to ^{3}He, and of ^{16}O relative to ^{15}N? (*Hint*: You need to find a nuclear mass table to obtain the masses of the involved nuclei. Then use Eq. (3.11).)
11. Find the energy in MeV released in the fusion of two deuterons. (*Hint*: You need to use a nuclear mass table.)
12. From the uncertainty principle $\Delta p \Delta x \sim \hbar$, and the fact that a nucleon is confined within the nucleus, what can be concluded about the energies of nucleons within the nucleus?
13. Evaluate the total binding energy of $^8_4$Be. According to this result the $^8_4$Be would be a stable nucleus but, in reality, it is extremely unstable. Try to justify this discrepancy. (*Hint*: You need to use a nuclear mass table.)
14. For which kinetic energy does the proton have velocity equal to half that of light? Compare with the result for the electron.
15. Assuming the proton to be a sphere of radius 1.5×10^{-13} cm, of uniform density, and with angular momentum $\hbar\sqrt{\frac{1}{2}(\frac{1}{2}+1)}$, evaluate the angular velocity in revolutions/s and the tangential velocity at the proton equator. (*Hint:* Use the relation between angular momentum of a rigid body with moment of inertia I and its angular velocity ω, $L = I\omega$. The moment of inertia of a sphere of mass M is $I = 2MR^2/5$).
16. Suppose that the difference between the square of the mass of the electron neutrino and that of the muon neutrino has the value $[m(\nu_\mu)^2 - m(\nu_e)^2]c^4 = 5 \times 10^{-5}\,\mathrm{eV}^2$, and that the difference between the square of the mass of the muon neutrino and that of the tau neutrino has the value $[m(\nu_\mu)^2 - m(\nu_e)^2]c^4 = 3 \times 10^{-3}\,\mathrm{eV}^2$. What values of $m(\nu_e)$, $m(\nu_\mu)$, and $m(\nu_\tau)$ minimize the sum $m(\nu_e) + m(\nu_\mu) + m(\nu_\tau)$, given these constraints?

17. What is the minimum photon energy required to dissociate the deuteron? Take the binding energy to be 2.22 MeV.
18. A nuclear power reactor generates 3.0×10^9 W of power. In one year, what is the change in the mass of the nuclear fuel due to energy being taken from the reactor?
19. Because pions had not been discovered in 1936 when Yukawa proposed the meson theory of the nuclear force, it was suggested that the muon was Yukawa's particle. What would the range of the nuclear force be if this were true?
20. Show that $^{238}_{92}$U is stable with respect to the decay by p, n, e^-, e^+ emission, but is unstable with respect to the emission of an α-particle.
21. Using relativistic expressions for momentum and energy conservation, show that a proton must have energy greater than 5.6 GeV to produce a proton–antiproton pair in a collision with another proton at rest.
22. The spin and parity of ^{9}Be and ^{9}B are $3/2^-$ for both nuclei. Assuming that these values are given by the last nucleon, justify the observed value, 3^+, of ^{10}B. What other combinations of spin-parity can appear? Verify in a nuclear chart the presence of excited states of ^{10}B that could correspond to those combinations.
23. The group of stable odd–odd nuclei consists of very light nuclei. Is it possible to find a justification for this?
24. The ground state of ^{137}Ba possesses spin-parity $3/2^+$. The first two excited states possess spin-parity 12^+ and $11/2^-$. According to the shell model, which levels would be expected for these excited states?
25. Using Eq. (3.31), calculate the density of electrons (in kg/m^3) so that the Fermi energy $E_F = m_e c^2$. How does this compare to the density of water on Earth?
26. ^{13}C and ^{13}N both have a ground state $1/2^-$ and three excited states below 4 MeV, of spin-parity $1/2^+$, $3/2^-$, and $5/2^+$. The other states are located above 6 MeV. Interpret these four states using the shell model.

Chapter 4

Radiation

4.1. Introduction

Photons are massless particles and accordingly (see Appendix B) individual photons travel at a constant speed c. Quantum mechanics further relates their frequency (and wavelength), energy, and momentum by

$$\nu = \frac{c}{\lambda}, \quad E = h\nu, \quad p = \frac{E}{c}, \quad \text{or, equivalently,}$$

$$\omega = 2\pi\nu, \quad E = \hbar\omega, \quad \hbar = \frac{h}{2\pi}. \tag{4.1}$$

To describe the radiative properties of a source of multiple photons, such as a star for instance, one needs to consider the distribution E_ν (or E_λ) of photon energies as well as the direction Ω in which the photons are propagating.

Photons are constantly being produced and destroyed in atomic and nuclear decays and excitations. In this chapter we discuss the interaction of photons with matter, multipole radiation, propagation of radiation, and stellar atmospheres.

4.2. Interaction of Radiation with Matter

4.2.1. *Lifetime of unstable states*

A quantum system, described by a wavefunction that is an Hamiltonian eigenfunction, is in a well defined energy state and, if it does not suffer external influences, it will remain indefinitely in that state. But this ideal situation does not prevail in excited atoms or excited nuclei, or in the ground state of *unstable atoms or nuclei.* Interactions of several types can add a perturbation to the Hamiltonian and the pure energy eigenstates no longer exist. In this situation a transition to a lower energy level of the same or of another nucleus can occur.

An unstable state normally lives a long time compared to the time spent by a particle with velocity near that of light in crossing an atomic or nuclear diameter. In this way we can admit that the particle is in an approximately stationary state, and write its wave function as

$$\Psi(\mathbf{r},t) = \psi(\mathbf{r})e^{-iWt/\hbar}. \tag{4.2}$$

$|\Psi(\mathbf{r},t)|^2 dV$ is the probability of finding the particle in the volume dV and, if the state described by Ψ decays with a decay constant Λ, it is reasonable to write

$$|\Psi(\mathbf{r},t)|^2 = |\Psi(\mathbf{r},0)|^2 e^{-\Lambda t}. \tag{4.3}$$

To obey Eqs. (4.2) and (4.3) simultaneously, W must be a complex quantity with imaginary part $-\Lambda\hbar/2$. We can write

$$W = E_0 - \frac{\hbar\Lambda}{2}i, \tag{4.4}$$

which shows that the wave function Eq. (4.2) does not represent a well defined stationary state since the exponential contains a real part $-\Lambda t/2$. However, we can write the exponent of Eq. (4.2) as a superposition of values corresponding to well defined energies E (for $t \geq 0$),

$$e^{-(iE_0/\hbar+\Lambda/2)t} = \int_{-\infty}^{+\infty} A(E)e^{-iEt/\hbar}\, dE. \tag{4.5}$$

Functions connected by a Fourier transform relate to each other as

$$f(t) = \frac{1}{\sqrt{2\pi}}\int_{-\infty}^{+\infty} g(\omega)e^{-i\omega t}\, d\omega, \quad g(\omega) = \frac{1}{\sqrt{2\pi}}\int_{-\infty}^{+\infty} f(t)e^{i\omega t}\, dt. \tag{4.6}$$

This allows us to establish the form of the amplitude $A(E)$,

$$A(E) = \frac{1}{2\pi\hbar}\int_0^{+\infty} e^{[i(E-E_0)/\hbar-\Lambda/2]t}\, dt, \tag{4.7}$$

where the lower limit indicates that the stationary state was created at time $t = 0$. The integral in Eq. (4.7) has an easy solution, giving

$$A(E) = \frac{1}{h\Lambda/2 - 2\pi i(E - E_0)}.$$

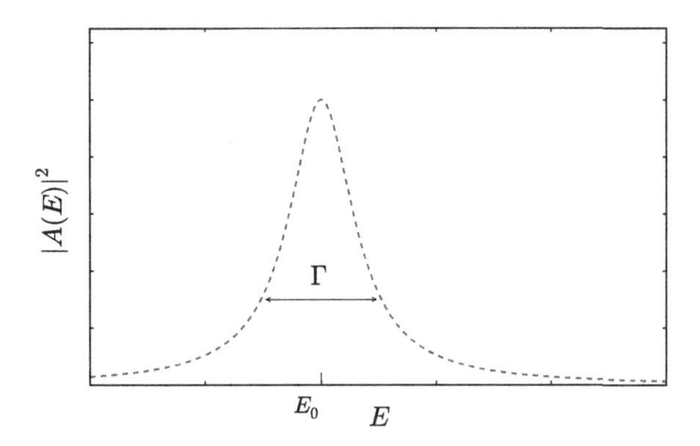

Figure 4.1. Form of the energy distribution given by Eq. (4.8).

The probability of finding a value between E and $E + dE$ in an energy measurement is given by the product

$$A^*(E)A(E) = \frac{1}{h^2\Lambda^2/4 + 4\pi^2(E - E_0)^2}, \tag{4.8}$$

and this function of energy has the form of a *Lorentzian*, with the aspect shown in Fig. 4.1. Its width at half-maximum is $\Gamma = \hbar\Lambda = \hbar/\tau$. The relationship

$$\tau\Gamma = \hbar \tag{4.9}$$

between the half-life and the width of a state is directly connected to the uncertainty principle and shows that the longer a state survives the greater the precision to which its energy can be determined. In particular, only to stable states one can attribute a single value for the energy.

4.2.2. *Decay rate of unstable states*

The decay constant Λ was presented at the beginning of the chapter as the probability per unit time of occurrence of a transition between quantum states, and its values were presumed known from experimental data. In this section we shall show that a formula to evaluate the decay constant can be obtained from the postulates of *perturbation theory*.

The previous section has described an unstable state by the addition of a perturbation to an stationary state. Formally we can write

$$H = H_0 + V, \tag{4.10}$$

where H is the Hamiltonian of the unstable state, composed by the non-perturbed Hamiltonian H_0 and a small perturbation V. The Hamiltonian H_0 satisfies an eigenvalue equation

$$H_0\psi_n = E_n\psi_n, \tag{4.11}$$

whose eigenfunctions form a complete basis in which the total wavefunction Ψ, obeying

$$H\Psi = i\hbar\frac{\partial\Psi}{\partial t}, \tag{4.12}$$

can be expanded as

$$\Psi = \sum_n a_n(t)\psi_n e^{-iE_nt/\hbar}. \tag{4.13}$$

Using Eqs. (4.10) and (4.13) in Eq. (4.12), and with the aid of Eq. (4.11), we obtain

$$i\hbar\sum_n \dot{a}_n\psi_n e^{-iE_nt/\hbar} = \sum_n V a_n\psi_n e^{-iE_nt/\hbar}, \tag{4.14}$$

with $\dot{a}_n \equiv \partial a_n(t)/\partial t$. Using the orthogonalization properties of the ψ_n, let us multiply Eq. (4.14) on the left by ψ_k^* and integrate it in coordinate space. From this, we get

$$\dot{a}_k = -\frac{i}{\hbar}\sum_n a_n V_{kn} e^{i\frac{E_k-E_n}{\hbar}t}, \tag{4.15}$$

where we introduced the matrix element

$$V_{kn} = \int \psi_k^* V\psi_n \, d^3r. \tag{4.16}$$

Let us make the following assumptions about the perturbation V: it begins to act at time $t = 0$, when the unperturbed system is described by an eigenstate ψ_m. It stays at a very low value, and after a short time interval, becomes zero at $t = T$. These assumptions allow us to say that the conditions

$$\begin{cases} a_m = 1 \\ a_n = 0, \quad \text{if } n \neq m, \end{cases} \tag{4.17}$$

are rigorously verified for $t < 0$ and also work approximately for $t > 0$. Thus, Eq. (4.15) has only one term, and the value of the amplitude is obtained from

$$a_k = -\frac{i}{\hbar}\int_0^T V_{km} e^{i\frac{E_k - E_m}{\hbar}t}\, dt, \tag{4.18}$$

whose value must necessarily be small by the assumption that followed Eq. (4.17). The above approach is also known as first order perturbation theory. The integral of Eq. (4.18) gives

$$a_k = \frac{V_{km}\left(1 - e^{i\frac{E_k - E_m}{\hbar}T}\right)}{E_k - E_m}. \tag{4.19}$$

We need now to interpret the meaning of the amplitude a_k. The quantity $a_k^* a_k$ measures the probability of finding the system in the state k. This characterizes a transition occurring from the initial state m to the state k, and the value of $a_k^* a_k$ divided by the interval T should be a measure of the decay constant Λ_k relative to the state k. The total decay constant is obtained by the sum over all states:

$$\Lambda = \sum_{k \neq m} \Lambda_k = \frac{\sum |a_k|^2}{T}. \tag{4.20}$$

Let us now suppose that there is a large number of available states k. We can, in this case, replace the summation in Eq. (4.20) by an integral. Defining $\rho(E)$ as the density of available states around the energy E_k, we write

$$\begin{aligned}\Lambda &= \frac{1}{T}\int_{-\infty}^{+\infty} |a_k|^2 \rho(E_k)\, dE_k \\ &= \frac{4}{T}\int_{-\infty}^{+\infty} |V_{km}|^2 \frac{\sin^2\left[\left(\frac{E_k - E_m}{2\hbar}\right)T\right]}{(E_k - E_m)^2} \rho(E_k)\, dE_k.\end{aligned} \tag{4.21}$$

The function $\sin^2 x/x^2$ only has significant amplitude near the origin. In the case of Eq. (4.21), if we suppose that V_{km} and ρ do not vary strongly in a small interval of the energy E_k around E_m, both these quantities can be taken outside of the integral, and we obtain the final expression

$$\Lambda = \frac{2\pi}{\hbar}|V_{km}|^2 \rho(E_k) \tag{4.22}$$

that we have been looking for. Eq. (4.22) is known as the *golden rule no.* 2 (also known as *Fermi golden rule*), and allows us to determine the decay constant if we know the wavefunctions of the initial and final states. The result (4.22) can also be used to obtain, to first order, the cross-sections of processes induced by particle interaction through a potential.

4.2.3. *Multipole radiation*

We limit ourselves to explaining the general framework of this theory. Classically, the electromagnetic radiation emitted by a system is the result of the variation in time of the charge density or of the distribution of charge currents in the system. The energy is emitted in two types of multipole radiation: the electric and the magnetic. Each one of them is expressed as function of the corresponding multipole moments, being the quantities which contain the variables (charge and current) of the system. If the wavelength of the emitted radiation is long in comparison to the dimensions of the system the power emitted by each multipole is given by ([Jac98], Chapter 16):

$$P_E(LM) = \frac{8\pi(L+1)c}{L[(2L+1)!!]^2}\left(\frac{\omega}{c}\right)^{2L+2}|\mathcal{M}(\mathrm{EL}, M)|^2 \tag{4.23}$$

for electric multipole radiation and

$$P_M(LM) = \frac{8\pi(L+1)c}{L[(2L+1)!!]^2}\left(\frac{\omega}{c}\right)^{2L+2}|\mathcal{M}(\mathrm{ML}, M)|^2 \tag{4.24}$$

for the corresponding magnetic radiation, with frequency $\nu = \omega/2\pi$. L is the angular momentum carried away by the radiation and M is its projection along the its propagation direction. The multipole moments $\mathcal{M}(\mathrm{M/EL}, M)$ depend on the initial and final states of the decaying system. They are defined below.

In quantum mechanics, the energy is not emitted continually but in packets of energy $\hbar\omega$. In a quantum calculation the disintegration constant is the same as the number of *quanta* emitted per unit of time when the power is given by the classical expressions (4.23) and (4.24). Thus,

$$\lambda_E(LM) = \frac{P_E(LM)}{\hbar\omega} = \frac{8\pi(L+1)}{\hbar L[(2L+1)!!]^2}\left(\frac{\omega}{c}\right)^{2L+1}|\mathcal{M}(\mathrm{EL}, M)|^2, \tag{4.25}$$

and

$$\lambda_M(ML) = \frac{P_M(LM)}{\hbar\omega} = \frac{8\pi(L+1)}{\hbar L[(2L+1)!!]^2}\left(\frac{\omega}{c}\right)^{2L+1}|\mathcal{M}(\mathrm{ML}, M)|^2. \tag{4.26}$$

The expressions for the multipole moments are

$$\mathcal{M}_{fi}(\mathrm{EL}, M) = \sum_{k=1} e_k \int r_k^L Y_{LM}^*(\theta_k, \phi_k)\Psi_f^*\Psi_i \, d\tau, \tag{4.27}$$

where e_k are the charges composing the system, r_k describes their positions, and

$$\mathcal{M}_{fi}(\text{ML}, M) = \frac{\hbar}{2mc(L+1)} \sum_{k=1} e_k \int r_k^L Y_{LM}^*(\theta_k, \phi_k) \boldsymbol{\nabla} \cdot [\Psi_f^* \mathbf{L}_k \Psi_i] \, d\tau, \tag{4.28}$$

where the argument $i(f)$ denotes the initial (final) state described by the wavefunction $\Psi_i(\Psi_f)$. Equations (4.27) and (4.28) refer to a single nucleon with mass m that emits radiation in its passage from the state i to the state f. Thus, a sum over all the protons should be incorporated to the result. The spins of both protons and neutrons also give a contribution to Eq. (4.28), but this will not be discussed here.

Note that for $L = 1$, the "operator" rY_{1m} appearing in Eq. (4.27) is proportional to the components of the vector $\mathbf{r}$. In fact, one can also write the electric dipole operator, $\mathbf{d}$, as

$$\mathbf{d} \equiv \mathcal{M}(E1) = \sum_k (e_k \mathbf{r}_k). \tag{4.29}$$

Similarly, the electric quadrupole operator $r^2 Y_{2m}$ for $L = 2$ is proportional to the products of components of $\mathbf{r}$. Often, these terms are rearranged and the quadrupole operator, Q_{ij} (with $i, j = 1, 2, 3$ meaning the cartesian components), is written as

$$Q_{ij} \equiv \mathcal{M}(E2) = \sum_k e_k [3(r_k)_i (r_k)_j - \delta_{ij} r_k^2]. \tag{4.30}$$

The magnetic dipole operator in Eq. (4.28) can also be expressed in a way that relates to the angular momentum of the particles in the system. Including the intrinsic spin $\mathbf{s}$, the M1 operator can be written as

$$\boldsymbol{\mu} \equiv \mathcal{M}(M1) = \frac{\hbar}{2c} \sum_k \frac{e_k}{m_k} (g_l \mathbf{l} + g_s \mathbf{s})_k, \tag{4.31}$$

where g_l (g_s) is the orbital (spin) *gyromagnetic* ratio, or *g-factor*. In principle, g_l should be equal to one, but the intrinsic g-ratio is $g_e = 2.0$ for the electron and $g_p = 5.59$ for the proton. For the neutron, $g_n = -3.83$.

4.2.4. *Selection rules*

Conservation of angular momentum and parity can block certain γ transitions between two states. Selection rules for the γ-radiation are easy to establish if we accept that a *quantum* of radiation carries an angular momentum $\mathbf{l}$ of magnitude $\sqrt{l(l+1)}\hbar$ and component z equal to $m\hbar$, where l is the multipole order. Thus, in transitions between an initial $\mathbf{I}_i$ and a final spin $\mathbf{I}_f$ conservation of angular momentum imposes $\mathbf{I}_i = \mathbf{I}_f + \mathbf{l}$. Then the possible values for the multipole order l obeys the relation

$$|I_i - I_f| \leq l \leq I_i + I_f, \tag{4.32}$$

where $|\mathbf{I}_i| = \sqrt{I_i(I_i+1)}\hbar$, etc. A special case is the transition $0^+ \to 0^+$. As multipole radiation of order zero, these transitions do not exist and they are effectively impossible through the emission of a γ-ray. But in this case a process of *internal conversion* can happen, where the energy from a nuclear transition is released by the ejection of an atomic electron.

Transitions between states of same parity can only be accomplished by electric multipole radiation of even number or by magnetic radiation of odd number. The inverse is valid for transitions where there is parity change. This can be understood examining the Eqs. (4.27) and (4.28). It is necessary that the integrand has even parity, otherwise the contribution in $\mathbf{r}$ cancels with the contribution in $-\mathbf{r}$ and the integral over the whole space vanishes. Let us look at the case of Eq. (4.27): r^l is always positive and the spherical harmonic $Y_l^m(\theta, \phi)$ is even if l is even. For Eq. (4.27) to be non-zero, Ψ_a should have the same parity as Ψ_b for l even and opposite parity for l odd. This justifies the transition rule for the electric multipole radiation. A similar procedure applied to (4.28) justifies the selection rules for the magnetic multipole radiation.

A very simple estimate of electromagnetic transition probabilities can be made by assuming that a system (atom or nucleus) decaying from an excited state is described by the wavefunction Ψ_a to the final state b of $l = 0$ (the restriction $l = 0$ makes the calculation simpler). The state b is described by a wavefunction Ψ_b in agreement with the schematic diagram in Fig. 4.2. For Ψ_a and Ψ_b one can use, in an approximate calculation,

$$R_a(r) = const. = R_a(r < R), \quad \text{and} \quad R_a(r) = 0(r > R),$$

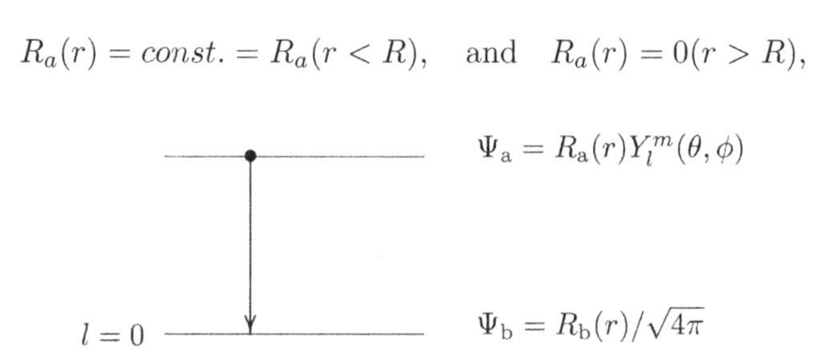

Figure 4.2. De-excitation of a system to a level with $l = 0$.

and to use the same approach for $R_b(r)$. The normalization yields immediately the values for the constants R_a and R_b:

$$R_a = R_b = \sqrt{\frac{3}{R^3}},$$

where R is the system radius. In this way, it is not difficult to calculate the electric multipole moment $\mathcal{M}(Elm)$ in Eq. (4.27):

$$\mathcal{M}(Elm) = e \int r^l Y_l^{m*}(\theta, \phi) \frac{3}{R^3} \frac{Y_l^m(\theta, \phi)}{\sqrt{4\pi}} r^2 \, dr \, d\Omega,$$

which yields

$$\mathcal{M}(Elm) = \frac{3eR^l}{\sqrt{4\pi}(l+3)}.$$

Thus, in this approximation, the disintegration constant (4.25) is

$$\lambda_E = \frac{2(l+1)}{l[(2l+1)!!]^2} \left(\frac{3}{l+3}\right)^2 \frac{e^2}{\hbar} \left(\frac{E_\gamma}{\hbar c}\right)^{2l+1} R^{2l},$$

where we wrote explicitly the disintegration energy $E_\gamma = \hbar\omega$. A similar calculation for the magnetic disintegration constant (4.26) yields

$$\lambda_M = \frac{20(l+1)}{l[(2l+1)!!]^2} \left(\frac{3}{l+3}\right)^2 \frac{e^2}{\hbar} \left(\frac{E_\gamma}{\hbar c}\right)^{2l+1} R^{2l} \left(\frac{\hbar}{mcR}\right)^2,$$

where m is the mass of a nucleon. For a typical nucleus of intermediate mass, $A = 120$, it is easy to see that $\lambda_E/\lambda_M \cong 100$, independently of the multipolar order l. Evidently, both constants decrease rapidly when the value of l increases.

The values of the magnetic multipole moments are small in comparison to the electric moments of same order. Also, the transition probability decreases quickly with increasing l, restricting the multipole orders that give significant contribution. These facts imply, in principle, that the electric dipole is always the dominant radiation. But the selection rules can modify this situation.

4.3. Propagation of Radiation

4.3.1. *Radiation intensity*

We define the *specific radiation intensity* $I_\nu(r, \Omega, t)$ as the radiative energy per frequency interval, per normally incident area, per solid angle, per unit

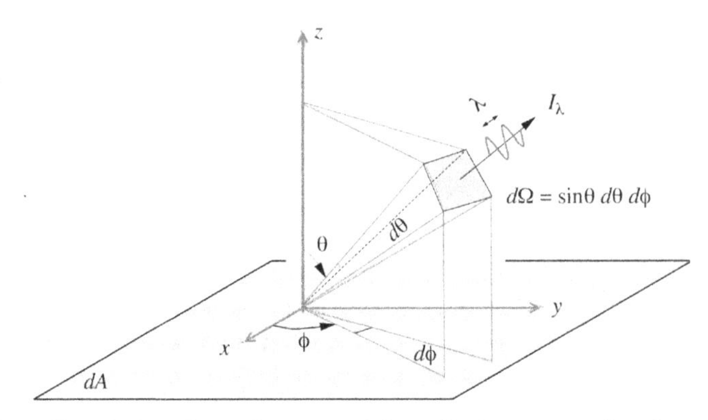

Figure 4.3. Description of coordinates used for radiation propagating along the radial direction.

time. Then, the radiative energy per unit frequency from an elemental source area dA propagating in a direction $\mathbf{\Omega} = (\theta, \phi)$ (see Fig. 4.3), with respect to the normal $\hat{\mathbf{n}}$ to dA, into an elemental solid angle $d\Omega$, per increment in frequency $d\nu$ is given by

$$dE = I_\nu(\mathbf{\Omega}) \cdot (\hat{\mathbf{n}} \cdot \hat{\mathbf{\Omega}})\, dA\, d\Omega\, d\nu\, dt, \tag{4.33}$$

where

$$(\hat{\mathbf{n}} \cdot \hat{\mathbf{\Omega}})dA = \hat{\mathbf{z}} \cdot \hat{\mathbf{\Omega}}dA = \cos\theta dA, \quad \text{and} \quad d\Omega = \sin\theta d\theta d\phi = -d(\cos\theta)d\phi. \tag{4.34}$$

The mean intensity, $\langle I_\nu \rangle$, is the average of I_ν, over all directions,

$$\langle I_\nu \rangle = \frac{1}{4\pi} \int I_\nu d\Omega = \frac{1}{4\pi} \int_0^{2\pi} \int_0^{\pi} I_\nu \sin\theta d\theta d\phi. \tag{4.35}$$

For most stars, we can assume that the atmosphere is thin compared to the radius, so that a planar approximation can be used, where I_ν is independent of ϕ. Then $I_\nu = I_\nu(z, \xi)$, where z is measured relative to some fixed reference level in the stellar atmosphere and $\xi = \cos\theta$, with $0 < \theta < \pi$. Thus,

$$\langle I_\nu \rangle = \frac{1}{2} \int_{-1}^{+1} I_\nu(z, \xi) d\xi. \tag{4.36}$$

The flux, $\mathbf{\Phi}_\nu$, is the radiative energy through a unit area. Formally, it is defined as the specific intensity from solid angles in all directions, i.e.,

$$\mathbf{\Phi}_\nu = \int I_\nu \hat{\mathbf{\Omega}} d\Omega. \tag{4.37}$$

The integral represents integrating all possible angles on the fixed surface (see Fig. 4.3). At the end a net direction exists.

The flux in the normal direction $\hat{\mathbf{n}}$ is

$$\Phi_\nu = \mathbf{\Phi}_\nu \cdot \hat{\mathbf{n}} = \int I_\nu \hat{\mathbf{\Omega}} \cdot \hat{\mathbf{n}} d\Omega = \int_0^{2\pi} \int_0^{\pi} I_\nu \cos\theta \sin\theta d\theta d\phi, \tag{4.38}$$

which implies that the net flux from an isotropic radiation field (I_ν independent of direction) is zero. This is because by definition an isotropic field has zero net in/out flux. This also implies that $\Phi \propto 1/r^2$ for a spherical surface of radius r surrounding an isotropic source, since $\hat{\mathbf{n}} \cdot \hat{\mathbf{\Omega}} = 1$, yielding $4\pi r_1^2 \Phi(r_1) = 4\pi r_2^2 \Phi(r_2)$ by conservation of energy.

Radiation pressure, $P_{rad,\nu}$, is the momentum flux of the radiation field, i.e. the rate at which momentum is transported through a unit area per unit time. The momentum of a pencil beam of radiative energy is dE_ν/c. Formally, this is dependent upon direction. The component normal to the area dA then gives the momentum flux per unit area as

$$\frac{dE_\nu \cos\theta}{cdA} = \frac{I_\nu dA \cos\theta d\Omega}{cdA}. \tag{4.39}$$

This leads to the following definition for the radiation pressure:

$$P_{rad,\nu} = \frac{1}{c} \int I_\nu \cos^2\theta d\Omega = \frac{2\pi}{c} \int_{-1}^{1} I_\nu \xi^2 d\xi. \tag{4.40}$$

Thus, radiation pressure is related to the second-order moment of the radiation field.

The *radiation energy density*, $U_{rad,\nu}$, is defined as the total energy in the radiation field per unit volume. Since the increment in total radiative energy is $dE_\nu = I_\nu dA \cos\theta d\Omega dt$, and $dt = ds/c$ for a path segment ds along a light ray, this implies (since $dV = dA \cos\theta ds$)

$$U_{rad,\nu} = \frac{1}{c} \int I_\nu d\Omega = \frac{4\pi}{c} \langle I_\nu \rangle. \tag{4.41}$$

Thus, the radiation energy density is related to the 0th-order moment of the radiation field. For an isotropic radiation field, we can use $\int_{-1}^{1} \cos^2\theta$ $d\cos\theta = 2/3$ to perform the angular integral in Eq. (4.40). Comparing to the results obtained from integrating Eq. (4.41), one gets

$$U_{rad,\nu} = 3P_{rad,\nu}, \tag{4.42}$$

a relation that will be used later.

4.3.2. *Radiative energy transport*

The energy in a beam of radiation propagating through a medium will be modified by emission, absorption, and scattering. These effects are described by emission and extinction coefficients. The *emission coefficient*, j_ν, is defined so that the energy added to a beam of radiation by an elemental volume $dV = dAds$ of matter intercepting the beam is

$$dE = j_\nu \, dAdsd\Omega d\nu dt. \tag{4.43}$$

In general, j_ν also contains a scattering component. Note also that sometimes j_ν is used to represent a mass emission coefficient.

The combined effects of absorption and scattering of radiation can be studied by using the definition of the *opacity*, or *extinction coefficient*, κ_ν. It is defined so that the energy removed from a beam of radiation by an elementary volume of cross-section dA and length ds is

$$dE = \kappa_\nu I_\nu dAdsd\Omega d\nu dt. \tag{4.44}$$

The quantity $1/\kappa_\nu$ is known as the mean free path of a photon. It is the mean distance travelled by a photon in a medium before it is absorbed and/or scattered. Both j_ν and κ_ν are properties of the medium, not of the radiation field (generally, they are functions of the temperature, T, and mass density, ρ, of the medium).

It is useful to define the *optical depth* of a medium as a dimensionless quantity that gives a measure of how much radiation is attenuated as it passes through the medium along a geometrical path s:

$$\tau_\nu = \int_{z_1}^{z_2} \kappa_\nu(z) ds = \frac{1}{\xi} \int_{z_1}^{z_2} \kappa_\nu(z) dz, \tag{4.45}$$

where

$$ds = \frac{dz}{\cos\theta} = \frac{dz}{\xi}, \tag{4.46}$$

for plane-parallel geometry. For applications to stellar atmospheres, the optical depth is defined such that it decreases with height (increasing z), as shown in Fig. 4.4, i.e. $d\tau_\nu = -\kappa_\nu dz/\xi$ for (plane-parallel) stellar atmospheres.

4.3.3. *The equation of radiative transfer*

The net change in the intensity of a radiation beam as a result of its interaction with matter is obtained by combining the effects of emission

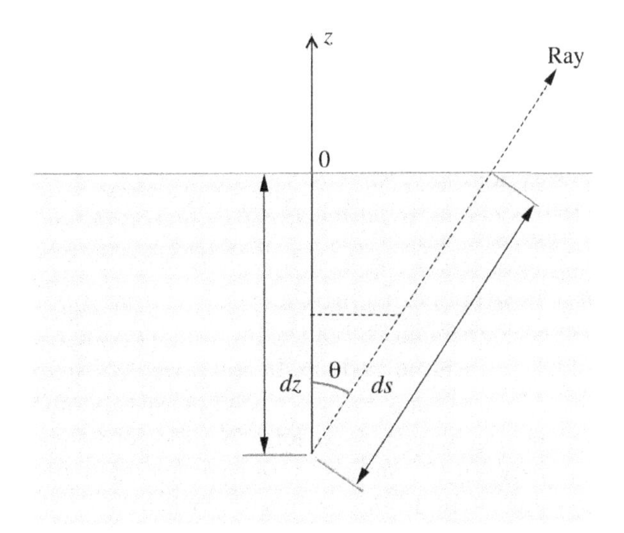

Figure 4.4. Schematic view of variables used to describe the calculation of optical depth.

and extinction. The net change in radiative energy is equal to emission minus extinction, i.e.,

$$I_\nu \, dAd\Omega d\nu dt = j_\nu \, dAdsd\Omega d\nu dt - \kappa_\nu I_\nu \, dAdsd\Omega d\nu dt. \tag{4.47}$$

This implies

$$\frac{dI_\nu}{ds} = j_\nu - \kappa_\nu I_\nu, \tag{4.48}$$

and using the definition of optical depth gives

$$\frac{dI_\nu}{d\tau_\nu} = \frac{j_\nu}{\kappa_\nu} - I_\nu. \tag{4.49}$$

Introducing the definition of the source function, $S_\nu = j_\nu/\kappa_\nu$, we get

$$\frac{dI_\nu}{d\tau_\nu} = S_\nu - I_\nu, \tag{4.50}$$

which is known as the *equation of radiative transfer.*

A few simple solutions of the above equations are:

1. if there is no material present, we have the solution $j_\nu = 0 = \kappa_\nu$ and $dI_\nu/ds = 0$, i.e., $I_\nu = const.$;
2. if there is a uniform medium in thermodynamic equilibrium, the radiation field does not change with time or position. Thus, $dI_\nu/d\tau_\nu = 0$,

i.e. the rate of extinction equals the rate of emission. Therefore, $I_\nu = S_\nu$ and $j_\nu = \kappa_\nu I_\nu$, known as *Kirchoff's law*;

3. if there is emission only, $\kappa_\nu = 0$, which means $\xi dI_\nu/dz = j_\nu$, or $I_\nu = I_\nu(0) + \int_0^z j_\nu dz/\xi$, where $I_\nu(0)$ is the incident specific intensity;
4. if there is absorption only, $j_\nu = 0$, yielding $\xi dI_\nu/dz = -\kappa_\nu I_\nu$. For a finite slab, the emergent radiation is $I_\nu = I_\nu(0)\exp(-\int_0^z \kappa_\nu dz/\xi) = I_\nu(0)\exp(-\tau_\nu)$.

In the limit of large optical depths, the specific intensity of the radiation field approaches the source function of the medium, i.e., the radiation field comes into equilibrium with the medium. In real astrophysical sources, S_ν is a sensitive function of the temperature, T, and density, ρ, of the medium. In stellar and planetary atmospheres, for example, both these quantities vary significantly between the outer surface layer and the innermost layer. Strict thermodynamic equilibrium prevails only when $S_\nu = I_\nu$ everywhere. A thermal source emitting a spectrum that is not quite of Planck distribution form at all frequencies (i.e., exhibits lines and edges) is referred to as being in non-thermodynamic equilibrium. The cosmic microwave background (CMB) radiation is the best example of a real astrophysical source in strict thermodynamical equilibrium. Generally, however, most real astrophysical sources are far from being in an ideal thermodynamical equilibrium state.

4.4. Blackbody Radiation

We already had a taste of a quantum distribution when we discussed the electron gas, leading to the *Fermi–Dirac distribution*, given by Eq. (3.34). Next we discuss the *Boltzmann distribution*, valid for classical particles. Then we will show how the quantum aspect of the photon leads to the *Bose–Einstein distribution* for them. For photons, the Bose–Einstein distribution is known as the *Planck*, or *blackbody radiation* law.

4.4.1. *Boltzmann law*

Let us assume that we have a closed system, containing a large number N of distinguishable particles, in thermal equilibrium with its surrounding at temperature T and under constant volume conditions. We also suppose that only the two lowest (non-degenerate) energy levels $\epsilon_0 = 0$ and $\epsilon_1 = \epsilon$ are occupied and contain n_0 and n_1 particles, respectively (see Fig. 4.5). The number of ways in which this configuration can be realized is given by

$$W = \frac{N!}{n_0! n_1!}, \tag{4.51}$$

ε_1 — n_1 particles

ε_0 — n_0 particles

Figure 4.5. $N(= n_0 + n_1)$ particles distributed in two states with energies ϵ_0 (n_0 particles) and ϵ_1 (n_1 particles), respectively.

with $N = n_0 + n_1$. The microscopic statement of the *Boltzmann distribution* relates the entropy of the system, S, to the number of possible configurations by means of

$$S = k \ln W, \tag{4.52}$$

k being, of course, the Boltzmann constant. Equation (4.52) implies that

$$S = k\left(\ln N! - \ln n_0! - \ln n_1!\right). \tag{4.53}$$

Now assume that a small amount of energy ϵ (corresponding to the separation energy between the levels ϵ_0 and ϵ_1) is added to the system and promotes one particle from the lower to the upper level (i.e. $n_0 \to n_0 - 1$ and $n_1 \to n_1 + 1$). Then,

$$S' = k(\ln(N!) - \ln(n_0 - 1)! - \ln(n_1 + 1)!). \tag{4.54}$$

The change in entropy, ΔS, is thus

$$\Delta S = S' - S = k \ln\left(\frac{n_0}{n_1 + 1}\right) = k \ln\left(\frac{n_0}{n_1}\right), \quad \text{for } n_1 \gg 1. \tag{4.55}$$

From the laws of thermodynamics, we have a connection between the entropy and the total energy, U, of the system,

$$\Delta S = \frac{dU}{T}, \tag{4.56}$$

since under constant volume, the heat transfer, dq, and the system energy change are the same, i.e., $dU = dq$. Since we also assumed that $dU = \epsilon$, we have $\Delta S = \epsilon/T$ and

$$k \ln\left(\frac{n_0}{n_1}\right) = \frac{\epsilon}{T}. \tag{4.57}$$

This implies that

$$\frac{n_1}{n_0} = \exp\left(-\frac{\epsilon}{kT}\right). \tag{4.58}$$

This is the *Boltzmann distribution*, which may be generalized to an arbitrary number of levels. That is, for a system with a large number of particles at a temperature T, the number of particles with energy E is given by

$$n(E) = A\exp\left(-\frac{E}{kT}\right), \tag{4.59}$$

where A is a normalization constant. This equation tell us that for a system of particles the probability of finding a particle with energy E decreases exponentially with the particle energy. But, as the temperature increases, it becomes more probable to find particles with high energies in the system.

4.4.2. *Planck's law*

Consider a photon of frequency ν propagating in a direction $\hat{\mathbf{n}}$. The propagation vector is given by $\mathbf{k} = (2\pi/\lambda)\hat{\mathbf{n}} = (2\pi\nu/c)\hat{\mathbf{n}} = (\omega/c)\hat{\mathbf{n}}$. Suppose the photon is inside a box of volume V and dimensions L_x, L_y, L_z, each of which is longer than λ, so that the photon can be represented as a standing wave in the box. The number of nodes in the wave in each direction is, e.g., $n_x = L/\lambda_x = k_x L_x/2\pi$ (i.e., 1 node per integral number of λ_x). The wave changes state when n_x changes by one or more. This means that the number of node changes per wave number interval is $\Delta n_x = L_x \Delta k_x/2\pi$. The number of state changes in the three-dimension wave vector element $\Delta k_x \Delta k_y \Delta k_z = d^3k$ is

$$dn = dn_x dn_y dn_z = \frac{2L_x L_y L_z d^3k}{(2\pi)^3} = \frac{2Vd^3k}{(2\pi)^3}, \tag{4.60}$$

where the factor 2 takes into account two independent polarization states per wave vector $\mathbf{k}$. Now we use the identity $d^3k = k^2 dk d\Omega = (2\pi)^3\nu^2 d\nu d\Omega/c^3$ to find that the number of states per volume per solid angle per frequency is given by

$$\rho_s = \frac{dn}{d\Omega d\nu} = \frac{2\nu^2}{c^3}, \tag{4.61}$$

which is known simply as the *density of states*.

Next we consider the average energy of each state. Each state may contain n photons of energy $h\nu$, so the energy of each state can be written as $E_n = nh\nu$. The probability to populate a state of energy E_n is $\exp(-\beta E_n)$, where $\beta = (kT)^{-1}$, so that the average energy is

$$\bar{E} = \frac{\sum_{n=0}^{\infty} E_n e^{-\beta E_n}}{\sum_{n=0}^{\infty} e^{-\beta E_n}} = -\frac{\partial}{\partial \beta} \ln \left(\sum_{n=0}^{\infty} e^{-\beta E_n} \right). \tag{4.62}$$

The formula for the sum of a geometric series implies that

$$\sum_{n=0}^{\infty} e^{-\beta E_n} = \sum_{n=0}^{\infty} e^{-\beta n h\nu} = \sum_{n=0}^{\infty} (e^{-\beta h\nu})^n = (1 - e^{-\beta h\nu})^{-1}, \tag{4.63}$$

which yields for the average energy of each state

$$\bar{E} = \frac{h\nu e^{-h\nu/kT}}{1 - e^{-h\nu/kT}} = \frac{h\nu}{e^{h\nu/kT} - 1}. \tag{4.64}$$

Since the energy per photon is $h\nu$, then $\bar{E}/h\nu$ gives the average number of photons of frequency ν in each state,

$$n_\nu = \frac{1}{e^{h\nu/kT} - 1}. \tag{4.65}$$

This is the general expression for the mean occupation number of any system of particles (such as photons) that obey *Bose–Einstein statistics*. Only for large energies, $E_\nu = h\nu \gg 1$, can the factor 1 in the denominator be neglected and then this distribution becomes equal to the Boltzmann distribution. For low energies the distribution deviates from Boltzmann and tends to favor the occupation of low energy states. That is, *bosons* (particles which obey the Bose–Einstein statistics) have a preference to occupy the lowest available energy states.

We can combine the density of states with the occupation number and energy per photon to get an expression for the energy per unit volume per solid angle per frequency. But this is just the energy density per solid angle and from Eq. (4.41), we can use $U_{rad,\nu}/\Delta\Omega = I_\nu/c$. This gives the *Planck distribution* function, which is the *blackbody radiation* at a temperature T. It is isotropic with specific intensity, $I_\nu = c\rho_s n_\nu$, given by

$$I_\nu = \frac{2h\nu^3}{c^2}[e^{h\nu/kT} - 1]^{-1}. \tag{4.66}$$

The Planck distribution for different temperatures and as a function of the wavelength is shown in Fig. 4.6. One can clearly see a shift of the

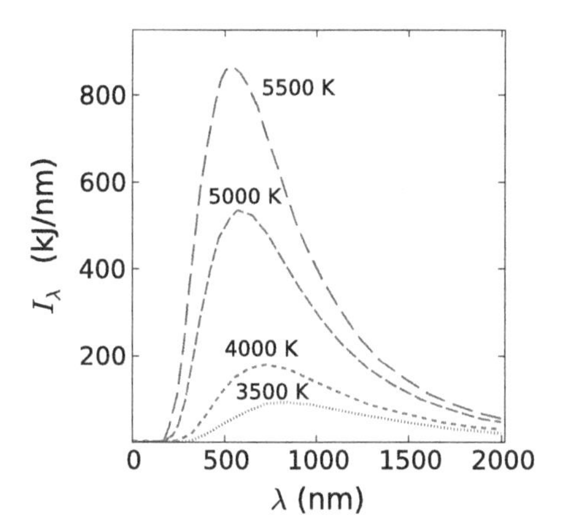

Figure 4.6. Planck distribution for different temperatures and as a function of the wavelength.

distribution maximum to smaller wavelengths (or larger frequencies) as the temperature increases. The peak is also more pronounced, i.e., has a smaller width, as the temperature increases.

The number of photons per unit energy can be obtained by dividing the energy density $U_{rad,\nu}$ (Eq. (4.41)) by the photon energy, $E_\gamma = h\nu$. That is,

$$n(E_\gamma) = \frac{8\pi}{c^3h^3}\frac{E_\gamma^2}{e^{E_\gamma/kT}-1}, \tag{4.67}$$

known as the *Bose–Einstein distribution* of photons in thermal equilibrium.

4.4.3. *Stefan–Boltzmann law*

Integrating the Planck distribution over all frequencies one obtains the *Stefan–Boltzmann law*. We consider the radiation passing by an infinite plane on the side of a blackbody radiation source. Then the energy flux from the radiation needs only to be integrated in the interval $0 \le \theta \le \pi/2$ for the outgoing radiation. Thus, we have

$$\Phi = \int \Phi_\nu d\nu = \int I_\nu d\nu \int_0^{\pi/2} \cos\theta \sin\theta d\theta \int_0^{2\pi} d\phi = \pi \int_0^\infty I_\nu d\nu = \sigma T^4, \tag{4.68}$$

where the last equality has been obtained by changing the integration variable to $x = h\nu/kT$ and using the integral $\int_0^\infty x^3[e^x-1]^{-1}dx = \pi^4/15$.

The above equation has a temperature dependence given by a T^4 law. The Stefan–Boltzmann constant arising from the integration is

$$\sigma = \frac{2\pi^5 k^4}{15h^3c^2} = 5.67 \times 10^{-8}\mathrm{W.m^{-2}K^{-4}}. \tag{4.69}$$

The *Stefan–Boltzmann law* is used to define the *effective temperature* of a source by means of

$$\sigma T_{eff}^4 = \Phi. \tag{4.70}$$

The effective temperature of a body such as a star is the temperature of a blackbody that would emit the same total amount of electromagnetic radiation. The pressure of a photon gas is obtained from Eq. (4.41), and performing the appropriate angular integral, which yields

$$P_{rad} = \frac{4\sigma T^4}{3c}. \tag{4.71}$$

The Planck spectrum is monotonic with temperature at each and every frequency, so a higher T gives a stronger I_ν. The frequency at which I_ν peaks is

$$h\nu_{peak} \simeq 2.82kT. \tag{4.72}$$

This linear shift of the peak with T is known as the *Wien displacement law.*

Two important limits of the Planck spectrum are $h\nu \ll kT$, leading to the *Rayleigh–Jeans law,*

$$I_\nu^{RJ} \simeq \frac{2\nu^2}{c^2}kT. \tag{4.73}$$

This was originally derived by assuming the classical equipartition relation $\bar{E} = kT$ for the energy of a photon. However, if it is applied to all frequencies, then the total energy is proportional to $\int \nu^2 d\nu$, which diverges. This is known as the *ultraviolet catastrophe.*

Because the Rayleigh–Jeans limit almost always applies at radio-frequencies, it is sometimes convenient to use the definition of *brightness temperature*, T_b, which is the temperature of a blackbody having the same brightness at a particular frequency,

$$kT_b = \frac{c^2}{2\nu^2}I_\nu. \tag{4.74}$$

This relation allows one to deduce the brightness temperature of a star. For example, for the human body, $T = 37^\circ$ C $= 310\,$K, so that $\lambda_{peak} = 9.4\,\mu$m (infrared light). For the Sun's surface, $\lambda_{peak} = 500\,\text{nm}$ (yellow light) so that $T = 5770\,\text{K}$.

In the other limit $h\nu \gg kT$, one obtains the *Wien limit*

$$I_\nu^W \simeq \frac{2\nu^3}{c^2} \exp\left(-\frac{h\nu}{kT}\right). \tag{4.75}$$

The cosmic microwave background (CMB) radiation is an astrophysical source that emits a perfect (ideal) blackbody spectrum. Generally, however, most real astrophysical sources are far from being in an ideal thermal equilibrium state. The accuracy with which the CMB obeys the Planck spectrum is a very strong physical constraint in favor of the quantum statistics obeyed by photons. An experimental verification of the CMB temperature is $T_{CMB} = 2.725 \pm 0.002\,\text{K}$.

The *luminosity of a star* is related to the radiation flux as $L = A\Phi$, or according to the Stefan–Boltzmann law, $L = A\sigma T^4$, where A is the surface area of the star. Our Sun produces $\Phi = L/A = 6.27 \times 10^7\,\text{W/m}^2$ of energy at the solar surface. Hence, if the surface solar temperature is $T = 5770\,\text{K}$ then Eq. (4.68) yields $\Phi = 5.67 \times 10^{-8} \times (5770)^4\,\text{W/m}^2 = 6.28 \times 10^7\,\text{W/m}^2$, which is the experimentally observed value. Thus, the energy emitted from a simple star such as our Sun seems to be consistent with the blackbody radiation.

The Stefan–Boltzmann law implies that for a star with luminosity L, one has

$$\frac{L}{L_\odot} = \left(\frac{R}{R_\odot}\right)^2 \left(\frac{T}{T_\odot}\right)^4, \tag{4.76}$$

which suggests that stars of a similar structure as our Sun might lie along a one-parameter path (corresponding to $R/R_\odot$) in the luminosity (or magnitude) vs. temperature (or color) plane. In fact, there is a dominant location in the Hertzsprung–Russell color-magnitude diagram along which roughly 80% of the stars reside (see Fig. 1.2).

As mentioned in the introduction, fluxes in astronomy are usually quoted in terms of magnitudes. Magnitudes are related to fluxes Φ via

$$m = -\frac{5}{2} \log(\Phi) + \text{constant},$$

where log denotes the logarithm with base 10. The *apparent magnitude* m is the flux we observe here on Earth, whereas the *absolute magnitude* $\mathcal{M}$ is the flux emitted at the source. They are related by

$$m - \mathcal{M} = 5 \log \left(\frac{d_L}{10\text{pc}} \right) + K,$$

where

$$d_L \simeq \frac{cz}{H_0}$$

is the approximate luminosity distance to an object with a redshift z in a flat ($k = 0$) Universe containing non-relativistic matter and vacuum energy.

4.5. Radiation in Stellar Plasmas

4.5.1. *Stellar atmospheres*

The atmosphere of a star is defined as those layers sufficiently close to the surface that some photons can escape, i.e., the atmosphere is as deep as one can see into the star. The primary layers in a stellar atmosphere, as shown in Fig. 4.7, are:

1. *Photosphere* — In this layer the energy transfer is dominated by radiation processes. The Sun's photosphere is $\sim 400\,\text{km}$ thick, and has $T \lesssim 5800\,\text{K}$.
2. *Chromosphere* — Here, the non-radiative energy dissipation heats up the gas to above the radiative equilibrium temperature. However, the density is sufficiently high that most of the dissipated energy can be

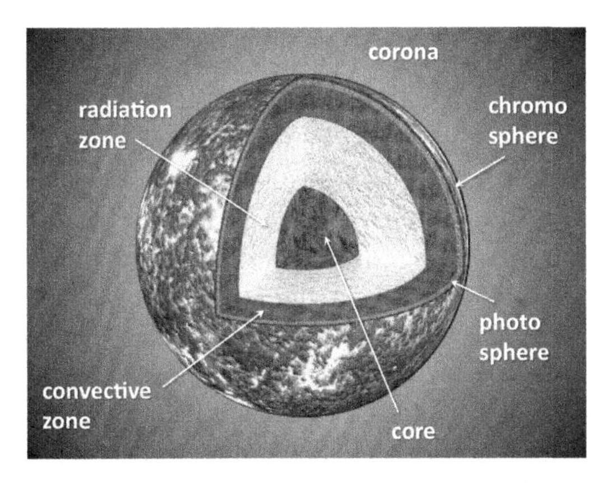

Figure 4.7. Schematic view of a cross-section of the Sun.

radiated away and the resultant temperature is not too high. The Sun's chromosphere is $\sim 10^4$ km thick and has $T \sim 1.5 \times 10^4$ K.

3. *Corona* — A layer containing significant non-radiative energy transport and dissipation. The density is too low for the dissipated energy to be efficiently radiated away, so the corona is very hot. The solar corona extends out to $> 10^6$ km and has $T > 2 \times 10^6$ K.
4. *Stellar wind* — Is the outermost layer of a stellar atmosphere. It is so tenuous that it is no longer gravitationally bound; most stars, including our Sun, lose mass through a stellar wind.

A theory of stellar atmospheres involves constructing models of the photosphere that take into account the effective temperature, T_{eff}, the hydrostatic equilibrium, and elemental abundances. Numerical models can compute the continuum spectrum as well as detailed profiles of the spectral lines.

Some typical assumptions and limitations in modeling stellar atmospheres are:

1. *Plane-parallel geometry* — The atmosphere is vertically stratified into homogeneous plane-parallel layers, with $d\tau_\nu(z, \xi) = -\kappa_\nu dz/\xi$ and with small overall geometric thickness. The optical depth is conventionally defined this way so that $\tau_\nu \to \infty$ deep in the atmosphere ($z \to \infty$) and $\tau_\nu \simeq 0$ at the photospheric surface ($z \simeq 0$). The plane-parallel approximation breaks down for luminous, low surface gravity stars, with extended envelope atmospheres. Spherically symmetric radiative transfer models show significant differences in the emergent spectra. Also, homogeneity is not a good assumption in stellar atmospheres with strong convective motions.
2. *Steady-state and local thermal equilibrium* — All variations in time are ignored, and hence, it is static. Under local thermal equilibrium, all the microscopic processes are in detailed balance and the densities of energy states are calculated from statistical relations. The steady-state assumption is not appropriate for pulsating stars and stars in closely interacting binary systems. Non-local thermal equilibrium effects are important in low surface gravity stellar atmospheres, where the densities are insufficient to maintain thermal equilibrium.
3. *Hydrostatic equilibrium* — The outward pressure gradients exactly balance gravity. Thus, dynamical effects are ignored. A net outward flow due to radiation pressure gradients cannot be neglected in hot stars such as giants, supergiants, and early-type stars ($\sim 10\%$ at 30,000 K).

4. *Radiative equilibrium* — The energy flux is transported primarily in the form of radiation. Energy conservation implies that

$$\nabla \cdot \mathbf{\Phi}_{rad} = 0, \tag{4.77}$$

where $\mathbf{\Phi}_{rad}$ is the total energy flux vector. This means that $d\Phi/dz = 0$, i.e., the radiative flux is constant with height. Using the definition of flux, namely, $\Phi = 2\pi \int_{-1}^{+1} I\xi d\xi$, together with the radiative transfer equation, $dI/ds = \xi dI/dz = j - kI$, we get

$$\frac{d\Phi}{dz} = 2\pi \int_{-1}^{+1} \frac{dI}{dz}\xi d\xi = 2\pi \int_{-1}^{+1} \kappa_\nu (S_\nu - I_\nu) d\xi d\nu = 0. \tag{4.78}$$

Now the assumption of local thermal equilibrium implies that locally S_ν is isotropic, and using the definition $J_\nu = (1/2) \int_{-1}^{+1} I_\nu d\xi$, we find

$$\int_0^\infty \kappa_\nu S_\nu d\nu = \int_0^\infty \kappa_\nu J_\nu d\nu, \tag{4.79}$$

i.e., $S_\nu = J_\nu$. The assumption of local thermal equilibrium is valid for atmospheres of hot stars, which are dominated by radiation, but breaks down for increasingly cooler stellar atmospheres, where convective energy transport is important.

A simple model, called *grey atmosphere* model, uses the simplifying assumption that $\kappa \equiv \kappa_\nu$, i.e., that absorption is independent of frequency. This is not a bad approximation for most stellar atmospheres, where the opacity is dominated by H^-, which has an absorption coefficient virtually independent of ν. Since $\kappa_\nu = k$ implies $\tau_\nu = \tau$, the radiative transfer equation integrated over ν becomes

$$\frac{dI}{d\tau} = I - S, \tag{4.80}$$

where we used $\tau = -\kappa ds$ and $S = \langle I \rangle$ for radiative equilibrium. One can derive using the approximation of isotropic radiation field, and the fact that radiation emerges uniformly outward but is zero inward, the grey atmosphere temperature equation

$$T^4(\tau) = \frac{3}{4} T_{eff}^4 \left(\tau + \frac{2}{3}\right). \tag{4.81}$$

This equation shows that the 4th power of T increases linearly with $\tau(z)$, meaning that T decreases with height. The temperature at the stellar surface is $T^4(0) = T_{eff}^4/2$, i.e., $T(0) \simeq 0.841 T_{eff}$. At $\tau = 1$, the temperature is $T(1) \simeq 1.057 T_{eff}$. One also concludes that $T(\tau) = T_{eff}$ for $\tau = 2/3$, i.e., the effective depth of continuum formation is $\tau = 2/3$. Thus, a photon emitted from this depth has a probability of $\exp(-2/3) \simeq 0.5$ of emerging from the surface.

4.5.2. *Einstein coefficients*

In considering the transition of an atom from one excited state to another, Einstein defined a set of coefficients that denote the probability of specific transitions taking place. These are known as the *Einstein coefficients.* All radiative processes and the equation of radiative transfer itself may be formulated in terms of these coefficients.

The Einstein coefficients describe the probability of atomic transitions between two energy levels E_n and E_m (where $E_m > E_n$). The transition probability per unit time for the emission of a photon of energy $h\nu = E_m - E_n$ is defined as the *Einstein A-coefficient*, A_{mn}, and the emission rate per unit volume is $n_m A_{mn}$, where n_m is the number density of atoms in state m. In the absence of any other processes, the mean lifetime of particles in state m is $\Delta t = A_{mn}^{-1}$ (in seconds). According to the Heisenberg uncertainty principle, there is a corresponding spread in energy $\Delta E = \hbar/\Delta t$, implying $\Delta\nu = 1/(2\pi\Delta t)$. This quantum broadening gives rise to a Breit–Wigner profile for the distribution (*line profile*) over possible frequencies emitted by the atoms (see Eqs. (4.8–4.9)), i.e.,

$$\phi(\nu) = \frac{\Gamma/4\pi^2}{(\nu - \nu_0)^2 + (\Gamma/4\pi)^2}, \tag{4.82}$$

where ν_0 is the frequency centroid and the width of the distribution, $\Gamma = 1/\Delta t$, is the *radiative damping constant.* For $\Gamma \ll \nu_0$ this distribution is zero, except for the state with frequency ν_0 (a line). For larger values of Γ it gives a distribution of probabilities (a profile) to find states around ν_0 within an energy interval Γ.

The transition probability per unit time for absorption of a photon of energy $h\nu = E_m - E_n$ is given by $B_{nm}\bar{J}$, where B_{nm} is the *Einstein B-coefficient* for photoabsorption, and $B_{nm}\bar{J}$ is the *mean intensity* weighted over the line profile $\phi(\nu)$, i.e.,

$$\bar{J} = \int_0^\infty J_\nu \phi(\nu) d\nu. \tag{4.83}$$

The rate of photoabsorption per unit volume is $n_n B_{nm}\bar{J}$ [$\mathrm{s^{-1}\ m^{-3}}$].

In some circumstances, a photon can induce a transition from a high atomic energy state E_m to a lower state E_n and a photon of energy $h\nu = E_m - E_n$ is emitted in the process. The rate per unit volume of this stimulated emission is given by $n_m B_{mn} \bar{J}$, where B_{mn} is the *Einstein coefficient for stimulated emission.* An important property of stimulated emission is that it takes place into precisely the same photon state (direction and frequency). The emitted photon is precisely coherent with the photon that stimulated the emission. This is the mechanism by which lasers and masers work.

Under the condition of radiative equilibrium, the number of radiative transitions per unit time per unit volume out of state n must equal the number of radiative transitions into state n, i.e.,

$$n_n B_{nm} \bar{J} = n_m A_{mn} + n_m B_{mn} \bar{J}. \tag{4.84}$$

The Einstein coefficients are linked via the *detailed balance relations,* which connect any microphysical process to its inverse (i.e., emission and absorption). In practice this means solving Eq. (4.84) for $\bar{J}$. Using the Boltzmann distribution, we have

$$\frac{n_n}{n_m} = \frac{g_n \exp(-E_n/kT)}{g_m \exp(-E_m/kT)} = \frac{g_n}{g_m} \exp\left(\frac{h\nu}{kT}\right).$$

Now setting $\bar{J}$ equal to the Planck distribution, Eq. (4.66), one finds

$$g_n B_{nm} = g_m B_{mn}, \quad A_{mn} = \frac{2h\nu^3}{c^2} B_{mn}, \tag{4.85}$$

where g_n are *statistical factors*, i.e., the number of states with energy E_n. These relations are independent of T. Thus, they should be valid even under conditions where thermal equilibrium is not satisfied.

Let us assume that the emission is distributed with the same line profile as the absorption. For emission, we know that the total energy emitted in a volume dV, solid angle $d\Omega$, frequency range $d\nu$ and time interval dt is $j_\nu dV d\Omega d\nu dt$. Since each atom contributes an energy $h\nu$ distributed over 4π in solid angle and $\phi(\nu)$ in frequency for each transition, this total energy is equivalent to $(h\nu/4\pi)\phi(\nu) n_m A_{mn} dV d\nu d\Omega dt$. A similar argument follows for absorption, yielding

$$j_\nu = \frac{h\nu_{mn}}{4\pi} n_m A_{mn} \phi(\nu), \quad \kappa_\nu = \frac{h\nu_{mn}}{4\pi}(n_n B_{nm} - n_m B_{mn})\phi(\nu), \tag{4.86}$$

where κ_ν now represents an absorption coefficient corrected for stimulated emission. Using Eqs. (4.85), one gets

$$\kappa_\nu = \frac{h\nu_{mn}}{4\pi} n_n B_{nm} \left(1 - \frac{n_m g_n}{n_n g_m}\right) \phi(\nu). \tag{4.87}$$

From the definition of the radiative thermal equilibrium, $dI_\nu/ds = j_\nu - \kappa_\nu I_\nu$, substituting the above relations gives

$$\frac{dI_\nu}{ds} = \frac{h\nu}{4\pi} n_m A_{mn} \phi(\nu) - \frac{h\nu}{4\pi} n_n B_{nm} \left(1 - \frac{n_m g_n}{n_n g_m}\right) \phi(\nu) I_\nu. \tag{4.88}$$

This now represents a radiative transfer equation for line (bound–bound) transitions.

Using Eqs. (4.85), the source function $S_\nu = j_\nu/\kappa_\nu$ can also be written as

$$S_\nu = \frac{2h\nu^3}{c^2} \left(\frac{n_n g_m}{n_m g_n - 1}\right)^{-1}. \tag{4.89}$$

Although it was derived from consideration of only line transitions, this expression is general and applies to all radiative transitions.

4.5.3. *Maxwell's velocity distribution*

As we discussed in Section 4.4.1, the Boltzmann distribution is the probability that any particle in an assembly of particles in thermodynamical equilibrium will be found with energy E, and is given by $n(E) = A\exp(-E/kT)$, where A is a normalization constant. If this distribution is applied to one direction of velocity for a particle in an ideal gas, so that $E = mv_z^2/2$, it becomes

$$n(v_z) = n_0 \sqrt{\frac{m}{2\pi kT}} e^{-mv_z^2/2kT}, \tag{4.90}$$

where the normalization factor ensures that $\int n(v_z) dv_z = n_0$ is equal to the number of particles, n_0.

One can readily extend the above relation to three dimensions by simply multiplying the distributions for the three cartesian components, i.e.,

$$n(v) = n_0 \left[\frac{m}{2\pi kT}\right]^{3/2} e^{-mv^2/2kT}, \tag{4.91}$$

where $v^2 = v_x^2 + v_y^2 + v_z^2$. However, the normalization of Eq. (4.91) is not right if we want $n(v)dv$ to represent the probability to find a particle with velocity v, i.e., between v and $v + dv$. This is obtained by expressing the volume element in spherical coordinates, $dv_x dv_y dv_z = v^2 dv d\Omega$ and integrating it over angles, yielding a factor 4π. The normalized distribution is

$$n(v) = 4\pi n_0 \left[\frac{m}{2\pi kT}\right]^{3/2} v^2 e^{-mv^2/2kT}, \tag{4.92}$$

which is the Maxwells velocity distribution of n_0 thermal particles of mass m at temperature T. Now

$$\int_0^\infty n(v)dv = n_0, \tag{4.93}$$

as can be easily verified.

Notice that as a function of the kinetic energy E of the particles,

$$n(E) \sim E \exp\left(-\frac{E}{kT}\right), \tag{4.94}$$

The form of the Maxwell–Boltzmann distribution can be seen in Fig. 4.8. The value $E = kT$ represents the energy at which any given particle has the highest probability of being found in.

4.5.4. *Saha equation*

In *Local Thermodynamical Equilibrium* (LTE) the distribution of atoms over various states of excitation and ionization is described by Boltzmann's relation for the relative populations of the ground atomic level n_1 and the

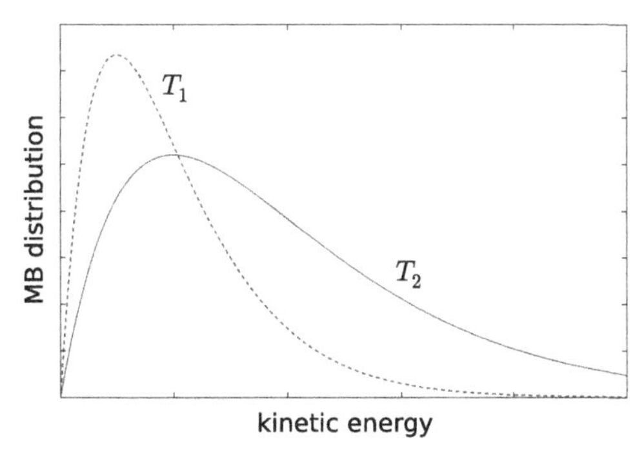

Figure 4.8. Maxwell–Boltzmann energy distribution for two different temperatures. The peak of the distribution occurs at $E \sim kT$. In this case $T_2 > T_1$.

n^{th} level n_n (per unit volume), with corresponding energies E_1 and E_n and statistical weights g_1 and g_n:

$$\frac{n_1}{n_n} = \frac{g_1}{g_m} \exp\left(\frac{E_n - E_1}{kT}\right), \tag{4.95}$$

where the statistical factors g_i represent, e.g., the possible number of spin orientations of the particles, which also count for the number of possible available states.

Saha's equation is a relation, based on Boltzmann's distribution, for the ratio of number densities of two different ionization states of an atomic species. For free electrons and large enough temperatures, we will assume a Boltzmann distribution. The *Saha equation* follows as

$$\frac{n_{i+1} n_e}{n_i} = \frac{g_{i+1} g_e}{g_i} \exp\left(-\frac{\chi_i}{kT}\right) \exp\left(-\frac{p^2}{2mkT}\right) d^3p, \tag{4.96}$$

which, when integrated over the free electron momentum p, yields

$$\frac{n_{i+1} n_e}{n_i} = \frac{g_{i+1} g_e}{g_i} \left(\frac{2\pi m_e kT}{h^2}\right)^{3/2} \exp\left(-\frac{\chi_i}{kT}\right), \tag{4.97}$$

where n_i is the number density of atoms in the ith ionization state, n_{i+1} is the number density in the next ionization state, n_e is the number density of free electrons, χ_i is the ionization energy difference between the ith and $i+1$ states, and the factors g_i, g_{i+1}, and g_e are the corresponding statistical weights (degeneracy) of the discrete energy states. For free (unbound) electrons, $g_e = 2$, owing to two spin states. Generally, one can use $g_n = 2n^2$ (n is the principal quantum number) for bound energy states. For the more complex atomic systems, a quantum treatment is usually necessary to calculate the statistical weights.

Let us see what we obtain for the ionization fraction of hydrogen in the solar photosphere, where $E = 13.6\,\text{eV}$, $T_{ioniz} = E/k \simeq 160,000\,\text{K}$, $T_{gas} \simeq 6,000\,\text{K}$, and $n_H \simeq 10^{17}\,\text{cm}^{-3}$. Thus, the actual gas temperature is much smaller than the temperature equivalent of the ionization energy. Naively one would expect a very low ionization fraction of hydrogen in the solar photosphere. Hydrogen is the dominating element, hence the electron density should be related to the hydrogen density by

$$n_e = \frac{n_{i+1}}{n_i + n_{i+1}} n_H. \tag{4.98}$$

The ratio of level partition sums is approximately 0.5, because the first excited level of atomic hydrogen is high compared to the thermal energy. This means that, approximately, n_i is the number of atoms in the ground state and n_{i+1} is the number of atoms either ionized or in the first excited state. Inserting the numbers in Saha's equation then gives the ionization fraction

$$\xi = \frac{n_{i+1}}{n_i + n_{i+1}} \longrightarrow \frac{n_{i+1}n_e}{n_i n_H} = \frac{\xi^2}{1-\xi} \simeq 10^{-7} \longrightarrow \xi \simeq 3 \times 10^{-4}, \tag{4.99}$$

where the intermediate step was obtained from Eq. (4.97) with $\chi_i = 13.6\,\text{eV}$. The result above seems small indeed, but repeating the calculation for a slightly hotter star with photospheric temperature $T = 12,000\,\text{K}$ and the same gas density yields an ionization fraction $\xi(T = 12,000\,\text{K}) \simeq 0.3$, much higher than what the ionization energy would suggest.

4.6. Opacities in Stars

The *opacity* k measures the impenetrability to electromagnetic or other kinds of radiation, especially visible light, through matter. The processes that result in emission, absorption, and scattering of photons in stars which contribute to the opacity are:

1. *Bound–bound transitions*, which correspond to bound electrons being excited or de-excited to another bound state as a result of absorption or emission of a photon; this results in *line radiation.*
2. *Bound–free transitions* correspond to photoionization and recombination, depending on whether a photon is absorbed or emitted and an electron is excited to a free state or a bound state, respectively.
3. *Free–free transitions* correspond to free electrons being excited or de-excited to another free state following absorption or emission of a photon; this results in continuum radiation.
4. *Scattering processes* can occur between atoms, free electrons and photons; they result in a redistribution of photon energies (i.e., scattering processes modify the radiation spectrum).

4.6.1. *Absorption and emission coefficients*

Recall the generalized expression for the source function S_ν, given by Eq. (4.89), derived from the Einstein coefficients. The Boltzmann relation implies $n_n g_m/(g_n n_m) = \exp(E_m - E_n)/kT$. Substituting this into Eq. (4.89) immediately gives

$$S_\nu = \frac{2h\nu^3}{c^2}\left[\exp\left(\frac{h\nu}{kT}\right) - 1\right]^{-1}. \tag{4.100}$$

Thus, we arrive at the result that the source function equals the Planck function under local thermal equilibrium, $S_\nu = I_\nu$. This equation confirms what we already knew. In fact, it confirms that Eqs. (4.85) are correct and that the transition rate for stimulated emission is the same as for absorption (i.e., $g_n B_{nm} = g_m B_{mn}$). But it was an astonishing result in 1917. Indeed, Einstein was forced to "invent" stimulated emission in order to reproduce Planck's formula.

Similarly, we can simplify the absorption coefficient under local thermal equilibrium conditions. Substituting the Boltzmann relation, $n_2/n_1 = (g_2/g_1)\exp[-(E_2 - E_1)/kT]$, into the expression (4.87) for the absorption coefficient yields

$$\kappa_\nu = \frac{h\nu}{4\pi} n_n B_{nm}\left[1 - \exp\left(-\frac{h\nu}{kT}\right)\right]\phi(\nu). \tag{4.101}$$

The factor $1 - \exp(-h\nu/kT)$ is the correction factor due to stimulated emission. Under local thermal equilibrium, the emission and absorption coefficients are related via Kirchoff's law, $j_\nu = \kappa_\nu B_\nu(T)$. Thus, it is only necessary to determine either κ_ν or j_ν.

Any population that does not satisfy the Boltzmann relation, i.e., with

$$\frac{n_m g_n}{n_n g_m} \neq \exp\left(-\frac{h\nu}{kT}\right),$$

is referred to as non-thermal. By definition, such populations cannot be in local thermal equilibrium. An example of when non-thermal populations become important is when stimulated emission is important. The correction factor for stimulated emission is $1 - n_m g_n/(n_n g_m)$. Under local thermal equilibrium, $n_m g_n/(n_n g_m) = \exp(-h\nu/kT)$, which is always less than unity (i.e., stimulated emission is negligible under local thermal equilibrium conditions). Under some conditions, however, it is possible to attain a state such that the higher energy levels are populated more than the lower energy levels, i.e., $n_m g_n/(n_n g_m) > 1$. This is referred to as an *inverted population* and gives a net negative absorption, which is formally equivalent to emission.

4.6.2. *Bound–bound transitions — oscillator strength*

The transition rates can be calculated by using the quantum theory of radiation discussed in Section 4.2. Let us consider the electric dipole radiation as the dominant multipole radiation. As an atom tries to absorb a photon, it simultaneously tries to decay back spontaneously to state i. To take account of this effect, define the rate of spontaneous emission A_{fi} (Einstein coefficient). For spontaneous emission in the $m \to n$ transition this coefficient is found to be

$$A_{mn} = \frac{4\omega_{mn}^3}{3c^3\hbar}|\mathbf{d}_{mn}|^2, \tag{4.102}$$

where $\omega_{mn} = (E_m - E_n)/\hbar$ is the angular frequency for the transition, and

$$\mathbf{d}_{mn} = \int \Psi_m^*(\mathbf{r}_1, \ldots, \mathbf{r}_N)\left(\sum_i e_i\mathbf{r}_i\right)\Psi_n(\mathbf{r}_1, \ldots, \mathbf{r}_N)d^3r_1 \cdots d^3r_N \tag{4.103}$$

is the electric dipole matrix element, and the sum runs over all charges in the system. The calculation depends on the knowledge of the quantum-mechanical wave functions $\Psi_\alpha(\mathbf{r}_1, \ldots, \mathbf{r}_N)$ for the system of N charges in positions $\mathbf{r}_i$. This equation can be used to estimate A_{mn} for the electric dipole transition corresponding to Ly-α emission from a hydrogen atom ($\nu_{12} \simeq 2.5 \times 10^{15}$ Hz). Ly-α is the Lyman-alpha line, which occurs when the electron falls from the orbital with principal quantum number $n = 2$ to that with $n = 1$. Using $d_{21} \simeq ea_H$, where $a_H = 0.529 \times 10^{-10}$ m is the Bohr radius, we find $A_{21} \simeq 5 \times 10^8\ s^{-1}$.

Oscillator strengths are dimensionless quantities defined by comparing the emission rate or absorption rate of an atom with the emission or absorption rate of a single-electron oscillator (with angular oscillation frequency ω_{mn}). We define an emission oscillator strength f_{mn} by the relation

$$f_{mn} = \frac{1}{3}\frac{A_{mn}}{\gamma_{cl}}, \quad \text{where } \gamma_{cl} = \frac{2e^2\omega_{mn}^2}{3m_ec^3}, \tag{4.104}$$

or, using Eq. (4.102),

$$f_{mn} = \frac{2m_e\omega_{mn}}{3e^2\hbar}|\mathbf{d}_{mn}|^2. \tag{4.105}$$

From Eq. (4.85), the Einstein B-coefficent is given by

$$B_{mn} = \frac{4\pi^2 e^2}{h\nu_{mn} m_e c} f_{mn}. \tag{4.106}$$

The oscillator strength has been defined for emission processes. For absorption processes the following relation is used

$$f_{mn} = -\frac{g_m}{g_n} f_{nm}. \tag{4.107}$$

which follows from Eq. (4.85). Thus, the bound–bound emission coefficient can be re-written as follows

$$\begin{aligned} j_\nu &= n_m \frac{h\nu}{4\pi} A_{mn}\phi(\nu) \\ &= n_m \frac{h\nu}{4\pi}\frac{2h\nu^3}{c^2} B_{mn}\phi(\nu) = n_m \frac{h\nu}{4\pi}\frac{2h\nu^3}{c^2}\frac{g_n}{g_m} B_{nm}\phi(\nu), \end{aligned} \tag{4.108}$$

yielding

$$j_\nu = n_m \frac{2h\nu^3}{c^2}\frac{g_n}{g_m}\frac{\pi e^2}{m_e c} f_{mn}\phi(\nu). \tag{4.109}$$

Only transition rates for hydrogen (and hydrogen-like ions such as He^+ and Li^{++}) can be calculated exactly. In this case, the energy of a photon absorbed or emitted in a transition between two discrete levels with principal quantum numbers n and m is

$$h\nu = Ry\left(\frac{1}{n^2} - \frac{1}{m^2}\right), \quad Ry = \frac{e^2}{2a_H} = 13.6\,\mathrm{eV}, \tag{4.110}$$

where $a_H = \hbar^2/m_e e^2$ is the Bohr radius. Ry is known as the Rydberg constant. When the upper level is in the continuum, so that there is a free electron with energy $E = m_e v^2/2$, then

$$h\nu = \frac{Ry}{n^2} + E = I_n + E. \tag{4.111}$$

where I_n is the ionization energy, also known as the *ionization potential.*

The oscillator strength for hydrogen bound–bound transitions can be calculated analytically using hydrogen wave functions for the states n and

m in $\mathbf{d}_{mn}$. One gets

$$f_{mn} = \frac{32C}{3\pi\sqrt{3}n^5m^3(1/n^2 - 1/m^2)^3}, \tag{4.112}$$

where C is a constant of the order of unity. For multi-electron atoms the calculations can become very difficult.

Continuum transitions

Bound–free transition rates are more complicated to calculate than bound–bound (or line) transition rates since the energy of the free electron must be taken into account. The oscillator strength can be used, but now it involves an integral over the free electron velocity distribution (a Maxwell distribution under local thermal equilibrium). What is usually calculated instead is the absorption cross-section. For bound–free and free–bound transitions, the energy of the photon absorbed or emitted must balance the ionization potential $I_n = E - E_n$ of the energy level involved and also the kinetic energy of the free electron.

For free–free transitions, both electron and photon have continuum energies, so the above restriction is no longer relevant, although an integration over the electron energy distribution is still necessary.

The *bound–free cross-section for absorption* of a photon and ionization from energy level n is approximately given by

$$\sigma_{\text{bf}} = \frac{64\pi^4 Z^4 m_e e^{10}}{3\sqrt{3}ch^6\nu^3 n^5}, \tag{4.113}$$

where Z is the atomic number. This expression is only valid for $h\nu > I_n$.

The inverse of photoionization is recombination and corresponds to a free–bound transition, i.e., the capture of a free electron to a bound energy state n and the emission of a photon. The volume emissivity for this process under local thermal equilibrium is

$$j_{\text{fb}} = \frac{2h\nu^3}{c^2} n_e n_i \left(\frac{h^2}{2\pi m_e kT}\right)^{3/2} \sigma_{\text{bf}} \exp\left[-\frac{(h\nu - I_n)}{kT}\right]. \tag{4.114}$$

The rates of electronic transitions are determined by assuming that electrostatic (Coulomb) interactions dominate. For bound–bound transitions, Coulomb interactions between close-neighbor bound electrons are the most important. For free electrons, the distant Coulomb interaction with the ionized atoms electrostatic potential is relevant. The emission coefficient (volume emissivity) for *free–free emission* (Bremsstrahlung) is given by

$$j_{\mathrm{ff}} = \sqrt{\frac{32\pi}{3}} C Z^2 n_e n_i \sigma_T r_0 m_e c^2 \left(\frac{kT}{m_e c^2}\right)^{1/2} \exp\left(-\frac{h\nu}{kT}\right), \tag{4.115}$$

where C is a constant of the order unity,

$$r_0 = \frac{e^2}{m_e c^2} \tag{4.116}$$

is the *classical electron radius*, and

$$\sigma_T = \frac{8\pi}{3} r_0^2 \simeq 6 \times 10^{-25}\,\mathrm{cm}^2 \tag{4.117}$$

is the *Thomson cross-section.* The frequency dependence of j_{ff} is only in the exponential factor. Thus, the emission spectrum is constant at frequencies $h\nu \ll kT$, and declines exponentially at $h\nu \gg kT$.

There are other related processes. Scattering is not an intrinsic radiative process, but it redistributes photon energies to produce a radiation spectrum that is different from one due to radiation processes alone. One of the most important type of scattering is electron scattering. The simplest treatment is to assume isotropic scattering. For non-relativistic electrons, scattering off photons is approximately elastic, or coherent: the energy absorbed is approximately the energy emitted.

Electronic transitions can also proceed via collisions between atoms and free electrons. When T is sufficiently high that a substantial fraction of atoms are in an ionized state and there are many free electrons available, collisional transition rates exceed radiative transition rates for densities above a critical density, n_{crit}. Collisional transition rates (per unit time) are parametrized by coefficients analogous to the Einstein coefficients. The critical density can be written in terms of the mean intensity, Eq. (4.83),

$$n_{crit} = \frac{A_{mn}}{\langle \sigma_{mn} v \rangle}(1 + c^2 2h\nu^3 \bar{J}). \tag{4.118}$$

Collisional rates can become extremely important when transitions cannot proceed radiatively, i.e., when they are forbidden or semi-forbidden. In many such cases the collisional rates are comparable to the allowed radiative transition rates, thus allowing the forbidden transitions to proceed collisionally.

Total opacities

The number of processes involved in the propagation of radiation in a stellar environment is very large, and can only be managed within a statistical theory. For a solar composition, main sequence star, (a) hydrogen and helium bound–free and free–free transitions, (b) heavy element (carbon to iron) bound–free, free-free and bound–bound transitions, and (c) electron scattering make roughly equal contributions. At lower heavy element abundance, source (b) decreases somewhat. In the zone above shell sources in luminous red giants, source (c) dominates. Thermal energy transport in the cores of red giants is dominated by (d) electron conduction. However, neutrino emission actually eliminates the need to transport energy through the star by simply carrying the energy directly into space. The bulk of the energy from red giant depleted cores is converted into neutrino emission.

The bound–free opacities are the most important throughout most of the radiative envelopes of stars. The species which contribute the opacity that retards the energy flow are those for which the bound–free transitions from the ground state have energy equal to the energy of photons at the peak of the Planck curve. As long as the gas is non-degenerate, the phase space of the free electrons favors ionization of these species so that the ionization state of elements contributing to the opacity is one higher than the species with the proper ionization energy. Hydrogenic species are especially important in this regard. In fact, at the higher energies in stellar interiors, essentially all the important transitions are reasonably close to hydrogenic. Accurate opacity calculations take into account the details and do not make hydrogenic approximations for the cross-sections; however, the gross properties can be estimated this way.

4.7. Exercises

1. Consistent with our general $p_\mu p^\mu = -m_0^2c^2$ formula, it is the case that for photons (which have zero rest mass), $p_\mu p^\mu = 0$. Given this, can an electron absorb a single photon in free space? Demonstrate.
2. A person claims that he/she has invented a microchip 1 cm square in size which can run at a clock speed of 300,000 GHz. Would you invest in this person's company so that he/she can manufacture and market his/her new invention? Explain your answer.
3. A nucleus with decay constant λ exists at time $t = 0$. What is the probability that it disintegrates between t and $t + \Delta t$?

4. A particle of mass m and electric charge e is placed in a one-dimensional harmonic oscillator potential of frequency ω and uniform electric field $\mathcal{E}$ along the same axis.

   (a) Find the wave functions and the energy spectrum of the particle.
   (b) With a particle in the ground state of the problem, at time $t = 0$ the electric field is suddenly switched off. Find the probability of finding the particle at $t > 0$ in the nth stationary state of the oscillator.

5. Consider a spherical cloud of radius r, at a distance $d \gg r$ from the Earth. Suppose that this cloud has uniform density ρ, temperature T and thermal emissivity η_ν (power per unit volume per unit frequency interval per unit solid angle), i.e. η_ν is a function of T only.

   (a) Suppose the cloud is optically thick. Consider a small area element on the surface of the cloud, located at distance b from the line of sight passing through the center of the cloud, which subtends a solid angle $d\Omega$ as seen from the Earth. Find $I_\nu d\Omega$, the power per unit area per unit frequency interval, received from this area element at the Earth.
   (b) Find the flux Φ_ν at the Earth from the center cloud, by using the result of a).
   (c) What is the effective temperature T_{eff} of the cloud?

6. Using Eqs. (4.40) and (4.41), show that $U_{rad,\nu} = 3P_{rad,\nu}$ for an isotropic radiation field.
7. Prove Eq. (4.72).
8. Since you are made mostly of water, you are very efficient at absorbing microwave photons. If you were in intergalactic space, approximately how many CMB photons would you absorb per second? (If you like, you may assume you are spherical.) What is the approximate rate, in watts, at which you would absorb radiative energy from the CMB? Ignoring other energy inputs and outputs, how long would it take the CMB to raise your temperature by one nanoKelvin (10^{-9} K)? (You may assume your heat capacity is the same as pure water, $C = 4200\,\mathrm{J\,kg^{-1}\,K^{-1}}$.)
9. Assume a γ-ray detector on board of a satellite. Its effective area is $1500\,\mathrm{cm}^2$ and its angular position accuracy is $\Delta\theta \simeq 10^\circ$. After observing a Quasar for two weeks, it collects about 3000 photons with energy 1 GeV. Determine (a) $I = \int I_\nu d\nu$ and (b) $\Phi = \int \Phi_\nu d\nu$.

10. The activity of a given material decreases by a factor 8 in a time interval of 30 days. What is its half-life, mean lifetime and decay constant? If the sample initially had 10^{20} atoms, how many disintegrations have occurred in its second month of life?
11. The theories of grand unification predict that the proton is not a stable particle, although it has a long half-life. For a half-life of 10^{33} years, how many proton decays can we expect in one year in a mass of 10^3 tons of water?
12. Calculate the average velocity of a Maxwell–Boltzmann distribution.
13. Use the literature as a source to prove Eq. (4.85).
14. Calculate the fraction of hydrogen atoms in the $n = 2$ state with respect to the ground state for 3 different temperatures: $T_1 = 6000\,\mathrm{K}$, $T_2 = 10,000\,\mathrm{K}$, $T_3 = 30,000\,\mathrm{K}$.
15. Use the Saha equation to calculate the temperature of recombination, which you can define as the temperature when there is an equal amount of neutral hydrogen and free protons/electrons. You can simplify the hydrogen atom to be a single state at $-13.6\,\mathrm{eV}$.
16. An atom with a single electron is in a heat bath at a temperature of $T_6 = 1$ (T_6 means that the temperature is given in units of $10^6\,\mathrm{K}$). The atom is high Z, so the electron is bound at this temperature, and only three states have appreciable occupations. The ground state has spin 5/2. The first excited state, at 210 eV, has spin 3/2. The second excited state, at 380 eV, has spin 3/2. What are the occupation probabilities for these three states?
17. A nucleus is in a plasma at thermal equilibrium. Calculate the population probabilities of the ground state ($E_0 = 0$) and of the first three excited states ($E_1 = 0.1\,\mathrm{MeV}$, $E_2 = 0.5\,\mathrm{MeV}$, $E_3 = 1.0\,\mathrm{MeV}$) for two temperatures, $T = 1.0 \times 10^9\,\mathrm{K}$ and $3.0 \times 10^9\,\mathrm{K}$, and assume that all states have the same spin value.
18. Using Eq. (4.113) (a) calculate the cross-section (in fm^2) for the absorption of a photon with energy $h\nu = I$, where I is the ionization potential of a hydrogen atom at its ground state; (b) find the ratio of the value obtained in (a) and πa_H^2 where a_H is the Bohr radius.
19. As a mechanism for downward transitions (final energy smaller than initial energy), spontaneous emission competes with thermally stimulated emission (stimulated emission for which Planck radiation is the source). Show that at room temperature ($T = 300\,\mathrm{K}$) thermal stimulation dominates for frequencies much smaller than $5 \times 10^{12}\,\mathrm{Hz}$,

whereas spontaneous emission dominates for much larger frequencies. Which mechanism dominates for visible light?

20. Generalize the Saha equation to consider the temperature for a process in the early Universe where two protons and two neutrons combine directly to form ^{4}He. That is, find the temperature where half of the neutrons are bound in ^{4}He. Later we will compare this result with Big Bang nucleosynthesis predictions.
21. Negative hydrogen ions H^-, which have a dissociation energy of 0.754 eV, play an important role in the opacity of the outer layers of the Sun. Assume the electron density is $n_e = 10^{26}\,\mathrm{m}^{-3}$, and estimate the temperature (to one figure accuracy) at which the number densities of H and H^- are equal.

Chapter 5

Nuclear Reactions

5.1. Introduction

The collision of two nuclei can give rise to a nuclear reaction where, similarly to a chemical reaction, the final products can be different from the initial ones. We can sometimes have more than two final products in a reaction, as in the examples

$$\begin{aligned} \mathrm{p} + {}^{14}\mathrm{N} &\rightarrow {}^{7}\mathrm{Be} + 2\alpha, \\ \mathrm{p} + {}^{23}\mathrm{Na} &\rightarrow {}^{22}\mathrm{Ne} + \mathrm{p} + \mathrm{n}, \end{aligned} \tag{5.1}$$

or just one, as in the capture reaction

$$\mathrm{p} + {}^{27}\mathrm{Al} \rightarrow {}^{28}\,\mathrm{Si}^*, \tag{5.2}$$

where the asterisk indicates an excited state, which usually decays emitting γ-radiation. Under special circumstances, more than two reactants are possible. Thus, for example, the reaction

$$\alpha + \alpha + \alpha \rightarrow {}^{12}\mathrm{C} \tag{5.3}$$

can take place in the overheated plasma of the stellar interior.

Also, the final products can be identical to the initial ones. This case characterizes a process of nuclear scattering, which can be elastic, as in the simple example

$$\mathrm{p} + {}^{16}\mathrm{O} \rightarrow \mathrm{p} + {}^{16}\mathrm{O}, \tag{5.4}$$

where there is only transfer of kinetic energy between projectile and target, or inelastic, as in the example

$$\mathrm{n} + {}^{16}\mathrm{O} \rightarrow \mathrm{n} + {}^{16}\mathrm{O}^*, \tag{5.5}$$

where part of the kinetic energy of the system is used in the excitation of ^{16}O.

Nuclear reactions can involve any type of particle, and also radiation. Thus, the reactions

$$\gamma + {}^{63}\mathrm{Cu} \to {}^{62}\mathrm{Ni} + \mathrm{p},$$
$$\gamma + {}^{233}\mathrm{U} \to {}^{90}\mathrm{Rb} + {}^{141}\mathrm{Cs} + 2\mathrm{n}, \tag{5.6}$$

are examples of nuclear processes induced by gamma radiation. In the first a γ-ray knocks a proton off ^{63}Cu and in the second it induces the process of nuclear fission in ^{233}U, with the production of two fragments and the emission of two neutrons.

Starting from two or more reactants there can exist dozens of possibilities of composition of the final products with an unlimited number of available quantum states. As an example, the collision of a deuteron with a nucleus of ^{238}U can give rise, among others, to the following reactions:

$$\begin{aligned} \mathrm{d} + {}^{238}\mathrm{U} &\to {}^{240}\mathrm{Np}^* + \gamma, \\ &\to {}^{239}\mathrm{Np} + \mathrm{n}, \\ &\to {}^{239}\mathrm{U} + \mathrm{p}, \\ &\to {}^{237}\mathrm{U} + \mathrm{t}, \end{aligned} \tag{5.7}$$

where t represent the tritium, ^{3}H.

Each one of the branches of the reaction which can occur, with well defined quantum states of the participants, is referred to as *channel.* In Eq. (5.7), for the *entrance channel* d + ^{238}U, there are four possible *exit channels.* Notice that a different output channel would be reached if some of the final products (such as ^{239}U*) were in an excited state. The probability of a nuclear reaction taking place through a certain exit channel depends on the energy of the incident particle and is measured by the *cross-section* for that channel. The theory of nuclear reactions must, besides elucidating the mechanisms that determine the occurrence of the different processes, also evaluate the cross-sections corresponding to all exit channels.

Several conservation laws contribute to nuclear reactions: (a) baryonic number, (b) charge, (c) energy and linear momentum, (d) total angular momentum, (e) parity, and (f) isospin. In this chapter we discuss the basics of low energy scattering theory, the treatment of resonances and the meaning of optical potentials. Then we proceed to study the compound nucleus formation and decay, and the physics of intermediate and relativistic nuclear collisions.

5.2. Nuclear Scattering

When a wave of any type hits a small obstacle, secondary waves (circular or spherical) are produced and they move away from it, going to infinity. In the same way, a monoenergetic beam of particles, that can be represented by a plane wave, undergoes scattering when it finds a region in which there exists a potential $V(\mathbf{r})$ created by a nucleus (Fig. 5.1). The solution of scattering problems consists in finding the angular distribution of the scattered particles, where the total energy of the system target + projectile can have any positive value.

Given an incident plane wave, whose stationary part can be represented by

$$\Psi_i(\mathbf{r}) = e^{i\mathbf{k}\cdot\mathbf{r}} = e^{ikz}, \tag{5.8}$$

and a scattering potential $V(\mathbf{r})$, our problem reduces to finding the wavefunction of the scattered particles, or scattered wave function.

In problems of atomic and nuclear physics the detectors lie far away from the scattering centers compared to their dimensions, i.e., they are in a region where the particles no longer feel the action of the potential. Thus, our interest will be limited to the asymptotic part of the scattered wave function, namely, its form when $r \to \infty$. When a short range potential $V(r)$, supposed for simplicity to be spherically symmetric, acts on the particles of an incident beam, a detector placed in the asymptotic region will register

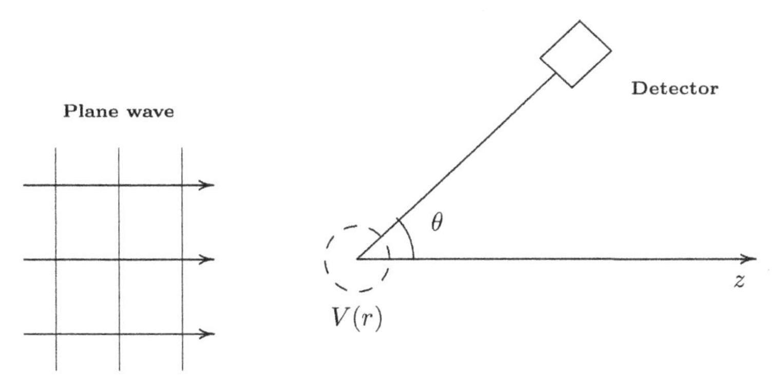

Figure 5.1. Scattering of a plane wave by a potential $V(r)$ limited to a small region of space.

not only the presence of the plane wave but also the particles scattered by the potential. That is, to the plane wave will be added an outgoing spherical wave created by the scattering center and we can write the wave function far from this center as

$$\Psi \sim e^{ikz} + f(\theta)\frac{e^{ikr}}{r}. \tag{5.9}$$

where the symbol $\sim$ means asymptotic value. The presence of the function $f(\theta)$ expresses the fact that the scattering directions do not have not the same probability. This function is called *scattering amplitude* and has, as we will see next, an essential role in the theory for the process.

The probability current, $\mathbf{j} = (\hbar/m)\,\mathrm{Im}(\Psi^*\nabla\Psi)$, will be now employed in the definition of a function that measures the angular distribution of the particles scattered by $V(r)$. For the incident plane wave the current is

$$j_i = \frac{\hbar}{m}\,\mathrm{Im}\left(e^{-ikz}\frac{d}{dz}e^{ikz}\right) = \frac{\hbar k}{m} = v, \tag{5.10}$$

and for the outgoing spherical wave

$$j_r \sim \frac{\hbar}{m}\mathrm{Im}\left\{f^*(\theta)\frac{e^{-ikr}}{r}\frac{\partial}{\partial r}\left[f(\theta)\frac{e^{ikr}}{r}\right]\right\} = \frac{v}{r^2}|f(\theta)|^2. \tag{5.11}$$

We define the *differential cross-section*, as a function of the angle θ (see Fig. 5.2), by

$$\frac{d\sigma}{d\Omega} = \frac{dN/d\Omega}{n\Phi}, \tag{5.12}$$

dN being the number of observed events in $d\Omega$ per unit time, n the number of target scattering centers comprised by the beam and Φ the incident flux (number of incident particles per unit area and per unit time). The solid

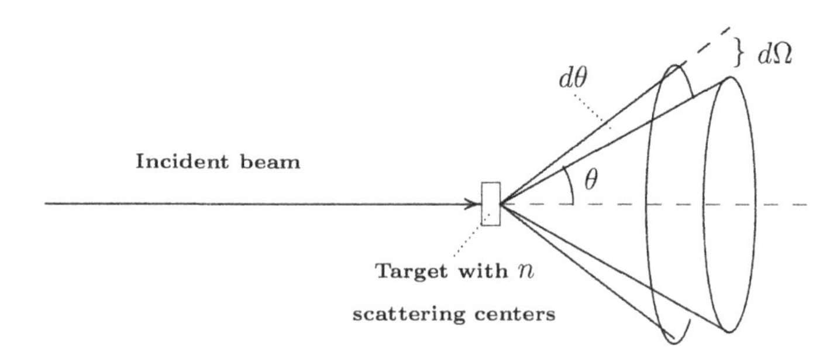

Figure 5.2. Quantities used in the definition of the differential cross-section.

angle $d\Omega = 2\pi \sin\theta\, d\theta$ is located between the cones defined by the directions θ and $\theta + d\theta$. If our assumption of spherical symmetry for the scattering potential is not valid, the solid angle is the one defined by the direction θ, ϕ, namely, $d\Omega = \sin\theta d\theta d\phi$.

Definition (5.12) is a general one and the *observed events*, in the present case, are particles scattered by the potential $V(r)$. $d\sigma/d\Omega$ has the dimension of area and its value is obtained from

$$\frac{d\sigma}{d\Omega} = \frac{j_r r^2}{j_i}, \tag{5.13}$$

using the fact that the number of particles that cross a given area per unit time is measured by the probability current flux through that area. Using (5.10) and (5.11) it is clear that

$$\frac{d\sigma}{d\Omega} = |f(\theta)|^2, \tag{5.14}$$

being thus the determination of the angular distribution reduced to the evaluation of the scattering amplitude $f(\theta)$.

The *total cross-section* is obtained by integrating (5.14),

$$\sigma = \int \frac{d\sigma}{d\Omega} d\Omega = 2\pi \int_{-1}^{+1} |f(\theta)|^2 d(\cos\theta), \tag{5.15}$$

and its meaning is obvious from the definition (5.12): the total cross-section measures the number of events per target nucleus per unit time divided by the incident flux defined above. It must include, in this way, events for which we cannot define a differential cross-section, such as the absorption of particles from the incident beam by the nucleus.

5.3. Partial Wave Expansion

When we study interactions governed by a central potential $V(r)$, solutions of the Schrödinger equation

$$\nabla^2\Psi + \frac{2m}{\hbar^2}[E - V(r)]\Psi = 0 \tag{5.16}$$

can be written as linear combinations of products of solutions separated into radial and angular parts,

$$\Psi = \sum_{l,m} a_{lm} \frac{u_l(r)}{r} Y_l^m(\theta, \phi), \tag{5.17}$$

where $u_l(r)$ obeys the radial equation

$$\frac{d^2u_l}{dr^2} + \frac{2m}{\hbar^2}\left[E - V(r) - \frac{\hbar^2}{2m}\frac{l(l+1)}{r^2}\right]u_l = 0 \tag{5.18}$$

and the boundary condition

$$u_l(0) = 0. \tag{5.19}$$

The axial symmetry of our problem allows us to eliminate the dependence in ϕ of (5.17) so that

$$\Psi = \sum_l a_l P_l(\cos\theta)\frac{u_l(r)}{kr}, \tag{5.20}$$

where the constant $k = \sqrt{2mE}/\hbar$ was introduced to make later applications of the expansion easier.

The terms of (5.20) can be understood as *partial waves*, from which the general solution Ψ can be constructed. An expression like (5.20) is convenient: if $V(r)$ is spherically symmetric, the angular momentum is a constant of motion, and states of different values of the angular momentum contribute in an independent way to the scattering. Thus, it is also useful to present the plane wave by an expansion in Legendre polynomials

$$e^{ikz} = e^{ikr\cos\theta} = \sum_{l=0}^{\infty}(2l+1)i^l j_l(kr)P_l(\cos\theta), \tag{5.21}$$

where $j_l(x)$ are spherical Bessel functions and $P_l(\cos\theta)$ the Legendre polynomials.

Expression (5.21) has the form of (5.20). This means that the plane wave $e^{i\mathbf{k}\cdot\mathbf{r}}$ can be understood as the sum of a set of partial waves, each one with orbital angular momentum $\sqrt{l(l+1)}\hbar$. The terms $j_l(kr)P_l(\cos\theta)$ specify the radial and angular dependence of the partial wave l, the weight of the contribution of each term being given by the amplitude $(2l+1)$ and by the phase factor i^l.

At large distances from the origin the spherical Bessel functions reduce to the simple expression

$$j_l(kr) \sim \frac{\sin\left(kr - \frac{l\pi}{2}\right)}{kr} = \frac{e^{i(kr-l\pi/2)} - e^{-i(kr-l\pi/2)}}{2ikr}. \tag{5.22}$$

Using (5.22) in (5.21) results in

$$e^{ikr\cos\theta} \sim \frac{1}{2i}\sum_{l=0}^{\infty}(2l+1)i^l P_l(\cos\theta)\left[\frac{e^{i(kr-l\pi/2)} - e^{-i(kr-l\pi/2)}}{kr}\right], \qquad (5.23)$$

that represents the asymptotic form of a plane wave.

In (5.23) the first term inside the square brackets corresponds to an outgoing spherical wave and the second to an ingoing spherical wave. Thus, each partial wave in (5.23) is, at large distances from the origin, a superposition of two spherical waves, namely the ingoing and the outgoing components. The total radial flux for the wavefunction $\Psi_i = e^{ikr\cos\theta}$ vanishes, since the number of free particles that enters into a region is the same that exits.

Let us now consider Ψ in (5.20) as a solution of a scattering problem, the scattering being caused by a potential $V(r)$. The asymptotic form of Ψ can be obtained if we observe that the presence of the potential has the effect of causing a perturbation in the outgoing part of the plane wave, and such a perturbation can be represented by a unitary module function, $S_l(k)$.

From (5.23), this leads to

$$\Psi \sim \frac{1}{2i}\sum_{l=0}^{\infty}(2l+1)i^l P_l(\cos\theta)\frac{S_l(k)e^{i(kr-l\pi/2)} - e^{-i(kr-l\pi/2)}}{kr}, \qquad (5.24)$$

where the *S-matrix* function $S_l(k)$ can be represented by

$$S_l(k) = e^{2i\delta_l}. \qquad (5.25)$$

When we write the form (5.25) we admit that the scattering is elastic. The unitary magnitude of $S_l(k)$ keeps the same value for the probability current and does not allow the possibility that the presence of the potential removes or add particles to the elastic channel k. From a comparison of (5.24) and (5.20) we can obtain the expressions for a_l and for the asymptotic form of $u_l(r)$,

$$a_l = i^l(2l+1)e^{i\delta_l}, \qquad (5.26)$$

and

$$u_l(r) \sim \sin\left(kr - \frac{l\pi}{2} + \delta_l\right). \qquad (5.27)$$

$u_l(r)$ differs from the asymptotic form of the radial function of a free particle by the presence of the *phase shifts* δ_l; the presence of the scattering potential creates in each partial wave a phase shift δ_l. The scattering problem would be solved with the determination of these phase shifts for a given potential $V(r)$. In fact, the use of (5.24) and (5.23) in (5.9) results in

$$f(\theta) = \frac{1}{k} \sum_{l=0}^{\infty} (2l+1) e^{i\delta_l} \sin \delta_l P_l(\cos\theta) \tag{5.28}$$

and the differential cross-section (5.14) is obtained from the knowledge of the phase shifts δ_l. Figure 5.3 shows the straightforward relation between the sign of the phase-shift and that of an attractive or a repulsive potential.

The phase shifts are evaluated by solving Eq. (5.18) for each l and comparing the phase of $u_l(r)$, for some large r, with the phase of $j_l(kr)$ for the same value of r. The total cross-section, in turn, has the expression

$$\sigma = \frac{4\pi}{k^2} \sum_l (2l+1) \sin^2 \delta_l, \tag{5.29}$$

obtained by the integration (5.15).

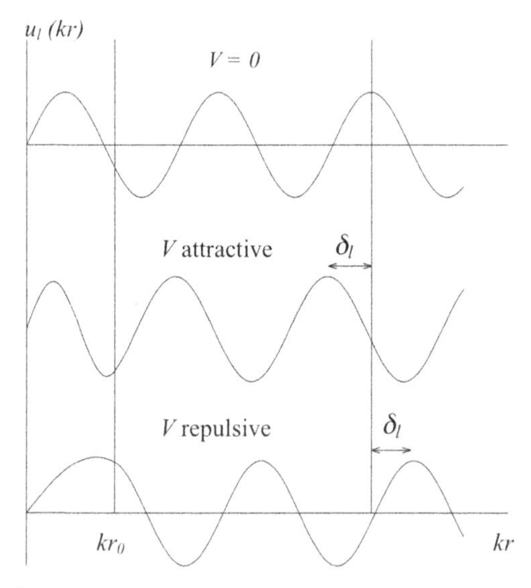

Figure 5.3. Radial part of the wavefunction for three different potentials, showing how the sign of the phase shifts are determined by the wave function behavior in the region $r < r_0$ where the potential acts.

From (5.28) and (5.29) one extracts an important relation. For this, it is enough to observe in (5.28) that

$$\text{Im } f(0) = \frac{1}{k} \sum_{l=0}^{\infty} (2l+1) \sin^2 \delta_l \tag{5.30}$$

and to compare this result with (5.29) to obtain

$$\sigma = \frac{4\pi}{k} \text{Im } f(0). \tag{5.31}$$

This relation is known as the *optical theorem.* It connects the total cross-section with the scattering amplitude at angle zero. This is physically understandable: the cross-section measures the removal of particles from the incident beam that arises, in turn, from the destructive interference between the incident and scattered beams at zero angle [Sc54]. The optical theorem is not restricted to elastic scattering, being also valid for inelastic processes.

5.3.1. *Low energy scattering*

If the energy is low enough, the sum (5.28) reduces to the term with $l = 0$. We have, in this case,

$$f(\theta) = \frac{1}{k} e^{i\delta_0} \sin \delta_0 \tag{5.32}$$

and

$$\sigma = \frac{4\pi}{k^2} \sin^2 \delta_0. \tag{5.33}$$

The differential cross-section that results from (5.32) is independent of θ: the scattering is isotropic. This is easily understandable since at low energies the incident particle wavelength is much greater than the dimension of the target nucleus; during its passage all points in the nucleus are with the same phase at each time and it is impossible to identify the direction of incidence.

In the extreme case $E \to 0$ the scattering amplitude (5.32) remains finite only if $\delta_0 \to 0$ together with the energy. In this case the phase difference is no more the main scattering parameter. A better parameter is the *scattering length* a, defined as the limit

$$\lim_{E\to 0} f(\theta) = \lim_{k\to 0} \frac{\delta_0}{k} = -a, \tag{5.34}$$

yielding the equation

$$\sigma = 4\pi a^2 \tag{5.35}$$

as the expression for the total cross-section at the zero energy limit.

The physical meaning of the scattering length can be obtained by observing that for $l = 0$, and in the limit $E \to 0$, Eq. (5.18) in the region outside the potential reduces to its first term. Hence, if $d^2u_0/dr^2 = 0$, we see that the wave function u_0 tends to a straight line and the abscissa at the point where this line crosses the r axis is the scattering length a. This can be easily seen if we rewrite (5.27) in the limit $E \to 0$:

$$u_0(r) \cong kr + \delta_0 = k(r - a). \tag{5.36}$$

When $V_0 = 0$, Eq. (5.18) has the simple form $u'' = 0$, with the trivial solution for u as a straight line crossing the origin. When V_0 is small and forms a well, the form of the wave function looks like Fig. 5.4(a). There is no bound state yet and the scattering length a is negative [see Eq. (5.36)]. When V_0 is deep enough to allow the existence of the first bound state, the form of the wave function is shown in Fig. 5.4(b), with a maximum inside the well. This is an expected behavior: the internal part of the wave function is not sensitive to the fact that E is a little positive or negative and a bound

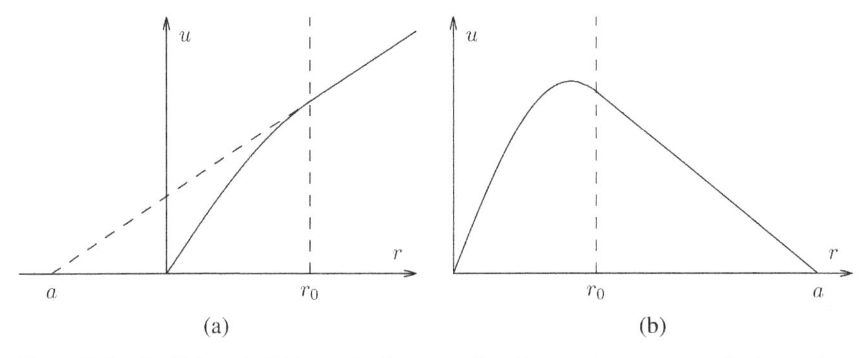

Figure 5.4. Radial part of the scattering wave function u at zero energy for two values of the well depth, where r_0 is the potential range. In (a) the depth is not large enough for the existence of a bound state and the scattering length is negative. In (b) there already exists a bound state, the function u has a maximum inside the potential range and a has a positive value.

state wave function should have in r_0 a negative derivative to match to the exponential decay of the external part. The essential result is that the sign of a is, in this case, positive. Thus, the sign of the scattering length can show us if a resonance occurs with a negative (bound) state or a positive (sometimes called "virtual") energy.

A higher-order expansion of the phase shift at low energies leads to

$$k \cot \delta_0 \simeq -\frac{1}{a} + \frac{1}{2} k^2 r_0, \tag{5.37}$$

where r_0 is called by *effective range*. It measures the effective interval in coordinate space where the wave function deviates from a free wave. It is convenient to re-write Eq. (5.32) in the form

$$f = \frac{1}{k \cot \delta_0 - ik}, \tag{5.38}$$

from which we get, as $k \to 0$,

$$f \simeq \frac{1}{-1/a + k^2 r_0/2 - ik}. \tag{5.39}$$

5.3.2. *Scattering and reaction cross-sections*

Elastic scattering is just one of the channels for which the reaction can be processed and we refer to it as the *elastic channel.* The inelastic scattering channel and all the other channels are grouped in the *reaction channel.*

The occurrence of a nuclear reaction proceeding through a given reaction channel leads to a modification of the outgoing part of (5.24), now not only by a real phase factor, but also by an imaginary part the magnitude of the scattering matrix S_l, indicating that there is a loss of particles in the elastic channel. To calculate the elastic cross-section we need to place Ψ in the form of Eq. (5.9); using Eqs. (5.23) and (5.24) we get

$$f(\theta) = \frac{1}{2k} \sum_{l=0}^{\infty} (2l+1) i (1 - S_l) P_l(\cos\theta), \tag{5.40}$$

which results in the differential scattering cross-section

$$\frac{d\sigma_e}{d\Omega} = |f(\theta)|^2 = \frac{1}{4k^2} \left| \sum_{l=0}^{\infty} (2l+1)(1 - S_l) P_l(\cos\theta) \right|^2. \tag{5.41}$$

The total scattering cross-section is calculated using the orthogonality of the Legendre polynomials, which results in

$$\sigma_e = \pi\lambdabar^2 \sum_{l=0}^{\infty}(2l+1)|1-S_l|^2, \tag{5.42}$$

with $\lambdabar = \lambda/2\pi = 1/k$.

To calculate the reaction cross-section it is necessary to compute initially the number of particles that disappear from the elastic channel, measured by the flux of the probability current vector through a spherical surface of large radius centered at the target, calculated with the total wave function of Eq. (5.24):

$$j_r = \frac{i\hbar}{2m}\int\left(\Psi^*\frac{\partial\Psi}{\partial r} - \Psi\frac{\partial\Psi^*}{\partial r}\right) r^2\, d\Omega. \tag{5.43}$$

The negative sign indicates that absorption corresponds to the incoming flux in the sphere. The cross-section will be the ratio between j_r and the current of probability for the incident plane wave, $j_i = \hbar k/m$. From this, one finds

$$\sigma_r = \pi\lambdabar^2 \sum_{l=0}^{\infty}(2l+1)(1-|S_l|^2). \tag{5.44}$$

From Eqs. (5.40), (5.42) and (5.44) we can easily see that the total cross-section, $\sigma_t = \sigma_e + \sigma_r$, is linked to the scattering amplitude at zero scattering angle through the relationship (5.31), showing that the optical theorem is still valid in the presence of reaction channels different from the elastic channel.

Let us examine equations (5.42) and (5.44): when $|S_l| = 1$ the reaction cross-section is zero and we have pure scattering. The contrary (pure absorption without elastic scattering), however, cannot happen, as the vanishing of σ_e also implies the vanishing of σ_r. In general there is a region of allowed values of S_l for which the two cross-sections can coexist.

The maximum of σ_r happens for $S_l = 0$, corresponding to total absorption. Let us suppose that the absorption potential is limited to the surface of a nucleus with radius $R \gg \lambdabar$, that is, that all the particles with impact parameter smaller than the radius R are absorbed. This is equivalent to say that all particles are absorbed for $l \leq R/\lambdabar$. In this case

$$\sigma_r = \pi\lambdabar^2 \sum_{l=0}^{R/\lambdabar}(2l+1) = \pi(R+\lambdabar)^2. \tag{5.45}$$

This is the value that from an intuitive approach would be adequate for the total cross-section, i.e., equal to the geometric cross-section (the part $\lambda\!\!\!^{-}$ can be understood as an uncertainty in the position of the incident particle). But, we saw above that the presence of scattering is always obligatory. As $S_l = 0$, the scattering cross-section is identical to the reaction one, producing a total cross-section

$$\sigma = \sigma_r + \sigma_e = \pi(R + \lambda\!\!\!^{-})^2 + \pi(R + \lambda\!\!\!^{-})^2 = 2\pi(R + \lambda\!\!\!^{-})^2, \tag{5.46}$$

that is twice the geometric cross-section.

5.4. Potential Scattering

5.4.1. *Coulomb scattering*

An essential condition for the scattering wave function to have the asymptotic behavior of Eq. (5.27) is that the potential $V(r)$ goes to zero faster than $1/r$ as $r \to \infty$. Otherwise, the incident wave function is distorted even at asymptotic distances. This is the case of the Coulomb potential between point particles (we use here the CGS unit system),

$$V_C(r) = \frac{Z_p Z_t e^2}{r}, \tag{5.47}$$

where Z_p and Z_t are respectively the projectile and target charges. Therefore, Coulomb scattering requires a special treatment.

It is convenient to introduce the *Sommerfeld parameter*

$$\eta = \frac{Z_p Z_t e^2}{\hbar \mathrm{v}} \tag{5.48}$$

and the *half distance of closest approach* in a head-on collision

$$a = \frac{Z_p Z_t e^2}{2E}. \tag{5.49}$$

These quantities are related as

$$\eta = ka. \tag{5.50}$$

One should solve the relative motion Schrödinger equation

$$-\frac{\hbar^2}{2\mu}\nabla^2\phi_c(\mathbf{r}) + \frac{Z_p Z_t e^2}{r}\phi_c(\mathbf{r}) = E\phi_c(\mathbf{r}),$$

or

$$\left[\nabla^2 + k^2 - \frac{2\eta k}{r}\right]\phi_c(\mathbf{r}) = 0, \tag{5.51}$$

where ϕ_c is the wavefunction for the two-body system with reduced mass μ in their center of mass. The solution is found in terms of the *confluent hypergeometric function* [As72], ${}_1F_1(d; c; s)$,

$$\phi_c(r, z) = C\, e_1^{ikz} F_1(-i\eta; 1; ik(r - z)). \tag{5.52}$$

Choosing the normalization constant as

$$C = \Gamma(1 + i\eta)e^{-\pi\eta/2},$$

where Γ is the gamma-function [AS72], and using the asymptotic form of ${}_1F_1(d; c; s)$, we obtain

$$\phi_c(r, z)_{|r-z|\to\infty} \to e^{ikz} \cdot [e^{-i\eta \ln(k(r-z))}] + f_c(\theta)\frac{e^{ikr}}{r} \cdot [e^{-i\eta \ln(2kr)}]. \tag{5.53}$$

Above, $f_c(\theta)$ is the *Coulomb scattering amplitude*,

$$f_c(\theta) = -\frac{\eta}{2k \sin^2(\theta/2)} e^{-i\eta \ln(\sin^2 \theta/2)} e^{2i\sigma_0}, \tag{5.54}$$

where, σ_0 is the s-wave *Coulomb phase shift*

$$\sigma_0 = \arg \Gamma(1 + i\eta). \tag{5.55}$$

Note that Eq. (5.53) is not valid at $\theta = 0$. In this case, $|r - z| = 0$ and the condition $|r - z| \to \infty$ cannot be satisfied.

Inspecting Eq. (5.53), we conclude that the first and second terms correspond respectively to the incident plane wave and a scattered wave, as in Eq. (5.9). However, due to the long range of the interaction, the waves have a logarithmic distortion given by the factors within square brackets.

Now we turn to the elastic cross-section. It can easily be checked that the distortion factors in Eq. (5.53) lead to additional terms in the incident and in the scattered currents. However, these terms are proportional to $1/r$, so that they vanish asymptotically. Therefore, the elastic cross-section is given by

$$\frac{d\sigma_c(\Omega)}{d\Omega} = |f_c(\theta)|^2.$$

Using Eq. (5.54), we get the *Rutherford scattering cross-section*

$$\frac{d\sigma_c(\Omega)}{d\Omega} = \frac{a^2}{4}\left[\frac{1}{\sin^4(\theta/2)}\right]. \tag{5.56}$$

A very interesting feature of Coulomb scattering is that the above quantum mechanical cross-section is identical to the classical cross-section, Eq. (5.56).

The wave function $\phi_c(\mathbf{r})$ can be expanded in partial waves as in Eq. (5.17). Making the change of variable, $r \to \rho = kr$, the radial equation takes the form

$$u_l''(\rho) + \left[1 - \frac{l(l+1)}{\rho^2} - \frac{2\eta}{\rho}\right] u_l(\rho) = 0\,. \tag{5.57}$$

One identifies above the Coulomb wave equation [AS72]. Its solutions can be expressed as linear combinations of the independent pair of real solutions $F_l(\eta,\rho)$ and $G_l(\eta,\rho)$. $F_l(\eta,\rho)$ is the regular solution and $G_l(\eta,\rho)$ diverges at the origin (except for $l = 0$). Explicit forms of the Coulomb functions F_l and G_l are given in Refs. [Mes61,AS72].

When $\rho \to \infty$ one has

$$F_l(\eta, \rho \to \infty) \to \sin\left[\rho - \frac{l\pi}{2} - \eta\ln(2\rho) + \sigma_l\right], \tag{5.58}$$

$$G_l(\eta, \rho \to \infty) \to \cos\left[\rho - \frac{l\pi}{2} - \eta\ln(2\rho) + \sigma_l\right], \tag{5.59}$$

where

$$\sigma_l = \arg\Gamma(1 + l + i\eta) \tag{5.60}$$

are the *Coulomb phase shifts.* Usually, one evaluates $\sigma_0 = \arg\Gamma(1 + i\eta)$ and obtain Coulomb phase shifts for higher partial waves from the series

$$\sigma_l = \sigma_0 + \sum_{s=0}^{l} \tan^{-1}\left(\frac{\eta}{s}\right).$$

When $V_c \to 0 \Rightarrow \eta \to 0$, one has

$$F_l(\eta \to 0, \rho) \to \frac{j_l(\rho)}{\rho}, \quad G_l(\eta \to 0, \rho) \to \frac{n_l(\rho)}{\rho}. \tag{5.61}$$

The solutions of the Coulomb wave equation can also be expressed as linear combinations of incoming and outgoing Coulomb waves, respectively $H_l^{(-)}(\rho)$ and $H_l^{(+)}(\rho)$. They are defined as

$$H_l^{(\pm)}(\eta,\rho) = G_l(\eta,\rho) \pm iF_l(\eta,\rho). \tag{5.62}$$

Using Eqs. (5.58) and (5.59) we find their asymptotic forms

$$H_l^{(\pm)}(\eta,\rho \to \infty) \to e^{\pm\, i[\rho - \frac{l\pi}{2} - \eta \ln(2\rho) + \sigma_l]}. \tag{5.63}$$

As the strength of the Coulomb field goes to zero ($\eta \to 0$), $H_l^{(\pm)}$ converge to the Haenkel functions,

$$H_l^{(\pm)}(\eta \to 0,\rho) \to \frac{h_l^{(\pm)}(\rho)}{\rho}. \tag{5.64}$$

The partial wave expansion of the wave function ϕ_c is given by

$$\phi_c(\mathbf{r}) = A \sum_{l=0}^{\infty} (2l+1) i^l e^{i\sigma_l}\, P_l(\cos\theta) \frac{F_l(\eta,\rho)}{kr}. \tag{5.65}$$

In the $\eta \to 0$ limit, $\sigma_l \to 0$, $F_l(\eta,\rho) \to \jmath_l(\rho)/\rho$ and Eq. (5.65) reduces to the expansion for the plane wave, Eq. (5.21), except for the normalization factor A.

It is also useful to write the partial wave expansion in terms of incoming and outgoing Coulomb waves. Using Eq. (5.62), Eq. (5.65) becomes

$$\phi_c(\mathbf{r}) = A\left(\frac{1}{2kr}\right) \sum_{l=0}^{\infty} (2l+1) i^{l+1}\, P_l(\cos\theta) e^{i\sigma_l} [H_l^{(-)}(\eta,\rho) - H_l^{(+)}(\eta,\rho)]. \tag{5.66}$$

Let us consider the scattering from the potential

$$V(r) = V_c(r) + V_N(r), \tag{5.67}$$

where $V_c(r)$ is the Coulomb potential of Eq. (5.47) and $V_N(r)$ is a short-range potential, which vanishes for $r > R$. The equation for $u_l(k,r)$ then reduces to the Coulomb wave equation. Therefore, $u_l(k,r)$ can be expressed as the linear combination

$$u_l(k,r) = \frac{i}{2}[H_l^{(-)}(\eta,\rho) - S_l H_l^{(+)}(\eta,\rho)]e^{-i\delta_l}, \tag{5.68}$$

where $S_l = e^{2i\delta_l}$ are the partial wave components of the S-matrix associated with V_N, and δ_l are the corresponding phase shifts. The normalization of

u_l is such that it reduces respectively to $F_l(\eta, \rho)$ and j_l/ρ in the $V_N \to 0$ and $\eta \to 0$ limits, respectively.

To get the cross-section, we consider that the wave function can be written, as in Eq. (5.14), with

$$f(\theta) = f_c(\theta) + f_N(\theta), \tag{5.69}$$

with $f_c(\theta)$ given by Eq. (5.54) and $f_N(\theta)$ by Eq. (5.28).

Usually one is interested in investigating the action of the short-ranged potential. For this purpose, the cross-sections are usually normalized with respect to the corresponding pure Coulomb (or Rutherford) ones.

5.4.2. *Resonances*

For simplicity, let us consider neutrons with $l = 0$ and ignore the spins of the neutron and of the target. In this case,

$$\frac{d^2u_0}{dr^2} + k^2u_0 = 0 \quad (r \geq R) \tag{5.70}$$

is valid for the radial wave function u_0 at distances r larger than the *channel radius* $R = R_a + R_A$, with R_a and R_A being the radii of the projectile and of the target, respectively. The solution of Eq. (5.70) is, using (5.20) and (5.24),

$$u_0 = S_0 e^{ikr} - e^{-ikr} \quad (r \geq R). \tag{5.71}$$

A radial wave function inside the nucleus should connect the external function (5.71) with a continuous function and its derivative at $r = R$. The *logarithmic derivative*

$$\mathcal{L}_l \equiv R\left[\frac{du_l/dr}{u_l}\right]_{r=R} \tag{5.72}$$

must have identical values if calculated with the internal or the external function and this condition creates a relationship between $\mathcal{L}_l$ and S_l. Thus, knowledge of $\mathcal{L}_l$ leads to knowledge of the cross-sections. For our example of neutrons with $l = 0$, the application of Eq. (5.72) results in

$$\mathcal{L}_0 = ikR\frac{S_0 + e^{-2ikR}}{S_0 - e^{-2ikR}}, \tag{5.73}$$

from which we extract

$$S_0 = \frac{\mathcal{L}_0 + ikR}{\mathcal{L}_0 - ikR} e^{-2ikR}. \tag{5.74}$$

If $\mathcal{L}_0$ is a real number, then $|S_0|^2 = 1$. The reaction cross-section (5.44) will be zero and we shall have pure scattering.

Using Eq. (5.74), the scattering cross-section (5.42) can be written as

$$\sigma_{e,0} = \pi ƛ^2 |A_{res} + A_{pot}|^2, \tag{5.75}$$

with

$$A_{res} = \frac{-2ikR}{\mathcal{L}_0 - ikR}, \tag{5.76}$$

and

$$A_{pot} = \exp(2ikR) - 1. \tag{5.77}$$

Separation of the cross-section into two parts has a physical justification: A_{pot} does not contain $\mathcal{L}_0$ and thus it does not depend on the conditions inside the nucleus. It represents the situation where the projectile does not penetrate the nucleus, being scattered by its external potential. This is clearly seen in the idealized situation where the nucleus is considered to be an impenetrable hard sphere. In this case the wave function is zero inside the nucleus and u_0 vanishes at $r = R$, implying $\mathcal{L}_0 \rightarrow \infty$ and $A_{res} \rightarrow 0$. In this way, A_{pot} is the only contribution responsible for the scattering.

Applying now Eq. (5.74) to the reaction cross-section (5.44) and using

$$\mathcal{L}_0 = \mathcal{L}_R + i\mathcal{L}_I \tag{5.78}$$

we have

$$\sigma_{r,0} = \pi ƛ^2 \frac{-4kR\mathcal{L}_I}{\mathcal{L}_R^2 + (\mathcal{L}_I - kR)^2}, \tag{5.79}$$

an equation that will be useful when we study the presence of resonances in the excitation functions (cross-section as a function of the energy).

The second parenthesis in the denominator of Eq. (5.79) is never zero, since the numerator forces $\mathcal{L}_I$ to be always negative. If for a certain energy $\mathcal{L}_R$ vanishes, $\sigma_{r,0}$ passes by a maximum in that energy. We can tentatively identify these energies as being the energy of the resonances. Let us take the extreme case of a single resonance at the energy E_R, that is, $\mathcal{L}_R = 0$

for $E = E_R$. We can expand $\mathcal{L}_R$ in a Taylor series in the neighborhood of a resonance,

$$\mathcal{L}_R(E) = (E - E_R)\left(\frac{d\mathcal{L}_R}{dE}\right)_{E=E_R} + \cdots . \tag{5.80}$$

Keeping just the first term of the expansion and applying it to Eqs. (5.75) and (5.79), we get

$$\sigma_{e,0} = \pi\lambda\!\!\!^{-2}\left|\exp(2ikR) - 1 + \frac{i\Gamma_\alpha}{(E - E_R) + i\frac{\Gamma}{2}}\right|^2, \tag{5.81}$$

$$\sigma_{r,0} = \pi\lambda\!\!\!^{-2}\frac{\Gamma_\alpha(\Gamma - \Gamma_\alpha)}{(E - E_R)^2 + \left(\frac{\Gamma}{2}\right)^2}, \tag{5.82}$$

where we define

$$\Gamma_\alpha = -\frac{2kR}{(d\mathcal{L}_R/dE)_{E=E_R}} \quad \text{and} \quad \Gamma = \frac{2kR - 2\mathcal{L}_I}{(d\mathcal{L}_R/dE)_{E=E_R}}. \tag{5.83}$$

The cross-section for formation of a compound nucleus with an energy E close to a resonance should be proportional to the probability of the resonant state existing at that energy. This probability is given by Eq. (4.8), and we get:

$$\sigma_{R,0} \propto \frac{1}{(E - E_R)^2 + \left(\frac{\Gamma}{2}\right)^2},$$

from which it is deduced that the energy Γ, defined in Eq. (5.83), that appears in Eq. (5.82) is the total width of the resonance, $\Gamma = \Gamma_\alpha + \Gamma_\beta + \ldots$, i.e., the sum of the widths for all the possible processes of decay of the nucleus, starting from the resonant state. The interpretation of Γ_α is obtained by a comparison of Eq. (5.82) with Eq. (5.81). From the latter, the contribution of the resonant part to the scattering cross-section contains the factor $\Gamma_\alpha\Gamma_\alpha$ and to Eq. (5.82) the factor $\Gamma_\alpha(\Gamma - \Gamma_\alpha)$. It is fair to interpret Γ_α as the entrance channel width, with $\Gamma - \Gamma_\alpha$ being the sum of the widths of all the exit channels except α. If we restrict the exit channels to a single channel β or, to put it in another way, we designate β as the group of exit channels except α, Eq. (5.82) is rewritten as

$$\sigma_{\alpha,\beta} = \pi\lambda\!\!\!^{-2}\frac{\Gamma_\alpha\Gamma_\beta}{(E - E_R)^2 + \left(\frac{\Gamma}{2}\right)^2}, \tag{5.84}$$

which is the usual way of presenting the *Breit–Wigner formula*, a formula that describes the form of the cross-section close to a resonance. Let us recall that Eq. (5.84) refers to an incident particle of $l = 0$, without charge and without spin. If the spins of the incident and target particles are s_a and s_A, respectively, and the incident beam is described by a single partial wave l_0, the cross-section (10.60a) should be multiplied by corresponding statistical factors.

The evaluation of the phase shifts δ_l when $V = V_c + V_N$ is carried out in the same way as in the absence of Coulomb potential. The boundary R divides the r-axis into an internal and an external region. The radial equation is solved numerically in the internal region and the logarithmic derivative, $\mathcal{L}_l$, is calculated. Matching it to the one calculated with the radial wave function at the external region, one finds the following expression for the phase shift δ_l,

$$\tan \delta_l = \frac{kRF_l'(kR) - F_l(kR)\mathcal{L}_l}{kRG_l'(kR) - G_l(kR)\mathcal{L}_l}, \qquad (5.85)$$

where the primes mean derivatives with respect to the arguments.

5.4.3. *Optical potentials*

The complications involved in the inclusion of all relevant reaction channels is avoided by the use of empirical optical potentials. In this model the interaction between the nuclei in a reaction is described by a potential $U(r)$, r being the distance between the center of mass of the two nuclei. This idea is similar to the one of the shell model. It replaces the complicated interaction that a nucleon has with the rest of the nucleus with a potential that acts on the nucleon. The potential $U(r)$ includes a complex part that takes into account the absorption effects, i.e., the inelastic scattering.

In its most commonly used form, the optical potential is written as the sum

$$U(r) = U_R(r) + U_I(r) + U_D(r) + U_S(r) + U_C(r), \qquad (5.86)$$

which contains parameters that can vary with the energy and the masses of the nuclei and that should be chosen by an adjustment to the experimental data. The optical potential $U(r)$ will only make sense if these variations are small for close masses or neighboring energies.

The first part of (5.86),

$$U_R(r) = -Vf(r, R, a), \tag{5.87}$$

is real and represents a nuclear well with depth V, being multiplied by a Woods–Saxon form factor

$$f(r, R, a) = \{1 + \exp[(r - R)/a]\}^{-1}, \tag{5.88}$$

where R is the radius of the nucleus and a measures the diffuseness of the potential, i.e., the width of the region where the function f is sensibly different from 0 or 1. This produces a well with round borders, closer to reality than a square well. In (5.87) and (5.88), V, R and a are treated as adjustable parameters.

The absorption effect or, in other words, the disappearance of particles from the elastic channel, is taken into account including the two following imaginary parts,

$$U_I(r) = -iWf(r, R_I, a_I), \tag{5.89}$$

and

$$U_D(r) = 4ia_I W_D \frac{d}{dr} f(r, R_I, a_I). \tag{5.90}$$

The expression (5.89) is responsible for the absorption in the whole volume of the nucleus, but (5.90), built from the derivative of the function f, acts specifically in the region close to the nuclear surface, where the form factor f suffers its largest variation. These two parts have complementary goals: at low energies there are no available unoccupied states for nucleons inside the nucleus and the interactions are essentially at the surface. In this situation $U_D(r)$ is important and $U_I(r)$ can be ignored. On the other hand, at high energies the incident particle has larger penetration and in this case the function $U_I(r)$ is important.

As with the shell model potential, a *spin–orbit interaction* term is added to the optical potential. This term, which is the fourth part of (5.86), is usually written in the form

$$U_S(r) = \mathbf{s} \cdot \mathbf{l} \left(\frac{\hbar}{m_\pi c^2} \right)^2 V_s \frac{1}{r} \frac{d}{dr} f(r, R_S, a_S), \tag{5.91}$$

incorporating a normalization factor that contains the mass of the pion m_π. $\mathbf{s}$ is the spin operator and $\mathbf{l}$ the angular orbital momentum operator. As with $U_D(r)$, the part $U_S(r)$ is also only important at the surface of the nucleus since it contains the derivative of the form factor (5.88). The values of V_S, R_S and a_S must be adjusted by the experiments. The spin–orbit interaction leads to asymmetric scattering due to the different signs of the product $\mathbf{s} \cdot \mathbf{l}$ as the projectile passes by one or the other side of the nucleus.

Finally, a term corresponding to the Coulomb potential is added to (5.86) whenever the scattering involves charged particles. It has the form

$$U_C(r) = \frac{Z_1 Z_2 e^2}{2R_c}\left(3 - \frac{r^2}{R_c^2}\right) \quad (r \leq R_c)$$
$$= \frac{Z_1 Z_2 e^2}{r} \quad (r > R_c), \tag{5.92}$$

where it is assumed that the nucleus is a homogeneously charged sphere of radius equal to the *Coulomb barrier radius* R_c, which defines the region of predominance of each one of the forces — nuclear or Coulomb.

The optical model has a limited group of adjustable parameters and is not capable of describing abrupt variations in the cross-sections, as it happens for isolated resonances. However, it can provide a good description of the cross-sections in the continuous region, as it treats these as an wave phenomenon.

5.5. Compound Nucleus

5.5.1. *Compound nucleus formation*

The compound nucleus model is a description of atomic nuclei to explain nuclear reactions as a two-stage process comprising the formation of a relatively long-lived intermediate nucleus and its subsequent decay. First, a bombarding particle loses all its energy to the target nucleus and becomes an integral part of a new, highly excited, unstable nucleus, called a *compound nucleus*. The formation stage takes a period of time approximately equal to the time interval for the bombarding particle to travel across the diameter of the target nucleus:

$$\Delta t \sim \frac{R}{c} \sim 10^{-21}\text{s}.$$

Second, after a relatively long period of time (typically from 10^{-19} to 10^{-15} s) and independent of the properties of the reactants, the compound nucleus disintegrates, usually into an ejected small particle and a product nucleus. For the calculation of properties for the decay of this system by *particle evaporation* one may thus borrow from the techniques of statistical mechanics. The energy distribution of the evaporated nucleons have sharp resonances, whose width is much smaller than those known from potential scattering. A qualitative way to understand this is through the time–energy uncertainty relation: the relaxation to equilibrium and the subsequent "evaporation" of a nucleon takes a long time, much longer than the period needed for a nucleon to move across the average potential. In other words, the emission of the nucleon with energy equal to that of the incident one is a concentration of the excitation energy on a single particle through a complicated process, which needs a long time. Hence the associated energy width will be quite narrow. For very slow neutrons the widths of these resonances are in the range of a few eV!

We shall assume initially that the incident particle is a neutron of low energy ($< 50\,\mathrm{MeV}$). When such a neutron enters the field of nuclear forces it can be scattered or begin a series of collisions with the nucleons. The products of these collisions, including the incident particle, will continue along their course, leading to new collisions and new changes of energy. During this process one or more particles can be emitted and they form with the residual nucleus the products of a reaction that is referred to as *pre-equilibrium*. But, at low energies, the largest probability is the continuation of the process so that the initial energy is distributed to all the nucleons, with no emitted particle. The final nucleus with $A + 1$ nucleons has an excitation energy equal to the kinetic energy of the incident neutron together with the binding energy that the neutron has in the new, highly unstable, nucleus. It can, among other processes, emit a neutron with the same or smaller energy to that which was absorbed. In the final stage the compound nucleus can evaporate one or more particles, fission, etc... In our notation, for the most common process in which two final products are formed (the evaporated particle plus the residual nucleus or two fission fragments, etc.) we write:

$$\mathrm{a} + \mathrm{A} \rightarrow \mathrm{C}^* \rightarrow \mathrm{B} + \mathrm{b}, \tag{5.93}$$

the asterisk indicating that the compound nucleus C is in an excited state.

The compound nucleus lives long enough to "forget" the way it was formed and the de-excitation to the final products b and B only depends on the energy, angular momentum and parity of the quantum state of the compound nucleus.

Another particularity that should happen in reactions in which there is formation of a compound nucleus refers to the angular distribution of the fragments, or evaporated particles: it should be isotropic in the center of mass, and this is verified experimentally. We know, however, that the total angular momentum is conserved and cannot be "forgotten". Reactions with large transfer of angular momentum, as happens when heavy ions are used as projectiles, can show a non-isotropic angular distribution in the center of mass system.

The occurrence of a nuclear reaction in two stages allows the cross-section for a reaction A(a,b)B to be written as the product,

$$\sigma(\mathrm{a}, \mathrm{b}) = \sigma_{\mathrm{CN}}(\mathrm{a}, \mathrm{A}) P(\mathrm{b}), \tag{5.94}$$

where $\sigma_{\mathrm{CN}}(\mathrm{a,A})$ is the cross-section of formation of the compound nucleus starting from the projectile a and the target A and $P(\mathrm{b})$ is the probability of the compound nucleus to emit a particle b leaving a residual nucleus B. If not only the particles but the quantum numbers of entrance and exit channels are well specified, i.e., if the reaction begins at an entrance channel α and ends at an exit channel β, (5.94) can be written as

$$\sigma(\alpha, \beta) = \sigma_{\mathrm{CN}}(\alpha) P(\beta). \tag{5.95}$$

We can associate the probability $P(\beta)$ to the width Γ_β of the channel β and write

$$P(\beta) = \frac{\Gamma(\beta)}{\Gamma}, \tag{5.96}$$

where Γ is the total width, that is, $\tau = \hbar/\Gamma$ is the half-life of disintegration of the compound nucleus. $\Gamma(\beta)$ is the partial width for the decay through channel β, and

$$\Gamma = \sum_\beta \Gamma(\beta).$$

In the competition between the different channels β, nucleon emission is preferred over γ-radiation whenever there is available energy for nucleon emission and among the nucleons the neutrons have preference as they do not have the Coulomb barrier as an obstacle. Thus, in a reaction where there is no restriction for neutron emission we can say that

$$\Gamma \cong \Gamma_n, \tag{5.97}$$

where Γ_n includes the width for the emission of one or more neutrons.

In some cases it is also useful to define the *reduced width*

$$\gamma_\beta^2 = \frac{\Gamma_\beta}{2P(\beta)}, \tag{5.98}$$

where the penetrability $P(\beta)$ is the imaginary part of the logarithmic derivative of the outgoing wave in the channel β, evaluated at the surface $r = R$ and multiplied by the channel radius R. For s-wave neutrons the outgoing wave is just e^{ikr} multiplied by the appropriate S-matrix element and hence $P = kR$. Thus the reduced width determines the probability that the components specified by β will appear at the surface so that the compound system can decay through this mode.

The study of the function $P(\beta)$ is done in an evaporation model that leads to results in many aspects similar to the evaporation of molecules of a liquid, with the energy of the emitted neutrons approaching the form of a Maxwell distribution

$$I(E) \propto E \exp\left(-\frac{E}{\theta}\right) dE, \tag{5.99}$$

with I measuring the number of neutrons emitted with energy between E and $E + dE$. The quantity θ, with dimension of energy, has the role of a *nuclear temperature*. It is related to the density of levels ω of the daughter nucleus B by

$$\frac{1}{\theta} = \frac{dS}{dE}, \tag{5.100}$$

with

$$S = \ln \omega(E), \tag{5.101}$$

where, to be used in (5.99), dS/dE should be calculated for the daughter nucleus B at the maximum excitation energy that it can have after the emission of a neutron, that is, in the limit of emission of a neutron with zero kinetic energy.

The *level density* $\omega(E)$ is a measure of the number of states of available energy for the decay of the compound nucleus in the interval dE around the energy E. In this sense, the relationship (5.101) is, neglecting the absence of the Boltzmann constant, identical to the thermodynamic relationship between the entropy S and the number of states available for the transformation of a system. The entropy S defined in this way has no dimension, and (5.100) is the well-known relation between the entropy and the temperature. The last one has in the present case the dimension of energy.

The cross-section for the formation of a compound nucleus $\sigma_{\rm CN}(\alpha)$ can be determined in a simple way if some additional hypotheses about the reaction can be done. We can still write the wave function inside the nucleus as just an incoming wave

$$u_0 \cong \exp(-iKr), \quad \text{for } (r < R), \tag{5.102}$$

where $K = \sqrt{2m(E - V_0)}/\hbar$ is the wave number inside the nucleus, and it is assumed that the neutron with total energy E is subject to a negative potential V_0. The expression (5.102) is clearly a crude simplification for a situation where the incident neutron interacts in a complicated way with the other nucleons in the nucleus. It allows us, however, to explain the average behavior of the cross-sections at low energies. Starting from (5.102) one determines the value of $\mathcal{L} = R[du_0/dr]_{r=R}/u_0$,

$$\mathcal{L} = -iKR. \tag{5.103}$$

This means that $\mathcal{L}_R = 0$ and $\mathcal{L}_I = -KR$ in Eq. (5.78). Inserting this result in Eq. (5.79) we get the expression

$$\sigma_{\text{CN}} = \frac{\pi}{k^2} \frac{4kK}{(k+K)^2} \tag{5.104}$$

for the cross-section of compound nucleus formation for neutrons with $l = 0$. At low energies, $E \ll |V_0|$, thus $k \ll K$. Under these conditions, $\sigma_{\text{CN}} = 4\pi/kK$. Thus, σ_{CN} varies with $1/k$, that is,

$$\sigma_{\text{CN}} \propto \frac{1}{\text{v}}, \tag{5.105}$$

where v is the velocity of the incident neutron. This is the well-known $1/v$ *law* that governs the behavior of the capture cross-section of low energy neutrons (Wigner's law). Figure 5.5 exhibits the *excitation function* (cross-section as function of the energy) for the reaction n + ^{235}U. The cross-section decays with $1/v$ up to 0.3 eV, where a series of resonances start to appear. This abrupt behavior of the cross-sections does not belong to the theory of the compound nucleus and the resonances appear exactly when is not possible to sustain the hypothesis that there is no return to the entrance channel, a hypothesis that was used in (5.102).

We have already arrived at an expression for the cross-section that describes a resonance, Eq. (5.81). That formula describes elastic scattering in a resonant situation. In the case of compound nuclei we shall call it the *compound elastic* cross-section. A similar expression can be obtained for the reaction, or absorption cross-section. Let us briefly review the concepts.

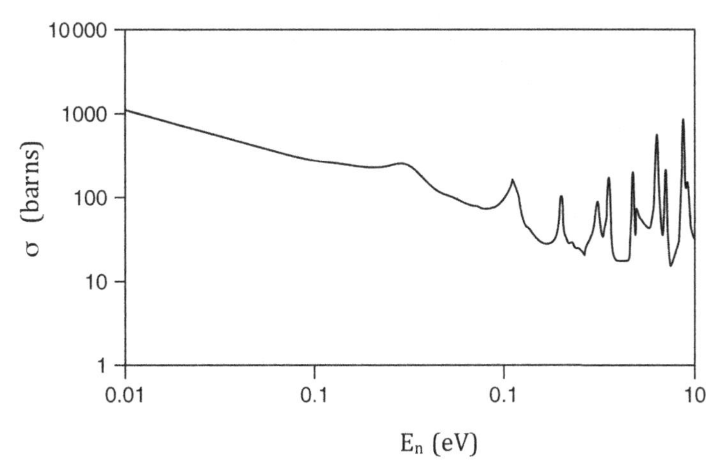

Figure 5.5. Total cross-section for neutrons of low energy hitting ^{235}U. There are resonances at energies 1–10 eV (epithermal neutrons) and at very low energies, $E < 0.1$ eV (thermal neutrons). The cross-section varies as $1/v$, as expected (see Eq. (5.105)).

We will study resonance processes involving s-wave neutron scattering and assume that the nucleus has a well-defined surface. The nucleon does not interact with the nucleus at separation distances larger than the *channel radius* R. Using the logarithmic derivative $\mathcal{L}_0 = R[du_0/dr]_{r=R}/u_0$ for a neutron wave function of the form

$$u_0 = \frac{i}{2k}[e^{-ikr} - S_0 e^{ikr}], \quad r \geq R, \tag{5.106}$$

we have

$$S_0 = \frac{\mathcal{L}_0 + ikR}{\mathcal{L}_0 - ikR} e^{-i2kR}. \tag{5.107}$$

If $\mathcal{L}_0$ is real, $|S_0|$ is unity and there is no reaction, but if $\Im\mathcal{L}_0 < 0$ then $|S_0| < 1$.

The scattering amplitudes and the total and reaction widths were obtained earlier. The total width is given by

$$\Gamma = -\frac{2(b + kR)}{a'(E_r)}, \tag{5.108}$$

so that the *reaction* (or absorption) *width* is

$$\Gamma_r = \sum_{\beta \neq \alpha} \Gamma_\beta = \Gamma - \Gamma_\alpha = -\frac{2b}{a'(E_r)}. \tag{5.109}$$

The compound elastic scattering cross-section is given in our present notation as

$$\sigma_{\mathrm{ce},\alpha} = \frac{\pi}{k^2} \frac{\Gamma_\alpha^2}{\left(E - E_r\right)^2 + \Gamma^2/4}, \tag{5.110}$$

and the absorption cross-section is

$$\sigma_{\mathrm{abs}} = \frac{\pi}{k^2} \frac{\Gamma_r \Gamma_\alpha}{(E - E_r)^2 + \Gamma^2/4}. \tag{5.111}$$

The cross-section for compound nucleus formation is obtained by adding the cross-sections for those processes which involve formation of the compound nucleus through channel α, i.e.

$$\sigma_{\mathrm{CN}} = \sigma_{\mathrm{abs}} + \sigma_{\mathrm{ce},\alpha} = \frac{\pi}{k^2} \frac{\Gamma \Gamma_\alpha}{(E - E_r)^2 + \Gamma^2/4}, \tag{5.112}$$

and from Eqs. (5.95–5.96) the cross-section for the process $\alpha \longrightarrow \beta$ is

$$\sigma_{\mathrm{abs}} = \frac{\pi}{k^2} \frac{\Gamma_\beta \Gamma_\alpha}{(E - E_r)^2 + \Gamma^2/4}, \tag{5.113}$$

as in Eq. (5.84)

If the exit channel is the same as the entrance channel α, the cross-section should be obtained from (5.81) and its dependence on energy is more complicated because in addition to the resonant scattering there is potential scattering, and the cross-section will contain, beyond these two terms, an interference term between both. The presence of these three terms results in a peculiar aspect of the scattering cross-section, differing from the simple form (5.112) for the compound nucleus cross-section.

The region of energy where resonances show up can extend to 10 MeV in light nuclei but it ends well before this in heavy nuclei. Starting from this upper limit the increase in the density of levels with the energy implies that the average distance between the levels is smaller than the width of the levels and individual resonances cannot be observed experimentally. They form a continuum and that region is referred to as the *continuum region*. In the continuum region there are no characteristic narrow peaks due to individual resonances. But the cross-section does not vary monotonically; peaks of large widths are seen. Their presence is mainly due to interference phenomena between the part of the incident beam that passes through the nucleus and the part that passes around it [Mv67].

For the decay of the compound nucleus through the channel β we can write Eq. (5.104) as

$$\sigma_{\rm CN} = \frac{\pi}{k^2}\frac{4kK}{(k+K)^2}\frac{\Gamma_\beta}{\Gamma}. \qquad (5.114)$$

The quantity $4kK/(k+K)^2$ is called the *transmission coefficient* T_0 for s-wave neutrons. If $\langle\Gamma_\alpha\rangle$ is the mean width for resonances due to particles in channel α and D is the mean spacing of levels within an energy interval I we may write

$$\begin{aligned}\sigma_{\rm CN}(\alpha) &= \frac{1}{I}\int_{E-I/2}^{E+I/2}\frac{\pi}{k^2}\sum_r \frac{\Gamma^r\Gamma^r_\alpha}{(E-E_r)^2+(\Gamma^r)^2/4}dE\\ &= \frac{\pi}{k^2}\frac{2\pi}{I}\sum_r \Gamma^r_\alpha = \frac{\pi}{k^2}2\pi\frac{\langle\Gamma_\alpha\rangle}{D}, \qquad (5.115)\end{aligned}$$

where the energy interval is chosen so that the variation of k^2 can be neglected.

Combining with Eq. (5.104) we have

$$\frac{\langle\Gamma_\alpha\rangle}{D} = \frac{1}{2\pi}\frac{4kK}{(k+K)^2} \approx \frac{2k}{\pi K}. \qquad (5.116)$$

This quantity is called the s-*wave strength.* To study the A dependence of this quantity, the comparison must be made at a fixed energy and since the strength function is measured for the different nuclei at different incident energies, one has to transform the values thus measured to those corresponding to a fixed energy. The fixed energy is conventionally chosen to be $E_0 = 1\,\mathrm{eV}$. From Eq. (5.116) we see that the energy dependence at low energies is

$$\frac{\langle\Gamma_\alpha\rangle}{D} \propto \sqrt{E}.$$

The strength function at the conventional energy $E_0 = 1\,\mathrm{eV}$ is thus related to that measured at an energy E by the relation

$$\frac{\langle\Gamma_{0,\alpha}\rangle}{D} = \left(\frac{E_0}{E}\right)^{1/2}\frac{\langle\Gamma_\alpha\rangle}{D}.$$

5.5.2. *Compound nucleus decay*

At low incident energies the compound nucleus states are excited individually and each produces a resonance in the cross-section that may be described by the Breit–Wigner theory (Eq. (5.112)). As the incident energy increases compound nucleus states of higher energy are excited and these are closer together and of increasing width. Eventually they overlap and it is no longer possible to identify the individual resonances. The cross-section then fluctuates.

This fluctuating behavior is due to the interference of the reaction amplitudes corresponding to the excitation of each of the overlapping states which vanish in the energy average of the cross-section since these amplitudes are complex functions with random modulus and phase. The energy average of the cross-sections thus shows a weak energy dependence and it is predictable by theory. To develop such a theory we consider a reaction that proceeds from the initial channel c through the compound nucleus to the final channel c'. If we forget, for the moment, that the compound nucleus may be created in states of different angular momentum J, the hypothesis of the independence of formation and decay of the compound nucleus, according to Eq. (5.95) and Eq. (5.96), then gives for the cross-section

$$\sigma_{cc'} \sim \sigma_{CN}(c)\frac{\Gamma_{c'}}{\Gamma}, \tag{5.117}$$

where $\sigma_{CN}(c)$ is the cross-section for formation of the compound nucleus and $\Gamma_{c'}$ and Γ are, respectively, the energy-averaged width for the decay of the compound nucleus in channel c' and the energy-averaged total width.

We now use the *reciprocity theorem*, derived in the next section, that relates the cross-section $\sigma_{cc'}$ to the cross-section for the time-reversed process $c' \to c$:

$$g_c k_c^2\, \sigma_{cc'} = g_{c'} k_{c'}^2 \sigma_{c'c}, \tag{5.118}$$

where $g_c = 2I_c + 1$ and $g_{c'} = 2I_{c'} + 1$ are the statistical weights of the initial and final channels, I_c and $I_{c'}$ are the spin of the projectile and the ejectile, and k_c and $k_{c'}$ their wave numbers. This gives

$$g_c k_c^2\, \sigma_{CN}(c)\Gamma_{c'} = g_{c'} k_{c'}^2 \sigma_{CN}(c')\Gamma_c \tag{5.119}$$

or, equivalently

$$\frac{\Gamma_c}{g_c k_c^2 \sigma_{CN}(c)} = \frac{\Gamma_{c'}}{g_{c'} k_{c'}^2\, \sigma_{CN}(c')}. \tag{5.120}$$

Since the channels c and c' are chosen arbitrarily, this relation holds for all possible channels. So,

$$\Gamma_c \propto g_c k_c^2 \sigma_{CN}(c). \tag{5.121}$$

Since the total width is obtained by summing the $\Gamma_{c'}$s over all open channels

$$\Gamma = \sum_c \Gamma_c\,, \tag{5.122}$$

the cross-section (5.117) becomes

$$\sigma_{cc'} = \sigma_{CN}(c)\frac{g_{c'}k_{c'}^2\sigma_{CN}(c')}{\sum_c g_c k_c^2 \sigma_{CN}(c)}. \tag{5.123}$$

Ejectiles with energy in the range $E_{c'}$ to $E_{c'} + dE_{c'}$ leave the residual nucleus with energy in the range $U_{c'}$ to $U_{c'} + dU_{c'}$ where

$$U_{c'} = E_{CN} - B_{c'} - E_{c'}, \tag{5.124}$$

and E_{CN} and $B_{c'}$ are respectively the compound nucleus energy and the binding energy of the ejectile in the compound nucleus. Introducing the density of levels of the residual nucleus $\omega(U_{c'})$, Eq. (5.123) becomes

$$\sigma_{cc'}dE_{c'} = \sigma_{CN}(c)\frac{g_{c'}k_{c'}^2\sigma_{CN}(c')\omega(U_{c'})dU_{c'}}{\sum_c g_c k_c^2 \sigma_{CN}(c)\omega(U_c)dU_c}, \tag{5.125}$$

or, since $k^2 = 2\mu E$,

$$\sigma_{cc'}(E_{c'})dE_{c'} = \sigma_{CN}(c)\frac{(2I_{c'}+1)\mu_{c'}E_{c'}\sigma_{CN}(c')\omega(U_{c'})dU_{c'}}{\sum_c \int_0^{E_c^{\max}}(2I_c+1)\mu_c E_c \sigma_{CN}(c)\omega(U_c)dU_c}, \tag{5.126}$$

where μ_c is the reduced mass of the ejectile c. This is the *Weisskopf–Ewing formula* for the angle-integrated cross-sections.

To a good approximation, the level density $\omega(U) \propto \exp(U/T)$, so the ejectile spectrum given by the Weisskopf–Ewing theory is Maxwellian. It rises rapidly above the threshold energy, attains a maximum and then falls exponentially as shown in Fig. 5.6.

Since (5.121) is a proportionality relation it does not allow one to evaluate the absolute values of the decay widths. This may be obtained by use of the *detailed balance principle* which in addition to the invariance for time reversal leading to the reciprocity theorem (5.118) implies the

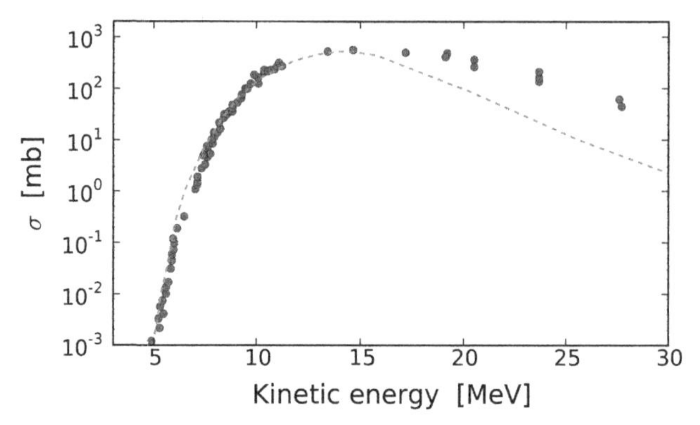

Figure 5.6. Excitation function for the ^{58}Ni(α,p) reaction compared with Weisskopf–Ewing calculations. The disagreement between experimental data and theoretical calculation at the higher energies is due to the emission of *pre-equilibrium* protons.

existence of a long-lived compound nucleus state. This principle states that two systems a and b, with state densities ρ_a and ρ_b, are in statistical equilibrium when the depletion of the states of system a by transitions to b equals their increase by the time-reversed process $b \longrightarrow a$.

If $W_{ab} = \Gamma_{ab}/\hbar$ is the decay rate (probability per unit time) for transitions from a to b and $W_{ba} = \Gamma_{ba}/\hbar$ is the decay rate for the inverse process, this equality occurs when

$$\rho_a \Gamma_{ab} = \rho_b \Gamma_{ba}. \tag{5.127}$$

In the case we are interested in, a is the compound nucleus with energy E_{CN} and, if one neglects the spin dependence, its state density $\rho_{CN}(E_{CN})$ coincides with its density of levels $\omega_{CN}(E_{CN})$ (this is not true for a system with spin J since $(2J+1)$ states correspond to each level). Γ_{ab} is the width $\Gamma_{c'}$ for decay in channel c'. System b is constituted by the ejectile c' with energy from $E_{c'}$ to $E_{c'} + dE_{c'}$ and a residual nucleus with excitation energy from $U_{c'}$ to $U_{c'} + dU_{c'}$. Thus ρ_b is the product of the density of continuum states of c' and the density of levels of the residual nucleus:

$$\rho_b = \rho_{c'}(E_{c'})\omega\left(U_{c'}\right). \tag{5.128}$$

The density of the continuum states of c' is given by the Fermi gas model expression (3.29) i.e.,

$$\rho_{c'}(E_{c'}) = \frac{\mu_{c'} E_{c'} V}{\pi^2 \hbar^3 v_{c'}} g_{c'} dE_{c'}, \tag{5.129}$$

where $\mu_{c'}$ and $v_{c'}$ are, respectively, the reduced mass and the velocity of c' and V is the space volume. The decay rate for the inverse process is

$$W_{c'c} = \frac{v_{c'}\sigma_{c'}(E_{c'})}{V}, \tag{5.130}$$

where $\sigma_{c'}(E_{c'})$ is the cross-section for the inverse process (formation of the compound nucleus with energy E_{CN} from channel c') which may be evaluated with the optical potential model.

From relations (5.127–5.130) one finally gets

$$\Gamma_{c'} = \frac{1}{\omega_{CN}(E_{CN})} \frac{(2I_{c'}+1)\mu_{c'}E_{c'}}{\pi^2\hbar^2} \sigma_{c'}(E_{c'})\omega(U_{c'})dE_{c'}. \tag{5.131}$$

The Weisskopf–Ewing theory provides a simple way of estimating the energy variation, at low incident energies, of the cross-sections of all available final channels in a particular reaction, and an example of such a calculation is shown in Fig. 5.6.

5.5.3. *Reciprocity theorem*

Extending the partial wave expansion theory to include specific channels, the cross-section for the reaction $\alpha \longrightarrow \beta$ is, as usual, given by the ratio of the outgoing flux in channel β and the incident flux in channel α. This gives for the angle-integrated cross-section of a reaction $\alpha \longrightarrow \beta$ $(\alpha \neq \beta)$

$$\sigma_{\alpha\beta} = \frac{\pi}{k^2} \sum_l (2l+1) \left| S_l^{\alpha\beta} \right|^2, \tag{5.132}$$

where $S_l^{\alpha\beta}$ is the S-matrix for the particle scattering from channel α to channel β. The angle-integrated elastic $(\alpha = \beta)$ cross-section is

$$\sigma_{\alpha\alpha} = \frac{\pi}{k^2} \sum_l (2l+1)|1 - S_l^{\alpha\alpha}|^2. \tag{5.133}$$

The two expressions can be unified by writing

$$\sigma_{\alpha\beta} = \frac{\pi}{k^2} \sum_l (2l+1)|\delta_{\alpha\beta} - S_l^{\alpha\beta}|^2. \tag{5.134}$$

This generalization explains why we called S_l by S-matrix, as it depends on two indices, (α, β), representing initial and final channels.

The scattering amplitude can be written as

$$f_{\alpha\beta}(\theta) = \frac{1}{2ik}\sum_l (2l+1)P_l(\cos\theta)|\delta_{\alpha\beta} - S_l^{\alpha\beta}|. \tag{5.135}$$

In terms of its partial wave components, the unitarity of the S-matrix, Eq. (5.25), can be written as

$$\sum_\alpha S_l^{\beta\alpha} S_l^{\alpha\gamma\dagger} = \sum_\alpha S_l^{\beta\alpha} S_l^{\gamma\alpha *} = \delta_{\gamma\beta}, \tag{5.136}$$

and so

$$S_l^{\gamma\alpha} = S_l^{\alpha\gamma}. \tag{5.137}$$

From Eqs. (5.132) and (5.137) one immediately gets

$$\frac{\sigma_{\alpha\beta}}{\lambda_\alpha^2} = \frac{\sigma_{\beta\alpha}}{\lambda_\beta^2}, \quad \text{or} \quad k_\alpha^2\sigma_{\alpha\beta} = k_\beta^2\sigma_{\beta\alpha}, \tag{5.138}$$

which is known as the reciprocity theorem.

If we consider the scattering of spin particles, the reciprocity theorem has to account for the statistical weights of the channels α and β. The relation (5.138) becomes then

$$g_\alpha k_\alpha^2\sigma_{\alpha\beta} = g_\beta k_\beta^2\sigma_{\beta\alpha}, \tag{5.139}$$

where g_α and g_β are the total number of spin states in channels α and β, respectively, i.e., $g_\alpha = (2I_a+1)(2I_A+1)$ and $g_\beta = (2I_b+1)(2I_B+1)$.

5.5.4. *The Hauser–Feshbach theory*

The Weisskopf–Ewing theory depends only on the nuclear level density and on the compound nucleus formation cross-section which may be easily obtained from optical model potentials. It is thus simple to use, but it has the disadvantage that it does not explicitly consider the conservation of angular momentum and does not give the angular distribution of the emitted particles. This is provided by the *Hauser–Feshbach theory.* This theory takes into account the formation of the compound nucleus in states of different J and parity π. Let us consider the case of a reaction leading from the initial channel c to a final channel c'. Since J and parity are good quantum numbers, the cross-section $\sigma_{cc'}$ is a sum of terms each corresponding to a given J^π

$$\sigma_{cc'} = \sum_{J\pi} \sigma_{cc'}^{J\pi}. \tag{5.140}$$

When one averages over the energy one further assumes that the independence hypothesis holds for each $\sigma_{cc'}^{J\pi}$. Thus,

$$\sigma_{cc'}^{J\pi} = \sigma_{CN}^{J\pi}(c)\frac{\Gamma_{c'}^{J\pi}}{\Gamma^{J\pi}}. \tag{5.141}$$

Repeating the procedure used to obtain the Weisskopf–Ewing expression gives

$$\Gamma_c^{J\pi} \propto g_c k_c^2\, \sigma_{CN}^{J\pi}(c), \tag{5.142}$$

where now $g_c = (2i_c + 1)(2I_c + I)$ and i_c and I_c are, respectively, the projectile and target spin in channel c.

If there is no pre-equilibrium emission, one may identify the compound nucleus formation cross-section

$$\sigma_{CN} = \sum_{J,\pi} \sigma_{CN}^{J\pi} \tag{5.143}$$

with the optical model reaction cross-section

$$\sigma_R = \frac{\pi}{k^2}\sum_l (2l+1)T_l, \tag{5.144}$$

which, if the transmission coefficients $T_l = 1 - |\langle S_l\rangle|^2$ do not depend on J, may also be written as[1]

$$\sigma_{CN} = \frac{\pi}{k^2}\sum_{l=0}^{\infty}\ \sum_{s=|I-i|_{\min}}^{I+i}\ \sum_{J=|l-s|_{\min}}^{l+s} \frac{(2J+1)}{(2i+1)(2I+1)}T_l, \tag{5.145}$$

where $\mathbf{s}$ is the channel spin $\mathbf{s} = \mathbf{i} + \mathbf{I}$. Inverting the order of summation over the various indexes, a change which does not affect the result of the sum, one gets

$$\sigma_{CN} = \frac{\pi}{k^2}\sum_J \frac{(2J+1)}{(2i+1)(2I+1)} \sum_{s=|I-i|_{\min}}^{I+i}\ \sum_{l=|J-s|_{\min}}^{J+s} T_l. \tag{5.146}$$

Thus, comparing (5.146) with (5.143), one may write

$$\sigma_{CN} = \frac{\pi}{k^2}\frac{(2J+1)}{(2i+1)(2I+1)}\sum_{s,l} T_l. \tag{5.147}$$

[1] $\langle S_l\rangle$ is the average value of the scattering amplitude over several overlapping resonances.

Using relations (5.142) and (5.147) one easily gets

$$\frac{\Gamma_c^{J\pi}}{\Gamma^{J\pi}} = \frac{\Gamma_c^{J\pi}}{\sum_c \Gamma_c^{J\pi}} = \frac{\sum_{s,l} T_l(c)}{\sum_c \sum_{s,l} T_l(c)}. \tag{5.148}$$

Thus the cross-section (5.140) for transition from the initial channel c to the final channel c' may be written

$$\sigma_{cc'} = \frac{\pi}{k^2} \sum_J \frac{(2J+1)}{(2i_c+1)(2I_c+1)} \frac{\sum_{s,l} T_l(c) \sum_{s',l'} T_{l'}(c')}{\sum_c \sum_{s,l} T_l(c)}. \tag{5.149}$$

This is the *Hauser–Feshbach* expression for the energy-averaged angle-integrated cross-section of statistical reactions. The compound nucleus states may be both of positive and negative parity. Since parity is conserved, in evaluating (5.149), one must take into account that the parity of compound nucleus states and the parity of the residual nucleus states may impose restrictions on the values of the emitted particle angular momentum. Thus, positive parity compound nucleus states decay to positive parity states of the residual only by even angular momenta and to negative parity residual nucleus states by odd angular momenta.

5.6. Heavy Ion Collisions

For intermediate energy nuclear collisions, $E_{lab} \sim 100-1000$ MeV/nucleon, the nucleons and the products of their collisions can be described individually and their propagation can be calculated by semiclassical equations. Hadronic transport theories have been quite successful in describing a multitude of measured particle spectra. The *nuclear equation of state* (EOS), $\epsilon(\rho, \delta)$, expresses the energy per nucleon of nuclear matter as a function of the nucleon density ρ and the relative neutron excess $\delta = (\rho_n - \rho_p)/\rho$. It is a fundamental quantity in theories of *neutron stars* and *supernova explosions*. The main measured quantities which can provide information about the EOS are binding energies and other data for finite nuclei. As the finite nuclei are in states near the standard nuclear matter state with normal nucleon density $\rho_0 \sim 0.16\,\text{fm}^{-3}$ and zero neutron excess, $\delta = 0$, our knowledge about the EOS can be confirmed experimentally only in a small region around $\rho \sim \rho_0$ and $\delta \sim 0$. With very neutron-rich nuclei and energetic heavy-ion collisions, the nuclear EOS can be tested well beyond this region.

If one assumes that the system of nucleons forms a dilute system (or gas) of particles, such that the total volume of the gas particles is small compared to the volume available to the gas, then $na^3 \ll 1$, where n is the number density of particles and a is the (interaction) radius of a particle. Since the particles in a neutral gas do not have long-range forces like the particles in a plasma, they are assumed to interact only when they collide, i.e., when the separation between two particles is not much larger than $2a$. The term "collision" normally means the interaction between two such nearby particles. A particle moves in a straight line between collisions. The average distance travelled by a particle between two collisions is known as the mean free path. The mean free path depends on the cross-section $\sigma \sim a^2$, and is given by $\lambda = 1/n\sigma$. One consequence of the requirement that the gas be dilute is that $\lambda \gg a$. If the gas is dilute, the probability of three-body collisions is much lower than that for two-body collisions and they can be neglected.

Assuming that these conditions are valid, a practical equation describing nucleus–nucleus collisions in terms of nucleon–nucleon collisions can be deduced. A popular transport equation is the *Boltzmann–Uehling–Uhlenbeck* (BUU) equation,

$$\frac{\partial f}{\partial t} + \left(\frac{\mathbf{p}}{m} + \nabla_{\mathbf{p}} U\right) \cdot \nabla_{\mathbf{r}} f - \nabla_{\mathbf{r}} U \cdot \nabla_{\mathbf{r}} f$$
$$= \int d^3p_2 \int d\Omega \sigma_{NN}(\Omega) |\mathbf{v}_1 - \mathbf{v}_2| \{f_1' f_2' [1 - f_1][1 - f_2]$$
$$- f_1 f_2 [1 - f_1'][1 - f_2']\}. \tag{5.150}$$

When the right-hand side is taken as zero, the resulting equation is known as the *Vlasov equation.*

This equation can be understood as follows. If dN is the number of particles in the volume element d^3r and whose momenta fall in the momentum element d^3p at time t, then the distribution function $f(\mathbf{r}, \mathbf{p}, t)$ is given by $dN = f(\mathbf{r}, \mathbf{p}, t) d^3r d^3p$. Thus the BUU equation is an equation for the distribution function $f(\mathbf{r}, \mathbf{p}, t)$. To account for the effect of each particle interacting with all others, one introduces the concept of *mean-field,* $U(\mathbf{r}, \mathbf{p}, t)$. This mean-field exerts a force on each particle, given by $-\nabla_{\mathbf{r}} U(\mathbf{r}, \mathbf{p}, t)$. Also, the momentum dependence of the potential introduces a dependence through the derivative $-\nabla_{\mathbf{p}} U(\mathbf{r}, \mathbf{p}, t)$.

Due to the nucleon–nucleon collisions, the distribution function within $d^3r d^3p$ can also be modified by nucleons leaving (or entering) this volume. This is taken care by the *collision term,* i.e. the right-hand-side of the BUU equation. σ_{NN} is the nucleon–nucleon differential scattering cross-section, while $\mathbf{v}_1$ and $\mathbf{v}_2$ are the velocities of two colliding nucleons. The first factor inside braces are for collisions populating the volume element and the second term for those depleting it. The factors $(1 - f)$ account for Pauli blocking of final occupied states. The prime and non-prime notations for f denote the distributions before and after the binary collisions. The integrals average over scattering angle and over all collisions within d^3rd^3p. The BUU equation falls in the category of what one calls *quantum transport theories*. Hadronic transport theories have been quite successful in applications, describing a multitude of measured particle spectra.

Eq. (5.150) needs as basic ingredients the mean field U and the cross-section σ_{NN}. However, in practice the simulations are often done with a phenomenological mean field and free nuclear cross-sections. Another important ingredient in the transport theory calculations is the *compressibility K of nuclear matter*, which refers to the second derivative of the compressional energy E with respect to the density,

$$K = k_F^2 \frac{\partial(E/A)}{\partial k_F^2}\bigg|_{k_{F0}} = 9\rho^2 \frac{\partial^2(E/A)}{\partial \rho^2}\bigg|_{\rho_0}. \tag{5.151}$$

This is an important quantity, e.g., for nuclear astrophysics.[2] In fact, the mechanism of supernova explosions is strongly dependent on the value K. Supernova models might or might not lead to explosions depending on the value of K. The central collisions of heavy nuclei are one of the few probes of this quantity in the laboratory. The dependence of the calculations on K follow from the dependence of the mean field potential U on the particle density ρ.

Following an initial interpenetration of projectile and target densities, the NN collisions begin to thermalize matter in the overlap region making the momentum distribution there centered at zero momentum in the c.m. system. The density in the overlap region rises above normal and a disk of excited and compressed matter forms at the center of the system. More and more matter dives into the region with compressed matter that begins to expand in transverse directions. At late stages, when the whole matter is excited, transverse expansion dominates. A further view of the situation is illustrated in Fig. 5.7. The measurement of the matter distribution in these collisions and the comparison with transport theories allows one to

[2]We have defined this quantity in Eq. (6.105) in a slightly different way.

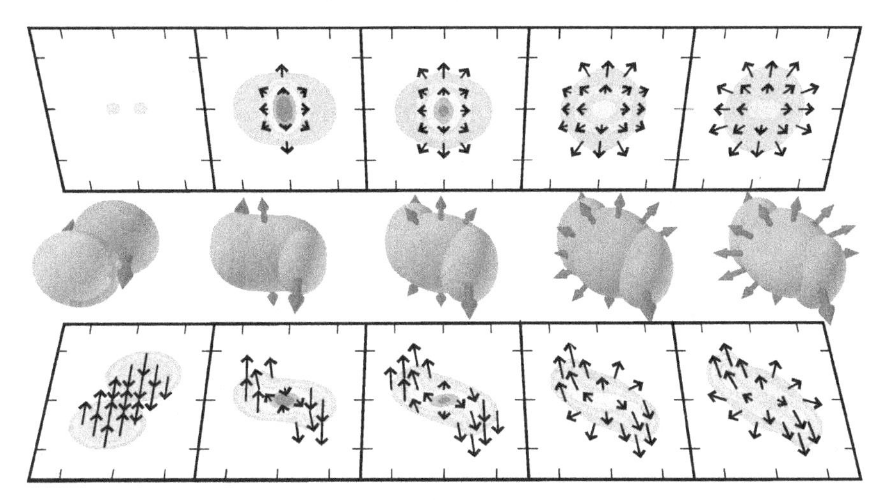

Figure 5.7. Matter distribution at several time stages of a nearly central nucleus–nucleus collision. (Image courtesy of P. Danielewicz).

deduce the incompressibility of nuclear matter. With neutron-rich nuclei, one will also able to extract the EOS dependence on the asymmetry properties of nuclear matter. A valuable contribution by using neutron-rich projectiles arises from measurements of particle production and their kinematic properties.

Using a transport theory based on the BUU equation together with available experimental data, one was able to determine that the maximum pressures attained in central collisions at 2 GeV/nucleon are in the range of $P = 80$ to $130\,\text{MeV/fm}^3$ [DLL02]. This corresponds to 1.3×10^{34} to $2.1 \times 10^{34}\,\text{Pa}$. A similar analysis for collisions at 6 GeV/nucleon yields the respective values of $P = 210$ to $350\,\text{MeV/fm}^3$ (3.4×10^{34} to $5.6 \times 10^{34}\,\text{Pa}$, respectively). These correspond to pressures 23 orders of magnitude larger than the maximum pressure attained in a laboratory, being 19 orders of magnitude larger than pressures within the core of the Sun. They are, in fact, only a comparable to pressures within neutron stars. The analysis seems to be consistent with bounds for K of Eq. (5.151) within the range $K = 170 - 380\,\text{MeV}$ [DLL02]. Figure 5.8 shows the pressure for neutron matter as a function of the density. The different curves are based on theoretical models. The shadow bands display the range of possibilities for the EOS based on the analysis of experiments using a soft (lower shadow region) and stiff (upper shadow region) EOS (adapted from [DLL02]).

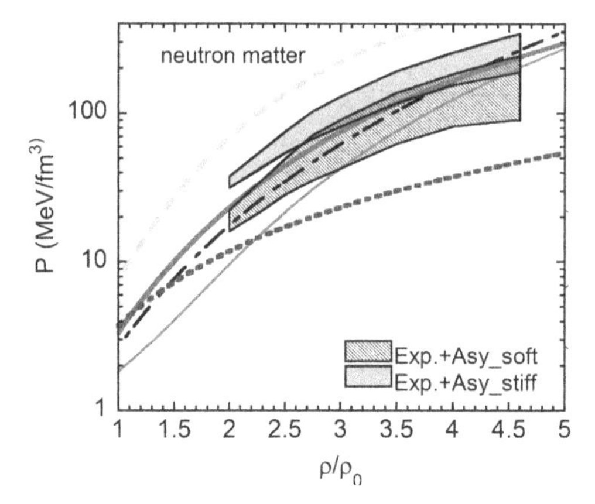

Figure 5.8. Pressure for neutron matter as a function of density. The different curves are based on theoretical models. The shadow bands display the range of possibilities for the EOS based on the analysis of experiments using a soft (lower shadow region) and stiff (upper shadow region) EOS. (Adapted from [DLL02]).

5.7. Relativistic Nuclear Collisions

In relativistic nucleus–nucleus collisions ($E_{lab} \gtrsim 1\,\text{GeV/nucleon}$), *rapidity* is a variable frequently used to describe the behavior of particles in inclusively measured reactions. It is defined by (compare with definition of ξ in the Appendix B)[3]

$$\xi = \frac{1}{2}\ln\left(\frac{E + p_{\parallel}}{E - p_{\parallel}}\right), \tag{5.152}$$

which corresponds to $\tanh\xi = p_{\parallel}/E$, where $p_{\parallel}$ is the longitudinal momentum along the direction of the incident particle and E is the energy, both defined for a given particle. The accessible range of rapidities for a given reaction is determined by the available center-of-mass energy and all participating particles' rest masses. One usually gives the limit for the incident particle, elastically scattered at zero angle:

$$|\xi_{\max}| = \ln\left[\frac{E + p}{m}\right] = \ln(\gamma + \gamma\beta), \tag{5.153}$$

where $\beta \equiv v$ is the velocity, $\gamma = (1 - \beta^2)^{-1/2}$, and all variables referring to the through-going particle given in the desired frame of reference (e.g., in the center of mass).

[3]In this section we use $\hbar = c = 1$, for simplicity.

Note that $\partial\xi/\partial p_{\|} = 1/E$. A Lorentz boost β along the direction of the incident particle adds a constant, $\ln(\gamma + \gamma\beta)$, to the rapidity. Rapidity differences, therefore, are invariant under a Lorentz boost, as discussed in Appendix B. Statistical particle distributions are flat in y for many physics production models. Frequently, the simpler variable *pseudorapidity* η is used instead of rapidity (and sloppy language mixes up the two variables).

The pseudorapidity is a handy variable to approximate the rapidity if the mass and momentum of a particle are not known. It is an angular variable defined by

$$\eta = -\ln\left[\tan\left(\frac{\theta}{2}\right)\right], \tag{5.154}$$

whose inverse function is $\theta = 2\arctan(e^{-\eta})$, where θ is the angle between the particle being considered and the undeflected beam. η is the same as the rapidity y if one sets $\beta = 1$ (or $m = 0$). Statistical distributions plotted in pseudorapidity rather than rapidity undergo transformations that have to be estimated by using a kinematic model for the interaction.

The primary motivation for studying ultra-relativistic heavy ion collisions is to gain an understanding of the equation of state of nuclear matter under extreme circumstances. Displayed in Fig. 5.9 is a schematic phase diagram of nuclear matter. The behavior of nuclear matter as a function of temperature and density (or pressure) is governed by its equation of state.

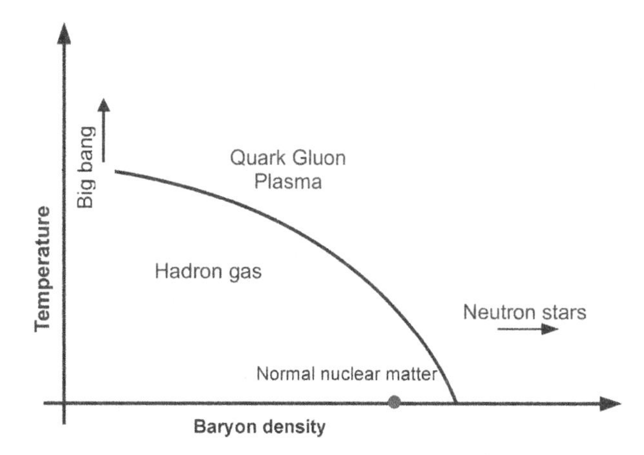

Figure 5.9. As water comes in different phases (solid, liquid, gas), so nuclear matter can come in its normal hadronic form and, at sufficiently high temperature and density, in the form of a deconfined state of quarks and gluons. The diagram shows how nuclear matter should behave as a function of density and temperature.

Conventional nuclear physics is concerned primarily with the lower left portion of the diagram at low temperatures and near normal nuclear matter density. Here normal nuclei exist and at low excitation a liquid–gas phase transition is expected to occur. This is the focus of experimental studies using low energy heavy ions. At somewhat higher excitation, nucleons are excited into baryonic resonance states, along with accompanying particle production and hadronic resonance formation. In relativistic heavy ion collisions, such excitation is expected to create hadronic resonance matter.

We now briefly discuss the QGP signatures in nucleus–nucleus collisions [Won94]. One group of such signatures can be classified as *thermodynamic variables.* This class involves determination of the energy density ϵ, pressure P, and entropy density s of the interacting system as a function of the temperature T and the baryochemical potential μ_B. Experimental observables can be identified with these variables and thus their relative behavior can be determined. If a phase transition to QGP occurs, a rapid rise in the effective number of degrees of freedom, expressed by ϵ/T^4 or s/T^3, should be observed over a small range of T. The variables T, s, and ϵ can be identified with the average transverse momentum $\langle p_T \rangle$, the hadron rapidity density dN/dy, and the transverse energy density dE_T/dy, respectively. The transverse energy produced in the interaction is $E_T = \sum_i E_i \sin\theta_i$, where E_i and θ_i are the kinetic energies of the ejectiles and the emission angles.

Electromagnetic (EM) probes, such as photons and leptons, provide information on the various stages of the interaction without modification by final state interactions. These probes may provide a measure of the thermal radiation from a QGP, if a region of photon energy or equivalently lepton pair invariant mass can be isolated for emission from a QGP relative to other processes. However, the yields for EM probes are small relative to background processes, which are primarily EM decays of hadrons and resonances. Lepton pairs from the QGP are expected to be identifiable in the 1–10 GeV invariant mass range. The widths and positions of the ρ, ω, and ϕ peaks in the lepton pair invariant mass spectrum are expected to be sensitive to medium-induced changes of the hadronic mass spectrum.

The production of J/Ψ particles in a quark–gluon plasma is predicted to be suppressed. This is a result of the *Debye screening* of a $c\bar{c}$ pair, initially formed in the QGP by fusion of two incident gluons. Less tightly bound excited states of the $c\bar{c}$ system, such as Ψ' and χ_c, are more easily dissociated and will be suppressed even more than the J/Ψ.

A long-standing prediction for a signature of QGP formation is the enhancement of strange hadrons. The production of strange hadrons relative to nonstrange hadrons is suppressed in hadronic reactions. This suppression increases with increasing strangeness content of the hadron. In a QGP the strange quark content is rapidly saturated by $s\bar{s}$ pair production in gluon–gluon reactions, resulting in an enhancement in the production of strange hadrons. Thus, multi-strange baryons and strange antibaryons are predicted to be strongly enhanced when a QGP is formed.

The connection between energy loss of a quark and the color-dielectric polarizability of the medium can be established in analogy with the theory of electromagnetic energy loss. Although radiation is a very efficient energy loss mechanism for relativistic particles, it is strongly suppressed in a dense medium by the *Landau–Pomeranchuk effect* [Won94]. Adding the two contributions, the stopping power of a quark–gluon plasma is predicted to be higher than that of hadronic matter. A quark or *gluon jet* propagating through a dense medium will not only lose energy but will also be deflected. This effect destroys the coplanarity of the two jets from a hard *parton–parton scattering*[4] with the incident beam axis. The angular deflection of the jets also results in an azimuthal asymmetry. The presence of a quark–gluon plasma is also predicted to attenuate the emission of jet pairs opposite to the trigger jet.

5.8. Exercises

1. Complete the reactions

$$\begin{aligned} \mathrm{p}+ &\rightarrow {}^{28}\mathrm{Si} + \mathrm{n} \\ {}^{197}\mathrm{Au} + {}^{12}\mathrm{C} &\rightarrow +\gamma \\ {}^{235}\mathrm{U} + \mathrm{n} &\rightarrow {}^{100}\mathrm{Mo} + +3\mathrm{n} \end{aligned}$$

2. What spin and parities can be expected in ^{20}Ne formed from the reaction $\alpha + {}^{16}$O?
3. Suppose that a meson π^- (spin 0 and negative parity) is captured from the orbit P in a pionic atom, giving rise to the reaction

$$\pi^- + \mathrm{d} \longrightarrow 2\mathrm{n}.$$

Show that the two neutrons should be in a singlet state.

[4]Parton is a generic name for either quarks or gluons.

4. A one-dimensional plane wave propagates freely with energy E until it reaches a thin slab of absorbing material with thickness a, represented by the imaginary potential $V(r) = iW_0$. In this expression, W_0 is a positive constant much smaller than E. Estimate the attenuation factor $\alpha = j_{in}/j_{out}$; where j_{in} and j_{out} are respectively the incident and the emergent currents.
5. When the reaction $\mathrm{p} + {}^{65}\mathrm{Cu} \to {}^{65}\mathrm{Zn} + \mathrm{n}$ is produced by protons of 15 MeV, a continuous energy distribution is observed for the emerging neutrons, in which a sharp peak at the energy $E_n = 5.57\,\mathrm{MeV}$ is superposed, associated with an excitation energy of 7.3 MeV of ^{65}Zn. What models can be used to justify this result?
6. Show that the loss of flux of a localized beam of particles in a region where a complex potential $U(\mathbf{r})$ exists is given by $\nabla \cdot \mathbf{j}(\mathbf{r}) = (2/\hbar)|\Psi(\mathbf{r})|^2 \mathrm{Im}[U(\mathbf{r})]$.
7. Build a reaction in which a projectile of $A = 20$ produces a compound nucleus $^{258}_{104}\mathrm{Rf}^*$. Using the approximate expression for the Coulomb barrier, $B_c = Z_1 Z_2 e^2/1.45(A_1^{1/3} + A_2^{1/3})$, determine the smallest possible excitation energy for the compound nucleus.
8. When the reaction $\mathrm{p} +^{65} \mathrm{Cu} \to^{65} \mathrm{Zn} + \mathrm{n}$ is produced by protons of 15 MeV, a continuous energy distribution is observed for the emerging neutrons, in which a sharp peak at the energy $E_n = 5.57\,\mathrm{MeV}$ is superposed, associated to an excitation energy of 7.3 MeV of ^{65}Zn. What models can be used to justify this result?
9. Show that in the capture of slow neutrons the total cross-section at the resonance energy has the value $\sigma = \lambda\sqrt{g\sigma_e/\pi}$, where σ_e is the scattering cross-section at the same energy and g the statistical factor of the Breit–Wigner formula.
10. ^{109}Ag has a neutron resonance absorption cross-section at 5.1 eV with a peak value of 7600 barn and a width of 0.19 eV. Evaluate the expected value of the compound elastic cross-section at the resonance.
11. The Breit–Wigner formula for the single-level resonant cross-section in nuclear reactions is

$$\sigma_{ab} = \frac{\pi}{k^2} g \frac{\Gamma_a \Gamma_b}{(E - E_0)^2 + \Gamma^2/4},$$

where

$$g = \frac{2J + 1}{(2I_A + 1)(2I_B + 1)},$$

J being the spin of the level, I_A the spin of the incident particle and I_B the spin of the target.
At neutron energies below 0.5 eV the cross-section in ^{235}U ($I_B = 7/2$) is dominated by one resonance ($J = 3$) at a kinetic energy of 0.29 eV with a width of 0.135 eV. There are three channels, which allow the compound state to decay by neutron emission, by photon emission, or by fission. At resonance the contributions to the neutron cross-sections are: i) elastic scattering (resonant) ($\ll$ 1 barn); ii) radiative capture (70 barns); and iii) fission (200 barns). (a) Calculate the partial widths for the three channels. (b) How many fissions per second will there be in a sheet of ^{235}U, of thickness 1 mg/cm^{-2}, traversed normally by a neutron beam of 10^5 per second with a kinetic energy of 0.29 eV? (c) How do you expect the neutron partial width to vary with the energy?

12. Evaluate the A dependence of the neutron s-wave strength function for a square well potential of depth 43 MeV and width 1.3 $A^{1/3}$ fm.
13. From the relations

$$\frac{1}{T} = \frac{dS}{dE}, \quad \text{and} \quad \rho(E) = \frac{\exp[S(U)]}{(2\pi T^2 dE/dT)^{1/2}},$$

and assuming that $U = aT^n$, show that

$$\rho(E) \propto \frac{E^{-1}}{(1+1/n)} \exp\left[\frac{n}{n-1} a^{1/n} E^{1-1/n}\right].$$

(Hint: Assume that E can be represented as a power series in T and $dE/dT \longrightarrow 0$, as $T \longrightarrow 0$, then if follows that the expansion must start with a term at best quadratic in T.).

14. From the statistical model show that

$$\frac{\Gamma_p}{\Gamma_n} = \frac{\int E_p \sigma(E_p) w(U_p) dE_p}{\int E_n \sigma(E_n) w(U_n) dE_n}.$$

15. Suppose that in an elastic scattering experiment between two structureless particles the center-of-mass differential cross-section may be represented by

$$\frac{d\sigma}{d\Omega} = A + BP_1(\cos\theta) + CP_2(\cos\theta) + \cdots$$

Express the coefficients A, B and C in terms of the phase shifts δ_l.

16. Show that in a compound nuclear reaction, the angular distribution of emitted particles with respect to the direction of incidence is symmetric about 90°.
17. There are several methods to associate a temperature with a nucleus in an ideal gas law statistically, such as
$$E = aT^2, \quad \text{and} \quad E = \frac{1}{11}AT^2 - T + \frac{1}{8}A^{2/3}T^{7/3},$$
where A is the mass number of the nucleus [LeC54]. Compare these using $A = 25$.
18. Prove that in the black-nucleus approximation the inverse cross-section for emission of a charged particle is given by $\sigma \approx \pi R^2(1 - V_C/E)$, where V_C is the particle Coulomb barrier. Say whether E is the laboratory particle energy or its center-of-mass energy (the energy of the particle relative to the nucleus). Give an expression for the maximum value of the impact parameter and the compound nucleus angular momentum.
19. Assume the interaction potential between the neutron and proton to be exponential, namely $V = V_0 e^{-r/r_0}$ where V_0 and r_0 are respectively the well depth and range of the nuclear potential. (a) Write down the Schrödinger equation (in the center-of-mass system) for the ground state of the deuteron for the case in which the angular momentum $l = 0$. (b) Let $x = e^{-r/r_0}$ and $\psi(r) = u(r)/r$. Show that the Schrödinger equation gives a Bessel function. Write down the general solution of this equation. (c) Applying the boundary conditions (ψ = finite at $r = 0$ and ∞), determine the relation between V_0 and r_0.
20. Evaluate the average angular momentum carried by neutrons, protons and α-particles emitted by ^{135}Tb at 22 MeV excitation energy. In this and in the following problems, when necessary, assume the nuclear temperatures to be approximately given by $T \approx \sqrt{8E/A}$ MeV and the radii for evaluating the Coulomb barriers are approximately given by $R \approx 1.5(A_R^{1/3} + A_p^{1/3})$, where A_R and A_p are the residual and the emitted particle mass.
21. Evaluate the mean energy of neutrons, protons and α-particles emitted by ^{55}V, ^{140}Cs and ^{227}Ra at 18 MeV excitation energy.
22. Assuming that the emission of neutrons, protons and α-particles are the dominant decay modes of the compound nuclei of the previous exercise, evaluate the compound nucleus total and partial decay widths.

Chapter 6

Stellar Structure

6.1. Introduction

Stars are held together by gravitation, an attraction exerted on each part of the star by all other parts, and their collapse is resisted by their internal thermal pressure. These two forces play the principal role in determining stellar structure and they must be (at least almost) in balance. Stars continually radiate energy into space. Thus, if thermal properties of stars are constant over a long period of time, a continual energy source must exist within the stars.

A theory for stars must describe the origin of the energy and its transport to surface. One can make two fundamental assumptions: a) neglect the rate of change of properties —assumed constant with time, and b) all stars are spherical and symmetric about their centers. Such assumptions lead to four basic equations to describe structure. All physical quantities depend upon the distance from the center of the star alone. These equations are:

1. equation of hydrostatic equilibrium: at each radius, r, forces due to pressure differences balance gravity;
2. conservation of mass;
3. conservation of energy: at each radius, r, the change in the energy flux to a volume of the star is equal to the local rate of energy release within this volume;
4. equation of energy transport: is a relation between the energy flux and the local gradient of temperature.

These basic equations are supplemented with

1. equation of state (EOS), which is the pressure of a gas as a function of its density and temperature;

2. opacity: i.e., how opaque the gas is to the radiation field;
3. nuclear energy generation rate.

It is the purpose of this chapter to deduce and discuss the properties of these equations and their application to stellar evolution.

6.2. Hydrostatic Support

The balance between gravity and internal pressure is known as *hydrostatic equilibrium*. Consider a mass element (see Fig. 6.1)

$$\delta m(r) = \rho(r)\delta s \delta r, \tag{6.1}$$

where $\rho(r)$ = density at distance r from the center. δs is the area size perpendicular to the radial coordinate and δr is the side of the volume along the radial coordinate. Now, consider the forces acting in the radial direction. They are, (1) the outward force: pressure exerted by stellar material on the lower face,

$$P(r)\delta s, \tag{6.2}$$

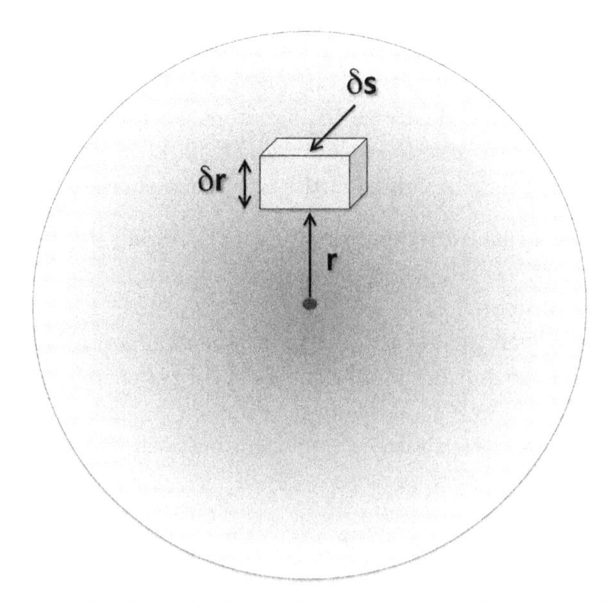

Figure 6.1. A small volume (evidently, highly exaggerated) element of a star at a distance r from its center.

and (2) the inward force due to the pressure exerted by the stellar material on the upper face, and (3) gravitational attraction of all stellar material lying within r. If they are in equilibrium, the total force is zero, hencc

$$P(r+\delta r)\delta s + \frac{GM(r)}{r^2}\delta m = P(r)\delta s, \tag{6.3}$$

or, since $\delta m = \rho \delta r \delta s$,

$$P(r+\delta r) - P(r) = -\frac{GM(r)}{r^2}\rho(r)\delta r. \tag{6.4}$$

If we consider an infinitesimal element, we can write

$$\frac{P(r+\delta r) - P(r)}{\delta r} \equiv \frac{dP(r)}{dr} = -\frac{GM(r)}{r^2}\rho(r), \tag{6.5}$$

which is the equation of *hydrostatic support.*

6.2.1. *Mass conservation*

The mass $M(r)$ contained within a spherical volume of radius r within the star is determined by the density of the gas $\rho(r)$. A thin shell inside the star with radius r and outer radius $r+\delta r$ has a volume $\delta V = 4\pi r^2 \delta r$ which contains as mass given by (see Fig. 6.2)

$$\delta M = \delta V \rho(r) = 4\pi r^2 \rho(r) \delta r, \tag{6.6}$$

which in the infinitesimal limit leads to the *equation of mass conservation*,

$$\frac{dM(r)}{dr} = 4\pi r^2 \rho(r). \tag{6.7}$$

Equations (6.5) and (6.7) are coupled. They assume that the gravity and pressure forces are balanced — but how valid is this assumption? In order to answer that, consider the case where the outward and inward forces are not equal. There will be a resultant force acting on the element which will give rise to an acceleration a. This force is

$$P(r+\delta r)\delta s + \frac{GM(r)}{r^2}\rho(r)\delta s \delta r - P(r)\delta s = a\rho(r)\delta s \delta r, \tag{6.8}$$

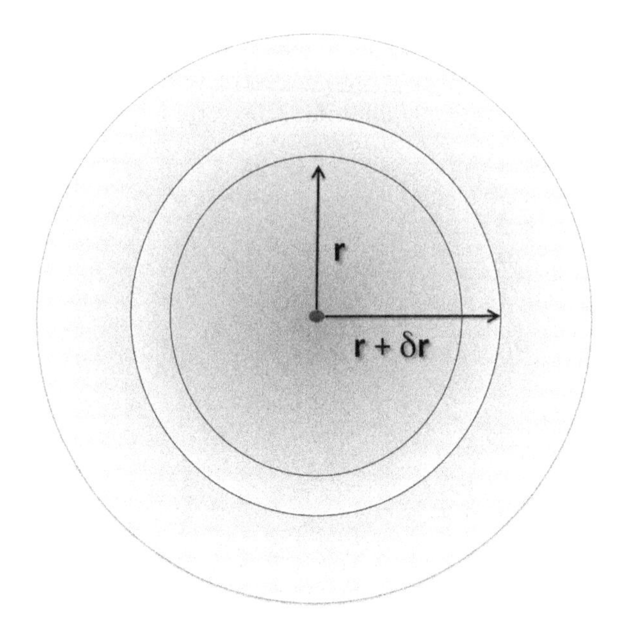

Figure 6.2. A spherical shell within a star.

or

$$\frac{dP(r)}{dr} + \frac{GM(r)}{r^2}\rho(r) = a\rho(r). \tag{6.9}$$

Now the local acceleration at position r due to gravity is $g = GM(r)/r^2$. Thus,

$$\frac{dP(r)}{dr} + g\rho(r) = \rho(r)a \tag{6.10}$$

which is the generalized form of the equation of hydrostatic support.

6.2.2. *Dynamical timescales and non-spherical symmetry*

The accuracy of the hydrostatic assumption can be tested by considering a resultant force on the element, i.e., the left-hand-side (LHS) of Eq. (6.10) different from zero. Suppose that their sum is small fraction β of the gravitational term, $\beta\rho(r)g = \rho(r)a$. Hence there is an inward acceleration of $a = \beta g$. Assuming that an accelerated volume element begins at rest, the spatial displacement d after a time t is

$$d = \frac{1}{2}at^2 = \frac{1}{2}\beta g t^2. \tag{6.11}$$

If we allow the star to collapse, i.e., if we set $d = r$, substitute $g = GM/r^2$, and assume $\beta = 1$, we get

$$t_d \approx \sqrt{\frac{2r^3}{GM}}, \tag{6.12}$$

where t_d is known as the *dynamical time.*

Stars are rotating gaseous bodies. So, they must be flattened at the poles. If so, departures from spherical symmetry must be accounted for. Consider a mass element δm near the surface of a star with mass M and radius r (see Fig. 6.3). This mass element will be acted upon by an additional inwardly acting force to provide circular motion. The centripetal force is given by $\delta m\omega^2 r$, where ω is the angular velocity of star. There will be no departure from spherical symmetry provided that

$$\frac{\delta m\omega^2 r}{GM\delta m/r^2} \ll 1, \quad \text{or} \quad \omega^2 \ll \frac{GM}{r^3}. \tag{6.13}$$

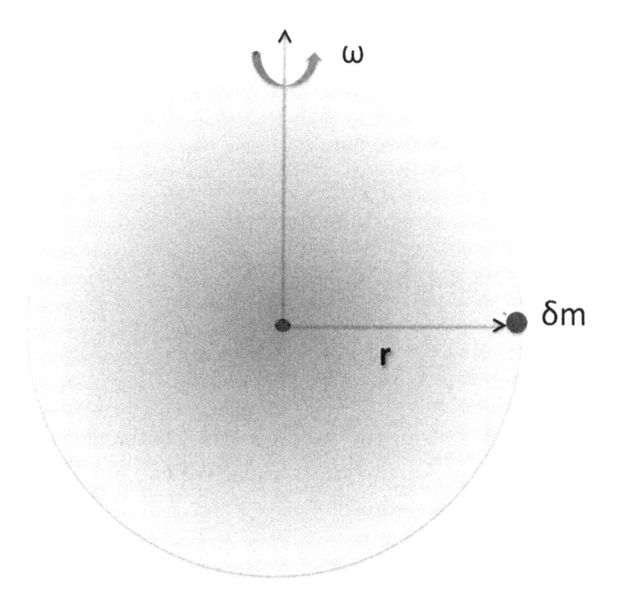

Figure 6.3. A small mass element, δm, at the equator of a rotating star.

Note that the above condition can be related to the dynamical time, Eq. (6.12). In fact, $GM/r^3 = 2/t_d^2$, and we can rewrite the condition for spherical symmetry as

$$\omega^2 \ll \frac{1}{t_d^2}. \tag{6.14}$$

Since the angular velocity is related to the rotation period, T, by means of $\omega = 2\pi/T$, spherical symmetry holds if $T \gg t_d$, i.e., if the period for rotation is much larger than the free-fall time.

As an example, for the Sun, we obtain $t_d(Sun) \sim 2000$ s, and $T \sim 1$ month. Indeed, for the majority of the stars, departures from the spherical symmetry can be ignored. But some stars do rotate rapidly and rotational effects must be included in the structure equations, which can change the output of the models.

6.2.3. *Pressure, temperature and virial theorem*

We have obtained only 2 of the 4 equations for stellar structure, and no knowledge yet of material composition or physical state. But we can already deduce very important results such as a minimum central pressure in the star. Why, in principle, there is need of a minimum pressure value? Given what we know, what is this likely to depend upon?

Minimum central pressure

Dividing Eqs. (6.5) and (6.7) we get

$$\frac{dP}{dM} = -\frac{GM}{4\pi r^4}, \tag{6.15}$$

and then we can integrate it to obtain

$$P_c - P_s = \int_0^{M_s} \frac{GM}{4\pi r^4} dM, \tag{6.16}$$

where the labels c and s mean the values at the center and at the surface of the star. Here we use R for the star radius. We can get a lower estimate of the right-hand side of this equation,

$$\int_0^{M_s} \frac{GM}{4\pi r^4} dM > \int_0^{M_s} \frac{GM}{4\pi R^4} dM = \frac{GM_s^2}{8\pi R^4}. \tag{6.17}$$

Hence we have

$$P_c - P_s > \frac{GM_s^2}{8\pi R^4}. \tag{6.18}$$

We can approximate the pressure at the surface of the star to be zero, leading to

$$P_c > \frac{GM_s^2}{8\pi R^4}. \tag{6.19}$$

For example, for the Sun,

$$P_\odot \geq \frac{GM_\odot^2}{8\pi R_\odot^4} = 4.5 \times 10^{13}\,\mathrm{N/m^2} = 4.5 \times 10^8 \text{ atmospheres}. \tag{6.20}$$

This seems rather large for a gaseous material. But we shall see later that this is not an ordinary gas.

Virial theorem

From Eq. (6.15) we obtain

$$4\pi r^3 dP = -\frac{GM}{r} dM. \tag{6.21}$$

Integrating over the whole star,

$$3\int_{P_c}^{P_s} V dP = -\int_0^{M_s} \frac{GM}{r} dM, \tag{6.22}$$

where V is the volume contained within radius r. Integrating the left-hand side by parts, and using $V_c = 0$ and $P_s = 0$, one obtains

$$3\int_0^{V_s} P dV - \int_0^{M_s} \frac{GM}{r} dM = 0. \tag{6.23}$$

Now the second term is the total gravitational potential energy of the star. It is the energy released in forming the star from its components dispersed at infinity. Thus, we can write the *virial theorem* as

$$3\int_0^{V_s} P dV + \Omega = 0. \tag{6.24}$$

This is of great importance in astrophysics and has many applications. We shall see that it relates the gravitational energy of a star to its thermal energy.

Mean temperature and source of pressure within a star

We have seen that pressure, P, is an important term in the equation of hydrostatic equilibrium and the virial theorem. We have derived a minimum value for the central pressure (for the Sun, $P_c > 4.5 \times 10^8$ atmospheres). But what physical processes give rise to this pressure and which are the most important? Is it the gas pressure, P_g, or the radiation pressure P_r? We shall show that P_r is negligible in stellar interiors and that the pressure is dominated by P_g. To do this, we first need to estimate the minimum mean temperature of a star.

Consider the Ω term, which is the gravitational potential energy:

$$\Omega = -\int_0^{M_s} \frac{GM}{r} dM. \tag{6.25}$$

We can obtain a lower bound on the right-hand side by noting that at all points inside the star $r < R$ and hence $1/r > 1/R$. Thus,

$$\int_0^{M_s} \frac{GM}{r} dM > \int_0^{M_s} \frac{GM}{R} dM = \frac{GM_s^2}{2R}. \tag{6.26}$$

Since $dM = \rho dV$, the virial theorem can be written as

$$-\Omega = 3\int_0^{V_s} PdV = 3\int_0^{M_s} \frac{P}{\rho} dM. \tag{6.27}$$

Pressure is the sum of radiation pressure and gas pressure, $P = P_g + P_r$. Assuming, for now, that stars are composed of ideal gas with negligible P_r, the relation between pressure and temperature is

$$P = nkT = \frac{k\rho T}{m}, \tag{6.28}$$

where n is the number of particles per cubic meter, m is the average mass of particles, and k is Boltzmann's constant. Hence we have

$$-\Omega = 3\int_0^{M_s} \frac{P}{\rho} dM = 3\int_0^{M_s} \frac{kT}{m} dM > \frac{GM_s^2}{2R}, \tag{6.29}$$

or,

$$\int_0^{M_s} TdM > \frac{GM_s^2 m}{6kR}. \tag{6.30}$$

We can think of the left-hand side as the sum of the temperatures of all the mass elements dM which make up the star. The mean temperature of the star, $\bar{T}$, is then just the integral divided by the total mass of the star, M_s. That is, $M_s\bar{T} = \int_0^{M_s} TdM$, or

$$\bar{T} > \frac{GM_s m}{6kR}. \tag{6.31}$$

As an example, for the Sun we have $\bar{T} > 4 \times 10^6 (m/m_H)$ K, where the mass of the hydrogen atom is $m_H = 1.67 \times 10^{-27}$ kg. We know that H is the most abundant element in stars and for a fully ionized hydrogen star $m/m_H = 1/2$ (as there are two particles, p^+ and e^-, for each H atom, and $m = (m_e + m_H)/2$). And for any other element m/m_H is greater than 1/2. We thus have $\bar{T}_\odot > 2 \times 10^6$ K.

We can also estimate the mean density of the Sun using

$$\rho_{av} = \frac{3M_\odot}{4\pi R_\odot^3} = 1.4 \times 10^3 \,\text{kg/m}^3. \tag{6.32}$$

Thus, the mean density of the Sun is only a little higher than water and other ordinary liquids. We know such liquids become gaseous at T much lower than $\bar{T}_\odot$. Also, the average kinetic energy of particles at $\bar{T}_\odot$ is much higher than the ionization potential of H. Thus the gas must be highly ionized, i.e., it is a plasma. It can thus withstand greater compression without deviating from an ideal gas. Note that an ideal gas demands that the distances between the particles are much greater than their sizes, and the nuclear dimension is 10^{-15} m compared to atomic dimension of 10^{-10} m.

Let us consider the issue of radiation versus gas pressure. We assumed that the radiation pressure was negligible. The pressure exerted by photons on the particles in a gas is, from Eq. (4.71),

$$P_{rad} = \frac{aT^4}{3}, \tag{6.33}$$

where a is the radiation density constant. If we compare gas and radiation pressure at a typical point in the Sun, we get

$$\frac{P_r}{P_g} = \frac{aT^4/3}{kT\rho/m} = \frac{maT^3}{3k\rho}. \tag{6.34}$$

Taking $T \sim T_{av} = 2 \times 10^6$ K, $\rho \sim \rho_{av} = 1.4 \times 10^3 \,\text{kg/m}^3$, and $m = 1.67 \times 10^{-27}/2$ kg, gives $P_r/P_s \sim 10^{-4}$. Therefore, the radiation pressure appears to be negligible at a typical (average) point in the Sun.

In summary, with no knowledge of how energy is generated in stars we have been able to derive a value for the Sun's internal temperature and deduce that it is composed of a near ideal gas plasma with negligible radiation pressure. However, P_r does become significant in higher mass stars. To give a basic idea of this dependency, replace ρ in the ratio equation above. We get

$$\frac{P_r}{P_g} = \frac{maT^4}{3kT(3M_s/4\pi R^3)} = \frac{4\pi ma}{9k}\left(\frac{R^3T^3}{M_s}\right). \tag{6.35}$$

From the virial theorem, $\bar{T} \sim M_s/R$, and we get from the above equation $P_r/P_g \propto M_s^2$. This shows that P_r becomes more significant in higher mass stars.

6.2.4. *Energy generation in stars*

So far we have only considered the dynamical properties of the star, and the state of the stellar material. We need to consider the source of the stellar energy. Let us consider the origin of the energy, i.e., the conversion of energy from some form in which it is not immediately available into some form that it can radiate. For example, how much energy does the Sun need to generate in order to shine with its measured flux? The luminosity of the Sun is $L_\odot = 4 \times 10^{26}$ J/s. The Sun has not changed its flux in 10^9 yr $= 3 \times 10^{15}$ s and during this time the Sun radiated 1.2×10^{43} J of energy. The mass lost during this time is $m_{lost} = 1.2 \times 10^{43}\,\text{J/c}^2 = 10^{-4} M_\odot$.

Source of energy generation

What is the source of the energy generation in stars? There are four possibilities: (a) cooling, (b) contraction, (c) chemical reactions, and (d) nuclear reactions.

Cooling and contraction are closely related, so lets consider them together. Cooling is the simplest idea of all. Suppose that the radiative energy of the Sun is due to the Sun being much hotter when it was formed, and has since been cooling down. We can test how plausible this is. Or is the Sun slowly contracting with consequent release of gravitational potential energy, which is converted to radiation?

In an ideal gas, the thermal energy of a particle (where n_f is the number of degrees of freedom $= 3$) is given by $(kT/2)n_f = 3kT/2$. The total thermal energy per unit volume is $3knT/2$, where n is the number of particles per unit volume. Now, according to the virial theorem $3\int_0^{V_s} PdV + \Omega = 0$. Assuming that the stellar material is an ideal gas (negligible P_r), and using

$P = nkT$, we get

$$3\int_0^{V_s} nkTdV + \Omega = 0. \tag{6.36}$$

If we define U as the integral over volume of the thermal energy per unit volume $(3knT/2)$, we get

$$2U + \Omega = 0. \tag{6.37}$$

That is, the negative gravitational energy of a star is equal to twice its thermal energy. This means that the time for which the present thermal energy of the Sun can supply its radiation and the time for which the past release of gravitational potential energy could have supplied its present rate of radiation differ by only a factor two. We can estimate the latter by recalling that negative gravitational potential energy of a star is related by the inequality $-\Omega > GM_s^2/2R$. As an approximation, we assume $-\Omega \sim GM_s^2/2R$. Then the total release of gravitational potential energy, that would have been sufficient to provide radiation energy at a rate given by the luminosity of the star L_s, occurs during a time

$$t_{th} \sim \frac{GM_s^2}{L_s R}. \tag{6.38}$$

t_{th} is known as the *thermal timescale* (or Kelvin–Helmholtz timescale). For the Sun this time would be $t_{\odot th} = 3 \times 10^7$ yrs.

We thus have shown that if Sun where powered by either contraction or cooling, it would have changed substantially in the last 10 million years, a factor of $\sim$100 too short to account for the constraints on age of the Sun imposed by fossil and geological records.

We can quickly rule out *chemical reactions* as possible energy sources for the Sun. We calculated above that we need to find a process that can produce at least 10^{-4} of the rest mass energy of the Sun. Chemical reactions such as the combustion of fossil fuels release $\sim 5 \times 10^{-10}$ of the rest mass energy of the fuel.

We conclude that the only known way of producing sufficiently large amounts of energy is through *nuclear reactions.* Two types of nuclear reactions are important for energy generation: fission and fusion. Fission reactions, such as those that occur in nuclear reactors or atomic weapons, can release $\sim 5 \times 10^{-4}$ of rest mass energy through fission of heavy nuclei (uranium or plutonium). Thus we can see that both fusion and fission could in principle power the Sun. Which is the more likely? As light elements are

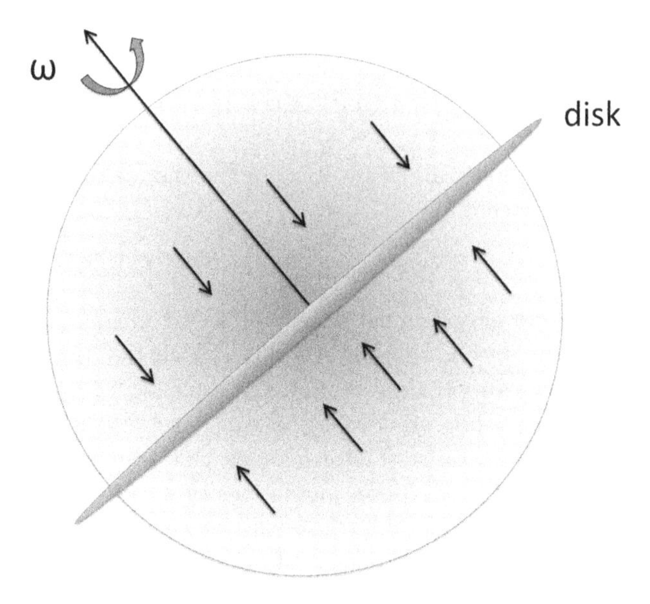

Figure 6.4. A fast rotating gas collapses to a disk.

much more abundant in the solar system than heavy ones, we expect that nuclear fusion is the dominant source.

6.2.5. *Galaxy and star formation*

Gaseous systems with a common angular momentum will collapse into a rotating disk to minimize internal friction, as shown in the Fig. 6.4. According to Eq. (6.12), the collapse time is

$$t_{coll} = \left(\frac{3}{2\pi G\rho}\right)^{1/2}. \tag{6.39}$$

Using the average density of our galaxy this yields $4 \times 10^{15}\,\mathrm{s} \sim 0.1$ Gy. Therefore a pretty quick formation of galaxies is possible, if initial fluctuations allow for it.

A gaseous cloud becomes unstable when the magnitude of its gravitational energy surpasses its thermal energy

$$E_G = \frac{GM^2}{R} > E_T = \frac{3}{2}kT\frac{M}{\mu}, \tag{6.40}$$

with M being the mass of the cloud, and μ its average molecular weight (defined later in this chapter). Taking $M = \rho 4\pi R^3/3$, and setting $E_G = E_T$ in Eq. (6.40) leads to the *Jeans mass*,

$$\begin{aligned} M_J &= \left(\frac{3}{4\pi}\right)^{1/2} \left(\frac{3kT}{2\mu G}\right)^{3/2} \rho^{-1/2} \\ &= 0.396 \times 10^{23}\,\mathrm{g} \left(\frac{T[\mathrm{K}]}{\mu[\mathrm{u}]}\right)^{3/2} \rho^{-1/2}[\mathrm{g/cm^3}]. \end{aligned} \tag{6.41}$$

We can also solve for the radius R, leading to the *Jeans radius*,

$$R_J = \left(\frac{9}{8\pi}\right)^{1/2} \left(\frac{kT}{\rho\mu G}\right)^{3/2} = 2.113 \times 10^7\,\mathrm{cm} \left(\frac{T[\mathrm{K}]}{\rho[\mathrm{g/cm^3}]\mu[\mathrm{u}]}\right)^{3/2}. \tag{6.42}$$

It is believed that stars form preferably in what is called a "molecular cloud". Inside a molecular cloud one finds about $T = 10\,\mathrm{K}$, $\mu = 2.5(\mathrm{H}^2)$. For $M = M_\odot$ this leads to $\rho = 5.4 \times 10^{-19}\,\mathrm{g/cm^3} \sim 1.3 \times 10^5$ atoms/cm^3 and $R_J = 9.6 \times 10^{16} \sim 0.1$ ly. This requires a compression of about 10^5 above the medium galactic density. The collapse time from Eq. (6.39) is about 40,000 years. However, in reality its more likely 10^6 years, as (a) the Jeans density has to be reached, and (b) the opacity of the cloud does not allow for effective energy loss by radiation. Indeed, a shockwave compression mechanism is necessary to produce an effective collapse. Sources of shock waves are e.g., supernova explosions.

One problem in modeling star formation is angular momentum. The angular momentum of a system of objects can only be changed by external torques. Any initial angular momentum of a cloud will lead to a fast spin up as the radius of the cloud decreases, and a disk is formed, as shown in Fig. 6.4. During the formation of a compact star (i.e., a neutron star), as the star collapses, its moment of inertia decreases dramatically but its spin angular momentum is conserved. If a solar-size star with a 100 day spin period collapses into a neutron star, its spin period will become about 1 ms. A large collapsing rotating cloud might also fission and produce binary or multiple star systems, which are about 50% of all star systems. As stars are born in these dense clouds the early history of a star can only be observed in the infrared. The star will have normally reached a stage close to main sequence, before the cloud clears.

6.3. Equation of State

Given the limits on $P(r)$ and $T(r)$ that we have just obtained we can ask if the central conditions are suitable for fusion. We will return to this later, but for now let's explore the consequences of a particular assumption for the pressure dependence on the density, *the equation of state*, $P(\rho)$. We will consider the pressure arising from classical gases, gases with a mixtures of particle species and quantum fluids.

6.3.1. *Classical gases with multiple particle species*

A plasma usually contains multiple types of particles, as a mixture of hydrogen and helium, for example, or a fully ionized gas, where the ions and electrons move separately, and obviously their masses are quite different. For a classical gas, the kinetic description makes the result obvious: each species follows the Boltzmann distribution, and the sum of the momentum transferred to a surface is simply the sum of the momenta transferred by the particles of each species, each of which is given by nkT. Thus, if we have N species present in the gas, then the total pressure is simply

$$P = \left(\sum_i^N n_i \right) kT, \tag{6.43}$$

where n_i is the number of particles of species i per unit volume. We can write this equivalently in terms of the mass fraction and mean mass. For distinction, in this section we denote mass and charge numbers of a nuclear species by $\mathcal{A}_i$ and $\mathcal{Z}_i$, respectively. If we let $\mathcal{A}_i m_H$ be the mass per particle of species i and X_i be the fraction of the mass (or *mass fraction*) at a given point that comes from species i, then, as before, we have

$$n_i = \frac{X_i}{\mathcal{A}_i m_H} \rho, \tag{6.44}$$

and therefore we can write the pressure as

$$P = \left(\sum_i^N \frac{X_i}{\mathcal{A}_i} \right) \rho \mathcal{R} T, \tag{6.45}$$

where $\mathcal{R} = k/m_H$ is the gas constant and ρ is the mass density.

For convenience we define

$$\frac{1}{\mu} = \sum_i^N \frac{X_i}{\mathcal{A}_i}, \tag{6.46}$$

where μ is the *mean mass* (measured in units of the hydrogen mass) per particle, so that the ideal gas law becomes[1]

$$P = \frac{\mathcal{R}}{\mu}\rho T. \tag{6.47}$$

If we only include ions (not electrons) in the sum, then we get the pressure due to ions alone, and we write μ in this case as μ_I, for the mean mass per particle of ions. Since the Sun is mostly hydrogen and helium, it is convenient to express its composition in terms of the fraction of the mass that is hydrogen, the fraction that is helium, and the fraction that is everything else — the everything else we call metals. To an astronomer, carbon, oxygen, and neon are all metals! We define X as the hydrogen mass fraction, Y as the helium mass fraction, and Z as the metal mass fraction, i.e., the abundance of everything else . For the Sun, $X = 0.707$, $Y = 0.274$, and $Z = 1 - X - Y = 0.019$.

We can write μ_I in terms of these definitions,

$$\frac{1}{\mu_I} = \frac{X}{1} + \frac{Y}{4} + \frac{Z}{\langle \mathcal{A} \rangle_{\text{metals}}}, \tag{6.48}$$

where $\langle \mathcal{A} \rangle_{\text{metals}}$ is the mean atomic mass of the metals, which is about 20 in the Sun. Thus for the Sun $\mu_I = 1.29$. We can similarly calculate the pressure due to (classical) electrons. In the outer layers of a star where it is cold there are none, but in the stellar interior the gas is close to being fully ionized. Thus there is one free electron per proton. If n_i is the number density of ions of species i, then the number density of electrons is

$$n_e = \sum_i \mathcal{Z}_i n_i = \frac{\rho}{m_H} \sum_i X_i \frac{\mathcal{Z}_i}{\mathcal{A}_i}. \tag{6.49}$$

Again, for convenience we give this sum a name,

$$\frac{1}{\mu_e} = \sum_i^N X_i \frac{\mathcal{Z}_i}{\mathcal{A}_i}. \tag{6.50}$$

The meaning of μ_e is that it is the average number of free electrons per nucleon, meaning per proton or neutron. In terms of our X, Y, and Z

[1] Caution: here μ denotes mean mass and previously we used it for chemical potential. Evidently, these are very different physical quantities.

numbers,

$$\frac{1}{\mu_e} = X + \frac{Y}{2} + Z \left\langle \frac{\mathcal{Z}}{\mathcal{A}} \right\rangle_{\text{metals}}, \tag{6.51}$$

where the term $\langle \mathcal{Z}/\mathcal{A} \rangle_{\text{metals}}$ represents the ratio of electrons (or protons) averaged over all the metal atoms. This is roughly 1/2, so to good approximation

$$\frac{1}{\mu_e} \simeq X + \frac{Y}{2} + \frac{Z}{2} = \frac{1}{2}(X+1), \tag{6.52}$$

since $Z = 1 - X - Y$. Thus for the Sun $\mu_e = 1.17$.

Thus the pressures of the ions and the electrons are $P_I = (\mathcal{R}/\mu_I)\rho T$ and $P_e = (\mathcal{R}/\mu_e)\rho T$, so the total pressure is $P = P_I + P_e = (\mathcal{R}/\mu)\rho T$, where

$$\frac{1}{\mu} = \frac{1}{\mu_I} + \frac{1}{\mu_e}. \tag{6.53}$$

According to Eq. (6.51), a conversion of hydrogen to helium causes μ to increase. Hence, we find that the luminosity is very sensitive to the balance between hydrogen and helium. Thus, the luminosity is sensitive to the initial helium mass fraction. If we knew the helium mass fraction, we could simply use this value to calculate the model luminosity and avoid any uncertainty. Unfortunately, the first excited state of helium is at 20 electron volts and since 1 eV corresponds to 11,000 K in the Boltzmann equation, at 20 eV, corresponding to the mean kinetic energy of particles in the Sun, there is a vanishingly small excitation at temperatures much below 10^4 K. With no atoms in the first excited state, the only absorption lines are from the ground state and are in the far ultraviolet. They cannot be observed in most stars much less used for an mass fraction determination. Accurate helium mass fractions can be obtained for hot stars and for gaseous nebulae but not for solar type stars. In the case of the Sun, helium can be observed directly in the solar wind but there is a problem due to the acceleration process which is known to produce mass fraction anomalies, so that the solar wind helium mass fraction does not help to constrain the solar helium mass fraction. In fact, the problem is usually turned around since the luminosity of the Sun is known to high accuracy, we can use the solar model to compute the solar helium mass fraction. For other stars, the helium to hydrogen ratio has to be treated as an unknown parameter which can be used to adjust the luminosity to match the observations.

Element abundance

It is common in nuclear astrophysics to define

$$Y_i = \frac{X_i}{\mathcal{A}_i}, \tag{6.54}$$

as the *abundance* of element i. Then it is better to change the notation X_i for the mass fraction of species i, avoiding the use of Z and Y, as defined previously. The abundance Y_i is proportional to number density but changes only if the nuclear species gets destroyed or produced. Changes in density are factored out. In terms of the number density

$$n_i = Y_i \frac{\rho}{m_H}, \tag{6.55}$$

and the *mean molecular weight* is

$$\mu = \frac{1}{\sum_i Y_i}. \tag{6.56}$$

Note that, since $\sum_i X_i = 1$, then $\sum_i Y_i < 1$ (as $Y = X/\mathcal{A} < X$).

As matter is usually electrically neutral, for each nucleus with charge number $\mathcal{Z}$ there are $\mathcal{Z}$ electrons, meaning that the electron abundance is

$$Y_e = \sum_i \mathcal{Z}_i Y_i. \tag{6.57}$$

Since the electron density is given by $n_e = Y_e \rho / m_H$, we get

$$Y_e = \frac{\sum_i \mathcal{Z}_i Y_i}{\sum_i \mathcal{A}_i Y_i}. \tag{6.58}$$

The numerator is proportional to the number of protons and the denominator is proportional to the number of nucleons. So Y_e is the ratio of protons to nucleons, counting all protons including the ones contained in nuclei (not just free protons as described by the "proton abundance"). Some special cases are (a) for 100% hydrogen, $Y_e = 1$, (b) for equal number of protons and neutrons ($N = Z$ nuclei), $Y_e = 0.5$, and (c) for pure neutron gas, $Y_e = 0$.

6.3.2. *Non-relativistic Fermi gas*

Above we have discussed the pressure of a system of classical particles. Here we will discuss the effects of quantum mechanics. In particular, we consider a quantum gas made of electrons which, due to the small electron mass, is the main source of quantum pressure at low temperatures and dense stellar environments. The results are easily extended to baryonic fluids, such as those within a neutron star, by replacing the electron mass with the nucleon mass. We have seen in Chapter 3, that the total energy of a degenerate electron gas is $E_{total} = N\bar{E} = 3NE_F/5$. To obtain the desired equation of state, we also need an expression for the pressure. From the first law of thermodynamics, $dE = dQ - PdV$, and the temperature T fixed at $T = 0$ (where $dQ = 0$ because $dT = 0$), we have (using $\mathcal{E} = E_{total}/V$ for the energy density),

$$P = -\left(\frac{\partial E_{total}}{\partial V}\right)_{T=0} = -\left(V\frac{\partial \mathcal{E}}{\partial V} + \mathcal{E}\right) = n_e \frac{d\mathcal{E}}{dn_e} - \mathcal{E} = n_e\mu - \mathcal{E}, \quad (6.59)$$

where the total energy is given by Eq. (3.33). The quantity

$$\mu = \frac{d\mathcal{E}}{dn_e} \quad (6.60)$$

is known as the *chemical potential* of the electrons. For the electron gas at $T = 0$, if we add one more electron to the system, the energy increases by E_F, which is the energy of the lowest available state for the particle. Thus, in this case, $\mu = E_F$.

The gas pressure of a non-relativistic electron gas is easier to calculate from $E_{total} = 3NE_F/5$ and Eq. (3.31). It reads

$$P = -\frac{\partial E_{total}}{\partial V} = \frac{2}{5} n_e E_F = \frac{\pi^3 \hbar^2}{15 m_e}\left(\frac{3n_e}{\pi}\right)^{5/3}. \quad (6.61)$$

since $E_F \propto V^{-2/3}$, from Eq. (3.31). Now, the pressure predicted by classical physics is NkT. A degenerate electron gas has a much higher pressure than predicted by classical physics at $T \sim 0$. This is an entirely quantum mechanical effect, and is due to the fact that identical fermions cannot get significantly closer together than a de Broglie wavelength without violating the Pauli exclusion principle. Note that, according to Eq. (3.30), the mean spacing between degenerate electrons is

$$d \sim n^{-1/3} \sim \frac{h}{\sqrt{m_e E}} \sim \frac{h}{p} \sim \lambda. \quad (6.62)$$

where λ is the de Broglie wavelength. Thus, an electron gas is non-degenerate when the mean spacing between the electrons is much greater than the de Broglie wavelength, and becomes degenerate as the mean spacing approaches the de Broglie wavelength.

The resistance to compression is usually measured in terms of a quantity known as the *bulk modulus*, which is defined as

$$B = -V\frac{\partial P}{\partial V}. \tag{6.63}$$

Now, for a fixed number of electrons, from Eqs. (3.30) and (6.61), $P \sim V^{-5/3}$. Hence,

$$B = \frac{5}{3}P = \frac{\pi^3\hbar^2}{9m_e}\left(\frac{3n_e}{\pi}\right)^{5/3}. \tag{6.64}$$

6.3.3. *Relativistic electron gas*

For a free electron gas at temperature $T = 0$ (lowest energy state), the electrons occupy all energy states up to the Fermi energy. The total density of the star can be calculated adding up the individual electronic energies. Since each phase-space cell $d^3p \cdot V$ (where V is the volume occupied by the electrons) contains $d^3p \cdot V/(2\pi\hbar)^3$ states, we get

$$\frac{E}{V} = 2\int_0^{p_F}\frac{d^3p}{(2\pi\hbar)^3}E(p) = 2\int_0^{p_F}\frac{d^3p}{(2\pi\hbar)^3}\sqrt{p^2c^2+m_e^2c^4} = \frac{m_e^4c^5}{8\pi^2\hbar^3}\epsilon(x), \tag{6.65}$$

where the factor 2 is due to the electron spin,

$$\epsilon(x) = 8x^3[(x^2+1)^{1/2}-1] - \Pi(x), \tag{6.66}$$

where $\Pi(x)$ is given below, and

$$x = \frac{p_F c}{m_e c^2} = \left(\frac{n}{n_0}\right)^{1/3} = \left(\frac{\rho}{\rho_0}\right)^{1/3}, \tag{6.67}$$

where

$$n_0 = \frac{m_e^3c^3}{\hbar^3} \quad \text{and} \quad \rho_0 = \frac{m_N n_0}{Y_e} = 9.79\times 10^5\, Y_e^{-1}\,\text{g/cm}^3. \tag{6.68}$$

In the above relations p_F is the Fermi momentum of the electrons, m_e (m_N) is the electron (nucleon) mass, $n = p_F^3/\hbar^3$ is the density of electrons, and ρ denotes the *mass density* within the star. Y_e is the number of electrons per nucleon.

The variable x characterizes the electron density in terms of

$$n_0 = 5.89 \times 10^{29}\,\text{cm}^{-3}. \tag{6.69}$$

At this density the Fermi momentum is equal to the inverse of the Compton wavelength of the electron, $(\hbar/m_e c)^{-1} = (386\,\text{fm})^{-1}$.

The pressure can be calculated from (6.59), yielding

$$P = -\frac{\partial E}{\partial V} = \frac{m_e^4 c^5}{24\pi^2\hbar^3}\Pi(x), \tag{6.70}$$

where

$$\Pi(x) = (2x^3 - 3x)\sqrt{x^2+1} + 3\sinh^{-1} x. \tag{6.71}$$

For large densities (ultra-relativistic electron gas), $x \gg 1$, and $P \simeq x^4$, i.e., the pressure is proportional to $n^{4/3}$, as mentioned in the previous section.

6.4. Polytropic Models

Some of the adopted equations of state are purely phenomenological, with parameters fitted to adapt to different stellar scenarios. Here we discuss some of these equations.

Lane–Emden equation

A popular equation of state is the *polytropic model*, which assumes that

$$P = \kappa\rho^\Gamma, \tag{6.72}$$

where ρ is the mass density (we now consider any mixture of particles, not only electrons), while κ and Γ are constants adjusted to the proper stellar scenario.

There are many physical situations for which this equation occurs. For example, for a non-relativistic electron gas ($\rho < 10^6\,\text{g/cm}^3$), we have $\Gamma = 5/3$, as given in Eq. (6.61). In terms of the electron fraction Y_e, the electron number density is given by $n_e = Y_e \rho N_A$ (only valid in cgs units, in which the atomic mass unit is $m_u = 1/N_A$), N_A is the Avogadro's number, and

$$\kappa = \frac{\pi^3}{15}\left(\frac{3}{\pi}\right)^{5/3}\frac{\hbar^2(N_A Y_e)^{5/3}}{m_e}. \tag{6.73}$$

For a relativistic electron gas ($\rho > 10^6\,\mathrm{g/cm^3}$), as we show later, $\Gamma = 4/3$, and one has

$$\kappa = \left(\frac{3}{\pi}\right)^{1/3} \frac{\hbar c (N_A Y_e)^{4/3}}{8}. \tag{6.74}$$

With the help of the hydrostatic equilibrium and mass conservation equations, Eqs. (6.5) and (6.7), one gets

$$\frac{1}{r^2}\frac{d}{dr}\left(\frac{r^2}{\rho}\frac{dP}{dr}\right) = -4\pi G\rho. \tag{6.75}$$

Using $P = \kappa\rho^\Gamma$, with $\Gamma = 1 + 1/n$, we obtain

$$\frac{1}{\xi^2}\frac{d}{d\xi}\left(\xi^2\frac{d\theta}{d\xi}\right) = -\theta^n, \tag{6.76}$$

where $\rho = \rho_c\theta^n(\xi)$ and $\xi = ar$, with

$$a = \left[\frac{(n+1)\kappa\rho_c^{\frac{1}{n}-1}}{4\pi G}\right]^{-1/2}. \tag{6.77}$$

Eq. (6.76) is known as the *Lane–Emden equation.* The boundary conditions are $\theta(0) = 1$, and $\theta'(0) = 0$.

In few cases, it is possible to have analytical solutions of the Lane–Emden equation. They are

$$\begin{aligned} \theta(\xi) &= 1 - \frac{\xi^2}{6}, \quad \text{for } n = 0, \\ \theta(\xi) &= \frac{\sin\xi}{\xi}, \quad \text{for } n = 1, \\ \theta(\xi) &= \left(1 + \frac{\xi^2}{3}\right)^{-1/2}, \quad \text{for } n = 5. \end{aligned} \tag{6.78}$$

The solutions have a maximum at $\xi = 0$, or $r = 0$. For $n < 5$, $\xi_n = aR$ is finite at $r = R$ corresponding to the first root of θ_n, i.e., the first point where the solution crosses zero. For $n > 5$, ξ_n is infinite.

It is easy to see from the first two solutions that they go to zero at some point, which one can regard as the boundary of the star. For example,

$$\begin{aligned} 1 - \frac{\xi_n^2}{6} &= 0, \quad \text{if } \xi_n = \sqrt{6} \\ \frac{\sin\xi_n}{\xi_n} &= 0, \quad \text{if } \xi_n = \pi. \end{aligned} \tag{6.79}$$

The two cases most interesting for real stars have $n = 1.5$ and $n = 3$, which correspond to the non-relativistic and relativistic equation of state for an electron gas. Unfortunately, these do not have analytic solutions.

The value of r which corresponds to ξ_n is clearly the radius R of the star. For all values of the polytropic index n it is possible to develop a series solution which is useful for starting many numerical methods for the solution of the Lane–Emden equation. The first few terms in the solution are

$$\theta = 1 - \frac{1}{6}\xi^2 + \frac{n}{120}\xi^4 - \left(\frac{8n^2 - 5n}{42 \times 360}\right)\xi^6 + \cdots \tag{6.80}$$

Central density in polytropic models

The mass of a star can be calculated in polytropic models from

$$m(r) = \int_0^r 4\pi\rho r'^2 dr' = 4\pi\rho_c \frac{r^3}{\xi^3}\int_0^\xi \theta^n \xi'^2 d\xi' = 4\pi\rho_c r^3 \left(-\frac{1}{\xi}\frac{d\theta}{d\xi}\right). \tag{6.81}$$

Since $r/\xi = 1/a = R/\xi_n$, we get

$$M = 4\pi\rho_c R^3 \left(-\frac{1}{\xi}\frac{d\theta}{d\xi}\right)_{\xi=\xi_n}. \tag{6.82}$$

We now define the average density $\bar{\rho} = 3M/4\pi R^3$ and we get

$$\frac{\rho_c}{\bar{\rho}} = \left(-\frac{3}{\xi}\frac{d\theta}{d\xi}\right)^{-1}_{\xi=\xi_n}. \tag{6.83}$$

A few examples of the numerical solution of the Lane–Emden equation is presented in Table 6.1. Figure 6.5 shows a numerical solution for the Lane–Emden equation.

The construction of a polytropic model for a star is thus straightforward:

1. for a given n, solve the Lane–Emden equation to obtain $\theta(\xi)$, $\theta'(\xi)$, ξ_n and $-(3d\theta/\xi d\xi)_{\xi_n}$;
2. use M, R to obtain $\bar{\rho}$ and ρ_c;
3. use $a^{-1} = r/\xi = R/\xi_n$ to adjust the ξ scale to the r scale. Then we know $\rho(r) = \rho_c\theta^n(\xi)$;

Table 6.1. Numerical solutions of the Lane–Emden equation.

n	ξ_n	$\rho_c/\bar{\rho}$
0	2.4494	1.0
1	3.14159	3.28987
2	4.35287	11.40254
3	6.89685	54.1825
4	14.97155	622.408
5	∞	∞

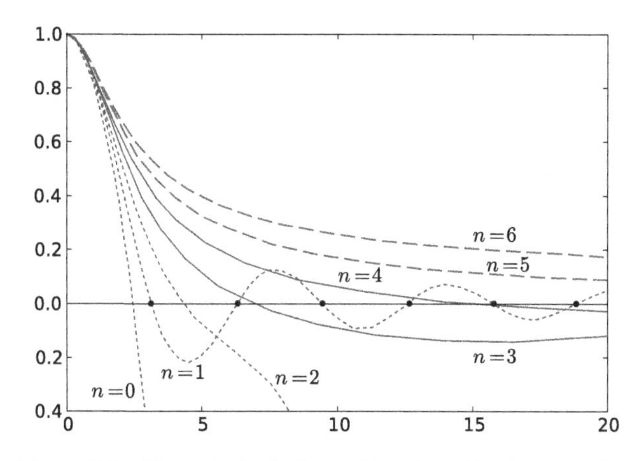

Figure 6.5. Numerical solutions of the Lane–Emden equation.

4. knowing a and ρ_c allows us to determine κ. Then

$$P(r) = \kappa\rho^{(n+1)/n} = \kappa\rho_c^{(n+1)/n}\theta^{n+1}; \tag{6.84}$$

5. get

$$m(r) = 4\pi\rho_c r^3 \left(-\frac{1}{\xi}\frac{d\theta}{d\xi}\right). \tag{6.85}$$

The above procedure assumes that κ is a free parameter. If κ is fixed, one can only construct models for given M or R, and fixed n.

As an application, we consider our Sun, with $M_\odot = 1.989 \times 10^{30}$ kg and $R_\odot = 6.96 \times 10^8$ m, and assume $\Gamma = 4/3$, i.e., $n = 3$. This is the *Eddington's model*. From Table 6.1 we get $\xi_{n=3} = 6.897$ and $\rho_c/\bar{\rho} = 54.18$. This leads to $\rho_c = 76.39\,\mathrm{g/cm^3}$, $\bar{\rho} = 1.41\,\mathrm{g/cm^3}$ and $a^{-1} = R/\xi_3 = 1.01 \times 10^{10}$. From this value of a we get $\kappa = 3.85 \times 10^{14}$ and $P_c = \kappa\rho_c^\Gamma = 1.24 \times 10^{17}$ dyn. cm^{-2}. Assuming a solar composition with mass fraction of hydrogen and helium given by $X_1 = 0.7$ (H) and $X_2 = 0.3$ (He), the *molecular weight* is (see Section 6.4)[2]

$$\mu = \left(\sum_i \frac{X_i(1+\mathcal{Z}_i)}{\mathcal{A}_i}\right)^{-1} = 0.62, \tag{6.86}$$

where $\mathcal{A}_i$ are the mass numbers of the nuclei, $\mathcal{A}_1 = 1$ and $\mathcal{A}_2 = 4$, and $\mathcal{Z}_i$ are their charges, $\mathcal{Z}_1 = 1$ and $\mathcal{Z}_2 = 2$. Using the ideal gas law in the form, $\mu P = \mathcal{R}\rho T$, one obtains $T_c = 1.2 \times 10^7$ K. A detailed calculation based on the Standard Solar Model [Bah89] yields $T_c = 1.57 \times 10^7$ K and $\rho_c = 156\,\mathrm{g/cm^3}$.

Polytropic model with fixed κ and n

We can also construct a stellar model for fixed κ, and n (and ρ_c). From the Lane–Emden equation we get $\theta(\xi)$, $\theta'(\xi)$ and $\rho = \rho_c\theta^n$ is a known function of ξ. One obtains

$$\left(\frac{r}{\xi}\right)^2 = a^{-2} = \frac{(n+1)\kappa\rho_c^{\frac{1}{n}-1}}{4\pi G} = \left(\frac{R}{\xi_n}\right)^2. \tag{6.87}$$

Thus the radius scales with central density as:

$$R \sim \rho_c^{\frac{1-n}{2n}}. \tag{6.88}$$

As long as $n > 1$, the radius becomes smaller with increasing ρ_c.

From Eq. (6.82) follows that

$$M = C_1\rho_c^{\frac{3-n}{2n}}, \quad \text{where } C_1 = 4\pi\left(\frac{-\theta'}{\xi}\right)_{\xi_n}\xi_n^3\left(\frac{n+1}{4\pi G}\right)^{3/2}\kappa^{3/2}. \tag{6.89}$$

Then one has the mass–radius relation

$$R \sim M^{\frac{1-n}{3-n}}, \tag{6.90}$$

which is useful for many estimates.

---

[2]Do not confuse μ used here with the chemical potential.

6.5. Compact Stars

For a few stars, nuclear fusion reactions do not provide enough energy to ensure a thermal pressure which can stop gravitational collapse. In this case, gravitational collapse is prevented by the pressure due to degenerated electron gas. As we discuss below in this section, for stars with masses $\lesssim 1.5\, M_\odot$ (*Chandrasekhar mass*) the internal pressure of the degenerated electron gas (i.e., when the electrons occupy all states allowed by the Pauli principle) does not allow star compression due to the gravitational attraction continuing indefinitely. Neutron stars and white dwarfs are the prototypes of stellar objects balanced by degenerate fermion gases.

The total energy density of a neutron star is

$$\mathcal{E} = \frac{E_{total}}{V} = n_p m_N c^2 \frac{A}{Z} + \frac{E_{total}^{(e)}}{V}, \tag{6.91}$$

where the last term is the total electron energy. The first term is the energy due to the nucleons and nuclei, with n_p equal to the number density of protons, and A/Z equal to the number of nucleons per electron.

White dwarfs

Since the pressure increases with the electron density, which increases with the decreasing volume of the star, we expect that the gravitational collapse stops when the electronic pressure equals the gravitational pressure. When this occurs the star cools slowly and its luminosity decreases. The star becomes a *white dwarf* and in some cases its diameter can become smaller than that of the Moon.

A white dwarf star is a low- or medium-mass star ($M < 8M_\odot$), being the remnant of a star near the end of its lifetime. Much of the mass of a medium mass star is lost by planetary nebula expulsion and stellar winds before a white dwarf is formed. At this stage, it has burned up through nuclear processes most of its hydrogen and helium, forming carbon, oxygen (C/O white dwarfs), neon (C/O/Ne white dwarfs), silicon, or perhaps iron. White dwarfs are dead stars. Typically white dwarfs have mass smaller than 1.4 times that of our Sun, $M_\odot = 1.989 \times 10^{30}$ kg. Some of them are also much smaller than our Sun ($\sim 1/100\ M_\odot$). Their radii are of the order of 10^4 km (note that $R_\odot = 6.96 \times 10^5$ km). In fact, the composition of white dwarfs is not well known. Most are clearly a mixture of carbon and oxygen, but the proportion of these two elements is not well constrained. A few massive white dwarfs in nova systems show evidence of heavy elements;

neon-oxygen-magnesium novae are relatively common. The surface layers of white dwarfs also vary greatly. These differences are reflected in their spectral types. It is an observed fact that the overwhelming majority of white dwarfs have masses near $\sim 0.6 M_\odot$, and there is very little dispersion about this mean.

Neutron stars

For stars with masses $M \sim 1.5 M_\odot$, the electron pressure is not sufficient to balance the gravitational attraction. The density increases to $2 \times 10^{14}\mathrm{g\,cm^{-3}}$ and the matter "neutronizes". This occurs via electron capture by the nuclei (inverse beta decay), transforming protons into neutrons. The final product is a *neutron star*, with a small radius (see Fig. 6.6). For example, if it were possible to form a neutron star from the Sun it would have a radius given by

$$R = \left(\frac{M_\odot}{\frac{4\pi}{3}\rho}\right)^{1/3} = \left(\frac{2 \times 10^{33}\,\mathrm{g}}{\frac{4\pi}{3} \times 2 \times 10^{14}\,\mathrm{g\,cm^{-3}}}\right)^{1/3} \simeq 14\,\mathrm{km}. \tag{6.92}$$

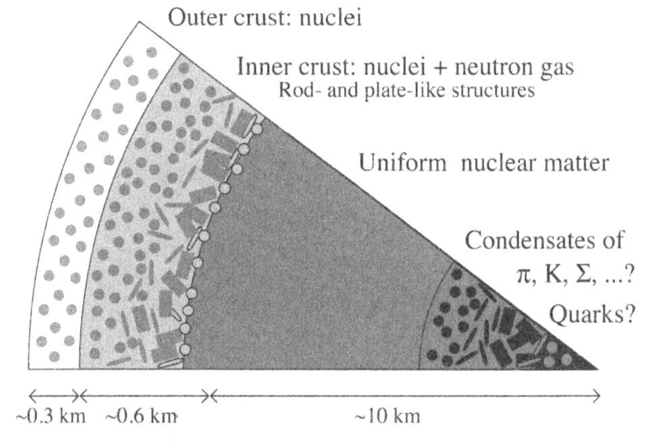

Figure 6.6. The structure of a neutron star. The outer regions of a neutron star may consist of thin layers of various elements that were produced by nuclear reactions during the star's lifetime. These outer layers are thought to have a rigid crystalline structure because of the intense gravitational field of the neutron star. The interior of the star contains a nuclear liquid of neutrons and about 10% protons at densities above nuclear matter density, n_0, increasing as it gets closer to the center. The composition of the inner core is unknown. It might be composed of quarks, or hyperon matter. Pion or kaon condensates might appear. Densities vary from 10^6 g/cm^3 at the crust to $\gtrsim 10^{15}$ g/cm^3 as one approaches the center [Hei02].

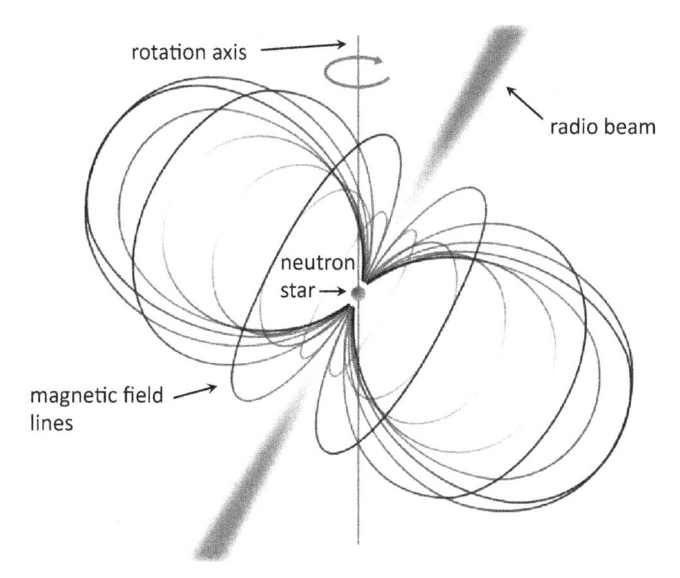

Figure 6.7. A rapidly rotating neutron star, or pulsar. At the magnetic poles, particles can escape and give rise to radio emission. If the magnetic axis is misaligned with the rotation axis of the neutron star (as shown), the star's rotation sweeps the beams over the observer as it rotates like a lighthouse, and one sees regular, sharp pulses of light (optical, radio, x-ray, etc.).

A cubic centimeter of the matter in neutron stars has 100 million tons! The neutron stars are at the limit of density that matter can have, the subsequent step being a black hole.

A *pulsar* is a rapidly rotating neutron star. Like a black hole, it is an endpoint to stellar evolution. The "pulses" of high-energy radiation we see from a pulsar are due to a misalignment of the neutron star's rotation axis and its magnetic poles (see Fig. 6.7). Today we know thousands of pulsars and they are formidable astrophysics laboratories, since: (a) their density is comparable to that of an atomic nucleus; (b) their mass and size give place to gravitational fields smaller only than those of the black holes, but easier to measure; (c) the fastest of the pulsars have at least 700 turns about its axis in one second. Thus, its surface rotates by 36,000 kilometers a second; (d) neutrons stars have more intense magnetic fields than any other known object in the Universe, million of times stronger than those produced in any terrestrial laboratory; (e) in some cases the regularity of their pulsations is the same or larger than the precision of the atomic clock, the latter being otherwise the best that we have.

Neutron stars (NS) are thought to consist of a massive dense core surrounded by a thin crust (of mass $\lesssim 0.01 M_\odot$ and thickness $\lesssim 1\,\text{km}$) [LP01]. The crust–core interface occurs at the density $\approx\rho/2$, where $\rho_0 = 2.8 \times 10^{14}\,\text{g}\,\text{cm}^{-3}$ is the standard density of saturated nuclear matter. The NS crust is composed of neutron-rich atomic nuclei, strongly degenerate electrons and (at $\rho \gtrsim 4 \times 10^{11}\,\text{g}\,\text{cm}^{-3}$) of free neutrons dripped from nuclei. NS cores contain degenerate matter of supranuclear density. Its composition and equation of state (EOS) are model dependent (being determined by still poorly known strong interactions in dense matter). On the other hand, these properties are almost not constrained by observations. The NS core is usually subdivided into the outer core $\rho \lesssim 2\rho_0$) and the inner core ($\rho \gtrsim 2\rho_0$). The outer core is found in all NS's, while the inner core is present only in massive, more compact NS's. The outer core is thought to be composed of neutrons, and an admixture of protons and electrons (and possibly muons). The composition and EOS of the inner core is still a mystery. It may be the same composition as in the outer core, with a possible addition of hyperons. Alternatively, the inner core may contain pion- or kaon condensates, or quark matter, or a mixture of these components. Another complication is introduced by superfluidity of neutrons, protons and other baryons in dense matter. Neutron stars usually rotate with such precision that they are known as the best timekeepers in the Universe, but every so often their rotation rate suddenly increases. It is thought that these *glitches* are related to superfluidity inside the star, which allows the neutrons to flow without friction.

Chandrasekhar mass

The equation of state of a fully degenerate electron gas takes a particularly simple form in the nonrelativistic limit (corresponding to $x \ll 1$), as well as in the extreme relativistic limit (corresponding to $x \gg 1$). Using Eqs. (6.70) and (6.61), we find

$$P = \frac{8\pi c}{15 h^3} \left[\frac{3\rho h^3 Y_e}{8\pi \mu_A m_u} \right]^{5/3}, \tag{6.93}$$

for $E_F \ll m_e c^2$, and

$$P = \frac{2\pi c}{3 h^3} \left[\frac{3\rho h^3 Y_e}{8\pi \mu_A m_u} \right]^{4/3}, \tag{6.94}$$

for $E_F \gg m_e c^2$. Here, μ_A is the molecular weight and m_u is the nuclear mass unit ($m_u \simeq m_N$, where m_N is the nucleon mass). Y_e is the fraction of electrons in the stellar medium (for a completely ionized hydrogen gas, $Y_e = 1/2$). For reasons that we will explain later, $\rho Y_e/\mu_A m_u$ replaces the electron density in the degenerate Fermi gas formulas.

Equation (6.93) is of the form of a polytrope, i.e., for ultrarelativistic electrons, $P \sim \rho^{4/3}$, which is a polytrope with $n = 3$. Then, the stellar mass is given by equation (6.89). But, for $n = 3$, the mass does not depend on ρ_c if κ is fixed. Thus, one has

$$M = C_1 = 4\pi \left(-\frac{\theta'}{\xi}\right)_{\xi_3} \xi_3^3 \left(\frac{\kappa}{\pi G}\right)^{3/2} = const. \tag{6.95}$$

Inserting the proper numerical values one has

$$M_{Ch} = 1.457(2Y_e)M_\odot. \tag{6.96}$$

This is the maximum mass for which a star with an ultrarelativistic electron EOS is stable.

We can also show that the main mass–radius relation for neutron stars are easily deduced from general arguments of quantum mechanics for an electron gas, without resort to elaborated equations. Suppose N fermions exist within a star with radius R. Then the number density is given by $n = N/V \sim N/R^3$. According to the Heisenberg principle, the momenta of the particles are given by $p \sim \hbar n^{1/3}$; the average energy is given by $E \sim \hbar n^{1/3} c \sim \hbar c N^{1/3}/R$. This is roughly the kinetic energy per fermion in the medium. The gravitational energy per fermion is $E_g \sim GNm_B^2/R$; with $m_B = M/N \simeq m_N$. In equilibrium,

$$E = E(R) = \frac{\hbar c N^{1/3}}{R} - \frac{GNm_B^2}{R} = \text{minimum}, \tag{6.97}$$

where the minimum occurs at $N = N_0$ at finite radius $R = R_0$ such that

$$\frac{\hbar c N_0^{1/3}}{R_0} = \frac{GN_0 m_B^2}{R_0}, \tag{6.98}$$

which yields

$$N_0 \sim \left(\frac{\hbar c}{G m_B}\right)^{3/2} \sim 2 \times 10^{57}. \tag{6.99}$$

Therefore, according to this estimate, there are about 10^{57} baryons within a neutron star, before reaching the limiting Chandrasekhar mass. Finally, one gets

$$M_{Ch} \sim N_0 m_B \sim 1.5 M_\odot, \tag{6.100}$$

which one can compare to Eq. (6.96).

Assume that equilibrium is related to the onset of relativistic degeneracy of particles of mass m ($E > mc^2$). Since $E = \hbar c N_0^{1/3}/R$, one gets

$$R_0 < \frac{\hbar}{mc}\left(\frac{\hbar c}{G m_B^2}\right)^{1/2}. \tag{6.101}$$

While pressure can come from electrons ($m = m_e$) or nucleons ($m = m_B$), the mass is always given by the baryons ($m_B \sim m_N$). Then

$$\begin{aligned} R_0 &\sim 5 \times 10^6\,\mathrm{m} \quad \text{for } (m = m_e), \\ R_0 &\sim 5 \times 10^3\,\mathrm{m} \quad \text{for } (m = m_B). \end{aligned} \tag{6.102}$$

These are typical radii for white dwarfs (upper value, degenerate electrons) or neutron stars (lower value, degenerate nucleons). Compare with Eq. (6.92).

For a given Y_e, a star balanced by a degenerate fermion gas model is totally determined by ρ_c. Figure 6.8 shows the mass versus central density of a white dwarf. The solution is obtained by numerically solving

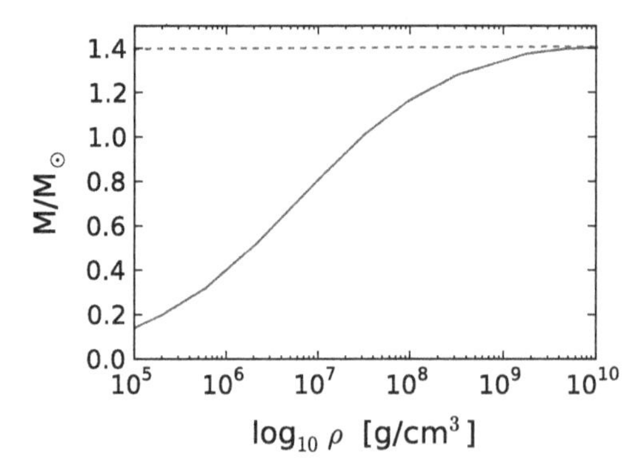

Figure 6.8. Mass of a white dwarf calculated as a function of ρ_c, the central density. With increasing ρ_c, the mass reaches a limiting value, the Chandrasekhar mass.

equations (6.5) and (6.7) together with the equation

$$\frac{dP}{dr} = \frac{d\rho}{dr}\frac{dP}{d\rho},$$

with $dP/d\rho$ obtained from the non-relativistic EOS, Eq. (6.61). The agreement with our simplified discussions is remarkable.

Equation of state

The equation of state (EOS) is a relation between the thermodynamic variables specifying the state of a physical system. The best known example of EOS is the ideal gas law, stating that the pressure of a collection of N noninteracting, point-like classical particles, enclosed in a volume V, grows linearly with the temperature T and the average particle density $n = N/V$.

In general, the EOS can be written as an expansion of the pressure, P, in powers of the density (here we use units such that Boltzmann's constant is $k = 1$),

$$P = nT[1 + nB(T) + n^2C(T) + \cdots]. \tag{6.103}$$

This expression is called *virial expansion* and the coefficients appearing in it are functions of temperature only. They describe the deviations from the ideal gas law and can be calculated in terms of the underlying elementary interaction. Therefore, the EOS carries a great deal of dynamical information, and its knowledge makes it possible to establish a link between measurable macroscopic quantities, such as pressure or temperature, and the forces acting between the constituents of the system at microscopic level.

If one uses the polytropic model, Eq. (6.72), where the adiabatic index $\Gamma = 1 + 1/n$ for a generic equation of state, one gets

$$\Gamma = \frac{d(\ln P)}{d(\ln \rho)}. \tag{6.104}$$

This is related to the *compressibility* χ characterizing the change of pressure with volume according to

$$\frac{1}{\chi} = -V\left(\frac{\partial P}{\partial V}\right)_N = \rho\left(\frac{\partial P}{\partial \rho}\right)_N \tag{6.105}$$

through

$$\Gamma = \frac{1}{\chi P}. \tag{6.106}$$

The compressibility is also simply related to speed of sound in matter, v_s, defined as

$$v_s = \frac{\partial P}{\partial \rho} = \frac{1}{\chi \rho}. \tag{6.107}$$

The magnitude of the compressibility reflects the so-called stiffness of the equation of state. Larger stiffness corresponds to more incompressible matter (smaller χ).

The dependence of the mass of the star upon its central density, obtained from integration of Eq. (6.5) using the equation of state of a fully ionized helium plasma, is illustrated in Fig. 6.8. The figure shows that the mass increases as the central density increases, until a limiting value $M \simeq 1.44\ M_\odot$ is reached at $\rho_0 \sim 10^{10}\,\mathrm{g/cm^3}$.

The white dwarf model can be improved by using more elaborated EOS. For example, models which include Coulomb corrections accounting for interaction of nuclei with electrons modify the results appreciably. High densities in the core of stars leads to electron Fermi energies that can be large enough to stimulate electron captures in nuclei such as ^{12}C, ^{12}B, ^{12}Be. The Fermi energy of the electron is high enough to energetically allow reactions such as $e^- + \mathcal{A}_\mathcal{Z} \longrightarrow \nu + \mathcal{A}_{Z-1}$ to occur. At low temperatures, $T \simeq 0$, the Fermi energy, or chemical potential, of the electron has to comply with $\mu + M(\mathcal{Z}, \mathcal{A}) \geq M(\mathcal{Z} - 1, \mathcal{A})$ or $\mu \geq M(\mathcal{Z} - 1, \mathcal{A}) - M(\mathcal{Z}, \mathcal{A})$. Electron capture reduces the electron-to-nucleon ratio Y_e and hence the Chandrasekhar mass. In Table 6.2 we show the critical densities for neutronization due to several electron capture reactions.

Other types of white dwarfs exist. *Red dwarfs* have between mass 8%–50% of the Sun and shine in red because of their low surface temperature of 2500–4000 K. They live very long, longer than the Universe is old now, and are consequently very numerous. They are not found in large number in groups of newly born stars, because the bigger stars are all still there. *Brown dwarfs* have mass below about $M = 0.1\ M_\odot$. From galactic lensing there is also evidence of a population of $0.5\ M_\odot$ massive objects (MACHO). MACHO stands for *Massive Astrophysical Compact Halo Objects*. Most of these objects are likely to be brown, white and red dwarfs. MACHOs could also be black holes, neutron stars or unassociated planets.

Table 6.2. Critical densities for onset of electron captures.

Reaction	Neutronization threshold (MeV)	ρ_0 (g cm^{-3})
$^1H \rightarrow n$	0.782	1.22×10^7
$^4_2He \rightarrow ^3_1H + n \rightarrow 4n$	20.596	1.37×10^{11}
$^{12}_6C \rightarrow ^{12}_5B \rightarrow ^{12}_4Be$	13.370	3.90×10^{10}
$^{16}_8O \rightarrow ^{16}_7N \rightarrow ^{16}_6C$	10.419	1.90×10^{10}
$^{20}_{10}Ne \rightarrow ^{20}_9F \rightarrow ^{20}_8O$	7.026	6.21×10^9
$^{24}_{12}Mg \rightarrow ^{24}_{11}Na \rightarrow ^{24}_{10}Ne$	5.513	3.16×10^9
$^{28}_{14}Si \rightarrow ^{28}_{13}Al \rightarrow ^{28}_{12}Mg$	4.643	1.97×10^9
$^{32}_{16}S \rightarrow ^{32}_{15}P \rightarrow ^{32}_{14}Si$	1.710	1.47×10^8
$^{56}_{26}Fe \rightarrow ^{56}_{25}Mn \rightarrow ^{56}_{24}Cr$	3.695	1.14×10^9

6.5.1. *Energy production*

The third equation of stellar structure is a relation between the energy release and the rate of energy transport. Consider a spherically symmetric star in which energy transport is radial and in which time variations are unimportant. $L(r)$ is the rate of energy flow across a sphere of radius r and $L(r + \delta r)$ is the rate of energy flow across a sphere of radius $r + \delta r$ (see Fig. 6.2). We define ϵ as the rate of energy release per unit mass per unit volume (W kg^{-1} s^{-1}). Hence, the energy release in shell is given by $4\pi r^2 \rho(r)\delta r \epsilon(r)$. Conservation of energy yields

$$L(r + \delta r) = L(r) + 4\pi r^2 \rho(r)\delta r \epsilon(r), \tag{6.108}$$

or,

$$\frac{L(r + \delta r) - L(r)}{\delta r} = 4\pi r^2 \rho(r)\epsilon(r), \tag{6.109}$$

which leads to, in the limit $\delta r \rightarrow 0$,

$$\frac{dL(r)}{dr} = 4\pi r^2 \rho(r)[\epsilon(r) - \epsilon_\nu(r)], \tag{6.110}$$

where an energy loss due to neutrino radiation was added by means of the term $\epsilon_\nu(r)$.

This is the *equation of energy production*. We now have three of the equations of stellar structure. However we have five unknowns $P(r), M(r)$,

$L(r), \rho(r), \epsilon(r)$. In order to make further progress we need to consider the energy transport equation.

6.5.2. *Energy transport*

There are three ways energy can be transported in stars. (a) *Convection* is the energy transport due to mass motions of the gas. (b) *Conduction* is the exchange of energy during collisions of gas particles (usually e^-). (c) *Radiation* is the energy transport by the emission and absorption of photons.

Conduction and radiation are similar processes. They both involve transfer of energy by direct interaction, either between particles or between photons and particles. Which is the more dominant in stars? The energy carried by a typical particle, $\sim 3kT/2$, is comparable to the energy carried by a typical photon, $\sim \hbar c/\lambda$. But the number density of particles is much greater than that of photons. This would imply conduction is more important than radiation. The mean free path of a photon is about $\sim 10^{-2}$ m. The mean free path of a particle is $\sim 10^{-10}$ m. Photons can move across temperature gradients more easily, hence they contribute to a larger transport of energy. In fact, conduction is negligible and radiation transport is the dominant process.

Convection

Convection is the mass motion of gas elements. It only occurs when the temperature gradient exceeds some critical value. We can derive an expression for this. Convective transport of energy refers to an exchange of energy between hotter and cooler layers in a dynamically unstable region through the exchange of macroscopic mass elements (blobs). The hotter blob moves upwards; the cooler downwards. The moving blobs will finally dissolve in their new surroundings and thereby deliver their excess or deficiency of heat. As stellar interiors are very dense, convective transport can be quite efficient. This mechanism requires driving processes such as *buoyancy forces* to operate.

Consider a convective element at distance r from the center of a star, as shown in Fig. 6.9. The element is in equilibrium with its surroundings. Now lets suppose that it rises to $r + \delta r$. As it expands, $P(r)$ and $\rho(r)$ are reduced to $P - \delta P$ and $\rho - \delta\rho$. But these may not be the same as the new surrounding gas conditions. Define those as $P - \Delta P$ and $\rho - \Delta\rho$. If gas element is denser than the surroundings at $r + \delta r$, it will sink (i.e., it is

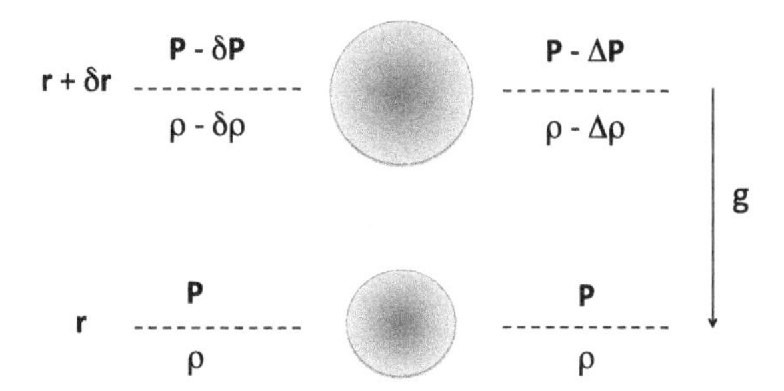

Figure 6.9. Adiabactic displacement of a volume element within a star.

convectively stable). If it is less dense then it will keep on rising, i.e., it is convectively unstable.

The condition for instability is therefore

$$\rho - \delta\rho < \rho - \Delta\rho. \tag{6.111}$$

Whether or not this condition is satisfied depends on two things: (a) the rate at which the element expands due to decreasing pressure, and (b) the rate at which the density of the surroundings decreases with height.

Let's make two assumptions: (1) the element rises adiabatically and (2) the element rises at a speed much smaller than the sound speed. During this motion, sound waves have time to smooth out the pressure differences between the element and the surroundings. Hence $\delta P = \Delta P$ at all times. The first assumption means that the element must obey the adiabatic relation between pressure and volume

$$PV^{\gamma} = const., \tag{6.112}$$

where $\gamma = c_p/c_v$ is the specific heat (i.e., the energy in J to raise the temperature of 1 kg of material by 1 K) at constant pressure, divided by the specific heat at constant volume. Given that V is inversely proportional to ρ, we can write

$$\frac{P}{\rho^{\gamma}} = const. \tag{6.113}$$

Thus, equating the term at r and $r + \delta r$,

$$\frac{P - \delta P}{(\rho - \delta\rho)^{\gamma}} = \frac{P}{\rho^{\gamma}}. \tag{6.114}$$

If $\delta\rho$ is small, we can expand $(\rho - \delta\rho)^\gamma \sim \rho^\gamma - \gamma\rho^{\gamma-1}\delta\rho$ and combine the last two expressions in

$$\delta\rho = \frac{\rho}{\gamma P}\delta P. \tag{6.115}$$

Now we need to evaluate the change in density of the surroundings, $\Delta\rho$. Let's consider an infinitesimal rise of δr. Then,

$$\Delta\rho = \frac{d\rho}{dr}\delta\rho. \tag{6.116}$$

Substituting these expressions for $\delta\rho$ and $\Delta\rho$ into the condition for convective instability derived above, i.e.,

$$\frac{\rho}{\gamma P}\delta P < \frac{d\rho}{dr}\delta r. \tag{6.117}$$

And this can be rewritten by recalling our second assumption that the volume element will remain at the same pressure as it surroundings, so that in the limit

$$\delta r \to 0, \quad \frac{dP}{\delta r} \equiv \frac{dP}{dr}, \tag{6.118}$$

or

$$\frac{\rho}{\gamma P}\frac{dP}{dr} < \frac{d\rho}{dr}. \tag{6.119}$$

The left-hand side of the equation above is the density gradient that would exist in the surroundings if there was an adiabatic relation between density and pressure. The right-hand side is the actual density gradient in the surroundings. We can convert this to a more useful expression, by first dividing both sides by dP/dr. Note that dP/dr is negative, hence the inequality sign must change. Thus,

$$\frac{\rho}{\gamma P} < \frac{d\rho/dr}{dP/dr}, \tag{6.120}$$

or

$$\left(\frac{P}{\rho}\right)\frac{d\rho}{dP} < \frac{1}{\gamma}, \tag{6.121}$$

and for an ideal gas in which radiation pressure is negligible (where m is the mean mass of particles in the stellar material),

$$P = \frac{\rho k T}{m}, \quad \text{and} \quad \ln P = \ln\rho + \ln T + const. \tag{6.122}$$

Differentiating it, we get

$$\frac{dP}{P} = \frac{d\rho}{\rho} + \frac{dT}{T}. \tag{6.123}$$

Combining this with the Equation (6.117) yields

$$\frac{P}{T}\frac{dT}{dP} > \frac{\gamma - 1}{\gamma}. \tag{6.124}$$

This is the condition for the occurrence of convection (in terms of the temperature gradient). A gas is convectively unstable if the actual temperature gradient is steeper than the adiabatic gradient. If the condition is satisfied, then large scale rising and falling motions transport energy upwards. The criterion can be satisfied in two ways: either the ratio of specific heats is close to unity or the temperature gradient is very steep. For example if a large amount of energy is released at the center of a star, it may require a large temperature gradient to carry the energy away. Therefore, where nuclear energy is being released, convection may occur.

Another way to describe the condition for dynamical stability is to compare the ratios of temperature and pressure gradients arising from adiabatic motion of the rising mass volumes with that of radiation transport, i.e.,

$$\nabla_{ad} \equiv \left.\frac{d\ln T}{d\ln P}\right|_{ad} = \left.\frac{d\ln T}{d\ln P}\right|_{rad}. \tag{6.125}$$

This is known as the *Schwarzschild criterion* for dynamical stability. As we are going to see later, one can write

$$\left.\frac{d\ln T}{d\ln P}\right|_{rad} = \frac{3}{16\pi ac}\frac{\bar{\kappa}L_{rad}P}{mT^4}, \tag{6.126}$$

where $\bar{\kappa}$ is the mean opacity and L_{rad} is the luminosity. From this condition one finds that convection occurs for large L_{rad}/T^4, e.g., in the cores of stars with $M > M_\odot$, or for large $\bar{\kappa}$, e.g., in the solar envelope.

Alternatively in the cool outer layers of a star, gas may only be partially ionized, hence much of the heat used to raise the temperature of the gas goes into ionization. Therefore, the specific heat of the gas at constant V is nearly the same as the specific heat at constant P, and $\gamma \sim 1$. In such a case, a star can have a cool outer convective layer.

Convection is an extremely complicated subject and it is true to say that the lack of a good theory of convection is one of the worst defects in our present studies of stellar structure and evolution. We know the conditions under which convection is likely to occur but do not know how much energy is carried by convection. Fortunately, we will see that we can often find occasions where we can manage without this knowledge.

Dynamical, thermal, and nuclear timescales

So far, we have identified 3 characteristic timescales that help us to understand stellar evolution:

1. The *dynamical timescale.* This defines the time-scale by which a slightly perturbed star can again reach hydrostatic equilibrium,

$$t_d = \left(\frac{2r^3}{GM}\right)^{1/2} \sim \frac{1}{2}\left(\frac{1}{G\bar{\rho}}\right)^{1/2}. \tag{6.127}$$

   For the Sun, $t_d \sim 2000\,\mathrm{s} \sim 27$ min.
2. The *thermal timescale.* This is the time for a star to emit its entire reserve of thermal energy upon contraction, provided it maintains constant luminosity (*Kelvin–Helmholtz timescale*). It is given by

$$t_{th} \sim \frac{GM^2}{Lr}. \tag{6.128}$$

   For the Sun, $t_{th} \sim 30$ million years.
3. The *nuclear timescale.* This is the time for a star to consume all its available nuclear energy (ϵ =typical nucleon binding energy/nucleon rest mass energy). It is given by

$$t_{nuc} \sim \frac{\epsilon M c^2}{L}. \tag{6.129}$$

   For the Sún, we can make a simple estimate. We assume that the source of energy generation is the conversion of four protons into a helium nucleus, i.e., 4p $\to$ ^{4}He. This releases an amount of energy $Q = 6.3 \times 10^{18}$ erg/g. Thus, $\epsilon = Q$, if the Sun consists entirely of hydrogen. With $L_\odot = 4 \times 10^{33}$ erg/s, we obtain $t_{nuc} \sim 3 \times 10^{18} s \sim 10^{11}$ y, which is larger than the age of the Universe.

Another time scale arises from the diffusion of radiation, a topic to be discussed in the next section. Here we make a simple estimate of this time scale.

Photons created in the core of the Sun scatter many times on their way out to the surface; actually they are absorbed and re-emitted more than scattered, but never mind that for now. The mean-free path between scatterings is $\lambda = (n_e \sigma)^{-1}$, where n_e is the number of electrons per unit volume, and σ is the scattering cross section per electron. Since the Sun is mostly hydrogen, $n_e \simeq \rho/m_H$. The mass density averaged over the volume of the Sun is $1.4\,\mathrm{g/cm^3}$ (similar to that of the human body!), hence $n_e \simeq 10^{24}\,\mathrm{cm^{-3}}$. The cross-section is generally of order of the Thomson cross-section, Eq. (4.117), so $\bar{\lambda} \simeq 1\,\mathrm{cm}$. The corresponding collision time is

$$t_{e\gamma} = \frac{\bar{\lambda}}{c} \simeq 10^{-10}\,\mathrm{s}. \tag{6.130}$$

It can be shown that the proton–electron and proton–proton collision times are similarly short, or even shorter, compared to t_d and the other relevant macroscopic timescales discussed above. Importantly therefore, all regions of the interior quickly relax to local thermodynamic equilibrium (LTE). Thermal equilibrium can never be perfect, however, since the surface, which radiates freely to space, is inevitably colder than the core. The temperature gradient drives an outward flux of heat.

Photons escape by a random walk of average step length equal to λ. The root-mean-square distance after N steps is $d_N = N^{1/2}\lambda$. Setting this distance equal to the Sun's radius $R_\odot$ gives the typical number of steps needed to travel from center to surface: $N \simeq (R_\odot/\lambda)^2 \simeq 10^{22}$. The corresponding *photon diffusion time* is $t_{diff} = N t_{e\gamma} \simeq 10^{12}\,\mathrm{s} \simeq 3 \times 10^4$ yr.

Thus, in general,

$$t_d \ll t_{th}, \quad t_{diff} \ll t_{nuc}. \tag{6.131}$$

When analyzing processes associated with one of these timescales, one can usually ignore the slower processes and assume that the more rapid ones are at equilibrium. Most unsolved problems in stellar theory have to do with breakdown in these assumptions: for example, convection and wind-borne mass loss are departures from dynamical equilibrium that have thermal and nuclear consequences.

6.6. Radiative Transport

We assume for the moment that the condition for absence of convection is satisfied, and we will derive an expression relating the change in temperature with radius in a star assuming that all energy is transported by radiation. Hence we ignore the effects of convection and conduction.

The equation of radiative transport assumes a spherically symmetric geometry, i.e., the gas conditions are a function of only one coordinate, in this case r. Photons interact with free electrons, cause atomic transitions and ionize atoms. The mean free path of photons is given by $l_{ph} = 1/\kappa\rho$. κ is the *opacity to radiation* and is a measure of the *effective area* (cross section) for photon interacting with matter. Typical values for the Sun are $\bar{\kappa} \equiv \langle\kappa\rangle = 0.4\,\mathrm{cm^2\,g^{-1}}$ and $\bar{\rho} = 1.4\,\mathrm{g\,cm^{-3}}$, which leads to $l_{ph} = 2\,\mathrm{cm}$. Hence, $l_{ph}/R_\odot \sim 3 \times 10^{-11}$, which means that the transport of radiation occurs by means of a diffusion process.

If D is the diffusion coefficient, then the diffusive flux is given by $\mathbf{F} = -D\boldsymbol{\nabla} U_{rad}$, where U_{rad} is the radiation density, and $D = cl_{ph}/3$. Thus, along the radial direction, the radiation flux is given by

$$F_{rad} = -\frac{c}{3\kappa\rho}\frac{dU_{rad}}{dr} = -\frac{c}{\kappa\rho}\frac{dP_{rad}}{dr}. \tag{6.132}$$

A few more steps are necessary to make this equation useful. We define

$$\beta = \frac{P_{gas}}{P}, \quad \text{where } P = P_{gas} + P_{rad} \tag{6.133}$$

for a star composed of a gas–radiation mixture.

Using the Stefan–Boltzmann law, $P_{rad} = aT^4/3 = 4\sigma T^4/3c$, we get

$$1 - \beta = \frac{P_{rad}}{P} = \frac{aT^4}{3P}. \tag{6.134}$$

Then, the luminosity, $L = 4\pi r^2 F_{rad}$, is given by

$$L = -4\pi r^2 \left(\frac{ac}{3\rho\kappa}\right)\frac{dT^4}{dr} = -4\pi r^2 \left(\frac{c}{\rho\kappa}\right)\frac{dP_{rad}}{dr}. \tag{6.135}$$

Now we average over all radiation frequencies and obtain

$$L = -4\pi r^2 \left(\frac{ac}{3\rho\bar{\kappa}}\right)\frac{dT^4}{dr} = -4\pi r^2 \left(\frac{c}{\rho\bar{\kappa}}\right)\frac{dP_{rad}}{dr}. \tag{6.136}$$

where $\bar{\kappa}$ is the *Rosseland mean opacity*, defined by

$$\frac{1}{\bar{\kappa}} = \left(\int \frac{1}{\kappa_\nu}\frac{dI(\nu,T)}{dT}d\nu\right)\left(\int \frac{dI(\nu,T)}{dT}d\nu\right)^{-1}, \tag{6.137}$$

where $I(\nu, T)$ is the Planck function for the intensity of blackbody radiation, Eq. (4.66).

We thus can write

$$F_{rad} = \frac{L(r)}{4\pi r^2} = -\frac{4}{3}\frac{ac}{\bar{\kappa}\rho}T^3\frac{dT}{dr}, \tag{6.138}$$

or

$$\frac{dT}{dr} = -\frac{3}{4ac}\frac{\bar{\kappa}\rho}{T^3}\frac{L(r)}{4\pi r^2}, \tag{6.139}$$

which is the *equation for radiation transport.*

As a by-product, we can now show how the power $n = 3$ appears in the Eddington's polytropic model, mentioned after Eq. (6.85). The ideal gas law is given in terms of the total pressure by

$$P_g = nkT = \frac{\rho}{\mu_A}\frac{kT}{m_u} \equiv \beta P, \tag{6.140}$$

and the radiation pressure is

$$P_{rad} = \frac{aT^4}{3} \equiv (1-\beta)P. \tag{6.141}$$

Eliminating T from the first equation and replacing in the second equation yields

$$T = \frac{\beta \mu_A m_u}{\rho k}P = \left[\frac{3(1-\beta)}{a}P\right]^{1/4}, \tag{6.142}$$

leading to

$$P = \left(\frac{k}{\mu_A m_u}\right)^{4/3}\left[\frac{3(1-\beta)}{a\beta^4}\right]^{1/3}\rho^{4/3}, \tag{6.143}$$

thus proving Eddington's model. This model assumes that β is a constant to obtain $n = 3$. But β needs not to be constant in general. Note that the same power, $\rho^{4/3}$, appears in the equation of state of a non-relativistic degenerate electron gas, which is not related to the Eddington's model.

The *Eddington luminosity* in a star is defined as the point where the gravitational pressure inwards equals the continuum radiation pressure outwards. When exceeding the Eddington luminosity, a star would initiate a very intense continuum-driven stellar wind from its outer layers. Since most massive stars have luminosities far below the Eddington luminosity, however, their winds are mostly driven by the less intense diffusion process. Assuming that the opacities are due to Thompson scattering only, Eq. (6.135) becomes

$$\frac{dP}{dr} = -\frac{\kappa\rho}{c}F_{rad} = -\frac{\sigma_T\rho}{m_p c}\frac{L}{4\pi r^2} \tag{6.144}$$

where σ_T is the Thomson scattering cross section for the electron, m_p is the proton mass, and the gas is assumed to be purely made of ionized hydrogen. Equating this pressure with Eq. (6.5) and solving for the luminosity gives the Eddington luminosity,

$$L_{\rm Edd} = \frac{4\pi G M m_{\rm p} c}{\sigma_{\rm T}} = 3.3 \times 10^4 \left(\frac{M}{M_\odot}\right) L_\odot, \tag{6.145}$$

where M is the mass of the central object.

Mean opacities

The opacity parameter alters the luminosity of a star. When the star is more transparent, it tends to have a higher luminosity. For much of the radiative zone of stars, the elements which make up the heavy element abundance Z are the determining factor in the opacity. A star with a lower abundance of the heavy elements has a lower opacity and a higher luminosity. This effect causes the globular clusters to have a higher turnoff luminosity than is found for clusters having similarly old stars but having a higher heavy element abundance. Both of the luminosity effects tend to cancel when plotted on the HR diagram. However, when one looks at the mass–luminosity diagram, one sees an offset. Unfortunately, the only way to get a mass is from binary stars. For these stars you need to know the orbital velocities of both components as well as the inclination of the orbit.

Calculations of radiation transport require the knowledge of $\bar{\kappa}$, or its frequency average, as function of ρ and T. Nowadays, there exist numerical opacity tables for different chemical mixtures. The basic processes involved in the opacity calculations are free–free transitions due to interactions with free electrons, bound–bound transitions excitations of bound electrons in an ion, and bound–free transitions excitation of bound electrons into continuum.

Summarizing our previous discussion on opacities, as a function of the photon frequency, ν, one obtains

1. Electron scattering by a free charge can be described by the Thomson scattering formula, Eq. (4.117),

$$\kappa_\nu = \frac{8\pi}{3}\frac{r_e^2}{\mu_e}, \tag{6.146}$$

   where r_e is the classical electron radius.
2. Free–free transitions yield the result $\kappa_\nu \sim Z^2 \rho T^{-1/2} \nu^{-3}$.
3. Bound–free transitions yield $\kappa_\nu \sim \nu^3$, different for different bound states.
4. Bound–bound transitions. In this case, the photon excites electrons only for certain frequencies. The energy can later be re-emitted, but in arbitrary directions. This leads to a weakening of directed photon beams. It is important to notice that the absorption lines in stars are strongly broadened by collisions. Bound–bound absorption can contribute significantly for $T < 10^6$ K (e.g., in our Sun).

H^- is very important for the opacity and convection. In the H^- ion, the proton binds two electrons. The binding energy is 0.75 eV (compared to 13.59 eV for H atom). As a consequence any photon with energy in excess of the ionization energy can be absorbed by $H^- + \gamma \rightarrow H + e^-$. The process liberates the extra electron; the remaining energy becomes kinetic energy. As the binding energy corresponds to a photon wavelength of 17000 Å, H^- contributes to the continuum opacity in the infrared and at shorter wavelength. The abundance of H^- in the Sun is quite rare (10^{-7} relative to H in the solar photosphere), but H is not capable of contributing significantly to the continuum opacity. If H^- appears in stellar environments, it increases the opacities greatly which leads to convection. Figure 6.10 shows a typical numerical calculation of opacities under stellar conditions.

The stellar radius as determined from model calculations depends on the volume of the convective zone if one is present. An important equation to start with is derived from hydrostatic equilibrium, Eq. (6.4),

$$\frac{dP}{dr} = -g\rho,$$

and the equation for the optical depth (see Section 4.3),

$$\frac{d\tau_\lambda}{dr} = \rho\kappa_\lambda, \tag{6.147}$$

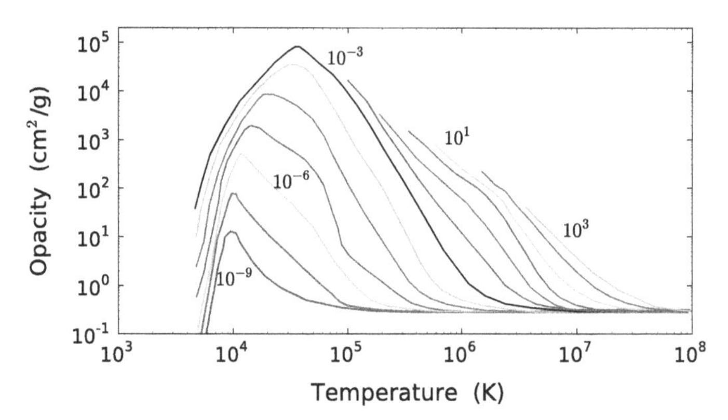

Figure 6.10. Numerical calculations of opacities (in cm^2/g) as a function of the temperature (in Kelvins) for several densities (in g/cm^3).

where κ_λ is the opacity for the radiation wavelength λ. If we consider a continuum wavelength near the peak of the Planck curve for a particular temperature T_{eff} (the temperature at the surface is the effective temperature of the star), the surface of the star is at $\tau_\lambda \sim 1$. We can take the ratio of the two above equations to find

$$\frac{dP}{d\tau_\lambda} = -g\kappa_\lambda. \tag{6.148}$$

Since the temperature is approximately T_{eff} at $\tau_\lambda = 1$ and the opacity can be considered to be a function of P and T, this equation gives a very specific relationship between P and g at the photosphere. Below this point, the pressure scales with the surface pressure and is thus fixed by this first point. Within the convection zone, the entropy is fixed so that specific volume $(1/\rho)$ as a function of pressure is fixed. One can thus calculate almost without ambiguity the volume of the convective envelope. The 'almost' comes from the fact that the entropy is not constant over the whole convective envelope and has a steep gradient at the levels just below the photosphere where the convection begins. The entropy at the photosphere is fixed by the above treatment of the optical depth and hydrostatic equilibrium. The uncertainty is the change in entropy between the photosphere and the adiabatic portion of the convection zone which encompasses all but the outer roughly 2000 km for the Sun. This difference is parameterized by a quantity called the mixing length parameter α. In a theory called the *mixing length theory*, the efficiency of convection is calculated in terms of a mean distance of travel for a typical convective element and this distance is taken to be αh_P where h_P is the pressure scale height [OC07].

6.6.1. *Solving the equations of stellar structure*

We now summarize the basic equations of stellar structure, in the absence of convection.

1. *Mass conservation.*

$$\frac{dm(r)}{dr} = 4\pi r^2 \rho(r). \tag{6.149}$$

2. *Hydrostatic support.*

$$\frac{dP(r)}{dr} = -\frac{Gm(r)}{r^2}\rho(r). \tag{6.150}$$

3. *Energy generation.*

$$\frac{dL(r)}{dr} = 4\pi^2 \rho(r)[\epsilon(r) - \epsilon_\nu(r)]. \tag{6.151}$$

4. *Radiation transport.*

$$\frac{dT}{dr} = -\frac{3}{4ac}\frac{\bar{\kappa}\rho}{T^3}\frac{L(r)}{4\pi r^2}. \tag{6.152}$$

Equation of state

The equations above are complemented by the equation of state $P \equiv P(\rho, T,$ chemical composition). We get the general equation of state for stellar matter by summing the expressions for radiation pressure, ion pressure, and electron pressure. Thus,

$$P = \frac{aT^4}{3} + \frac{\rho kT}{\mu_A m_u} + \frac{8\pi}{3h^3}\int_0^\infty \frac{p^3 \mathrm{v}(p)}{1+\exp(E/kT - \psi)}dp, \tag{6.153}$$

$$\rho = \frac{4\pi}{3}(2m_e)^{3/2}\mu_e m_u \int_0^\infty \frac{E^{1/2}}{1+\exp(E/kT - \psi)}dE, \tag{6.154}$$

where E is the electron kinetic energy, $\mathrm{v}(p) = p/(m_e^2 + p^2/c^2)^{1/2}$ is the electron velocity and ψ is the *degeneracy parameter*,

$$\psi = \ln\left[\frac{n_e h^3}{2(2\pi m_e kT)^{3/2}}\right], \tag{6.155}$$

which accounts for the degree of departure of the electron gas from the fully degenerate case. When $\psi \to \infty$, we have a completely degenerate electron gas, $\psi \simeq E_F/kT$, and when $\psi \to -\infty$ the electrons behave almost like an ideal gas.

Thus, we have implicit formulae for the equation of state (EOS). Given ρ and T, one can solve for ψ via (6.155), and then P. Similarly, the equation for each particle's internal energy is

$$u = \frac{aT^4}{\rho} + \frac{3}{2}\frac{kT}{\mu_A m_u} + \frac{8\pi}{h^3\rho}\int_0^\infty \frac{p^2 E(p)}{1+\exp(E/kT-\psi)}dp. \quad (6.156)$$

Unfortunately, there is no way to use these equations to generate analytical expressions for the thermodynamic quantities. Unless one of the terms dominates, these quantities must be computed numerically.

Crystallization

More complicated EOS are needed for some stellar environments. For example, in white dwarf stars one needs to abandon the assumption of the electron–nucleon fluid through the so-called "lattice" model which introduces the concept of the *Wigner–Seitz cell*: each cell contains a point-like nucleus of charge $+Ze$ with A nucleons surrounded by a uniformly distributed cloud of Z fully degenerate electrons. The global neutrality of the cell is guaranteed by the condition

$$Z = V_{WS} n_e = \frac{n_e}{n_{WS}}, \quad (6.157)$$

where $n_{WS} = 1/V_{WS}$ is the Wigner–Seitz cell density and $V_{WS} = 4\pi R_{WS}^3/3$ is the cell volume.

The total energy of the Wigner–Seitz cell is modified by the inclusion of the Coulomb energy, that is,

$$E_L = \mathcal{E} V_{WS} + E_C, \quad (6.158)$$

where

$$E_C = E_{e-N} + E_{e-e} = -\frac{9}{10}\frac{Z^2e^2}{R_{WS}}, \quad (6.159)$$

where $\mathcal{E}$ is the uniform energy density given by Eq. (6.91) and E_{e-N} and E_{e-e} are the electron–nucleus and the electron–electron Coulomb energies

$$E_{e-N} = -\int_0^{R_{WS}} 4\pi r^2 \left(\frac{Ze}{r}\right) en_e dr = -\frac{3}{2}\frac{Z^2e^2}{R_{WS}}, \tag{6.160}$$

$$E_{e-e} = \frac{3}{5}\frac{Z^2e^2}{R_{WS}}. \tag{6.161}$$

The pressure of the Wigner–Seitz cell is then given by

$$P_L = -\frac{\partial E_L}{\partial V_{WS}} = P + \frac{1}{3}\frac{E_C}{V_{WS}}, \tag{6.162}$$

where P is the pressure given by Eq. (6.153). The inclusion of the Coulomb interaction results in a decreasing of the pressure of the cell due to the negative lattice energy E_C.

The Wigner–Seitz cell chemical potential is

$$\mu_L = E_L + P_L V_{WS} = \mu + \frac{4}{3}E_C, \tag{6.163}$$

where the chemical potential for uniform matter is given by

$$\mu_L = Am_u c^2 + Z\mu_e, \tag{6.164}$$

with the electron chemical potential $\mu_e = \sqrt{m_e^2c^4 + p_F^2c^2}$.

Boundary conditions

Two of the boundary conditions are fairly obvious: at the center of the star ($r = 0$), $M = 0$, $L = 0$.

The conditions at the surface of the star its not so clear, but we can use approximations to allow for a solution. There is no sharp edge to a star, but for the Sun, $\rho(\text{surface}) \sim 10^{-4}\,\text{kg}\,\text{m}^{-3}$. This is much smaller than the mean density $\rho(\text{mean}) \sim 1.4 \times 10^3\,\text{kg}\,\text{m}^{-3}$ (which we derived). We know that the surface temperature ($T_{eff} = 5780\,\text{K}$) is much smaller than its mean temperature ($2 \times 10^6\,\text{K}$). Thus, we can make two approximations for the surface boundary conditions: $\rho = 0$ and $T = 0$ at $r = R$, so that the star has a sharp boundary with the surrounding vacuum.

Mass as an independent variable

The above formulas would, in principle, allow theoretical models of stars with a given radius. However, from a theoretical point of view it is the mass of the star which is chosen, the stellar structure equations solved, then the radius (and other parameters) are determined. We observe stellar radii to change by orders of magnitude during stellar evolution, whereas mass appears to remain constant. Hence, it is much more useful to rewrite the equations in terms of m rather than r.

If we divide the other three equations by the equation of mass conservation, and invert the latter, we obtain

$$\frac{dr}{dm} = \frac{1}{4\pi r^2 \rho}, \tag{6.165}$$

and (ϵ here contains all energy production and loss)

$$\frac{dL}{dm} = \epsilon. \tag{6.166}$$

Similarly, we get

$$\frac{dP}{dm} = -\frac{Gm}{4\pi r^4}, \tag{6.167}$$

and

$$\frac{dT}{dm} = -\frac{3\bar{\kappa} L_{rad}}{64\pi^2 r^4 acT^3}, \tag{6.168}$$

where $r \equiv r(m)$. We then use the boundary conditions $r = 0$ and $L = 0$ at $m = 0$. Also, $\rho = 0$ and $T = 0$ at $m = M$.

We specify M and the chemical composition and now have a well defined set of relations to solve. It is possible to do this analytically if simplifying assumptions are made, but in general these equations need to be solved numerically on a computer.

In a convective region we must solve the four differential equations, together with the equation

$$\frac{P}{T}\frac{dT}{dP} = \frac{\gamma - 1}{\gamma}. \tag{6.169}$$

And once the other equations have been solved, L_{rad} can be calculated. This can be compared with L (from $dL/dm = \epsilon$) and the difference gives the value of luminosity due to convective transport $L_{conv} = L - L_{rad}$. In solving the equations of stellar structure the equations appropriate to a convective region must be switched on whenever the temperature gradient reaches the adiabatic value, and switched off when all energy can be transported by radiation. This process can break down near the surface of a star.

The derivation of the above equations miss important aspects, which may can become important in particular cases. For example, the force on a mass element, $\rho d^3 r$, involves an extra term (*Euler or convective derivative*): $d/dt = \partial/\partial t + \mathbf{v}.\mathbf{\nabla}$, where $\mathbf{v}$ is the velocity at $\mathbf{r}$. The hydrostatic equilibrium equation at a fixed point r should therefore follow as

$$\rho \frac{\partial v}{\partial t}\bigg|_r + \rho v \frac{\partial v}{\partial r} = -\frac{\partial P}{\partial r} + f. \tag{6.170}$$

If one used a fixed mass point, instead of fixed point in space, Euler's derivative becomes

$$\frac{\partial}{\partial t}\bigg|_m = \frac{\partial}{\partial t}\bigg|_r + \frac{\partial}{\partial t}\bigg|_m \frac{\partial}{\partial r}. \tag{6.171}$$

Using $f = -Gm/r^2$ and $\delta m = 4\pi r^2 \delta r$, the equation for hydrostatic equilibrium becomes

$$\frac{dP}{dm} = -\frac{Gm}{4\pi r^4} - \frac{1}{4\pi r^2} \frac{\partial^2 r}{\partial t^2}\bigg|_m. \tag{6.172}$$

The last term can be safely neglected in most cases (e.g., our Sun), because of Eq. (6.131). But it might become important in other stellar environments.

Role of nuclear abundances

If there are no bulk motions in the interior of the star, then any changes of chemical composition are localized in the element of material in which the nuclear reactions occur. Thus, the star would have a chemical composition which is a function of mass M. In the case of no bulk motions, the set of equations we derived must be supplemented by equations describing the rate of change of abundances of the different chemical elements. Let $C_{X,Y,Z}$ be the chemical composition of the stellar material in terms of mass fractions of hydrogen (X), helium (Y), and metals (Z) (e.g., for solar system $X = 0.7$, $Y = 0.28$, $Z = 0.02$). Then

$$\frac{\partial (C_{X,Y,Z})_M}{\partial t} = f(\rho, T, C_{X,Y,Z}). \tag{6.173}$$

Such a model evolves as (from time t_0 to time $t_0 + \delta t$)

$$(C_{X,Y,Z})_{M,t_0+\delta t} = (C_{X,Y,Z})_{M,t_0} + \frac{\partial (C_{X,Y,Z})_M}{\partial t}. \tag{6.174}$$

At each time step the equation of energy production and energy transport (opacities) will depend on the chemical composition changes obtained from the equation above. Thus, the evolution of chemical composition is an important part of calculations of stellar evolution. This will be discussed in more detail later.

In summary, we have derived the equations to describe stellar structure, and explored the ways to solve these equations. As they are not time dependent, we must iterate with the calculation of changing chemical composition to determine short steps in the lifetime of stars. The crucial changing parameter is the H/He-ratio content of the stellar core (and afterwards, He burning will become important).

We have also discussed the boundary conditions applicable to the solution of the equations and made approximations that are not too far from realistic models. Finally, we have explored the influence of convection on energy transport within stars and have shown that it must be considered, but only in areas where the temperature gradient approaches the adiabatic value. In other areas, the energy can be transported by radiation alone and convection is not required.

6.6.2. *Luminosity of white dwarfs*

A simple way to model the cooling of a white dwarf is to use a two-zone model consisting of a degenerate, non-relativistic, isothermal core covered by a thin layer of ideal gas. One can show that the density at the transition region will be [Cha89]

$$\rho_t = \left(\frac{20m_e k}{\mu}\right)^{3/2}\left(\frac{\pi}{3N_A h^3}\right)\mu_e^{5/2}T^{3/2} = C_0\mu_e^{5/2}\mu^{-3/2}T_c^{3/2}, \qquad (6.175)$$

where N_A is Avogadro's number, k the Boltzmann constant, μ and μ_e are defined in Eqs. (6.46)–(6.53) and T_c is the core temperature. From the ideal gas law, the pressure at this location will be

$$P_t = \frac{N_A\rho_t}{\mu}kT_c = C_0\left(\frac{\mu_e}{\mu}\right)^{5/2}N_A kT_c^{5/2}. \qquad (6.176)$$

Now consider the behavior of a thin radiative atmosphere, which is neither a source nor sink of luminosity. Since this layer is thin, the mass of the atmosphere is negligible. Hence, if we adopt an opacity law of the form

$$\kappa = \kappa_0 \rho^s T^t = \kappa_0 \left(\frac{\mu}{N_A k}\right)^s P^s T^{t-s}, \tag{6.177}$$

then, from Eq. (6.126),

$$\nabla_{\rm rad} = \frac{P}{T}\frac{dT}{dP} = C_1 \mu^s \left(\frac{L}{M}\right) P^{s+1} T^{t-s-4}. \tag{6.178}$$

Following our discussion of the radiative atmospheres in Chapter 4, we can integrate this expression from the stellar photosphere down to the transition layer and, since the pressure and temperature at the transition region will be much larger than at the surface, one gets

$$T_c^{4+s-t} = \frac{4+s-t}{1+s} C_1 \mu^s \left(\frac{L}{M}\right) P_t^{1+s}, \tag{6.179}$$

or

$$P_t = \left[\frac{M(1+s)}{L\, C_1 \mu^s (4+s-t)}\right]^{1/1+s} T_c^{(4+s-t)/(1+s)}. \tag{6.180}$$

If we equate this equation to (6.176), we get an expression for the luminosity and temperature of the star in terms of the temperature of the central core,

$$L = \left[\frac{1+s}{C_1(4+s-t)}\right]\left[\frac{\mu_e^{-5/2}}{C_0 N_A k}\right]^{1+s} \mu^{(3s+5)/2} M\, T_c^{(3-3s-2t)/2}. \tag{6.181}$$

For an opacity dominated by bound–free absorption (see Section 6.6), $s = 1$, $t = -7/2$, and $\kappa_0 \approx 4 \times 10^{25}\,{\rm cm^2\,g^{-1}}$. Moreover, using $\mu_e \approx 2$, and $\mu \approx 1.75$, we get

$$L = C_2\ M\ T_c^{7/2}, \tag{6.182}$$

where $C_2 \sim 5 \times 10^{-30}$, if the mass and luminosity are in solar units.

The core is already degenerate, so gravitational contraction will not occur. However the core will cool, and the amount of this cooling will be given by the specific heat [Cha89],

$$L = -\frac{dE}{dt} = -c_V M \frac{dT_c}{dt}. \tag{6.183}$$

From this, we can compute the cooling curve of the star. If we take the derivative of Eq. (6.182) with respect to time, write it in terms of luminosity and mass, and then substitute in for the temperature derivative using Eq. (6.183), then

$$\frac{dL}{dt} = -\frac{7}{2c_V} C_2^{2/7} M^{-5/7} L^{12/7}. \tag{6.184}$$

This can be integrated easily to yield

$$t_{\text{cool}} = \frac{2}{5} C_2^{-2/7} c_V M^{5/7} [L^{-5/7} - L_0^{-5/7}]. \tag{6.185}$$

To calculate the specific heat, we can take advantage of the fact that under degenerate conditions, the specific heat of electrons is negligible. Thus c_V is given almost entirely by the ions

$$c_V = \left(\frac{dE}{dT}\right)_V = \frac{d}{dT}\left[\frac{3}{2}\frac{N_A k T}{\mu_I}\right] = \frac{3}{2}\frac{N_A k}{\mu_I}. \tag{6.186}$$

If we plug in the numbers, then in solar units,

$$t_{\text{cool}} = 2 \times 10^8 \, \mu_I^{-1} \, M^{5/7} [L^{-5/7} - L_0^{-5/7}] \text{ years}. \tag{6.187}$$

Note that $\mu_I \sim 12$, if the white dwarf is entirely carbon.

This simple cooling law reproduces the observed distribution of white dwarfs rather well. But, as a white dwarf cools, solid-state effects becomes more important and crystallization can occur. This greatly changes the equation of state. Also, many intermediate temperature white dwarf atmospheres do convect. Thus, simple radiative energy transport is not applicable to all white dwarfs. This changes the cooling curve somewhat, and can change the surface abundances.

6.7. Structure of Neutron Stars

6.7.1. *Relativistic corrections for neutron stars*

In the older astrophysical theories, non-relativistic mechanics and Newtonian gravitation theory were commonly used for stellar structure. This was apparently justified, since in many cases the crucial factor GM/c^2R was really small as compared to unity. When, however, this factor approaches unity, the use of the general theory of relativity is imperative. The general relativistic hydrostatic equilibrium equation is referred to as the *Tolman–Oppenheimer–Volkoff* (TOV) equation.

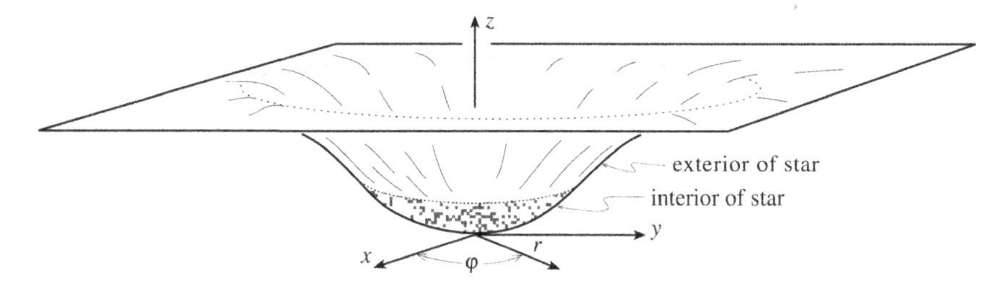

Figure 6.11. Embedding diagram for the warped space around a star.

The derivation of the TOV equation follows similar steps as we discussed before to obtain the FRW and Schwarzschild metrics. The equation is based on the following assumptions.

a) Matter constitutes a perfect fluid with energy–momentum tensor given by Eq. (D.123).
b) The star is spherically symmetric. This assumption allows us to write the metric inside the star in the form

$$ds^2 = e^{\nu(r)}c^2dt^2 - e^{\lambda(r)}dr^2 - r^2(d\theta^2 - \sin^2\theta d\phi^2). \qquad (6.188)$$

This metric describes the structure of space around a massive object, as depicted in Fig. 6.11.

A long and rather tedious calculation (best done on a computer), based on the metric coefficients (see Appendix C) produces the time–time and radial–radial components of the Einstein tensor, and thence of the Einstein field equation. For a massive star with a diagonal energy–momentum tensor (see Eq. 2.82), one can show that the Einstein's field equations yield

$$e^{\lambda(r)} = \frac{1}{1 - 2Gm(r)/c^2r}. \qquad (6.189)$$

c) The system is in equilibrium. This assumption leads to two conditions. First, it follows that the metric coefficients are independent of the time t. Second, the velocity has no space-like components in the chosen metric. With the use of these values it follows that the energy–momentum tensor has only diagonal components non-vanishing, given by Eq. (2.82).

Under these assumptions we obtain the TOV, which constrains the structure of a spherically symmetric body of isotropic material which is in static gravitational equilibrium, as modeled by general relativity. It reads

$$\frac{dp(r)}{dr} = -\frac{G}{r^2}\left[\rho(r) + \frac{p(r)}{c^2}\right]\left[m(r) + 4\pi r^3 \frac{p(r)}{c^2}\right]\left[1 - \frac{2Gm(r)}{c^2 r}\right]^{-1}. \tag{6.190}$$

where $m(r)$ is the total mass inside radius $r = r_0$, as measured by the gravitational field felt by a distant observer. The first two factors in the square brackets represent special relativity corrections of order $1/c^2$.

For a solution to the TOV equation, the function $\nu(r)$ is determined by the constraint

$$\frac{d\nu(r)}{dr} = -\frac{2}{p(r) + \rho(r)c^2}\frac{dp(r)}{dr}. \tag{6.191}$$

When supplemented with an equation of state, which relates density to pressure, the TOV equation completely determines the structure of a spherically symmetric body of isotropic material in equilibrium. If terms of order $1/c^2$ are neglected, it becomes the Newtonian hydrostatic equation (6.5), used to find the equilibrium structure of a spherically symmetric body of isotropic material when general-relativistic corrections are not important. These correction factors are all positive definite. It is as if Newtonian gravity becomes stronger for any value of r; relativity strengthens the pull of gravity.

Although the density of matter near the center of a neutron star is above that of an atomic nucleus, where the equation of state is ill-understood, we can be confident that there is an upper limit on the masses of neutron stars, a limit in the range $1.6M_\odot \lesssim M_{max} \lesssim 3M_\odot$. This mass limit cannot be avoided by postulating that a more massive neutron star develops an arbitrarily large central pressure and thereby supports itself against gravitational implosion. The reason is that an arbitrarily large central pressure is self-defeating: The "gravitational pull" which appears on the right-hand side of Eq. (6.190) is quadratic in the pressure at very high pressures (whereas it would be independent of pressure in Newtonian theory). This purely relativistic feature guarantees that if a star develops too high a central pressure, it will be unable to support itself against the resulting "quadratically too high" gravitational pull.

If the equation is used to model a bounded sphere of material in a vacuum, the zero-pressure condition $p(r = 0)$ and the condition $\exp[\lambda(r)] = 1/(1-2GM/rc^2)$ should be imposed at the boundary. The second boundary condition is imposed so that the metric at the boundary is continuous with the unique static spherically symmetric solution to the vacuum field equations, the Schwarzschild metric, Eq. (6.188). Computing the mass by integrating the density of the object over its volume will yield a larger value than obtained from Eq. (6.7). In fact,

$$M = \int_0^R \frac{4\pi\rho(r)r^2}{\sqrt{1 - 2GM(r)/rc^2}} dr. \tag{6.192}$$

The difference between this quantity and the non-relativistic counterpart, Eq. (6.7), yields

$$\delta M = \int_0^R 4\pi\rho(r)r^2 \left[\frac{1}{\sqrt{1 - 2GM(r)/rc^2}} - 1\right] dr, \tag{6.193}$$

which is an additional gravitational binding energy of the object divided by c^2.

The TOV equations must be integrated with the boundary conditions $M(0) = 0$, $\rho(R) = 0$ and $p(R) = 0$, where R is the radius of the star. We now introduce an independent variable η and the dimensionless functions $\epsilon(\eta)$ (energy density), $P(\eta)$ (pressure) and $m(\eta)$ (mass) by means of the transformations

$$r = a\eta, \quad \rho = \rho_c\epsilon(\eta), \quad p = \rho_c c^2 P(\eta), \quad \text{and} \quad M = M^* m(\eta). \tag{6.194}$$

Here a is a scale factor (a characteristic length), ρ_c the central density of the star, and M^* a characteristic mass. We then obtain the following non-dimensional form of the mass continuity and TOV equations:

$$\frac{dm(\eta)}{d\eta} = \eta^2\epsilon(\eta), \quad \frac{dP(\eta)}{d\eta} = -\frac{[\epsilon(\eta) + P(\eta)][P(\eta)\eta^3 + m(\eta)]}{\eta^2\left[1 - 2m(\eta)/\eta\right]}, \tag{6.195}$$

where we used

$$a^2 = \frac{c^2}{4\pi G\rho_c}, \quad \text{and} \quad M^* = 4\pi\rho_c a^3 = \frac{c^3}{\sqrt{4\pi G^3\rho_c}}. \tag{6.196}$$

In the new variables, the equation of state becomes $P(\eta)$, while the boundary conditions are given by

$$m(0) = 0, \quad \epsilon(\eta_s) = 0, \quad \text{and} \quad P(\eta_s) = 0, \tag{6.197}$$

where $\eta_s = R/a$ is the value of the non-dimensional radial co-ordinate η_s at the surface of the star. If we suppose that m is an increasing (non-decreasing) function of η, while ϵ and P are decreasing (non-increasing) functions of the same argument, then $0 \leq \epsilon \leq 1$ and $0 \leq P \leq p_c/\rho_c c^2$.

These equations can be integrated by assuming a central density ρ_c. This determines p_c, while $dP/d\eta$ and m vanish at $r = 0$. The mass continuity equation then determines m for an infinitesimal increase in η, and plugging in this value of m and ρ_c and p_c into the TOV equation determines the value of $dP/d\eta$, allowing a determination of P at the next step. For the P thus found the equation of state determines ρ and we go over the whole process once again to determine the values of the variables at the next step. In this way, the computation of p, ρ and m for increasing values of r goes until we arrive at $p = 0$. We identify this as the boundary of the star. The results in the model calculation may be represented in a number of ways. We can have a plot of the mass M against the central density ρ_c or against the radius R. The first case was already used in plotting Fig. 6.8. In Fig. 6.12 we plot several solutions for $M/M_\odot$ as a function for the radius.

The general feature of these curves is the occurrence of a maximum mass at a finite value of the central density ($\rho_c \sim 10^{15}\,\mathrm{g/cm^3}$) and approximately 10 km radius. The maximum mass has a value ranging from 0.7 $M_\odot$ for a non-interacting neutron gas to a value of around 3 $M_\odot$, corresponding to the stiff equation of state $p = \rho c^2$.

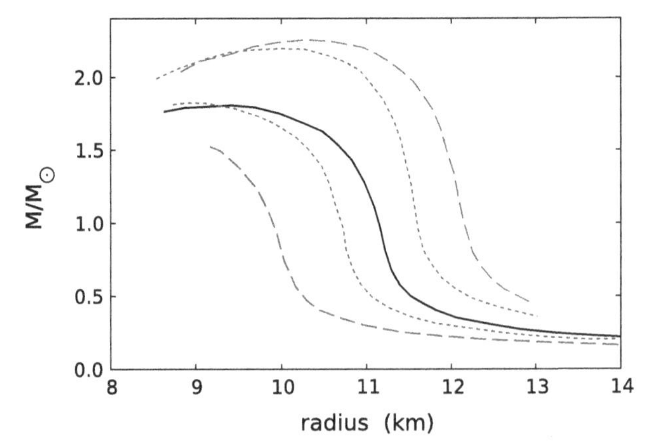

Figure 6.12. Nuclear masses (ratio with solar mass) of neutron stars for several equations of state as a function of their radii in kilometers.

A simple solution can be obtained by considering the structure and interior geometry of a homogeneous gaseous sphere, that is, we suppose that the energy density is constant throughout the star, $\rho = \rho_m = \rho_c$. In this case we have $\epsilon = 1$ for any $0 \leq \eta \leq \eta_s$. Using the boundary conditions $m(0) = 0$, $P(0) = P_c$, Eq. (6.195) can be solved analytically, yielding

$$m(\eta) = \frac{\eta^3}{3}, \quad P(\eta) = \frac{\frac{1}{3}(1+P_c) - \left(\frac{1}{3}+P_c\right)\left(1-\frac{2}{3}\eta^2\right)^{1/2}}{\left(\frac{1}{3}+P_c\right)\left(1-\frac{2}{3}\eta^2\right)^{1/2} - (1+P_c)}. \tag{6.198}$$

The radius $R = a\eta_s$ of the star can be obtained from the condition $P(\eta_s) = 0$ and is given by

$$R = a\frac{[6P_c(2P_c+1)]^{1/2}}{3P_c+1}. \tag{6.199}$$

In the dimensionless variables the total mass of the star is given by $m_s = \eta_s^3/3$, or $M_s = 4\pi\rho_c R_s^3/3$.

The radius R of the homogeneous stellar configuration relates to the gravitational (Schwarzschild) radius of the star $R_S = 2GM_s/c^2$ (with M_s the total mass of the star), by the relation

$$R = \frac{(3P_c+1)^2}{2P_c[2(2P_c+1)]}R_S. \tag{6.200}$$

The values of P_c depend on the physically allowed upper limit for the pressure. If we consider the classical restriction of general relativity, $p \leq \rho c^2/3$, as $p \geq 0$, it follows that in the case of the homogeneous sphere $0 \leq P_c \leq 1/3$. For the sake of generality we can consider the more general restriction $p \leq \rho c^2$, the upper limit being the stiff equation of state for the hot nucleonic gas. It is largely believed today that matter actually behaves in this manner at densities above about ten times nuclear, that is at densities greater than $10^{17}\,\mathrm{g/cm^3}$, namely at temperatures $T = (\rho/\sigma)^{1/4} > 10^{13}\,\mathrm{K}$, where σ is the radiation constant. In this case it follows that $0 \leq P_c \leq 1$. For the classical restriction of the pressure given by equation $p \leq \rho c^2$, we find the following simple stability criterion for a homogeneous star in the presence of a cosmological constant: $R \geq 9R_S/5$.

The expression of the metric tensor component e^{λ} in Eq. (6.188) inside the homogeneous star is given by

$$e^{\lambda(r)} = 1 - \frac{8\pi G\rho}{c^2}r^2, \tag{6.201}$$

while ν is given as a function of the dimensionless variable η by

$$e^{\nu(\eta)} = \frac{C}{[1+P(\eta)]^2} = C\left[\left(\frac{1}{3}+P_c\right)\left(1-\frac{2}{3}\eta^2\right)^{1/2} - (1+P_c)\right]^2, \tag{6.202}$$

where C is a constant of integration.

For $r > R$ the geometry of the space-time is described by the Schwarzschild line element, given by

$$e^{\nu} = e^{\lambda} = 1 - \frac{2GM_s}{r}. \tag{6.203}$$

In order to match the interior metric in the star smoothly on the boundary surface $r = R$ with the exterior Schwarzschild line element, we have to require the continuity of the gravitational potential across that surface. Matching Eq. (6.202) with the exterior Schwarzschild gravitational metric tensor component (6.203) at the boundary $r = R$ gives the value of the integration constant C. One gets

$$C = \frac{1 - 2GM_s/R}{\left[\left(\frac{1}{3}+P_c\right)\left(1-\frac{2}{3}\left(\frac{R}{a}\right)^2\right)^{1/2} - (1+P_c)\right]^2}. \tag{6.204}$$

Hence, the problem of the structure of the homogeneous general relativistic fluid sphere is completely solved.

Let us consider the speed of sound in nuclear matter. Starting from the elementary formula for the square of the speed of sound in terms of the bulk modulus B [You92], one can show that

$$\left(\frac{v_s}{c}\right)^2 = \frac{B}{\rho c^2} = \frac{dp}{d\epsilon} = \frac{dp/dn}{d\epsilon/dn}, \tag{6.205}$$

where p (ϵ) [n] is the pressure (energy) [number] density. To satisfy *relativistic causality* we must require that the sound speed does not exceed that of light. This can happen when the density becomes very large. Thus, causality sets restrictions on the possible equation of state.

6.7.2. *Neutron stars with nucleons and electrons*

Neutron stars (NS) are mostly composed of neutrons, but a small fraction of protons and electrons are also present. The neutron can decay into a proton and an electron via weak decay, $\mathrm{n} \to \mathrm{p} + e^- + \bar{\nu}_e$. The energy of such a decay ($m_n - m_p - m_e = 0.778\,\mathrm{MeV}$) is carried away by the light electron and (nearly massless) neutrino.[3] But when the low-energy levels for the proton are already filled, then the Pauli exclusion principle prevents the decay. Analogous effect occurs for the electrons. Electrons must be present within the star to cancel the positive charge of the protons, as a NS is electrically neutral. Therefore, equal numbers of electrons and protons imply that

$$k_{F,p} = k_{F,e}. \tag{6.206}$$

Within a NS the weak interaction must be in equilibrium, meaning that as many neutrons decay as electrons get captured via $p + e^- \to n + \nu_e$. This equilibrium can be expressed in terms of the chemical potentials for the three particle species,

$$\mu_n = \mu_p + \mu_e. \tag{6.207}$$

The chemical potential for particle i is $\mu_i(k_{F,i}) = \partial E_i/\partial n_i = (k_{F,i}^2 + m_i^2)^{1/2}$ and chemical equilibrium for the three species, $i = \mathrm{n, p, e}$, implies that

$$(k_{F,n}^2 + m_n^2)^{1/2} - (k_{F,p}^2 + m_p^2)^{1/2} - (k_{F,p}^2 + m_e^2)^{1/2} = 0. \tag{6.208}$$

For the densities within a NS, the electron mass is much smaller than the neutron Fermi energy, $m_e \ll k_{F,n}$, and Eq. (6.208) yields

$$k_{F,p}(k_{F,n}) \simeq \frac{k_{F,n}^2 + m_n^2 - m_p^2}{2(k_{F,n}^2 + m_n^2)^{1/2}}. \tag{6.209}$$

The total pressure and energy density in the star is the sum of the individual energy densities, of protons, neutrons and electrons. It can be calculated as in Eqs. (6.154) and (6.154). The liquid drop mass formula, Eq. (3.16), for nuclides with Z protons and N neutrons gives, for *symmetric nuclear matter* ($A = N + Z$ with $N = Z$), an equilibrium number density n_0 of 0.16 nucleons/fm^3. For this value of n_0 the Fermi momentum is $k_F = 263\,\mathrm{MeV}/c$ (compare to Eq. (3.42)). This momentum is small enough compared with $m_N = 939\,\mathrm{MeV/c^2}$ to allow a non-relativistic treatment of normal nuclear matter. At this density, the average binding energy per nucleon, $BE = -B(Z, A) = -16\,\mathrm{MeV}$ (see Section 3.4.1). A useful quantity for "measuring" nuclear pressure is the *nuclear compressibility*, K, defined in Eq. (5.151). This is a quantity which is not all that well established but is in the range of 200 to 400 MeV. The other is the so-called *symmetry*

[3]Here we use $\hbar = c = 1$.

energy term (Section 3.4.1), which, when $Z = 0$, contributes about 30 MeV of energy above the symmetric matter minimum at n_0.

Symmetric nuclear matter

In this case, $n_n = n_p$ and the total nucleon density $n = n_n + n_p = 2n_n$. We will relate the quantities n_0, BE, and K to the energy density for symmetric nuclear matter, $\epsilon(n)$, where $n = n(k_F)$ is the nuclear density (more details are found in Ref. [SR04]). The average energy per nucleon, E/A, for symmetric nuclear matter is related to ϵ by

$$E(n)/A = \epsilon(n)/n, \tag{6.210}$$

which includes the rest mass energy, m_N. As a function of n, $E(n)/A - m_N$ has a minimum at $n = n_0$ with a depth $BE = -16$ MeV. This minimum occurs when

$$\frac{d}{dn}\left(\frac{E(n)}{A}\right) = \frac{d}{dn}\left(\frac{\epsilon(n)}{n}\right) = 0 \quad \text{at } n = n_0. \tag{6.211}$$

Another constraint is

$$\frac{\epsilon(n)}{n} - m_N = BE, \quad \text{at } n = n_0. \tag{6.212}$$

The positive curvature at $n = n_0$ is given by

$$K(n) = 9\frac{dp(n)}{dn} = 9\left[n^2\frac{d^2}{dn^2}\left(\frac{\epsilon}{n}\right) + 2n\frac{d}{dn}\left(\frac{\epsilon}{n}\right)\right], \tag{6.213}$$

where

$$p(n) = n^2\frac{d}{dn}\left(\frac{\epsilon}{n}\right), \tag{6.214}$$

which defines the pressure in terms of the energy density. At $n = n_0$, $K(n_0) = K$, defined in Eq. (5.151).

For symmetric nuclear matter, one can model the energy density so that [Pra96]

$$\frac{\epsilon(n)}{n} = m_N + \frac{3}{5}\frac{\hbar^2 k_F^2}{2m_N} + \frac{A}{2}u + \frac{B}{\sigma+1}u^{\sigma}, \tag{6.215}$$

where $u = n/n_0$ and σ and A and B constants. The first term represents the rest mass energy and the second the average kinetic energy per nucleon, $\langle E_{F0}\rangle$. From the above three constraints, Eqs. (6.211)–(6.213), and noting that $u = 1$ at $n = n_0$, one can solve for the parameters A, B, and σ. Using

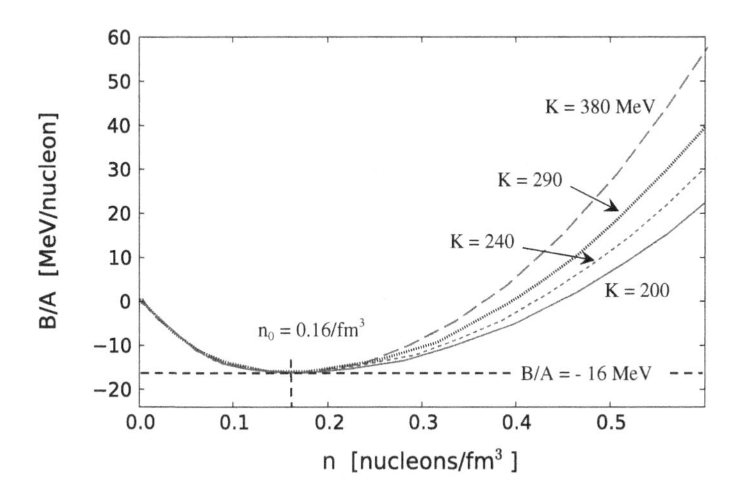

Figure 6.13. Binding energy per nucleon as a function of the nuclear matter density. The numerical solutions for different compressibilities are shown.

a particular value of K and the average value of the kinetic energy, the energy density is completely determined and so is the pressure,

$$p(n) = n^2 \frac{d}{dn}\left(\frac{\epsilon}{n}\right) = n_0 \left[\frac{2}{3}\langle E_{F0}\rangle u^{5/3} + \frac{A}{2}u^2 + \frac{B\sigma}{\sigma+1}u^{\sigma+1}\right]. \tag{6.216}$$

Many other parameterizations of the nuclear matter density are available in the literature [BP12]. An analogous procedure as above are followed to calculate the EOS.

Figure 6.13 shows the binding energy per nucleon as a function of the nuclear matter density. The position of the minimum is at $n = n_0 = 0.16\,\text{fm}^{-3}$, its depth there is $BE = -16\,\text{MeV}$, and its curvature (second derivative) there corresponds to the nuclear compressibility K.

Non-symmetric nuclear matter

We now represent the neutron and proton densities in terms of a parameter α as [Pra96,SR04]

$$n_n = \frac{1+\alpha}{2}n, \quad n_p = \frac{1-\alpha}{2}n, \tag{6.217}$$

where, for pure neutron matter, $\alpha = 1$. This definition implies that

$$\alpha = \frac{n_n - n_p}{n} = \frac{N-Z}{A}. \tag{6.218}$$

It is also usual to define the fraction of protons in the star as

$$x = \frac{n_p}{n} = \frac{1-\alpha}{2}. \tag{6.219}$$

For symmetric nuclear matter, $\alpha = 0$ (or $x = 1/2$).

The contributions to the kinetic energy part of ϵ from both neutrons and protons yield

$$\epsilon_{KE}(n,\alpha) = \frac{3}{5}\frac{k_{F,n}^2}{2m_N}n_n + \frac{3}{5}\frac{k_{F,p}^2}{2m_N}n_p = n\langle E_F\rangle\frac{1}{2}[(1+\alpha)^{5/3} + (1-\alpha)^{5/3}], \tag{6.220}$$

where

$$\langle E_F\rangle = \frac{3}{5}\frac{\hbar^2}{2m_N}\left(\frac{3\pi^2 n}{2}\right)^{2/3} \tag{6.221}$$

is the mean kinetic energy of symmetric nuclear matter at density n. For $n = n_0$ one has $\langle E_F\rangle = 3\langle E_{F0}\rangle/5$ [see Eq. (6.215)]. For non-symmetric matter, $\alpha \neq 0$, the excess kinetic energy is

$$\begin{aligned}\Delta\epsilon_{KE}(n,\alpha) &= \epsilon_{KE}(n,\alpha) - \epsilon_{KE}(n,0)\\ &= n\langle E_F\rangle\left\{\frac{1}{2}[(1+\alpha)^{5/3} + (1-\alpha)^{5/3}] - 1\right\}\\ &= n\langle E_F\rangle\{2^{2/3}[(1-x)^{5/3} + x^{5/3}] - 1\}.\end{aligned} \tag{6.222}$$

For pure neutron matter, $\alpha = 1$,

$$\Delta\epsilon_{KE}(n,\alpha) = n\langle E_F\rangle(2^{2/3} - 1). \tag{6.223}$$

To leading order in α,

$$\Delta\epsilon_{KE}(n,\alpha) = n\langle E_F\rangle\frac{5}{9}\alpha^2\left(1 + \frac{\alpha^2}{27} + \cdots\right) = n\,E_F\frac{\alpha^2}{3}\left(1 + \frac{\alpha^2}{27} + \cdots\right). \tag{6.224}$$

We can also assume the quadratic approximation in α works well for the total energy per particle, i.e.,

$$E(n,\alpha) = E(n,0) + \alpha^2 S(n). \tag{6.225}$$

The isospin-symmetry breaking is proportional to α^2, which reflects the pairwise nature of the nuclear interactions. The function $S(u)$, $u = n/n_0$, only depends on the density of the symmetric nuclear matter.

From the energy density, $\epsilon(n,\alpha) = n_0 u E(n,\alpha)$, the corresponding pressure is, from Eq. (6.214),

$$p(n,x) = u\frac{d}{du}\epsilon(n,\alpha) - \epsilon(n,\alpha) = p(n,0) + n_0\alpha^2 S(n), \qquad (6.226)$$

where $p(n,0)$ is defined by Eq. (6.216).

With these EOS we can now solve the TOV equation for a pure neutron star with nuclear interactions. There are substantial departures of the solutions for symmetric and asymmetric EOS [Hae07].

Quark phase EOS

The matter in the core of a NS possesses densities ranging from a few times n_0 ($0.16\,\mathrm{fm}^{-3}$, the normal nuclear matter density) to one order of magnitude higher. Therefore, a detailed knowledge of the EOS is required for densities $n \gg n_0$, where a description of matter only in terms of nucleons and leptons may be inadequate. In fact, at densities $n \gg n_0$ several species of other particles, such as hyperons and Δ isobars, may appear, and meson condensations may take place. At very high densities, nuclear matter is expected to undergo a transition to a quark–gluon plasma [Wit84,Bay85,Gle90]]. Besides the cold quark–gluon phase within a NS, a similar quark–gluon phase of matter possibly exists at extremely high temperatures and densities (see Section 5.7). It is believed to have existed in the first $10^{-5}\,\mathrm{s}$ after the Big Bang. The maximum mass of a NS changes when one takes into account the phase transition from hadronic matter to quark matter inside the neutron star.

One of the possible ways to handle the deconfined quark phase, is to use the MIT bag model [Cho74]. The total energy density is the sum of a non-perturbative energy shift B, the bag constant, and the kinetic energy for non-interacting massive quarks of flavors f with mass m_f and Fermi momentum $k_F^{(f)} = (3\pi^2 n_f)^{1/3}$, with n_f as the quarks' density of flavor f (see Section 3.2.4),

$$\epsilon = B + \sum_f \frac{3m_f^4}{8\pi^2}[x_f\sqrt{x_f^2+1}(2x_f^2+1)^2 - \sinh^{-1} x_f], \qquad (6.227)$$

where $x_f = k_F^{(f)}/m_f$. One usually considers massless u and d quarks, whereas the s quark mass is assumed to be about 150 MeV. The *bag constant* B is interpreted as the difference between the energy densities of the perturbative vacuum and the physical vacuum [Cho74]. Inclusion of a perturbative interaction among quarks introduces additional terms in the thermodynamic potential and hence in the number density and the energy density. In the original MIT bag model the bag constant has the value $B \sim 50\,\text{MeVfm}^{-3}$, which is quite small when compared with the ones ($\sim$200 MeV fm^{-3}) estimated from lattice calculations [Lep00].

The use of an energy density and a corresponding EOS from Eq. (6.227) generates stable solutions for compact stars, yielding small and really dense stars mainly composed of quark matter. But the description based on Eq. (6.227) is oversimplified. Several other models have been developed, with mixed phases (hybrid stars) including quarks, hyperons, the possibility of formation of quark nuggets, stars made up of strange quarks, and so on. The introduction of strangeness reduces Pauli repulsion by increasing the flavor degeneracy and ensures a lower charge-to-baryon ratio for strange quark matter compared to nuclear matter. Since the electron chemical potential, μ_e, is significantly smaller than the quark chemical potential μ_q, a parameterization of the EoS can be obtained by expanding in powers of μ_e/μ_q [JRS06],

$$p_{\text{QM}} = p_0(\mu_q, m_s) - n_q(\mu_q, m_s)\mu_e + \frac{1}{2}\chi_q(\mu_q, m_s)\mu_e^2 + \cdots \tag{6.228}$$

where p_0, n_q, and χ_q are well-defined and calculable functions of μ_q and the strange quark mass, m_s. This second-order expansion, which neglects the electron pressure $p_e \sim \mu_e^4$, can be used to model an EOS predicted by QCD. The mass–radius relation for stars composed of mixed and pure quark phases is diverse [JRS06].

Rotation

Neutron stars rotate very fast due to the conservation of angular momentum during their creation. As the NS absorbs orbiting matter from a companion star, it increases the rotation to several hundred times per second and they become more oblate. After creation neutron stars slow down because their rotating magnetic fields radiate energy. The rate at which the rotation of a NS slows down is usually constant and very small, about $-(d\omega/dt)/\omega = 10^{-10} - 10^{-21}$ seconds/rotation. Sometimes a NS spins up or undergoes a *glitch*, a sudden small increase of its rotation. Glitches are often linked to starquake. As the rotation of the star slows down, its shape becomes more spherical. Due to the stiffness of the NS crust, the crust might rupture, creating a starquake. After the starquake, the star will have a smaller equatorial radius, and since angular momentum is conserved, rotational

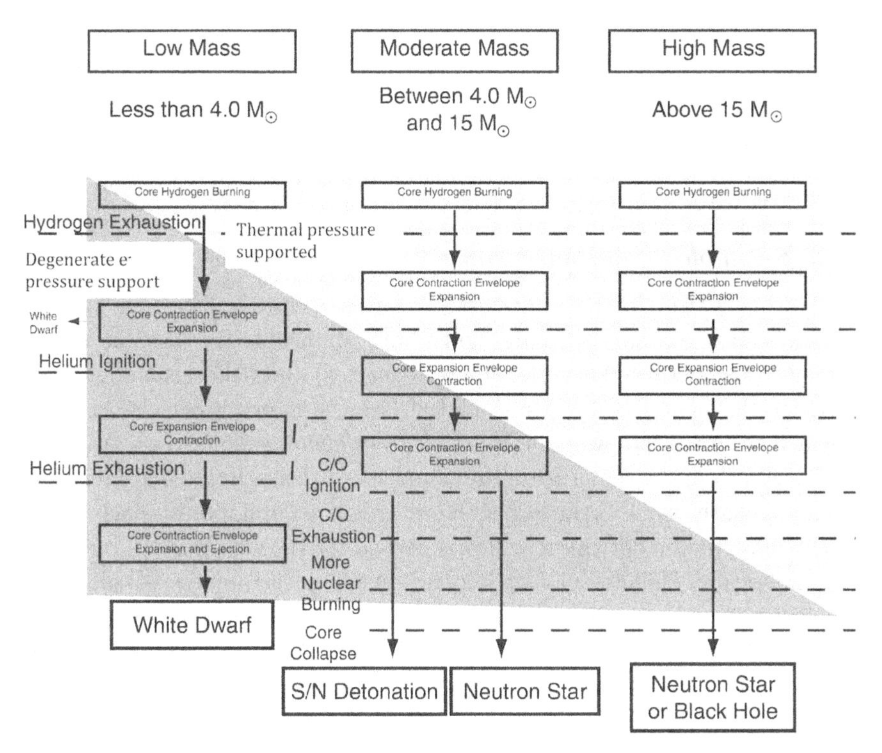

Figure 6.14. Schematic description of stellar evolution in terms of the environmental conditions. The path to formation of white dwarfs, neutron stars, supernova (SN) explosions and black holes is emphasized.

speed increases. Glithces might also be caused by transitions of vortices in the superfluid core of the star from one metastable energy state to a lower one [AI76].

6.8. Additional Stellar Properties

Figure 6.14 summarizes the several stages of stellar evolution and its dependence on the environment. Stellar evolution is driven by nuclear transformations. Stars with nuclear energy coming from a shell source evolve in two ways: (a) stellar surfaces expand when stellar centers contract, and (b) stellar surfaces contract when stellar centers expand. Adding mass to a depleted stellar core causes it to contract as its mass grows as long as there is no additional nuclear fuel in the core. Ignition of a nuclear fuel under conditions of degeneracy is unstable, leading to quick changes of stellar conditions.

In the following chapters we will discuss how nuclear reactions are experimentally or theoretically described, and how their magnitude influences the outcome of stellar evolution, having impact on cosmology models, astronomical observations, and cosmic ray composition and energy distribution.

6.8.1. *Seismology and pulsation*

In addition to the methods discussed in this chapter of testing stellar structure, there are two other methods, both applicable for the Sun: solar neutrinos (also observed in other stellar scenarios) and helioseismology. We will defer the neutrino discussion to the later chapters, when we discuss nuclear reactions. Here we point out that from helioseismology, we are able to deduce most of the structure of the sound speed through the solar interior. This is almost the same as the temperature structure but not quite since there is a thermodynamic derivative which is needed for the sound speed but not the temperature. Helioseismology depends on the measurement of coherent frequencies of oscillation for global modes. These are classified according to their spatial structure based on the locations and shapes of the nodal lines. The number of nodal surfaces of spherical shape between the center and surface gives the principal eigenvalue n. The total number of nodal surfaces in the angular dimension gives the degree ℓ of the mode and the number of nodal surfaces passing through the axis of rotation gives the angular order m of the mode. The frequencies depend primarily on n and ℓ but through rotation also on m. The resulting acoustic spectrum provides information on the sound travel time from center to surface and on the gradient of helium near the center.

Figure 6.15 shows the structure of coherent solar oscillation modes on an equatorial cut for a few values of ℓ and n. The direction of motion is indicated by the white/dark coding, but the amplitude grows toward the surface more than is indicated in the figure. The grey section in the center is not sampled. These are sectorial modes where $m = \pm\ell$.

Every star in the HR diagram (Chapter 1) has a natural (fundamental) frequency for pulsation. To understand this, consider a pulsation as the resonance of a sound wave. To first order, the speed at which a sound wave traverses a star is

$$v_s = \left(\frac{\gamma P}{\rho}\right)^{1/2}, \tag{6.229}$$

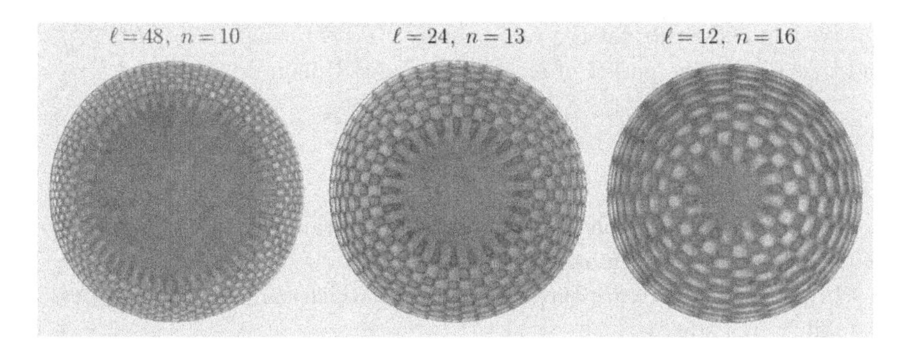

Figure 6.15. The structure of coherent solar oscillation modes on an equatorial cut.

where γ is the ratio of the specific heats. Now for simplicity, let's assume a constant density star. The pressure at any point in such a star can be found by directly integrating Eq. (6.150), which under constant density becomes $dP/dr = -4\pi G\rho^2 r/3$. One gets

$$P(r) = \frac{2}{3}\pi G\rho^2(R^2 - r^2), \tag{6.230}$$

where R is the stellar radius. The period of pulsation is then roughly the time it takes for this sound wave to cross the star, i.e.,

$$\Pi = 2\int_0^R \frac{dr}{v_s} \simeq \int_0^R \left[\frac{2}{3}\pi\gamma G\rho(R^2 - r^2)\right]^{-1/2} dr \simeq \left(\frac{3\pi}{\gamma G\rho}\right)^{1/2}. \tag{6.231}$$

There is only one variable in this equation — the density. To a good approximation, this holds true for all stars pulsating (radially) in the fundamental mode: the period of the star is inversely proportional to the square root of the average density. In other words,

$$\Pi\langle\rho\rangle^{1/2} = C, \tag{6.232}$$

where C is some constant.

An alternative way of seeing this is to consider the recovery time of a pulsating star if gravity is the restoring force. Under gravity, the infall time for a pulsating star is just the free-fall, or dynamical, timescale, given by Eq. (6.127). Since the free fall is one-half of the period, we get

$$\Pi^2 = \frac{R^3}{M} \propto \frac{1}{\langle\rho\rangle}, \tag{6.233}$$

and again $\Pi\langle\rho\rangle^{1/2} = C$.

We can substitute for the radius in Eq. (6.127) using $L = 4\pi R^2 \sigma T_{eff}^4$, and thus obtain the period of any star in the HR diagram

$$\Pi^2 = \frac{L^{3/4}}{T_{eff}^3 M^{1/2}}. \tag{6.234}$$

In other words, the temperature–luminosity diagram can just as easily be plotted as a period–luminosity diagram.

In principle, there are three mechanisms which can cause mechanical instability in a star.

(a) The ϵ *mechanism*: If the center of the star is compressed slightly, the nuclear reaction rates will go up, causing an increase in expansion. The expansion can then decrease the reaction rates, cool the central core, and cause contraction.

(b) The κ *mechanism*: Suppose the opacity in some region of a star were to increase with density. Upon compression, the material would absorb more energy, heat up, and expand. In the ensuing expansion, the opacity would decrease, heat would be lost from the system, and the material would fall back down. Pulsation would be driven by changes in the opacity.

(c) The γ *mechanism*: If, during compression, a region of the star were to heat up less than its surroundings, heat would flow into it. This heat could then cause the region to expand, and in the expansion, the excess heat could be returned to its surroundings. The specific heat of the gas would drive pulsation.

6.9. Exercises

1. Estimate the timescale for the Sun's radius to change by an observable amount (as a function of β in Eq. (6.11)). Assume that β is small. Is the timescale likely? (Assume $r = 7 \times 10^8$ m ; $g = 2.5 \times 10^2$ m/s^2).
2. What is the mean distance between particles in the Sun? Does it justify the assumption of an ideal gas?
3. We know from geological and fossil records that it is unlikely that the Sun has changed its flux output significantly over the last 10^9 y. Hence find an upper limit for β. What does this imply about the assumption of hydrostatic equilibrium?
4. What is the dynamical timescale of Eq. (6.12) for the Sun? (use $R_\odot = 7 \times 10^8$ m, $M_\odot = 1.99 \times 10^{30}$ kg).

5. Show that the statement following Eq. (6.23) is true. Note that work done = force × distance = mass × acceleration × distance.
6. Can you show that fusion reactions can release enough energy to feasibly power a star? Use the atomic weight of H as 1.008172 amu and that of ^{4}He as 4.003875 amu.
7. Show that the the helium mass fraction is

$$Y = 2\left(1 + \frac{n_p}{n_n}\right)^{-1}$$

   if all the baryons in the Universe are in H and He.
8. What would have to be the mass of a star compared to the Sun's mass so that the radiation pressure would be of the same order of magnitude as the ideal gas pressure?
9. Using Eq. (3.31), calculate the density of electrons (in kg/m^3) so that the Fermi energy $E_F = m_e c^2$. How does this compare to the density of water on Earth?
10. It turns out that the conduction (i.e., free) electrons inside metals are highly degenerate (since the number of electrons per unit volume is very large). Indeed, most metals are hard to compress as a direct consequence of the high degeneracy pressure of their conduction electrons. For example, the number density of free electrons in magnesium is $\rho \simeq 8.6 \times 10^{28}\,\mathrm{m^{-3}}$. Estimate the bulk modulus for this metal. The actual bulk modulus is $B = 4.5 \times 10^{10}\,\mathrm{N\,m^{-3}}$.
11. Show that in the low energy limit, $x \ll 1$, Equation (6.70) yields the non-relativistic pressure given by Eq. (6.61).
12. Calculate the bulk modulus for an ultrarelativistic electron gas.
13. Estimate the mass of a white dwarf at which the relativistic degenerate equation of state becomes essential for representing its structure.
14. Prove the Lane–Emden equation, Eq. (6.77), by substitution of the definition of θ and ξ on the preceding equation.
15. Show that the values of $\rho_c/\bar{\rho}$ for $n = 0$ and $n = 1$ in Table 6.1 are correct.
16. Calculate the percent difference between the values of ξ_n based on Eq. (6.80) with the exact values given in Table 6.1.
17. From the dynamical time calculate the speed of free-fall for a neutron star.
18. Suppose we have matter in which electrons, protons, and neutrons all have the same number density. For a low density, which has the highest Fermi energy?

19. What is the approximate value of the electron Fermi energy when $\rho = 10^6\,\mathrm{g\,cm^{-3}}$?
20. At the average density of a $1.4\,M_\odot$, $R = 10\,\mathrm{km}$ neutron star, what is the Fermi energy of a neutron?
21. Suppose that pure hydrogen falls on a neutron star surface, and that it merely settles without undergoing any fusion. Suppose also that at some point it fuses instantly and completely to helium. With these (incorrect) assumptions, make an energetic calculation to determine approximately what fraction of the accreted matter could be ejected to zero speed at infinity.
22. Do the same calculation as in problem above, except for a white dwarf of mass $0.6\ M_\odot$ and radius 10^9 cm. What do the results of your calculations suggest about some important differences between X-ray bursts on neutron stars, and classical novae from white dwarfs (both involving runaway fusion)?

Chapter 7

Reactions in the Cosmos

7.1. Introduction

Energy production in the stars is a well known process. The initial energy which ignites the process arises from the gravitational contraction of a mass of gas. The contraction increases the pressure, temperature, and density at the center of the star until values are achieved which enable thermonuclear reactions, initiating the star lifetime. The energy liberated in these reactions yield a pressure in the plasma, which opposes compression due to gravitation. Thus, an equilibrium is reached for the energy which is produced, the energy which is liberated by radiation, the temperature, and the pressure. The Sun is obviously a special case for us. The temperature in its surface is about 6000°C, while in its interior the temperature reaches 1.5×10^7 K, with a pressure given by 6×10^{11} atm and a density of $150\,\mathrm{g/cm^3}$. The present mass of the Sun is $M_\odot = 2\times10^{33}$ g and its main composition is hydrogen (70%), helium (29%) and less than 1% of more heavy elements, like carbon, oxygen, etc.

For reactions involving charged particles, nuclear physicists often encounter cross-sections near the Coulomb barrier of the order of millibarns. One can obtain a characteristic luminosity L_C based on this cross-section and the nuclear energy released per reaction

$$L_C \sim \frac{\epsilon N \Delta E}{\tau_C}, \tag{7.1}$$

where $\epsilon \approx 10^{-2}$ is the fraction of total number of solar nuclei $N \sim 10^{57}$ that take part in nuclear fusion reactions generating typically $\Delta E \sim 25$ MeV in hydrogen to helium conversion. Here, τ_C is the characteristic timescale for reactions, which becomes minuscule for the cross-sections at the Coulomb barrier and the ambient density and relative speed of the reactants, that is,

$$\tau_C \sim \frac{1}{n\sigma v}. \tag{7.2}$$

With $n = 10^{26}\,\mathrm{cm}^{-3}$, $\sigma = 1\,\mathrm{mb}$, and $v = 10^9\,\mathrm{cm\,s}^{-1}$, this would imply a characteristic luminosity of $L_C \approx 10^{20} L_\odot$, even for a small fraction of the solar material taking part in the reactions (i.e., $\epsilon \sim 10^{-2}$). If this were really the appropriate cross-section for the reaction, the Sun would have burned out very quickly. Instead, the cross-sections are much less than that at the Coulomb barrier penetration energy (say at proton energies of 1 MeV), to allow for a long lifetime of the Sun (in addition, weak-interaction process gives a smaller cross-section for some reactions than electromagnetic process).

Stellar nuclear reactions can be either: (a) charged particle reactions (both target and projectile are nuclei) or (b) neutral particle (neutron) induced reactions. Both sets of reactions can go through either a resonant state of an intermediate nucleus or can be a non-resonant reaction. In the former reaction, the intermediate state could be a narrow unstable state, which decays into other particles or nuclei. In general, a given reaction can involve both types of reaction channels. In charged particle induced reactions, the cross-section for both reaction mechanisms drops rapidly with decreasing energy, due to the effect of the Coulomb barrier (and thus it becomes more difficult to measure stellar reaction cross-sections accurately). In contrast, the neutron induced reaction cross-section is very large and increases with decreasing energy (here, resonances may be superposed on a smooth non-resonant yield which follows the $1/v \sim 1/\sqrt{E}$ dependence). These reaction rates and cross-sections can be then directly measured at stellar energies that are relevant (if such nuclei are long lived or can be generated). The schematic dependence of the cross-sections are shown in Fig. 7.1.

In the next sections we discuss how nuclear reaction rates are calculated for charged and neutral particles. The main input of these calculations are reaction cross-sections, many of which are beyond the possibility of measurement in terrestrial laboratories. Therefore, we also discuss the role that theory and indirect methods have on completing the information needed for stellar modeling.

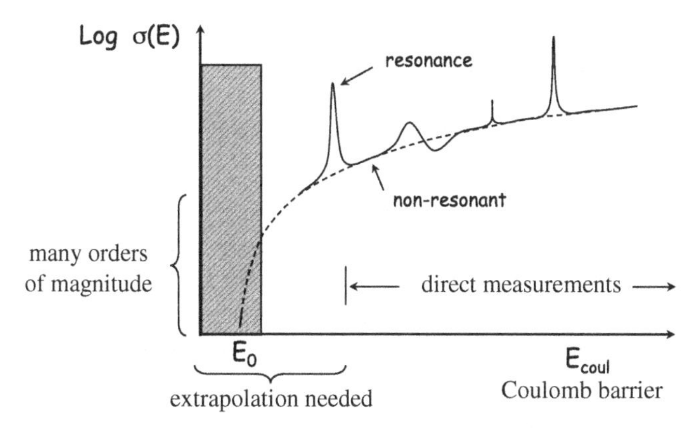

Figure 7.1. Schematic energy dependence of total cross-sections for fusions of charged particles. Note the presence of resonances (narrow or broad) superimposed on a slowly varying non-resonant cross-section. Cross-sections are needed at very low energy, often unaccessible with direct measurements.

7.2. Thermonuclear Cross-Sections

7.2.1. *Nuclear abundances*

Mass fractions and abundances have been discussed in Section 6.3. Here we summarize the main relations. We define N_i as the number of particles of nuclear species i, of one mole mass M_i, in a stellar plasma per unit volume. We can compare the number density to *Avogadro's number* N_A which is the number of atoms of species i which makes M_i grams, $N_A = M_i/m_i = 6.022 \times 10^{23}$ mol^{-1}, where m_i is the mass of a single atom of species i. The mass density is then given by $\rho = N_i m_i = N_i M_i/N_A$. If the mass density contains a mixture of species, then

$$\rho = \frac{1}{N_A}\sum_i N_i M_i. \tag{7.3}$$

Now we define the *mass fraction* of species i,

$$X_i = \frac{N_i M_i}{\rho N_A}, \tag{7.4}$$

in terms of which we can write

$$\frac{\sum_i N_i M_i}{\rho N_A} = \sum_i X_i = 1. \tag{7.5}$$

The *abundance* of species i is defined by (compare Eq. (6.54) — the difference is the choice of units in this section)

$$Y_i = \frac{X_i}{M_i} = \frac{N_i}{\rho N_A}. \tag{7.6}$$

The number density N_i changes in a stellar plasma, if nuclear transmutations take place, or due to variations of the mass density ρ caused by compression or expansion of the stellar gas. In these situations it is more convenient to express abundances in terms of Y_i which remains constant in the absence of nuclear reactions or mixing, whereas N_i is proportional to the mass density ρ. The matter density will also change even if no compression or expansion of the stellar gas occurs. That is because nuclear reactions transform a fraction of the nuclear mass into energy or other lighter particles. That is why the density is sometimes defined as $\rho_A = (1/N_A)\sum_i N_i A_i$ in terms of the mass number A_i instead of the atomic mass M_i, since the number of nucleons is always conserved in a nuclear reaction.

7.2.2. *Reaction rates in the laboratory*

Assume a beam of particles with velocity v incident on a thin target, so that beam intensity is unattenuated throughout target (see Fig. 7.2). The *reaction rate* per target nucleus is defined by the relation $\lambda = \sigma j$, where σ is the reaction cross-section and j is the current density (current per unit area) of the incident beam. If the number density of target nuclei is n_T, then the total reaction rate (reactions per second) is given by $r = \lambda A d n_T = \sigma I n_T$, where A is the cross-sectional area of the beam and d is the target thickness. In the second step, we have used the definition of current density $j = I/A$,

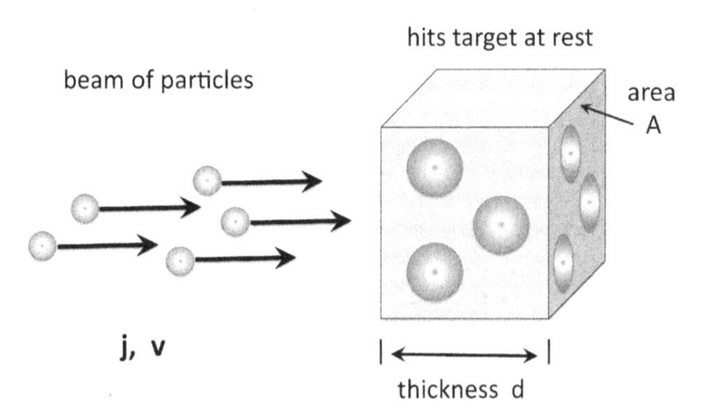

Figure 7.2. Beam of particles incident on a target.

where I is the beam number current (number of particles per second hitting the target). Note that n_T is in fact the number of target nuclei per unit area. Often the *target thickness* is specified in these terms.

7.2.3. *Reaction rates in stellar environments*

We generalize the above discussion by defining the nuclear cross-section for a reaction between a nuclear target j and a nuclear projectile k as

$$\sigma = \frac{\text{number of reactions target}^{-1}\text{sec}^{-1}}{\text{flux of incoming projectiles}} = \frac{r/n_j}{n_k v}, \tag{7.7}$$

where the target number density is given by n_j, the projectile number density is given by n_k, and v is the relative velocity between target and projectile nuclei. The number of reactions per unit volume and time can be expressed as $r = \sigma_{jk} v n_j n_k$, or, more generally, by

$$r_{j,k} = \int \sigma_{jk} |\mathbf{v}_j - \mathbf{v}_k| d^3n_j d^3n_k, \tag{7.8}$$

because what matters is the relative velocity $v = |\mathbf{v}_i - \mathbf{v}_k|$ between the particles. In terms of relativistic kinematics, the relative velocity is defined by

$$|\mathbf{v}_j - \mathbf{v}_k| = \frac{\sqrt{(p_j \cdot p_k)^2 - m_j^2 m_k^2}}{E_j E_k}, \tag{7.9}$$

where the dot product between the particle momenta four-vector is $p_j \cdot p_k = E_j E_k - \mathbf{p}_j \cdot \mathbf{p}_k$ (see Appendix A for the definition of four-momenta). It is left as an exercise to show that, in the rest frame of particle k, one has $|\mathbf{p}_j|/E_j = v_j$, and that in the non-relativistic case one obtains $m_j m_k |\mathbf{v}_j - \mathbf{v}_k|/E_j E_k \sim |\mathbf{v}_j - \mathbf{v}_k|$.

The evaluation of the integral in Eq. (7.8) depends on the type of particles and their distributions. For nuclei j and k in an astrophysical plasma, obeying a Maxwell–Boltzmann (MB) distribution,

$$d^3n_j = n_j \left(\frac{m_j}{2\pi kT}\right)^{3/2} \exp\left(-\frac{m_j^2 v_j}{2kT}\right) d^3v_j. \tag{7.10}$$

Equation (7.8) simplifies to $r_{j,k} = \langle \sigma v \rangle n_j n_k$, where the reaction rate $\langle \sigma v \rangle$ is the average of σv over the temperature distribution in (7.10). More specifically,

$$r_{j,k} = \frac{1}{V}\frac{dN}{dt} = \langle\sigma \mathrm{v}\rangle_{j,k}\frac{n_j n_k}{1+\delta_{jk}}, \tag{7.11}$$

where

$$\langle\sigma \mathrm{v}\rangle_{j,k} = \left(\frac{8}{m_{jk}\pi}\right)^{1/2}(kT)^{-3/2}\int_0^\infty E\sigma(E)\exp\left(-\frac{E}{kT}\right)dE. \tag{7.12}$$

Here m_{jk} (or μ) denotes the reduced mass of the target-projectile system. Notice that we have introduced the factor $1+\delta_{jk}$ in Eq. (7.11) to account for the case of identical particles, $i = k$. The rate should be proportional to the number of pairs of interacting particles in the volume. If the particles are distinct, that is just $\propto n_j n_k$. But if the particles are identical, the sum over distinct pairs is $\propto n_j n_k/2$.

7.2.4. *Photons*

When in Eq. (7.8) particle k is a photon, the relative velocity is always c and there is no need to integrate quantities over $d^3 n_j$. Thus, one obtains $r_j = \lambda_{j,\gamma} n_j$, where $\lambda_{j,\gamma}$ results from an integration of the photodisintegration cross-section over a Planck distribution for photons of temperature T,

$$d^3 n_\gamma = \frac{E_\gamma^2}{\pi^2(c\hbar)^3}\frac{1}{\exp(E_\gamma/kT)-1}dE_\gamma, \tag{7.13}$$

which leads to

$$r_j = \lambda_{j,\gamma}(T)n_j = \frac{1}{\pi^2(c\hbar)^3}\int d^3 n_j \int_0^\infty \frac{c\sigma(E_\gamma)E_\gamma^2}{\exp(E_\gamma/kT)-1}dE_\gamma. \tag{7.14}$$

There is, however, no direct need to evaluate photodisintegration cross-sections, because, due to detailed balance, they can be expressed by the capture cross-sections for the inverse reaction $l + m \to j + \gamma$. Using the reciprocity theorem, Eq. (5.139), together with the same arguments leading to the Saha Equation, (4.97), one obtains

$$\lambda_{j,\gamma}(T) = \left(\frac{\xi_l \xi_m}{\xi_j}\right)\left(\frac{A_l A_m}{A_j}\right)^{3/2}\left(\frac{m_u kT}{2\pi\hbar^2}\right)^{3/2}\langle\sigma \mathrm{v}\rangle_{l,m}\exp\left(-\frac{Q_{lm}}{kT}\right), \tag{7.15}$$

where $m_u = m_{^{12}C}/12$ is the *atomic mass unit*, Q_{lm} is the *reaction Q-value*, which measures the energy gained (or lost) due to the difference between the initial and final masses,

$$Q_{lm} = (m_l + m_m - m_j)c^2. \qquad (7.16)$$

In Eq. (7.15), $\langle \sigma \mathrm{v} \rangle_{l,m}$ is the inverse reaction rate, $\xi(T) = \sum_i (2J_i + 1)\exp(-E_i/kT)$ are partition functions, and A_i are the mass numbers of the participating nuclei in a thermal bath of temperature T.

7.2.5. *Electrons*

A procedure similar to Eq. (7.14) is used for electron captures by nuclei. Because the electron is about 2000 times less massive than a nucleon, the velocity of the nucleus j is negligible in the center of mass system in comparison to the electron velocity ($|v_j - v_e| \approx |v_e|$). The electron capture cross-section has to be integrated over a Boltzmann, partially degenerate, or Fermi distribution of electrons, depending on the astrophysical conditions. The electron capture rates are a function of T and $n_e = Y_e \rho N_A$ is the electron number density given in terms of the electron fraction Y_e. In a neutral, completely ionized plasma, the electron abundance is equal to the total proton abundance in nuclei $Y_e = \sum_i Z_i Y_i$, and

$$r_j = \lambda_{j,e}(T, \rho Y_e) n_j. \qquad (7.17a)$$

7.2.6. *Other particles*

This treatment can be generalized for the capture of positrons, which are in thermal equilibrium with photons, electrons, and nuclei. At high densities ($\rho > 10^{12}$ g.cm^{-3}), such as during supernovae explosions, the size of the neutrino scattering cross-section on nuclei and electrons ensures that enough scattering events occur to thermalize a neutrino distribution. Then also the inverse process to electron capture (neutrino capture) can occur and the neutrino capture rate can be expressed similarly to Eqs. (7.14) or (7.17a), integrating over the neutrino distribution. Also inelastic neutrino scattering on nuclei can be expressed in this form. Finally, for normal decays, like beta or alpha decays with half-life $\tau_{1/2}$, we obtain an equation similar to Eqs. (7.14) or (7.17a) with a decay constant λ_j and

$$r_j = \frac{\text{number of decays}}{\text{unit time . unit volume}} = \lambda_j n_j. \qquad (7.18)$$

Process of interest in stars may involve the decay of particle j to possible final states, which we might number k, m, n, etc. In this case k might stand for a final nucleus, an electron, and an antineutrino, in the case of β decay. The rate can be written

$$r_j = n_j(\lambda_{jk} + \lambda_{jm} + \lambda_{jn} + \cdots),$$

where each λ_{jk} describes a decay channel. The *mean lifetime* is defined as

$$\tau_{jk} = \langle t \rangle = \int_0^\infty e^{-\lambda_{jk} t} t dt = \frac{1}{\lambda_{jk}}.$$

Note that the *half-life* is defined by

$$\frac{1}{2} = e^{-\lambda_{jk}\tau_{jk}^{(1/2)}} \quad \Rightarrow \quad \tau_{jk}^{(1/2)} = \frac{\ln 2}{\lambda_{jk}} = (\ln 2)\tau_{jk}.$$

As the total decay rate is

$$\lambda_{\text{total}} = \lambda_{jk} + \lambda_{jm} + \cdots ,$$

it follows that

$$\frac{1}{\tau_{\text{total}}} = \frac{1}{\tau_{jk}} + \frac{1}{\tau_{jm}} + \cdots$$

7.2.7. *Reaction networks*

The time derivative of the number densities of each of the species in an astrophysical plasma (at constant density) is governed by the different expressions for r, the number of reactions per volume and time, as discussed above for the different reaction mechanisms which can change nuclear abundances. If we define by $N^i_{j,k,l,\ldots}$ the number of particles of nuclear species i created or destroyed by the reaction $j + k + l + \cdots \leftrightarrow i$, we have

$$\left(\frac{\partial n_i}{\partial t}\right)_{\rho=const} = \sum_j N^i_j r_j + \sum_{j,k} N^i_{j,k} r_{j,k} + \sum_{j,k,l} N^i_{j,k,l} r_{j,k,l}. \tag{7.19}$$

The reactions listed on the right hand side of the equation belong to the three categories of reactions: (1) decays, photodisintegrations, electron and positron captures and neutrino induced reactions ($r_j = \lambda_j n_j$), (2) two-particle reactions ($r_{j,k} = \langle\sigma v\rangle_{j,k} n_j n_k$), and (3) three-particle reactions ($r_{j,k,l} = \langle\sigma v\rangle_{j,k,l} n_j n_k n_l$) like the triple-alpha process ($\alpha + \alpha + \alpha \longrightarrow {}^{12}\text{C} + \gamma$). Note that the $N_i^\prime s$ can be positive or negative numbers.

In order to exclude changes in the number densities $\dot{n}_i$, which are only due to expansion or contraction of the gas, we introduce the nuclear abundances $Y_i = n_i/(\rho N_A)$. For a nucleus with atomic weight A_i, A_iY_i represents the mass fraction of this nucleus, therefore $\sum A_iY_i = 1$. In terms of nuclear abundances Y_i, a reaction network is described by the following set of differential equations

$$\dot{Y}_i = \sum_j N^i_j \lambda_j Y_j + \sum_{j,k} N^i_{j,k} \rho N_A < j,k > Y_j Y_k + \sum_{j,k,l} N^i_{j,k,l} \rho^2 N_A^2 < j,k,l > Y_j Y_k Y_l. \tag{7.20}$$

Equation (7.20) derives directly from Eq. (7.19) when the definition for the $Y_i^{\prime} s$ is introduced.

The above set of differential equations needs to be solved numerically. They can be rewritten as difference equations of the form $\Delta Y_i/\Delta t = f_i(Y_j(t + \Delta t))$, where $Y_i(t + \Delta t) = Y_i(t) + \Delta Y_i$. In this treatment, all quantities on the right hand side are evaluated at time $t + \Delta t$. This results in a set of non-linear equations for the new abundances $Y_i(t + \Delta t)$, which can be solved using, e.g., a multi-dimensional Newton–Raphson iteration procedure.

The *total energy generation per unit mass*, due to nuclear reactions in a time step Δt which changed the abundances by ΔY_i, is expressed in terms of the mass excess M_ic^2 of the participating nuclei

$$\Delta\epsilon = -\sum_i \Delta Y_i N_A M_i c^2, \quad \dot{\epsilon} = -\sum_i \dot{Y}_i N_A M_i c^2. \tag{7.21}$$

Therefore, the important ingredients for nucleosynthesis calculations are decay half-lives, electron and positron capture rates, photodisintegrations, neutrino induced reaction rates, and strong interaction cross-sections. The relative abundances of elements, as well as nuclide densities and temperature, are obtained by means of the above equations using a modeling of the initial conditions.

7.3. Astrophysical S-Factor

Nuclear reactions of various types can occur in stars. The first division is between charged reactions and neutron-induced reactions. The physics distinctions are the Coulomb barrier suppression of the former, and the need for a neutron source in the latter. The charged particles break up into two

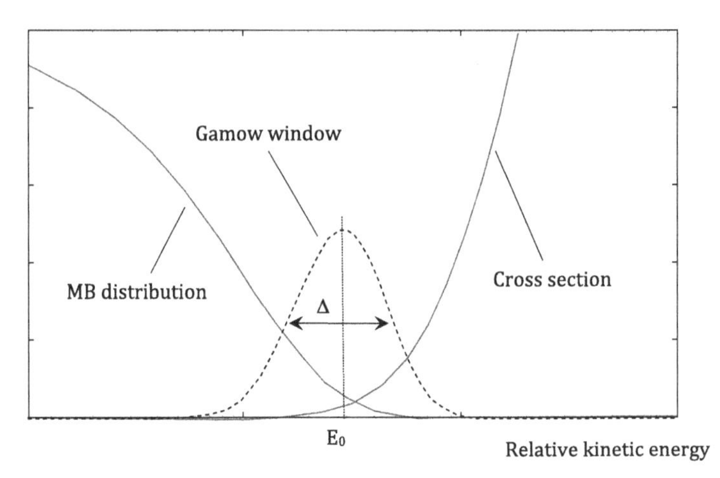

Figure 7.3. Schematic representation of the Gamow window as a result of the product of an exponentially falling MB distribution with a fast growing cross-section in energy. Units are arbitrary.

classes, resonant (where the incident energy corresponds to a resonance) and nonresonant.

7.3.1. *Charged particles: Non-resonant reactions*

Nuclear cross-sections for charged particles are strongly suppressed at low energies due to the Coulomb barrier. For particles having energies less than the height of the Coulomb barrier, the product of the penetration factor and the MB distribution function at a given temperature results in the so-called Gamow peak, in which most of the reactions will take place. Location and width of the Gamow peak depend on the charges of projectile and target, and on the temperature of the interacting plasma (see Fig. 7.3).

The *classical turning point radius* for a projectile of charge Z_j and kinetic energy E_p (in a Coulomb potential $V_c = Z_j Z_k e^2/r$, and effective height of the Coulomb barrier $E_c = Z_j Z_k e^2/R_N = 550\,\text{keV}$ for a p + p reaction[1]), is $R_c = Z_j Z_k e^2/E_p$. Thus, classically a p + p reaction would proceed only when the kinetic energy exceeds 550 keV. Since the number of particles traveling at a given speed is given by the MB distribution, only the tail of the MB distribution above 550 keV is effective when the typical thermal energy is 0.86 keV ($T_9 = 0.01$). The ratio of the tails of

[1] R_N is the strong interaction radius

the MB distributions $n_{MB}(550\,\text{keV})/n_{MB}(0.86\,\text{keV})$ is quite minuscule, and thus classically at typical stellar temperatures this reaction will be virtually absent.

According to the formalism developed in Chapter 5, we can obtain the fusion cross-section from a sum over partial waves, so that

$$\sigma(E) = \sum_l \sigma_l(E), \tag{7.22}$$

where E is the relative kinetic energy, and

$$\sigma_l(E) = \frac{2\pi}{k^2}(2l+1)\beta_l T_l(E), \tag{7.23}$$

where T_l is the *barrier transmission coefficient* and β_l is the probability that fusion occurs when nuclear contact occurs. For simplicity, we assume $\beta_l \sim 1$, but evidently this is an inadequate approximation in many cases, especially those involving rearrangement reactions. The barrier transmission is defined as the ratio of particles entering the nucleus per unit time to the number of particles incident on the barrier per unit time. One can show that the transmission coefficient scales with the nuclear potential depth as

$$T_l = \left(\frac{E}{U_0}\right)^{1/2} P_l(E), \tag{7.24}$$

which is the product of the barrier *penetration factor*, P_l, measuring the probability that the two nuclei touch their surfaces, and of a potential discontinuity factor, due to the difference between the wavelength of the free particle and that of the compound nucleus in the nuclear well $\lambda_0/\lambda \sim \sqrt{E/U_0}$.

Suppose we were interested in the reaction

$$j + k \to m \to n + \gamma,$$

where m is the intermediate nucleus formed in the fusion. To calculate the penetration factor and the cross-section, it will prove sufficient to ask the following question: given the nucleus m, what is the probability for it to decay into the channels $j + k$ and $n + \gamma$? The former will be related by time reversal to the probability for forming the compound nucleus. For definiteness we ask for the rate for decaying into $j + k$. This is

$$\Lambda(m) = \frac{1}{\tau} = \text{prob./sec for flux of } j/k \text{ through a sphere at very large } r,$$

where r is the relative coordinate of the j and k. This can be written in terms of the wave function Ψ of relative motion of j and k as

$$\begin{aligned}
&\lim(r \to \infty)\left[v \int r^2 \sin\theta d\theta d\phi |\Psi(r,\theta,\phi)|^2\right] \\
&\quad = \lim(r \to \infty)\left[v \int \left|\frac{u_l(r)}{r}\right|^2 |Y_{lm}|^2 r^2 \sin\theta d\theta d\phi\right] \\
&\quad = v|u_l(\infty)|^2,
\end{aligned} \tag{7.25}$$

where u_l is the radial part of the wave function. Note that $|u_l(\infty)|^2$ is a constant for very large r. We can rewrite this result as follows

$$\Lambda = vP_l|u_l(R_N)|^2, \quad \text{and} \quad P_l = \frac{|u_l(\infty)|^2}{|u_l(R_N)|^2}, \tag{7.26}$$

where R_N is the nuclear reaction radius and $u_l(R_N)$ is the wave function at the nuclear radius (see Fig. 7.4).

The penetration factor P_l is thus the square of the ratio of the wave function at infinity to that the nuclear radius. If the Coulomb barrier is high, this penetration factor will be very small because the tunneling probability is low. The simplest estimate of this would come from treating the wave function as a pure Coulomb wave function. As discussed in Section 5.4.1,

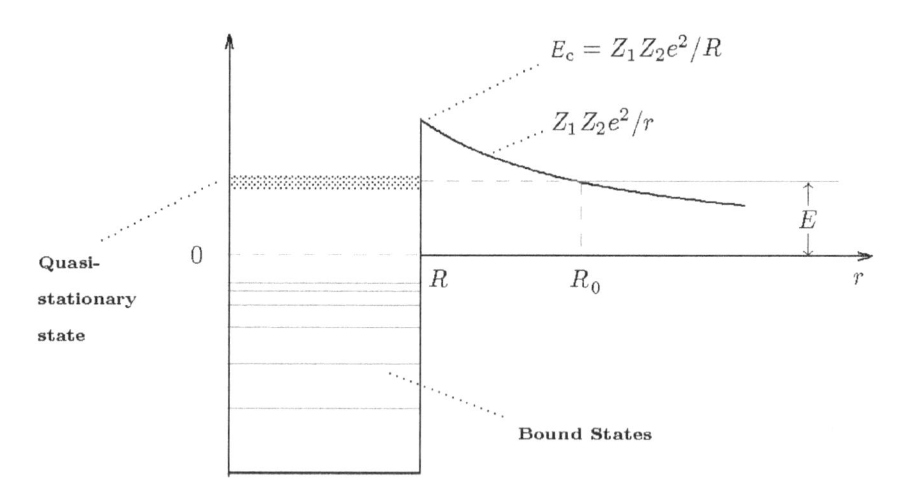

Figure 7.4. Potential barrier for charged particles.

the Coulomb radial equation is

$$\left[\frac{1}{2\mu}\frac{d^2}{dr^2} + \frac{l(l+1)}{2\mu r^2} + \frac{\alpha Z_j Z_k}{r} - E\right] u_l(r) = 0, \tag{7.27}$$

where r is the relative j-k coordinate. Defining

$$E = \frac{p^2}{2\mu}, \quad \rho = pr, \quad \eta = \frac{\alpha Z_j Z_k}{v} = \frac{\alpha Z_j Z_k \mu}{p} = \alpha Z_j Z_k \sqrt{\frac{\mu}{2E}}, \tag{7.28}$$

the outgoing solution of the Schrödinger equation corresponds to the following combination of the standard *Coulomb functions*

$$A[G_l(\rho) + iF_l(\rho)] \longrightarrow A(e^{i(pr - l\pi/2 - \eta \ln 2p + \sigma_l)} \sim Ae^{ipr},$$

where A is a normalization constant and the right-hand side is valid for $r \to \infty$. Thus we find the penetration factor

$$P_l = \frac{|\chi_l(\infty)|^2}{|\chi_l(R_N)|^2} = \frac{1}{|F_l(pR_N)|^2 + |G_l(pR_N)|^2}.$$

In this approximation, the values for the penetration can be obtained by looking up numerical values of the Coulomb functions.

In order to simplify the physics discussion, we consider $l = 0$ only and use approximations for the Coulomb functions at the nuclear radius. This approach leads to a penetration probability given by

$$P_0 = \exp\left(-2KR_c\left[\frac{\tan^{-1}(R_c/R_N - 1)^{1/2}}{(R_c/R_N - 1)^{1/2}} - \frac{R_N}{R_c}\right]\right), \tag{7.29}$$

with $K = [2\mu/\hbar^2(E_c - E)]^{1/2}$. This probability reduces to a much simpler relation at the low energy limit $E \ll E_c$, which is equivalent to the classical turning point R_c being much larger than the nuclear radius R_N. The probability is

$$P = \exp(-2\pi\eta) = \exp\left[-\frac{2\pi Z_j Z_k e^2}{\hbar v}\right] = \exp\left[-31.3 Z_j Z_k \left(\frac{\mu}{E}\right)^{1/2}\right], \tag{7.30}$$

where in the second equality, μ is the reduced mass in atomic mass units and E is the center of mass energy in keV.

The exponential quantity involving the square brackets in the second expression is called the *Gamow factor*. The reaction cross-section between particles of charge Z_j and Z_k has this exponential dependence due to the Gamow factor. In addition, because the cross-sections are essentially "areas" proportional to $\pi\lambda^2 \propto 1/E$, it is customary to write the cross-section, with these two energy dependencies filtered out

$$\sigma(E) = \frac{\exp(-2\pi\eta)}{E} S(E), \tag{7.31}$$

where the factor $S(E)$ is called the *astrophysical S-factor.*

Although this equation was derived for s-waves ($l = 0$), it is used widely as a definition of the astrophysical S-factor. The S-factor may contain degeneracy factors due to spin, e.g. $(2J+1)/(2J_j+1)(2J_k+1)$ as reaction cross-sections are summed over final states and averaged over initial states. Because the rapidly varying parts of the cross-section (with energy) are thus filtered out, the S-factor is a slowly varying function of center of mass energy, at least for the non-resonant reactions. It is thus much safer to extrapolate $S(E)$ to the energies relevant for astrophysical environments from the laboratory data, which is usually generated at higher energies (due to difficulties of measuring small cross-sections), than directly extrapolating the $\sigma(E)$, which contains the Gamow transmission factor (see Fig. 7.5).

One can easily see the two contributions of the velocity distribution and the penetrability in the integral

$$\langle\sigma\mathrm{v}\rangle = \left(\frac{8}{\pi\mu}\right)^{1/2} \frac{1}{(kT)^{3/2}} \int_0^\infty S(E) \exp\left[-\frac{E}{kT} - \frac{b}{E^{1/2}}\right], \tag{7.32}$$

where the quantity $b = 2\pi\eta E^{1/2} = (2\mu)^{1/2}\pi e^2 Z_j Z_k/\hbar$ arises from the barrier penetrability. Experimentally it is very difficult to take direct measurements of fusion reactions involving charged particles at very small energies. The experimental data can be guided by a theoretical model for the cross-section, which can then be extrapolated to the Gamow energy, as

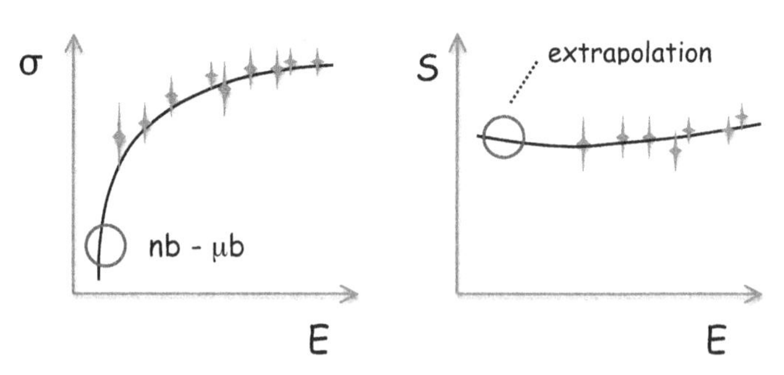

Figure 7.5. Left: Schematic representation of the energy dependence of a fusion reaction involving charged particles. Right: The astrophysical S-factor as defined by Eq. (7.31).

displayed in Fig. 7.5(b). The dots symbolize the experimental data points. The solid curve is a theoretical prediction, which supposedly describes the data at high energies. Its extrapolation to lower energies yields the desired value of the S-factor (and of σ) at the energy E_0. The extrapolation can be inadequate due to the presence of resonances and of subthreshold resonances, as shown schematically in Fig. 7.5.

The exponential in Eq. (7.32) is clearly small at small E and at large E. Now $S(E)$ is assumed to be a slowly varying function. The standard method for estimating such an integral, then, is to find the energy that maximizes the exponential, and expand around this peak in the integrand. This corresponds to solving

$$\frac{d}{dE}\left(\frac{E}{kT}+\frac{b}{\sqrt{E}}\right)=0.$$

The solution is $b = 2E_0^{3/2}/kT$. We now expand the argument of the exponential around this peak energy

$$f(E)=\frac{E}{kT}+\frac{b}{\sqrt{E}}\sim f(E_0)+f''(E_0)\frac{1}{2}(E-E_0)^2+\cdots,$$

as the first derivative $f'(E_0)$ vanishes by definition of E_0.

It follows that

$$\langle\sigma v\rangle\sim\left(\frac{8}{\pi\mu}\right)^{1/2}\frac{1}{(kT)^{3/2}}S(E_0)e^{-f(E_0)}\int_{-\infty}^{\infty}dEe^{-\frac{1}{2}(E-E_0)^2f''(E_0)}.$$

In deriving this result, we have assumed that $S(E)$ is slowly varying in the vicinity of the integrand peak at E_0, and thus can be replaced by its value at the peak.

The integral can be done, yielding $\sqrt{2\pi/f''(E_0)}$. Now $f''(E_o) = 3b/4E_0^{5/2} = 3/2E_0kT$ and $f(E_0) = E_0/kT + b/\sqrt{E_0} = 3E_0/kT$. Thus, we get,

$$\langle\sigma v\rangle=\frac{16}{9\sqrt{3}}\frac{1}{\mu}\frac{1}{2\pi\alpha Z_jZ_k}S(E_0)e^{-3E_0/kT}\left(\frac{3E_0}{kT}\right)^2.$$

Defining a quantity A by

$$Am_N=\frac{A_jA_k}{A_j+A_k}m_N\sim\frac{m_jm_k}{m_j+m_k}=\mu,$$

where m_N is the nucleon mass, evaluating some constants, and dividing out the dimensions of S (note S has the units of a cross-section times energy) yields

$$r_{jk} = \frac{n_j n_k}{1+\delta_{jk}}(7.21 \times 10^{-19}\text{cm}^3/\text{sec})\frac{1}{AZ_jZ_k}\frac{S(E_0)}{\text{keV barns}}e^{-3E_0/kT}\left(\frac{3E_0}{kT}\right)^2.$$

Note that the overall dimensions are clearly $1/(\text{cm}^3\text{sec})$, as the number densities have units $1/\text{cm}^3$. Also recall that a barn $= 10^{-24}$ cm^2.

E_0 is called the *Gamow energy*, which from its definition $E_0 = (kTb/2)^{2/3}$ yields

$$\frac{E_0}{kT} = \left(\frac{\pi\alpha Z_j Z_k}{\sqrt{2}}\right)^{2/3}\left(\frac{\mu c^2}{kT}\right)^{1/3}, \tag{7.33}$$

where the speed of light has been reinserted to make it explicit that this quantity carries no units. For example, in the center of our Sun $kT \sim 1.5 \times 10^7\text{K} \sim 1.3$ keV. So if we plug in the appropriate numbers for the ^{3}He+^{3}He reaction one finds

$$E_0 \sim 16.5kT \sim 21.5\,\text{keV}.$$

One could compare this to the average energy of a Maxwell–Boltzmann distribution of particles of $\langle E\rangle \sim 3kT$. Thus, indeed, the reactions are occurring far out on the Boltzmann tail, where nuclei have a better chance of penetrating the Coulomb barrier.

The location E_0 of the Gamow peak, and the effective width Δ of the energy window, as shown in Fig. 7.3, can also be written as

$$E_0 = 1.22(Z_j^2 Z_k^2 A T_6^2)^{1/3}\,\text{keV},$$

$$\Delta = \frac{16E_0kT}{3}^{1/2} = 0.749(Z_j^2 Z_k^2 A T_6^5)^{1/6}\,\text{keV}, \tag{7.34}$$

where the reduced mass A of the involved nuclei is given in units of m_u, and the temperature T_6 is given in units of 10^6 K.

Note that our formula can easily be improved by doing a Taylor expansion on $S(E)$,

$$S(E) = S(0) + S'(0)E + \frac{1}{2}S''(0)E^2 + \cdots \tag{7.35}$$

Using this expansion in Eq. (7.32) and approximating the product of the exponentials $\exp(-E/k_{\text{B}}T)$ and $\exp[2\pi\eta(E)]$ by a Gaussian centered at the energy E_0, Eq. (7.32) can be evaluated as

$$\langle \sigma v \rangle = \left(\frac{2}{m_{ab}}\right)^{1/2} \frac{\Delta}{(kT)^{3/2}} S_{\text{eff}}(E_0) \exp\left(-\frac{3E_0}{kT}\right), \tag{7.36}$$

with

$$S_{\text{eff}}(E_0) = S(0)\left[1 + \frac{5}{12\tau} + \frac{S'(0)}{S(0)}\left(E_0 + \frac{35E_0}{12\tau}\right) + \frac{S''(0)}{2S(0)}\left(E_0^2 + \frac{89E_0^2}{12\tau}\right)\right]. \tag{7.37}$$

The quantity E_0 defines the effective mean energy for thermonuclear fusion and is given by Eq. (7.34). The quantity τ is given by

$$\tau = \frac{3E_0}{kT}, \tag{7.38}$$

and Δ is given by Eq. (7.34). The maximum value of the integrand in the Equation (7.32) is:

$$\mathrm{I}_{\max} = \exp(-\tau). \tag{7.39}$$

The values of $\mathrm{E}_0, \mathrm{I}_{\max}, \Delta$, etc., for several reactions are tabulated in Table 7.1 for $T_6 = 15$.

As the nuclear charge increases, the Coulomb barrier increases, and the Gamow peak E_0 also shifts towards higher energies. Note how rapidly the maximum of the integrand I_{max} decreases with the nuclear charge and the Coulomb barriers. The effective width Δ is a geometric mean of E_0 and kT, and $\Delta/2$ is much less rapidly varying between reactions (for $kT \ll E_0$). The rapid variation of I_{max} indicates that of several nuclei present in the stellar core, those nuclear pairs which have the smallest Coulomb barrier will have the largest reaction rates. The relevant nuclei will be consumed most rapidly at that stage. (Note however that for the p + p reaction, apart

Table 7.1. Parameters of the thermally averaged reaction rates at $T_6 = 15$.

Reaction	Coulomb Barrier (MeV)	Gamow Peak (E_0) (keV)	I_{max} ($e^{-3E_0/kT}$)	Δ (keV)	$(\Delta)I_{max}$
p + p	0.55	5.9	1.1×10^{-6}	6.4	7×10^{-6}
p + N	2.27	26.5	1.8×10^{-27}	13.6	2.5×10^{-26}
$\alpha + C^{12}$	3.43	56	3×10^{-57}	19.4	5.9×10^{-56}
$O^{16} + O^{16}$	14.07	237	6.2×10^{-239}	40.4	2.5×10^{-237}

from the Coulomb barrier, the strength of the weak force, which transforms a proton to a neutron, also comes into play).

When nuclei of the smallest Coulomb barrier are consumed, there is a temporary dip in the nuclear generation rate, and the star contracts gravitationally until the temperature rises to a point where nuclei with the next lowest Coulomb barrier start burning. At that stage, further contraction is halted. The star therefore goes through well defined stages of different nuclear burning phases in its core at later epochs dictated by the height of the Coulomb barriers of the fuels. Note also from the Table 7.1, how far E_0, the effective mean energy of reaction, is below the Coulomb barrier at the relevant temperature. The stellar burning is so slow because the reactions are taking place at such a far sub-Coulomb region, and this is why the stars can last so long.

7.3.2. *Resonant reactions*

For the case of resonances, where E_r is the resonance energy, we can approximate $\sigma(E)$ by a Breit–Wigner resonance formula.

$$\sigma(E) = \frac{\pi}{k^2}\frac{(2J_r+1)}{(2J_j+1)(2J_k+1)}\frac{\Gamma_\mathrm{p}(E)\Gamma_\gamma}{(E_\mathrm{r}-E)^2+(\Gamma/2)^2}\,, \tag{7.40}$$

where J_r, J_j, and J_k are the spins of the resonance and the nuclei j and k, respectively, and the total width Γ is the sum of the particle decay partial width Γ_p and the γ-ray partial width Γ_γ. The particle partial width, or entrance channel width, Γ_p can be expressed in terms of the *spectroscopic factor* $\mathcal{S}_i$ and the width Γ_jk of the resonance state in the $j+k$ channel, i.e.,

$$\Gamma_\mathrm{p}(E) = \mathcal{S}_i \times \Gamma_\mathrm{jk}(E)\,. \tag{7.41}$$

The spectroscopic factor is a measure of how well a nucleus can be described in terms of a configuration of two clusters j and k: a nucleus is a complex many-body system and the probability (basically, $\mathcal{S}_i$) to find it in a $(j+k)$-configuration does not have to be unity.

Hence, one can re-express the S-factor as

$$S(E) = \frac{\pi}{2\mu}\frac{(2J_r+1)}{(2J_j+1)(2J_k+1)}\frac{\Gamma_\gamma}{(E-E_r)^2+(\Gamma/2)^2} \times\left[\Gamma_p(E)\exp\left(\frac{2\pi\alpha Z_j Z_k}{v}\right)\right].$$

Now the product of the exponential and Γ_p on the right should be roughly energy independent, as the exponential cancels the penetration probability buried in Γ_p. Thus the assumption that $S(E)$ is weakly energy dependent requires that one be not too close to the resonance. The width Γ_{jk} can be calculated from the scattering phase shifts of a scattering potential with the potential parameters being determined by matching the resonance energy. Simple estimates can also be done, as we show next.

From the definition (7.26) we obtain that, for a uniform probability density inside the nucleus,

$$|u_l(R)|^2 dr = \frac{4\pi R^2 dr}{4\pi R^3/3} = \frac{3}{R} dr. \tag{7.42}$$

One defines the *reduced width* θ_l by means of

$$|u_l(R)|^2 = \theta_l^2 \frac{3}{R}. \tag{7.43}$$

For realistic nuclear states

$$0.01 < \theta_l^2 < 1, \tag{7.44}$$

and θ_l^2 gives a measure of the degree to which a quasi-stationary nuclear state can be described by a relative motion of in a potential.

The *partial width* of that state is given by

$$\Gamma_l = \frac{3\hbar v}{R} P_l \theta_l^2. \tag{7.45}$$

One needs the tunneling probability for a potential V_l containing Coulomb, nuclear, and centrifugal terms. A simple estimate can be done by using *WKB wave functions* for $u_l(r)$. This procedure yields

$$P_l = \left[\frac{V_l(R) - E}{E}\right]^{1/2} \exp\left\{-\frac{2\sqrt{2\mu}}{\hbar}\int_R^{R_c} [V_l(r) - E]^{1/2} dr\right\}, \tag{7.46}$$

with $l(l+1) \longrightarrow (l+1/2)^2$. A numerical computation of the above equation shows that a good approximation consists in replacing $V_l(R) - E \simeq E_c = Z_1 Z_2 e^2/R$ in the factor preceding the exponential. This approximation is justified since at astrophysical energies $E \ll E_c$.

Calling the exponent in Eq. (7.46) as W_l, we can show that, to lowest order in E/E_c,

$$W_0 = 2\pi \frac{Z_1 Z_2 e^2}{\hbar v}\left[1 - \frac{4}{\pi}\left(\frac{E}{E_c}\right)^{1/2} + \frac{2}{3\pi}\left(\frac{E}{E_c}\right)^{3/2} - \cdots\right]. \tag{7.47}$$

To first order in E/E_c, the partial width Γ_0 of a state with $l = 0$, in terms of the reduced width θ_0^2 and W_0, is

$$\Gamma_0 \propto \exp(-bE^{-1/2}), \tag{7.48}$$

which displays a characteristic exponential decay with $E^{1/2}$.

For $l \neq 0$,

$$W_l = \frac{2\sqrt{2\mu}}{\hbar} \int_R^{R_c} \left[E_c \frac{R}{r} + E_l \left(\frac{R}{r} \right)^2 - E \right]^{1/2} dr, \tag{7.49}$$

where $E_l = (l + 1/2)^2 \hbar^2 / 2\mu R^2$. For astrophysically relevant cases, $R/R_c \lesssim 10^{-3}$ and the ratio R/r is quite small for most of the range of the integration. As a consequence the second term in the square-root bracket never dominates, and the integrand may be expanded. The leading terms become

$$W_l \simeq \frac{2\sqrt{2\mu}}{\hbar} \int_R^{R_c} \left(E_c \frac{R}{r} - E \right)^{1/2} dr + \frac{\sqrt{2\mu}}{\hbar} \int_R^{R_c} \frac{E_l R^{3/2}}{E_c^{1/2}} \frac{dr}{r^{3/2}}. \tag{7.50}$$

The first term is just equal to W_0, whereas the second term reflects the additional effects of the centrifugal barrier. Equation (7.50) becomes

$$W_l = W_0 + 2 \left[\frac{(l + 1/2)^2 E_l}{E_c} \right]^{1/2} \left[1 - \left(\frac{E}{E_c} \right)^{1/2} \right]. \tag{7.51}$$

Neglecting the correction in $(E/E_c)^{1/2}$ in the above equation, we obtain

$$P_l \approx \left(\frac{E_c}{E} \right)^{1/2} \exp \left[-2\pi \frac{Z_1 Z_2 e^2}{\hbar v} + 4 \left(\frac{2\mu R^2 E_c}{\hbar^2} \right)^{1/2} - 2(l + 1/2)^2 \left(\frac{\hbar^2}{2\mu R^2 E_c} \right)^{1/2} \right], \tag{7.52}$$

where the correction of order $(E/E_c)^{3/2}$ in W_0 has also been dropped. Finally, one obtains the following expression for partial width:

$$\Gamma_l = 6\theta_l^2 \left(\frac{\hbar^2 E_c}{2\mu R^2} \right)^{1/2} \exp(-W_l), \tag{7.53}$$

which is useful to obtain estimates of the particle decay partial width with angular momentum l.

The gamma partial widths Γ_γ are calculated from the electromagnetic reduced transition probabilities $B(J_i \to J_f;L)$ which carry the nuclear structure information of the resonance states and the final bound states (see Chapter 4). Later in this chapter we will discuss an example of a theoretical model for B(E/ML)-values. The reduced transition rates B(E/ML) are usually computed within the framework of the shell model. Most of the typical transitions are M1 or E2 transitions. For these the relations are

$$\Gamma_{\mathrm{E2}}[\mathrm{eV}] = 8.13 \times 10^{-7}\, E_\gamma^5\, [\mathrm{MeV}]\, B(E2)\, [\mathrm{e}^2\mathrm{fm}^4], \tag{7.54}$$

and

$$\Gamma_{\mathrm{M1}}[\mathrm{eV}] = 1.16 \times 10^{-2}\, E_\gamma^3\, [\mathrm{MeV}]\, B(M1)\, [\mu_N^2], \tag{7.55}$$

where the reduced matrix elements B(E/ML) are given in the units inside the brackets.

Back to Eq. (7.40), we obtain that, for the case of narrow resonances with width $\Gamma \ll E_r$, the Maxwellian exponent $\exp(-E/kT)$ can be taken out of the reaction rate integral, and one finds

$$\langle \sigma v \rangle = \left(\frac{2\pi}{\mu k T} \right)^{3/2} \hbar^2 (\omega\gamma)_r \exp\left(-\frac{E_r}{kT} \right), \tag{7.56}$$

where the *resonance strength* is defined by

$$(\omega\gamma)_r = \frac{2J_r + 1}{(2J_j + 1)(2J_k + 1)} (1 + \delta_{jk}) \frac{\Gamma_{\mathrm{p}}\Gamma_\gamma}{\Gamma}. \tag{7.57}$$

For broad resonances Eq. (7.32) is usually calculated numerically. An interference term has to be added. The total capture cross-section is then given by

$$\sigma(E) = \sigma_{\mathrm{nr}}(E) + \sigma_{\mathrm{r}}(E) + 2\left[\sigma_{\mathrm{nr}}(E)\sigma_{\mathrm{r}}(E)\right]^{1/2} \cos[\delta_{\mathrm{r}}(E)] \ , \tag{7.58}$$

where σ_{nr} (σ_r) is the non-resonant (resonant) cross-section and $\delta_{\mathrm{r}}(E)$ is the resonance phase shift. Only the contributions with the same angular momentum of the incoming wave interfere in Eq. (7.58).

Experimentally it is very difficult to perform direct measurements of fusion reactions involving charged particles at very small energies. The experimental data at higher energies can be guided by a theoretical model for the cross-section, which can then be extrapolated down to the Gamow energy. However, the extrapolation can be inadequate due to the presence of resonances and subthreshold resonances, for example.

Figure 7.5 outlines one of the main challenges in astrophysical reactions with charged particles. The experimental data can be guided by a theoretical model for the cross-section, which can then be extrapolated to the Gamow energy. The solid curve is a theoretical prediction, which supposedly describes the data at high energies. Its extrapolation to lower energies yields the desired value of the S-factor, or cross-section, at the Gamow energy E_0. The extrapolation can be complicated by the presence of unknown resonances.

7.3.3. *Neutron-induced reactions*

The low energy behavior of radiative capture cross-sections is fundamental in nuclear astrophysics because of the small projectile energies in the thermonuclear region. For example, for neutron capture near the threshold the cross-section can be written (see Chapter 5) as

$$\sigma_{if} = \frac{\pi}{k^2} \frac{-4kR\Im\mathcal{L}_0}{|\mathcal{L}_0|^2}, \tag{7.59}$$

where $\mathcal{L}_0$ is the logarithmic derivative for the s-wave, and $\Im$ means imaginary part. Since $\mathcal{L}_0$ is only weakly dependent on the projectile energy, one obtains for low energies the well-known $1/v$-behavior for low-energy neutron scattering.

With increasing neutron energy higher partial waves with $l > 0$ contribute more significantly to the radiative capture cross-section. Thus the product σv becomes a slowly varying function of the neutron velocity and one can expand this quantity in terms of v or $\sqrt{E}$ around zero energy:

$$\sigma v = S^{(n)}(0) + \dot{S}^{(n)}(0)\sqrt{E} + \frac{1}{2}\ddot{S}^{(n)}(0)E + \cdots. \tag{7.60}$$

The quantity $S^{(n)}(E) = \sigma v$ is the astrophysical S-factor for neutron-induced reactions and the dotted quantities represent derivatives with respect to $E^{1/2}$, i.e., $\dot{S}^{(n)} = 2\sqrt{E}\, dS^{(n)}/dE$ and $\ddot{S}^{(n)} = 4Ed^2S^{(n)}/dE^2 + 2dS^{(n)}/dE$. Notice that the above astrophysical S-factor for neutron-induced reactions is different from that for charged-particle induced reactions. In the astrophysical S-factor for charged-particle induced reactions the penetration factor through the Coulomb barrier also has to be considered (Eq. 7.31).

Inserting this into Eq. (7.12) we obtain for neutron-induced reactions the reaction rate

$$\langle \sigma v \rangle = S(0) + \left(\frac{4}{\pi}\right)^{\frac{1}{2}} \dot{S}(0)(k_{\rm B}T)^{\frac{1}{2}} + \frac{3}{4}\ddot{S}(0)k_{\rm B}T + \cdots. \tag{7.61}$$

In most astrophysical neutron-induced reactions, neutron s-waves will dominate, resulting in a cross-section showing a $1/v$-behavior (i.e., $\sigma(E) \propto 1/\sqrt{E}$). In this case, the reaction rate will become independent of temperature, $R = const$. Therefore it suffices to measure the cross-section at one temperature in order to calculate the rates for a wider range of temperatures. The rate can then be computed very easily by using

$$R = \langle \sigma v \rangle = \langle \sigma \rangle_T v_T = const, \tag{7.62}$$

with

$$v_T = \left(\frac{2kT}{m} \right)^{1/2} . \tag{7.63}$$

The mean lifetime $\tau_{\rm n}$ of a nucleus against neutron capture, i.e., the mean time between subsequent neutron captures, is inversely proportional to the available number of neutrons $n_{\rm n}$ and the reaction rate $R_{{\rm n}\gamma}$:

$$\tau_{\rm n} = \frac{1}{n_{\rm n} R_{{\rm n}\gamma}} . \tag{7.64}$$

If this time is shorter than the beta-decay half-life of the nucleus, it will be likely to capture a neutron before decaying (r-process). In this manner, more and more neutrons can be captured to build up nuclei along an isotopic chain until the beta-decay half-life of an isotope finally becomes shorter than $\tau_{\rm n}$. With the very high neutron densities encountered in several astrophysical scenarios, isotopes very far from stability can be synthesized.

In the case of neutron-induced reactions the effective energy window has to be derived in a slightly different way. For s-wave neutrons ($l = 0$) the energy window is simply given by the location and width of the peak of the MB distribution function. For higher partial waves the penetrability of the centrifugal barrier shifts the effective energy E_0 to higher energies. For neutrons with energies less than the height of the centrifugal barrier this is approximated by

$$E_0 \approx 0.172 T_9 \left(l + \frac{1}{2} \right) \text{ MeV}, \quad \Delta \approx 0.194 T_9 \left(l + \frac{1}{2} \right)^{1/2} \text{ MeV}, \tag{7.65}$$

The energy E_0 will always be comparatively close to the neutron separation energy.

7.4. Environment Electrons

The form of the astrophysical S-factor given in Eq. (7.31) assumes that the electric charges of nuclei are "bare" charges. However, neither at very low laboratory energies nor in stellar environments is this the case. In stars, the bare Coulomb interaction between the nuclei is screened by the electrons in the plasma surrounding them. If one measures reaction rates in the laboratory, using atomic targets (always), then atomic electrons screen as well. But the screening is different from the screening in the stellar plasma. Therefore we discuss these two problems separately in the following subsections.

7.4.1. *Stellar electron screening*

In astrophysical plasmas with high densities and/or low temperatures, effects of electron screening are very important, as will be discussed later. This means that the reacting nuclei, due to the background of electrons and nuclei, feel a different Coulomb repulsion than in the case of bare nuclei. Under most conditions (with non-vanishing temperatures), the generalized reaction rate integral can be separated into the traditional expression without screening (7.12) and a screening factor

$$\langle \sigma \mathrm{v} \rangle^*_{j,k} = f_{scr}(Z_j, Z_k, \rho, T, Y_i) \langle \sigma \mathrm{v} \rangle_{j,k}, \qquad (7.66)$$

in terms of the nuclear abundances, Y_i.

This screening factor is dependent on the charge of the involved particles, the density, temperature, and the composition of the plasma. At high densities and low temperatures, screening factors can enhance reactions by many orders of magnitude and lead to *pycnonuclear ignition* [Ga38,Wil40,SvH69]. The pycnonuclear regime occurs when the Coulomb barrier is penetrated due to zero-point vibrations of nuclei arranged, for instance, in a lattice. The thermonuclear regime is realized in a rather low-density and warm plasma, whereas the pycnonuclear regime operates at high densities and not too high temperatures. Pycnonuclear reactions are almost temperature independent and occur even at $T = 0$. Pycnonuclear reactions are extremely slow at densities typical for normal stars but intensify with increasing ρ. For example, carbon burns rapidly into heavier elements at $\rho \gtrsim 10^{10}\,\mathrm{g.cm}^{-3}$.

Consider a concentration of negative and positive charges with neutral total charge, that is, $\sum_i Z_i e c_{i0} = 0$, where c_{i0} is the spatially uniform concentration of positive ($i = +$) or negative ($i = -$) charges. Because of the interaction between the charges, these concentrations are no longer spatially uniform, with smaller charges tending to concentrate around larger charges.

The concentrations around the charges are populated according to the statistical distribution of the individual charge energies in the presence of a

Coulomb field $V(r)$, yet to be found. Assuming Boltzmann statistics, this argumentation implies that

$$c_+(r) = c_{+0} \exp\left[-\frac{Z_+ eV(r)}{kT}\right], \quad \text{and} \quad c_-(r) = c_{-0} \exp\left[-\frac{Z_- eV(r)}{kT}\right]. \tag{7.67}$$

If the ion close to which we are considering the screening is positive, then $V(r) > 0$ and $c_+(r) < c_{+0}$, or $c_-(r) > c_{-0}$, and the reverse is true if $V(r) < 0$.

The charge density at position r is given by

$$\rho(r) = \sum_i Z_i e c_i = \sum_i Z_i e c_{i0} \exp\left[-\frac{Z_i eV(r)}{kT}\right]. \tag{7.68}$$

If $Z_i eV(r)/kT \ll 1$ (*weak screening*), then $\rho(r) = -(e^2 V(r)/kT) \sum_i Z_i^2 c_{i0}$.

To obtain the potential $V(r)$ one has to solve the Poisson equation for the potential $V(r)$ which, for the above charge distribution, becomes

$$-\frac{1}{r^2}\frac{d}{dr}\left[r^2 \frac{dV}{dr}\right] = 4\pi\rho(r) = \left(\frac{1}{R_D}\right)^2 V,$$

where the *Debye radius* R_D is defined by

$$R_D^2 = \frac{kT}{4\pi e^2 \sum_i Z_i^2 c_{i0}}. \tag{7.69}$$

Since $V(r) \to 0$ as $r \to \infty$, the solution of this equation is $V(r) = (A/r)\exp(-r/R_D)$. The normalization constant is fixed by the condition $V(r) \to Z_i e/r$ as $r \to 0$. Thus,

$$V(r) = \frac{Z_i e}{r} \exp\left(-\frac{r}{R_D}\right). \tag{7.70}$$

Screening modifies the Coulomb potential between the nuclear radius R and the classical turning point R_0, and consequently modifies the barrier penetration. For weak screening $R_D \gg R, R_0$. In other words, we can expand $V(r)$ around $r = 0$. To first order, the barrier energy for an incoming projectile with charge $Z_2 e$ is $V(r) = Z_1 Z_2 e^2/r + U(r)$, where the Debye–Hückel *screening potential*, $U(r) = U(0) = const.$, is given by $U_0 = -Z_1 Z_2 e^2/R_D$.

The impact of the screening potential on the barrier penetrability and therefore on the astrophysical reaction rates can be approximated through

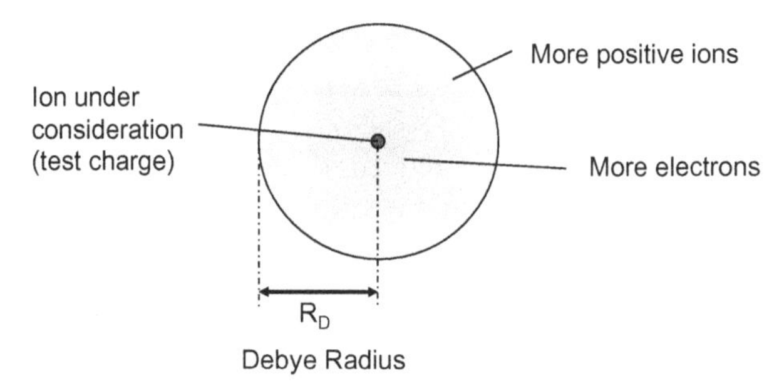

Figure 7.6. Schematic view of the Debye–Hückel sphere. An ion at the center of the sphere is surrounded by a cloud of ions, with the ions of opposite charge (electrons) agglomerating closer to it.

a screening factor $f = \exp(U_0/kT)$, which, in the weak screening limit, becomes $f \simeq 1 - U_0/kT$.

In summary, for the weak screening limit, the reaction rates are modified according to

$$\langle \sigma v \rangle_{screened} = f \langle \sigma v \rangle_{bare}, \tag{7.71}$$

with

$$f = 1 + 0.188 \frac{Z_1 Z_2 \rho^{1/2} \xi^{1/2}}{T_6^{3/2}}, \quad \text{and} \quad \xi = \sum_i (Z_i^2 + Z_i)^2 Y_i. \tag{7.72}$$

The Debye–Hückel model (Fig. 7.6) is an ideal plasma since the average interaction energy between particles is smaller than the average kinetic energy of a particle. In the case of *strong screening*, in low density plasmas, the potential energy cannot be described by the Debye–Hückel model since the probability of finding other charged particles in a Debye sphere almost vanishes. For a strongly coupled plasma, the ion-sphere model [Sal54] is more suitable. The ion-sphere model is equivalent to the *Wigner–Seitz cell* used in condensed-matter theory. As discussed in Chapter 6, the Wigner–Seitz cell around a lattice point is defined as the locus of points in space that are closer to that lattice point than to any of the other lattice points. In the case of a three-dimensional lattice, a perpendicular plane is drawn at the midpoint of the lines between the lattice points. The smallest volume is enclosed in this way and is called the Wigner–Seitz primitive cell. All space within the lattice will be filled by this type of primitive cell and will leave no gaps in the lattice.

The ion-sphere model assumes an ion having Z_b bound electrons, positioned at $r = 0$, and Z_f free electrons ($Z_b + Z_f = Z$) occupying the rest of the ion sphere volume. The plasma effects are taken into account by confining the ion and the Z_f electrons inside the ion sphere. To obtain the potential $V(r)$ one adds to the bare ion potential, $V = Ze^2/r$, the potential due to bound electrons, V_b, and that due to free electrons, V_f. A Slater type, or Kohn–Sham type, exchange-potential is also added. To obtain the bound and free electron densities one solves the Schrödinger equation, or Dirac equation, with $V(r)$. From this, one builds the bound and free electron densities which are then used to calculate the new potentials V_b and V_f. This process is done iteratively until convergence is reached.

The plasma density enters the ion-sphere model through the boundary conditions imposed on the potential, that is, through the neutrality conditions of the ion-sphere. Approximate schemes to obtain the ion-sphere potential have been developed. A widely used approximation for the potential energy of a single free electron electron inside the ion-sphere is given by

$$V(\mathbf{r}, \mathbf{r}') = \left(-\frac{Ze^2}{r} + \frac{e^2}{|\mathbf{r} - \mathbf{r}'|}\right)\left[1 - \frac{r}{2R_i}\left(3 - \frac{r^2}{R_i^2}\right)\right]\theta(R_i - r), \quad (7.73)$$

where $\mathbf{r}$ and $\mathbf{r}'$ are the positions of the bound electron and the projectile ion, respectively, and $\theta(R_i - r)$ is the step function. The ion-sphere radius $R_i(= [3(Z-1)/4\pi n_e]^{1/3})$ is given by the plasma electron density n_e since the total charge within the ion-sphere is neutral. This hydrogenic ion-sphere potential (7.73) can be generalized to Z_f free electrons inside the ion-sphere. The ion-sphere model also has its limitations. For instance, charge transfer processes in collisions between positive ions in strongly coupled plasmas can modify the range of validity of the model.

For screening in plasmas with intermediate densities, i.e., when $n_e R \approx 1$, where $R = R_i$ or $R = R_D$, more complicated models are necessary and are still under theoretical scrutiny. This is based on the simple observation that in the stars along the main sequence, there is only about 1–3 ions within the Debye sphere. Thus, in principle, the Debye screening model should not be applicable to screening in these environments. Also, static models such as the Debye–Hückel and ion-sphere models do not contain dynamical effects due to the fast motion of free electrons. Dynamical fluctuations, due to the fast motion of the electrons, and non-spherical effects could modify the screening in non-static models. In fact, mean field models might cease to be valid under the conditions prevailing in stellar cores in general and in the Sun, because particle fluctuations within the Debye–Hückel sphere are percent-wise large.

7.4.2. *Laboratory atomic screening*

Laboratory screening has been studied in more detail experimentally, as one can control different charge states of the projectile + target system in the laboratory [Ass87,Rol95]. Experimental techniques improve steadily and one can measure fusion cross-sections at increasingly lower energies where the screened Coulomb potential can be significantly smaller than the bare Coulomb potential. The deviation from the bare Coulomb potential is seen as an increase in the astrophysical S-factor extracted at the lowest energies (see Fig. 7.7). This enhancement has been experimentally observed for a large number of systems. The screening effects of the atomic electrons can be calculated [Ass87] in the adiabatic approximation at the lowest energies and in the sudden approximation at higher energies with a smooth transition in between.

In the adiabatic approximation one assumes that the velocities of the electrons in the target are much larger than the relative motion between the projectile and the target nucleus. In this case, the electronic cloud adjusts to the ground state of a "molecule" consisting of two nuclei separated by a time-dependent distance $R(t)$, at each time instant t. Since the closest-approach distance between the nuclei is much smaller than typical

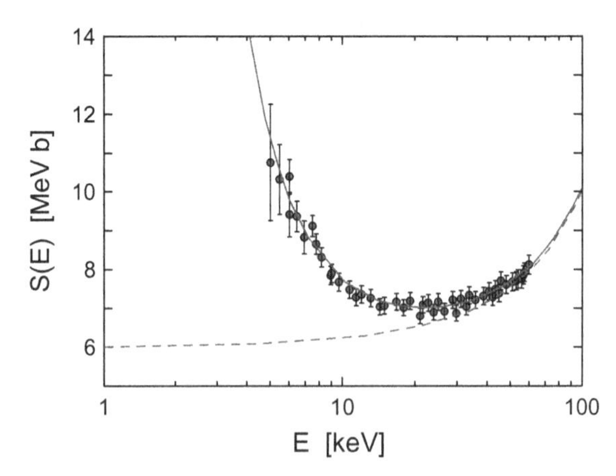

Figure 7.7. S-factor data for the $^3He(d,p)^4He$ reaction [Ali01]. The dashed curve represents the S-factor for bare nuclei and the solid curve that for screened nuclei with $U_e = 219\,\mathrm{eV}$.

atomic cloud sizes, the binding energy of the electrons will be given by the ground state energy of the $Z_P + Z_T$ atom, $B(Z_P + Z_T)$. Energy conservation implies that the relative energy between the nuclei increases by $U_e = B(Z_P + Z_T) - B(Z_T)$. This energy increment enhances the fusion probability because the tunneling probability through the Coulomb barrier between the nuclei increases accordingly. In other words, the fusion cross-section measured at laboratory energy E represents in fact a fusion cross-section at energy $E + U_e$, with U_e being known as the *screening potential.* Using Eq. (7.31), one gets for non-resonant reactions

$$\sigma(E + U_e) = \frac{S(E + U_e)}{(E + U_e)} \exp[2\pi\eta(E + U_e)] \approx \frac{S(E)}{E} \exp[2\pi\eta(E + U_e)]$$
$$\approx \exp[\pi\eta(E)\frac{U_e}{E}]\sigma(E), \qquad (7.74)$$

where we assumed that $U_e/E \ll 1$, and that the factor $S(E)/E$ varies much slower with E, as compared to the energy dependence of $\exp[-2\pi\eta(E)]$.

The exponential factor on the right-hand-side of Eq. (7.74) is the enhancement factor due to screening by the atomic electrons in the target. For light systems, the velocity of the atomic electrons is comparable to the relative motion between the nuclei. Thus, a dynamical calculation would be more appropriate to study the effect of atomic screening. However, the screening potential U_e obtained from a dynamical calculation cannot exceed that obtained in the adiabatic approximation because the dynamical calculation includes atomic excitations, which reduce the energy transferred from the electronic binding to the relative motion. The adiabatic approximation is thus the *upper limit* of the enhancement due to laboratory screening.

7.5. Reaction Models

Explosive nuclear burning in astrophysical environments produces unstable nuclei, which again can be targets for subsequent reactions. In addition, it involves a very large number of stable nuclei, which have not been fully explored by experiments. Thus, it is necessary to be able to predict reaction cross-sections and thermonuclear rates with the aid of theoretical models. Especially during the hydrostatic burning stages of stars, charged-particle induced reactions proceed at such low energies that a direct cross-section measurement is often not possible with existing techniques. Hence extrapolations down to the stellar energies of the cross-sections measured at the lowest possible energies in the laboratory are the usual procedures to apply. To be trustworthy, such extrapolations should have as strong a theoretical foundation as possible. Theory is even more mandatory when excited nuclei are involved in the entrance channel, or when unstable very neutron-rich or neutron-deficient nuclides (many of them being even impossible to produce with present-day experimental techniques) have to be considered. Such situations are often encountered in the modeling of explosive astrophysical scenarios.

Various models have been developed in order to complement the experimental information.

7.5.1. *Potential models*

This model assumes that the physically important degrees of freedom are the relative motion between the (structureless) nuclei in the entrance and exit channels, and by the introduction of spectroscopic factors and strength factors in the optical potential. The associated drawbacks are that the nucleus–nucleus potentials adopted for calculating the initial and final wave functions from the Schrödinger equation cannot be unambiguously defined.

In this model the bound state wave functions of $c = a+b$ are specified by

$$\Psi_{JM}(\mathbf{r}) = \frac{u_{lj}^{J}(r)}{r}\mathcal{Y}_{JM}^{l}, \tag{7.75}$$

where $\mathbf{r}$ is the relative coordinate of a and b, $u_{lj}^{J}(r)$ is the radial wave function and $\mathcal{Y}_{JM}^{l}$ is the spin-angle wave function

$$\mathcal{Y}_{JM}^{l} = \sum_{m,M_a} \langle jmI_aM_a|JM\rangle|jm\rangle|I_aM_a\rangle, \quad \text{with } |jm\rangle = \sum_{m_l,M_b} Y_{lm_l}(\hat{\mathbf{r}})\chi_{M_b}, \tag{7.76}$$

where χ_{M_b} is the spinor wave function of particle b and $\langle jmI_aM_a|JM\rangle$ is a Clebsch–Gordan coefficient. The ket $|I_aM_a\rangle$ denotes the angular momentum state of the core particle a.

The ground state wave function is normalized so that

$$\int d^3r|\Psi_{JM}(\mathbf{r})|^2 = \int_0^\infty dr|u_{lj}^{J}(r)|^2 = 1. \tag{7.77}$$

The wave functions are usually calculated with a central + spin–orbit + Coulomb potential of the form

$$V(\mathbf{r}) = V_0(r) + V_S(r)(\mathbf{l}.\mathbf{s}) + V_C(r), \tag{7.78}$$

which follows the same kind of parametrization of the real part of the optical potential, as described in Section 5.4. The potential parameters are adjusted so that the ground state energy E_B (or the energy of excited states) is reproduced.

The bound-state wave functions are calculated by solving the radial Schrödinger equation

$$-\frac{\hbar^2}{2m_{ab}}\left[\frac{d^2}{dr^2}-\frac{l(l+1)}{r^2}\right]u_{lj}^J(r)+[V_0(r)+V_C(r)+\langle \mathbf{s}.\mathbf{l}\rangle V_{S0}(r)]u_{lj}^J(r)$$
$$=E_i u_{lj}^J(r), \tag{7.79}$$

where $\langle \mathbf{s}.\mathbf{l}\rangle = [j(j+1)-l(l+1)-s(s+1)]/2$. This equation must satisfy the boundary conditions $u_{lj}^J(r=0) = u_{lj}^J(r=\infty) = 0$ which is only possible for discrete energies E corresponding to the bound states of the nuclear + Coulomb potential.

The continuum wave functions are calculated with the potential model as described above. The parameters are often not the same as the ones used for the bound states. The continuum states are now identified by the notation $u_{Elj}^J(r)$, where the (continuous) energy E is related to the relative momentum k of the system $a+b$ by $E=\hbar^2k^2/2m_{ab}$. They are normalized so as to satisfy the relation

$$\langle u_{Elj}^J | u_{E'l'j'}^{J'}\rangle = \delta(E-E')\delta_{JJ'}\delta_{jj'}\delta_{ll'}, \tag{7.80}$$

what means, in practice, that the continuum wave functions $u_{Elj}(r)$ are normalized to

$$-\sqrt{\frac{2m_{ab}}{\pi\hbar^2 k}}\, e^{i\delta_{lJ}}\sin(kr+\delta_{lJ}),$$

at large r.

The radial equation for the continuum wave functions can also be obtained with Eq. (7.79), but with the boundary conditions at infinity replaced by (see Chapter 5)

$$u_{Elj}^J(r\longrightarrow\infty) = i\sqrt{\frac{m_{ab}}{2\pi k\hbar^2}}[H_l^{(-)}(r)-S_{lJ}H_l^{(+)}(r)]e^{i\sigma_l(E)}, \tag{7.81}$$

where $S_{lJ}=\exp[2i\delta_{lJ}(E)]$, with $\delta_{lJ}(E)$ being the nuclear phase shift and $\sigma_l(E)$ the Coulomb one, and

$$H_l^{(\pm)}(r) = G_l(r) \pm iF_l(r). \tag{7.82}$$

F_l and G_l are the regular and irregular Coulomb wave functions. If the particle b is not charged (e.g., a neutron) the Coulomb functions reduce to the usual spherical Bessel functions, $j_l(r)$ and $n_l(r)$.

At a conveniently chosen large distance $r = R$, outside the range of the nuclear potential, one can define the logarithmic derivative (notice that this definition is different from the one used in Chapter 5)

$$\mathcal{L}_{lJ} = \left(\frac{du^J_{Elj}/dr}{u^J_{Elj}} \right)_{r=R} . \tag{7.83}$$

The phase shifts $\delta_{lJ}(E)$ are obtained by matching the logarithmic derivative with the asymptotic value obtained with the Coulomb wave functions. This procedure yields

$$S_{lJ} = \frac{G'_l - iF'_l - \mathcal{L}_{lJ}(G_l - iF_l)}{G'_l + iF'_l - \mathcal{L}_{lJ}(G_l + iF_l)}, \tag{7.84}$$

where the primes mean derivation with respect to the radial coordinate at the position R.

As discussed in Chapter 4, the operators for electric transitions of multipolarity $\lambda\pi$ are given by

$$\mathcal{M}_{E\lambda\mu} = e_\lambda r^\lambda Y_{\lambda\mu}(\widehat{\mathbf{r}}), \tag{7.85}$$

where the effective charge, which takes into account the displacement of the center-of-mass, is

$$e_\lambda = Z_b e \left(-\frac{m_a}{m_c} \right)^\lambda + Z_a e \left(\frac{m_b}{m_c} \right)^\lambda . \tag{7.86}$$

For magnetic dipole transitions

$$\begin{aligned} \mathcal{M}_{M1\mu} &= \sqrt{\frac{3}{4\pi}} \mu_N \left[e_M l_\mu + \sum_{i=a,b} g_i (s_i)_\mu \right], \quad \text{and} \\ e_M &= \left(\frac{m_a^2 Z_a}{m_c^2} + \frac{m_b^2 Z_b}{m_c^2} \right), \end{aligned} \tag{7.87}$$

where l_μ and s_μ are the spherical components of order μ ($\mu = -1, 0, 1$) of the orbital and spin angular momentum ($\mathbf{l} = -i\mathbf{r} \times \nabla$, and $\mathbf{s} = \sigma/2$) and g_i are the gyromagnetic factors of particles a and b. The nuclear magneton is given by $\mu_N = e\hbar/2m_N c$.

The matrix element for the transition $J_0M_0 \longrightarrow JM$, using the convention of [BM69], is given by

$$\langle JM|\mathcal{M}_{E\lambda\mu}|J_0M_0\rangle = \langle J_0M_0\lambda\mu|JM\rangle \frac{\langle J\|\mathcal{M}_{E\lambda}\|J_0\rangle}{\sqrt{2J+1}}. \tag{7.88}$$

From the single-particle wave functions one can calculate the reduced matrix elements $\langle lj\|\mathcal{M}_{E\lambda}\|l_0j_0\rangle_J$. The subscript J is a reminder that the matrix element depends on the channel spin J, because one can use different potentials in the different channels. The reduced matrix element $\langle J\|\mathcal{M}_{E\lambda}\|J_0\rangle$ can be obtained from a standard formula of angular momentum algebra, e.g., Eq. (7.17) of [Ed60]. One gets

$$\begin{aligned}\langle J\|\mathcal{M}_{E\lambda}\|J_0\rangle &= (-1)^{j+I_a+J_0+\lambda}[(2J+1)(2J_0+1)]^{1/2} \\ &\quad \times \begin{Bmatrix} j & J & I_a \\ J_0 & j_0 & \lambda \end{Bmatrix} \langle lj\|\mathcal{M}_{E\lambda}\|l_0j_0\rangle_J,\end{aligned} \tag{7.89}$$

where the $\{\begin{smallmatrix}\cdots\\\cdots\end{smallmatrix}\}$ denote the 6-j Wigner coefficients [BS93].

To obtain $\langle lj\|\mathcal{M}_{E\lambda}\|l_0j_0\rangle_J$ one needs the matrix element $\langle lj\|r^\lambda Y_\lambda\| l_0j_0\rangle_J$ for the spherical harmonics, e.g., Eq. (A2.23) of [La80]. For $l_0+l+\lambda =$ even, the result is

$$\begin{aligned}\langle lj\|\mathcal{M}_{E\lambda}\|l_0j_0\rangle_J &= \frac{e_\lambda}{\sqrt{4\pi}}(-1)^{l_0+l+j_0-j}\frac{\hat{\lambda}\hat{j}_0}{\hat{j}}\left\langle j_0\frac{1}{2}\lambda 0|j\frac{1}{2}\right\rangle \\ &\quad \times \int_0^\infty dr\, r^\lambda\; u_{lj}^J(r)u_{l_0j_0}^{J_0}(r),\end{aligned} \tag{7.90}$$

where we use here the notation $\widehat{k} = \sqrt{2k+1}$, and $\widetilde{k} = \sqrt{k(k+1)}$. For $l_0+l+\lambda = \text{odd}$, the reduced matrix element is null.

The multipole strength, or response functions, for a particular partial wave, summed over final channel spins, is defined by

$$\begin{aligned}\frac{dB(\pi\lambda;l_0j_0 \longrightarrow klj)}{dk} &= \sum_J \frac{|\langle kJ\|\mathcal{M}_{\pi\lambda}\|J_0\rangle|^2}{2J_0+1} \\ &= \sum_J (2J+1)\begin{Bmatrix} j & J & I_a \\ J_0 & j_0 & \lambda \end{Bmatrix}^2 |\langle klj\|\mathcal{M}_{\pi\lambda}\|l_0j_0\rangle_J|^2,\end{aligned} \tag{7.91}$$

where $\pi = E$, or M.

If the matrix elements are independent of the channel spin, this sum reduces to the usual single-particle strength $|\langle klj\|\mathcal{M}_{\pi\lambda}\|l_0j_0\rangle|^2/(2j_0+1)$.

For transitions between the bound states the same formula as above can be used to obtain the reduced transition probability by replacing the continuum wave functions $u^J_{klj}(r)$ by the bound state wave function $u^J_{lj}(r)$. That is,

$$B(\pi\lambda;\ l_0 j_0 J_0 \longrightarrow ljJ) = (2J+1)\begin{Bmatrix} j & J & I_a \\ J_0 & j_0 & \lambda \end{Bmatrix}^2 |\langle lj\|\mathcal{M}_{\pi\lambda}\|l_0 j_0\rangle|^2. \tag{7.92}$$

For bound state to continuum transitions the total multipole strength is obtained by summing over all partial waves,

$$\frac{dB(\pi\lambda)}{dE} = \sum_{lj} \frac{dB(\pi\lambda;\ l_0 j_0 \longrightarrow klj)}{dE}. \tag{7.93}$$

The photo-absorption cross-section for the reaction $\gamma + c \longrightarrow a + b$ is given in terms of the response function by

$$\sigma_\gamma^{(\lambda)}(E_\gamma) = \frac{(2\pi)^3(\lambda+1)}{\lambda[(2\lambda+1)!!]^2}\left(\frac{m_{ab}}{\hbar^2 k}\right)\left(\frac{E_\gamma}{\hbar c}\right)^{2\lambda-1}\frac{dB(\pi\lambda)}{dE}, \tag{7.94}$$

where $E_\gamma = E + |E_B|$, with $|E_B|$ being the binding energy of the $a+b$ system. For transitions between bound states, one has

$$\sigma_\gamma^{(\pi\lambda)}(E_\gamma) = \frac{(2\pi)^3(\lambda+1)}{\lambda[(2\lambda+1)!!]^2}\left(\frac{E_\gamma}{\hbar c}\right)^{2\lambda-1} \times B(\pi\lambda;\ l_0 j_0 J_0 \longrightarrow ljJ)\delta(E_f - E_i - E_\gamma), \tag{7.95}$$

where E_i (E_f) is the energy of the initial (final) state.

The cross-section for the radiative capture process $a + b \longrightarrow c + \gamma$ can be obtained by detailed balance (see Eq. (5.139)),

$$\sigma^{(\mathrm{rc})}_{(\pi\lambda)}(E) = \left(\frac{E_\gamma}{\hbar c}\right)^{2\lambda-1}\frac{2(2I_c+1)}{(2I_a+1)(2I_b+1)}\sigma_\gamma^{(\lambda)}(E_\gamma). \tag{7.96}$$

The total capture cross-section σ_{nr} is determined by the capture to all bound states with the single particle spectroscopic factors $\mathcal{S}_i$ in the final nucleus

$$\sigma_{\mathrm{nr}}(E) = \sum_{i,\pi,\lambda} \mathcal{S}_i\ \sigma^{(\mathrm{rc})}_{(\pi\lambda),i}(E). \tag{7.97}$$

Experimental information or detailed shell model calculations have to be performed to obtain the spectroscopic factors $\mathcal{S}_i$.

Spectroscopic factors

In a microscopic model, the wave function for an A-body system can be projected onto the wave function of an (A–1)-body system, yielding the *overlap function*

$$I(\mathbf{r}) = \int \Psi_A^*(\mathbf{r}_1, \mathbf{r}_2, \dots \mathbf{r}_A)\Psi_{A-1}(\mathbf{r}_1, \mathbf{r}_2, \dots \mathbf{r}_{A-1})d\mathbf{r}_1 d\mathbf{r}_2 \cdots d\mathbf{r}_{A-1}. \tag{7.98}$$

The spectroscopic factor is defined as the integral of $|I(\mathbf{r})|^2$ over all space, i.e.,

$$\mathcal{S} = \int |I(\mathbf{r})|^2 d\mathbf{r}. \tag{7.99}$$

In a loose sense, the spectroscopic factor is the probability of finding a certain configuration of (A–1) particles within the A-body system.

The spectroscopic factors $\mathcal{S}_i$ are usually obtained by adjusting the calculated cross-sections to reproduce the experimental ones.

Asymptotic normalization coefficients

In a microscopic approach, instead of the single-particle wave functions one often makes use of overlap integrals, $I(\mathbf{r})$, and a many-body wave function for the relative motion, $\Psi_{ab}(\mathbf{r})$. Both $I(\mathbf{r})$ and $\Psi_{ab}(\mathbf{r})$ might be very complicated to calculate, depending on how elaborated the microscopic model is. The variable $\mathbf{r}$ is the relative coordinate between a and b, with all the intrinsic coordinates of the nucleons in a and b being integrated out. The direct capture cross-sections are obtained from the calculation of $\sigma_{L,J}^{\text{d.c.}} \propto |\langle I(r)||r^L Y_L||\Psi_{ab}(r)\rangle|^2$.

The imprints of many-body effects will eventually disappear at large distances between the nucleon and the nucleus. One thus expects that the overlap function asymptotically matches the solution of the Schrödinger Equation (7.79), with $V = V_C$ for protons and $V = 0$ for neutrons. That is, when $r \to \infty$,

$$\begin{aligned} I(r) &= C_1 \frac{W_{-\eta, l_b+1/2}(2\kappa r)}{r}, \quad \text{for protons} \\ &= C_2 \sqrt{\frac{2\kappa}{r}} K_{l_b+1/2}(\kappa r), \quad \text{for neutrons,} \end{aligned} \tag{7.100}$$

where the binding energy of the $a + b$ system is related to κ by means of $E_b = \hbar^2\kappa^2/2m_{ab}$, $W_{p,q}$ is the Whittaker function and K_μ is the modified Bessel function. In Eq. (7.100), C_i is the *asymptotic normalization coefficient* (ANC) [MT90].

In the calculation of $\sigma^{\text{d.c.}}_{L,J}$ above, one often meets the situation in which only the asymptotic part of $I(r)$ and $\Psi_{ab}(r)$ contributes significantly to the integral over r. In these situations, $\Psi_{ab}(r)$ is also well described by a simple two-body scattering wave (e.g., Coulomb waves). Therefore the radial integration in $\sigma^{\text{d.c.}}_{L,J}$ can be done accurately and the only remaining information from the many-body physics at short-distances is contained in the asymptotic normalization coefficient C_i, i.e., $\sigma^{\text{d.c.}}_{L,J} \propto C_i^2$. We thus run into an effective theory for radiative capture cross-sections, in which the constants C_i carry all the information about the short-distance physics, where the many-body aspects are relevant. It is worthwhile to mention that these arguments are reasonable for proton capture at very low energies, because of the Coulomb barrier.

The asymptotic normalization coefficients, C_α, can also be obtained from the analysis of peripheral, transfer and breakup reactions. As the overlap integral, Eq. (7.100), asymptotically becomes a Whittaker function, so does the single particle bound-state wave function u_α, calculated with Eq. (7.79) (see Fig. 7.8). If we call the single particle ANC by b_i, then the relation between the ANC obtained from experiment, or a microscopic model, with the single particle ANC is given by $\mathcal{S}_i b_i^2 = C_i^2$. The values of $\mathcal{S}_i$ and b_i obtained with the simple potential model are useful telltales of

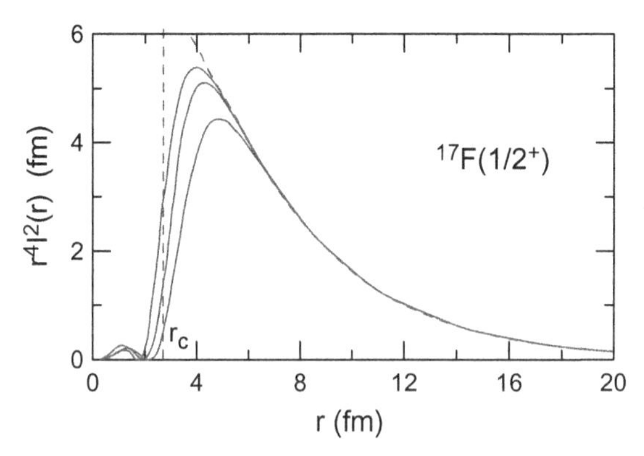

Figure 7.8. Comparison of various radial overlap integrals $r^4 I^2(r)$ for $^{17}F^*(1/2^+)$ with the normalized Whittaker function (dashed curve). Most of the contribution to the rms radius comes from the region outside the core, with radius r_c.

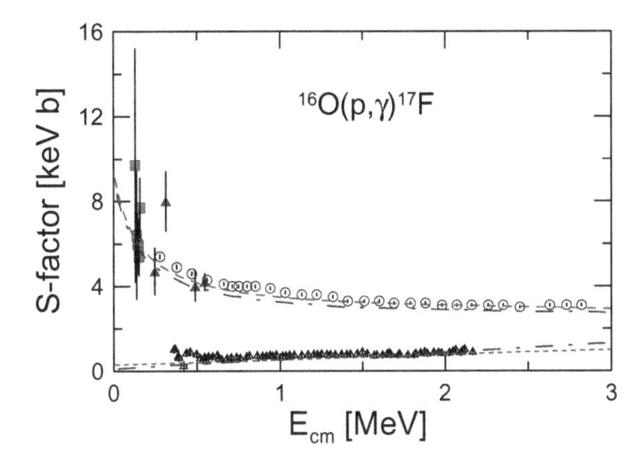

Figure 7.9. Potential model calculation ([Jun10] and references therein) for the reaction ^{16}O(p,γ)^{17}F. The dotted line and the dashed line are for the capture to the ground state and to the first excited state respectively. The experimental data are from several sources [Jun10]. The dotted-dashed lines are the result of shell model calculations [Mor97].

the complex short-range many-body physics of radiative capture reactions. One can also invert this argumentation and obtain spectroscopic factors if the C_i are deduced from a many-body model, or from experiment, and the b_i are calculated from a single particle potential model.

Figure 7.9 shows a potential model calculation [Jun10] for the S-factor of the $^{16}\text{O}(\text{p},\gamma)^{17}\text{F}$ reaction. The rate of this reaction influences sensitively the $^{17}\text{O}/^{16}\text{O}$ isotopic ratio predicted by models of massive ($\geq 4M_\odot$) asymptotic giant branch (AGB) stars, where proton captures occur at the base of the convective envelope (hot bottom burning). A fine-tuning of the $^{16}\text{O}(\text{p},\gamma)^{17}\text{F}$ reaction rate may account for the measured anomalous $^{17}\text{O}/^{16}\text{O}$ abundance ratio in small grains which are formed by the condensation of the material ejected from the surface of AGB stars via strong stellar winds. The agreement of the potential model calculation with the experimental data seen in Fig. 7.9 is very good and comparable to more microscopic calculations [Mor97].

Microscopic calculations of ANCs rely on obtaining the projection, or overlap, of the many-body wave functions of nuclei A and $A-1$. The overlap integral $\langle (A-1)|A\rangle \equiv I_b(r)$ must have correct asymptotic behavior with respect to the variable r which is the distance between the nucleon N and the c.m. of the nucleus $A-1$. The most common methods are: (a) the resonating group method (RGM), as described below, (b) the Fadeev method for three-body systems, (c) a combination of microscopic cluster method and $\mathcal{R}$-matrix approaches, to be discussed later, (d) Green's function Monte-Carlo method, (e) no-core shell model, and (f) hyperspherical functions method.

7.5.2. *Microscopic models*

In some microscopic models, nucleons are grouped into clusters and the completely antisymmetrized relative wave functions between the various clusters are determined by solving the Schrödinger equation for a many-body Hamiltonian with an effective nucleon–nucleon interaction. Typical cluster models are based on the *Resonating Group Method* (RGM) or the *Generator Coordinate Method* (GCM).

The many-body wave function Ψ is expanded in terms of Slater determinants, $\phi(\boldsymbol{\xi}, \mathbf{r})$, describing the system at a given mean distance $\mathbf{r}$ of the fragment nuclei

$$\Psi(\boldsymbol{\xi}) = \int \phi(\boldsymbol{\xi}, \mathbf{r}) f(\mathbf{r}) d^3r, \tag{7.101}$$

where $\boldsymbol{\xi}$ stands for all single particle coordinates of the nucleons, including spin and isospin. The clusters A and B are described by the functions $\phi(\boldsymbol{\xi}_A, \mathbf{r}_A)$ and $\phi(\boldsymbol{\xi}_B, \mathbf{r}_B)$ centered around points $\mathbf{r}_A$ and $\mathbf{r}_B$, respectively. The functions $\phi(\boldsymbol{\xi}, \mathbf{r})$ are then taken as the anti-symmetrized product

$$\phi(\boldsymbol{\xi}, \mathbf{r}) = \mathcal{A}\left\{\phi(\boldsymbol{\xi}_A, \mathbf{r}_A)\phi(\boldsymbol{\xi}_B, \mathbf{r}_B)\right\}, \tag{7.102}$$

where the operator $\mathcal{A}$ performs a full anti-symmetrization of all nucleons.

The expansion coefficients $f(\mathbf{r})$ are determined by solving the Schrödinger equation in the frozen density approximation

$$\langle \phi(\boldsymbol{\xi}, \mathbf{r}')|H - E|\Psi\rangle = 0, \quad \text{for all } \mathbf{r}'. \tag{7.103}$$

A partial wave decomposition of Eq. (7.103) yields a set of coupled integro-differential equations of the form

$$\int drr^2 [\mathcal{H}_l^{AB}(r, r') - E\mathcal{N}_l^{AB}(r, r')] f_l(r) = 0, \tag{7.104}$$

where the integral kernels are defined as

$$\mathcal{H}_l^{AB}(r, r') = \langle \phi_l(\boldsymbol{\xi}, r)|H|\phi_l(\boldsymbol{\xi}, r')\rangle \quad \text{and} \quad \mathcal{N}_l^{AB}(r, r') = \langle \phi_l(\boldsymbol{\xi}, r)|\phi_l(\boldsymbol{\xi}, r')\rangle.$$

Equations (7.104) can be solved numerically imposing scattering conditions on $f_l(r)$.

Methods based on Eq. (7.104) seem to be the best way to obtain scattering wave functions needed for astrophysical purposes. It is also worth mentioning that this approach has provided the best description of bound, resonant, and scattering states of nuclear systems.

As an example of applications of this method, we again give a radiative capture reaction. The creation and destruction of ^{7}Be in astrophysical environments is essential for the description of several stellar and cosmological processes and is not well understood. ^{8}B also plays an essential role in understanding our Sun. High energy ν_e neutrinos produced by ^{8}B decay in the Sun oscillate into other active species on their way to Earth. Precise predictions of the production rate of ^{8}B solar neutrinos are important for testing solar models, and for limiting the allowed neutrino mixing parameters. The most uncertain reaction leading to ^{8}B formation in the Sun is the ^{7}Be(p, γ)^{8}B radiative capture reaction. Figure 7.10 shows a comparison of microscopic calculations for the reaction ^{7}Be(p,γ)^{8}B with experimental data. The dashed-dotted line in Fig. 7.10 is the no-core shell-model calculation [Nav06] and the dotted line is for the resonant group method calculation of [Des04]. Experimental data are from several works [Ade11].

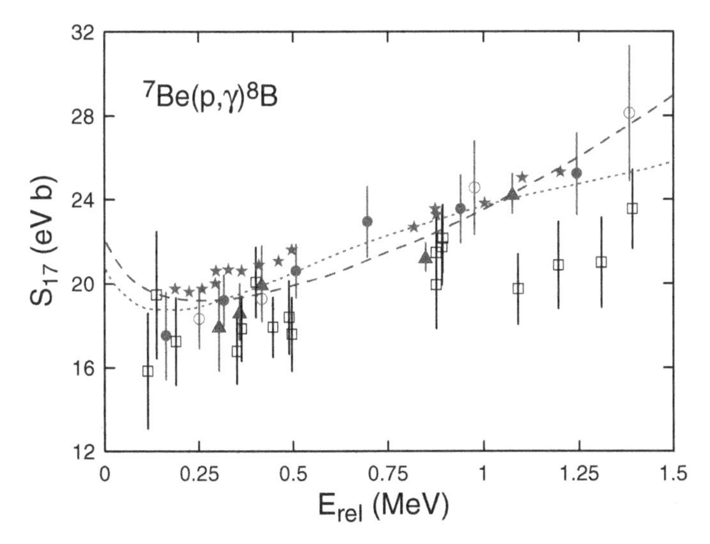

Figure 7.10. Microscopic calculations for the reaction ^{7}Be(p, γ)^{8}B. The dashed line is a no-core shell-model calculation [Nav06] and the dotted line is from a resonant group method calculation [Des04]. Experimental data are from several direct and indirect experiments [Ade11].

7.5.3. $\mathcal{R}$-matrix theory

Reaction rates dominated by the contributions from a few resonant or bound states are often extrapolated to energies of astrophysical interest in terms of *$\mathcal{R}$- or K-matrix* fits. The appeal of these methods rests on the fact that analytical expressions can be derived from underlying formal reaction theories that allow for a rather simple parametrization of the data. However, the relation between the parameters of the $\mathcal{R}$-matrix model and the experimental data (resonance energies and widths) is only quite indirect. The K-matrix formalism solves this problem, but suffers from other drawbacks [Bar94].

Elastic and inelastic scattering reactions

In the $\mathcal{R}$-matrix formalism [LT58,Brei59,DB10], the eigenstates of the nuclear Hamiltonian in the interior region of a nucleus are denoted by X_λ, with energy E_λ, and are required to satisfy the boundary condition

$$r\frac{dX_\lambda}{dr} + bX_\lambda = 0$$

at the channel radius $r = R$, where the constant b is a real number. The true nuclear wave function Ψ for the compound system is not stationary, but since the X_λ form a complete set, it is possible to expand Ψ in terms of X_λ, i.e.,

$$\Psi = \sum_\lambda A_\lambda X_\lambda, \quad \text{where } A_\lambda = \int_0^R X_\lambda \, \Psi \, dr.$$

The differential equations for Ψ and X_λ are (for s-wave neutrons in a potential V)

$$-\frac{\hbar^2}{2m}\frac{d^2\Psi}{dr^2} + V\Psi = E\Psi \tag{7.105}$$

$$-\frac{\hbar^2}{2m}\frac{d^2X_\lambda}{dr^2} + VX_\lambda = E_\lambda X_\lambda, \quad r \le R. \tag{7.106}$$

Multiplying Eq. (7.105) by Ψ and Eq. (7.106) by X_λ, subtracting and integrating, we have

$$A_\lambda = \frac{1}{E - E_\lambda}\frac{\hbar^2}{2mR}X_\lambda(R)[R\Psi'(R) + b\Psi(R)],$$

where the prime indicates the differentiation with respect to r. This result, together with the definition of Ψ, gives

$$\Psi(R) = \mathcal{R}[R\Psi'(R) + b\Psi(R)], \tag{7.107}$$

where the function $\mathcal{R}$ relates the value of $\Psi(R)$ at the surface to its derivative at the surface:

$$\mathcal{R} = \frac{\hbar^2}{2mR}\sum_{\lambda}\frac{X_\lambda(R)X_\lambda(R)}{E_\lambda - E}. \tag{7.108}$$

Rearranging Eq. (7.107) we have $R\Psi'(R)/\Psi(R) = (1 - b\mathcal{R})/\mathcal{R}$, which is just the logarithmic derivative $\mathcal{L}^I$ which can be used to determine the S-matrix element S_0 in terms of the $\mathcal{R}$ function. This gives

$$S_0 = \left[1 + \frac{2ikR\mathcal{R}}{1 - (b + ikR)\mathcal{R}}\right] e^{-2ikR}.$$

Finally, we assume that E is near a particular E_λ, say E_α, neglect all terms $\lambda \neq \alpha$ in Eq. (7.108), and define

$$\Gamma_\alpha = \frac{\hbar^2 k}{m}X_\alpha^2(R), \quad \text{and} \quad \Delta_\alpha = -\frac{b}{2kR}\Gamma_\alpha,$$

so that the S-matrix element becomes

$$S_0 = \left[1 + \frac{i\Gamma_\alpha}{(E_\alpha + \Delta_\alpha - E) - i\Gamma_\alpha/2}\right] e^{-2ikR}, \tag{7.109}$$

and the scattering cross-section is

$$\sigma_{\text{sc}} = \frac{\pi}{k^2}\left|e^{2ikR} - 1 + \frac{i\Gamma_\alpha}{(E_\alpha + \Delta_\alpha - E) - i\Gamma_\alpha/2}\right|^2. \tag{7.110}$$

We see that the procedure of imposing the boundary conditions at the channel radius leads to isolated s-wave resonances of Breit–Wigner form. If the constant b is non-zero, the position of the maximum in the cross-section is shifted. The level shift does not appear in the simple form of the Breit–Wigner formula because $E_\alpha + \Delta_\alpha$ is defined as the resonance energy. In general, a nucleus can decay through many channels and when the formalism is extended to take this into account, the $\mathcal{R}$-function becomes a matrix. In this $\mathcal{R}$-matrix theory the constant b is real and $X_\lambda(R)$ and E_λ can be chosen to be real so that the eigenvalue problem is Hermitian [WE47].

The $\mathcal{R}$-matrix theory can be easily generalized to account for higher partial waves and spin-channels. If we define the reduced width by

$\gamma_\lambda^2 = \hbar^2 X_\lambda^2(R)/2mR$, which is a property of a particular state and not dependent on the scattering energy E of the scattering system, we can write

$$\mathcal{R}_{\alpha\alpha'} = \sum_\lambda \frac{\gamma_{\lambda\alpha'}\gamma_{\lambda\alpha}}{E_\lambda - E},$$

where α is the channel label. $\gamma_{\lambda\alpha}$, E_λ, and b are treated as parameters in fitting the experimental data. If we write the wave function for any channel as $\Psi \sim I + S_\alpha O$, where I and O are incoming and outgoing waves, Eq. (7.107) means

$$R\frac{I'(R) + S_\alpha O'(R)}{I(R) + S_\alpha O(R)} = \frac{1 - b\mathcal{R}}{\mathcal{R}}.$$

Thus, as in Eq. (7.109), the S-matrix is related to the $\mathcal{R}$-matrix and from the above relation we obtain that,

$$S_\alpha = \frac{I(R)}{O(R)}\left[\frac{1 - (\mathcal{L}^I)^*\mathcal{R}}{1 - \mathcal{L}^I\mathcal{R}}\right]. \tag{7.111}$$

The total cross-sections for states with angular momenta and spins given by l, s and J is

$$\sigma_{\alpha\alpha'} = \frac{\pi}{k_\alpha^2}\sum_{JJ'll'ss'} g_J|S_{\alpha Jls,\alpha' J'l's'}|^2, \quad \alpha \neq \alpha', \tag{7.112}$$

where g_J are spin geometric factors.

In the statistical model, it can be argued that because the S-matrix elements vary rapidly with energy, the statistical assumption implies that there is a random phase relation between the different components of the S-matrix. The process of energy averaging then eliminates the cross terms and gives

$$\sigma_{\text{abs}} = \sum_{\alpha'\neq\alpha,J'l's'} \sigma_{\alpha\alpha'} = \frac{\pi}{k_\alpha^2}\sum_{Jls} g_J[1 - |S_{\alpha Jls}|^2] = \frac{\pi}{k_\alpha^2}\sum_{Jls} g_J T_{ls}^J(\alpha), \tag{7.113}$$

where the symmetry properties of the S-matrix in the form $\sum_{\alpha'\neq\alpha,\mathcal{J}'} S_{\alpha\mathcal{J},\alpha'\mathcal{J}'}\, S^*_{\alpha\mathcal{J},\alpha'\mathcal{J}'} = 1$ with $\mathcal{J} = Jls$, $\mathcal{J}' = J'l's'$ have been used, and we have introduced the general definition of the transmission coefficient

$$T_{ls}^J(\alpha) = 1 - |S_{\alpha Jls}|^2. \tag{7.114}$$

Radiative capture reactions

As an example of the application of $\mathcal{R}$-matrix theory, we will consider an $\mathcal{R}$-matrix calculation of the radiative capture reaction $n + x \rightarrow a + \gamma$ to a state of nucleus a with a given spin J_f. We only present the main formulas, without a detailed derivation. We refer the reader to Refs. [LT58,Brei59,DB10] for more details. The cross-section can be written as $\sigma_{J_f} = \sum_{J_i} \sigma_{J_i J_f}$, with

$$\sigma_{J_i J_f} = \frac{\pi}{k^2} \frac{2J_i + 1}{(2J_n + 1)(2J_x + 1)} \sum_{Il_i} |T_{Il_i J_f J_i}|^2. \tag{7.115}$$

Here, J_i is the total angular momentum of the colliding nuclei n and x in the initial state, J_n and J_x are their spins, and I, k, and l_i are their channel spin, wave number and orbital angular momentum in the initial state. $T_{Il_i J_f J_i}$ is the transition amplitude from the initial continuum state (J_i, I, l_i) to the final bound state (J_f, I). In the one-level, one-channel approximation, the resonant amplitude for the capture into the resonance with energy E_{R_n} and spin J_i, and subsequent decay into the bound state with the spin J_f can be expressed as

$$T^R_{Il_i J_f J_i} = -ie^{i(\sigma_{l_i} - \phi_{l_i})} \frac{[\Gamma^{J_i}_{bIl_i}(E)\Gamma^{J_i}_{\gamma J_f}(E)]^{1/2}}{E - E_{R_n} + i\frac{\Gamma_{J_i}}{2}}. \tag{7.116}$$

Here we assume that the boundary parameter is equal to the shift function at resonance energy and ϕ_{l_i} is the hard-sphere phase shift in the l_ith partial wave,

$$\phi_{l_i} = \arctan\left[\frac{F_{l_i}(k, r_c)}{G_{l_i}(k, r_c)}\right], \tag{7.117}$$

where F_{l_i} and G_{l_i} are the regular and irregular Coulomb functions, r_c is the channel radius, and σ_{l_i} is the Coulomb phase factor, $\sigma_{l_i} = \sum_{k=1}^{l_i} \arctan (\eta_i/k)$, where η_i is the Sommerfeld parameter. $\Gamma^{J_i}_{nIl_i}(E)$ is the observable partial width of the resonance in the channel $n + x$, $\Gamma^{J_i}_{\gamma J_f}(E)$ is the observable radiative width for the decay of the given resonance into the bound state with the spin J_f, and $\Gamma_{J_i} \approx \sum_I \Gamma^{J_i}_{nIl_i}$ is the observable total width of the resonance level. The energy dependence of the partial widths is determined by

$$\Gamma^{J_i}_{nIl_i}(E) = \frac{P_{l_i}(E)}{P_{l_i}(E_{R_n})} \Gamma^{J_i}_{nIl_i}(E_{R_n}), \tag{7.118}$$

and

$$\Gamma^{J_i}_{\gamma J_f}(E) = \left(\frac{E+\varepsilon_f}{E_{R_n}+\varepsilon_f}\right)^{2L+1} \Gamma^{J_i}_{\gamma J_f}(E_{R_n}), \tag{7.119}$$

where $\Gamma^{J_i}_{nIl_i}(E_{R_n})$ and $\Gamma^{J_i}_{\gamma J_f}(E_{R_n})$ are the experimental partial and radiative widths, ε_f is the binding energy of the bound state in nucleus a, and L is the multipolarity of the γ-ray transition. The penetrability $P_{l_i}(E)$ is expressed as

$$P_{l_i}(E) = \frac{kr_c}{F^2_{l_i}(k,r_c)+G^2_{l_i}(k,r_c)}. \tag{7.120}$$

The non-resonant amplitude can be calculated by

$$\begin{aligned} T^{NR}_{Il_iJ_fJ_i} &= -(2)^{3/2} i^{l_i+L-l_f+1} e^{i(\sigma_{l_i}-\phi_{l_i})} \frac{(\mu_{nx}k_\gamma r_c)^{L+1/2}}{\hbar k} \\ &\quad \times e_L\sqrt{\frac{(L+1)(2L+1)}{L[(2L+1)!!]^2}} C_{J_fIl_f} F_{l_i}(k,r_c) \\ &\quad \times G_{l_i}(k,r_c) W_{l_f}(2\kappa r_c)\sqrt{P_{l_i}}(l_i0L0|l_f0)U(Ll_fJ_iI;l_iJ_f)J'_L(l_il_f), \end{aligned} \tag{7.121}$$

where, e_L is the effective charge, U is a geometric coefficient, and

$$J'_L(l_il_f) = \frac{1}{r_c^{L+1}}\int_{r_c}^{\infty} dr\, r \frac{W_{l_f}(2\kappa r)}{W_{l_f}(2\kappa r_c)}\left[\frac{F_{l_i}(k,r)}{F_{l_i}(k,r_c)} - \frac{G_{l_i}(k,r)}{G_{l_i}(k,r_c)}\right]. \tag{7.122}$$

$W_l(2\kappa r)$ is the Whittaker hypergeometric function, $\kappa = \sqrt{2\mu_{nx}\varepsilon_f}$ and l_f are the wave number and relative orbital angular momentum of the bound state, and $k_\gamma = (E+\varepsilon_f)/\hbar c$ is the wave number of the emitted photon.

The non-resonant amplitude contains the radial integral ranging only from the channel radius r_c to infinity since the internal contribution is contained within the resonant part. Furthermore, the $\mathcal{R}$-matrix boundary condition at the channel radius r_c implies that the scattering of particles in the initial state is given by the hard sphere phase. Hence, the problems related to the interior contribution and the choice of incident channel optical parameters do not occur. Therefore, the direct capture cross-section only depends on the ANC and the channel radius r_c.

The $\mathcal{R}$-matrix method described above can be extended to the analysis of other types of reactions, e.g., transfer reactions [Des04]. The goal of the $\mathcal{R}$-matrix method is to parameterize some experimentally known quantities, such as cross-sections or phase shifts, with a small number of parameters,

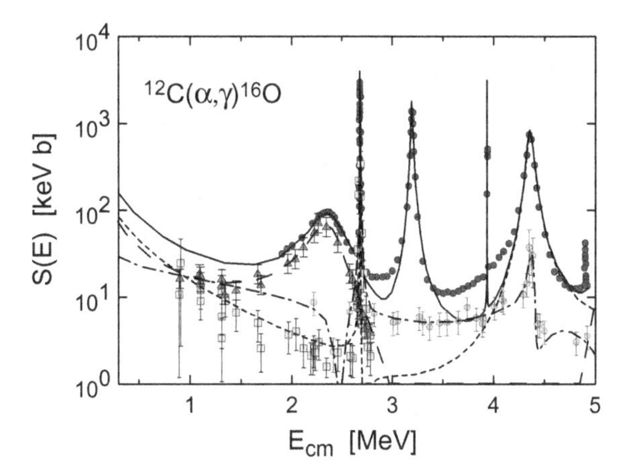

Figure 7.11. Total S factor data (filled-in circles) for $^{12}C(\alpha,\gamma)^{16}O$ compared with E1 (open triangles) and E2 (open squares) contributions. The solid line represents the sum of the single amplitudes of an $\mathcal{R}$-matrix fit (the dotted and dashed lines are the $E1$ and $E2$ amplitudes, respectively). In addition, the $\mathcal{R}$-matrix fit of to their data (dotted-dashed line) is shown. (Adapted from Ref. [Str08]).

which are then used to extrapolate the cross-section down to astrophysical energies. One example is given in Fig. 7.11 which shows the experimental data and $\mathcal{R}$-matrix fits for the cross-section of the reaction $^{12}C(\alpha,\gamma)^{16}O$ cross-section, of relevance to helium burning [Str08].

7.5.4. *Statistical models*

A large fraction of the reactions of interest proceed through compound systems that exhibit high enough level densities for statistical methods to provide a reliable description of the reaction mechanism. The theoretical treatment of nuclear reactions leading to formation and decay of compound nuclei was developed by Ewing and Weisskopf [WW40], based on two ideas: (a) the compound nucleus formation independence hypothesis as proposed by Niels Bohr [Bo36], and (b) the reciprocity theorem, or time-reversal properties of the underlying Hamiltonian. This allows one to relate capture and decay cross-sections, usually expressed in terms of transmission probabilities, defined in Eq. (7.114).

Later, the Ewing–Weisskopf theory was extended to include angular momentum dependence by Hauser and Feshbach [HH52] (see Section 5.5). The Hauser–Feshbach (HF) model has been widely used with considerable success in nuclear astrophysics. Explosive burning in supernovae involves in general intermediate mass and heavy nuclei. Due to a large nucleon number, they have intrinsically a high density of excited states. A high level density in the compound nucleus at the appropriate excitation energy allows for the use of the statistical-model approach for compound nuclear reactions [HH52] which averages over resonances.

A high level density in the compound nucleus also allows for the use of averaged transmission coefficients T, which do not reflect resonance behavior, but rather describe absorption via an imaginary part of the (optical) nucleon–nucleus potential as described in [MW79]. This leads to the expression (see Section 5.5)

$$\sigma_i^{\mu\nu}(j,o;E_{ij}) = \frac{\pi\hbar^2/(2\mu_{ij}E_{ij})}{(2J_i^\mu+1)(2J_j+1)}\sum_{J,\pi}(2J+1) \times \frac{T_j^\mu(E,J,\pi,E_i^\mu,J_i^\mu,\pi_i^\mu)T_o^\nu(E,J,\pi,E_m^\nu,J_m^\nu,\pi_m^\nu)}{T_{tot}(E,J,\pi)} \tag{7.123}$$

for the reaction $i^\mu(j,o)m^\nu$ from the target state i^μ to the excited state m^ν of the final nucleus, with a center of mass energy E_{ij} and reduced mass μ_{ij}. J denotes the spin, E the corresponding excitation energy in the compound nucleus, and π the parity of excited states. When these properties are used without subscripts they describe the compound nucleus, subscripts refer to states of the participating nuclei in the reaction $i^\mu(j,o)m^\nu$ and superscripts indicate the specific excited states. Experiments measure $\sum_\nu \sigma_i^{0\nu}(j,o;E_{ij})$, summed over all excited states of the final nucleus, with the target in the ground state. Target states μ in an astrophysical plasma are thermally populated and the astrophysical cross-section $\sigma_i^*(j,o)$ is given by

$$\sigma_i^*(j,o;E_{ij}) = \frac{\sum_\mu(2J_i^\mu+1)\exp(-E_i^\mu/kT)\sum_\nu\sigma_i^{\mu\nu}(j,o;E_{ij})}{\sum_\mu(2J_i^\mu+1)\exp(-E_i^\mu/kT)}. \tag{7.124}$$

The summation over ν replaces $T_o^\nu(E,J,\pi)$ in Eq. (7.123) by the total transmission coefficient

$$T_o(E,J,\pi) = \sum_{\nu=0}^{\nu_m} T_o^\nu(E,J,\pi,E_m^\nu,J_m^\nu,\pi_m^\nu) + \int_{E_m^{\nu_m}}^{E-S_{m,o}} \sum_{J_m,\pi_m} T_o(E,J,\pi,E_m,J_m,\pi_m)\rho(E_m,J_m,\pi_m)dE_m. \tag{7.125}$$

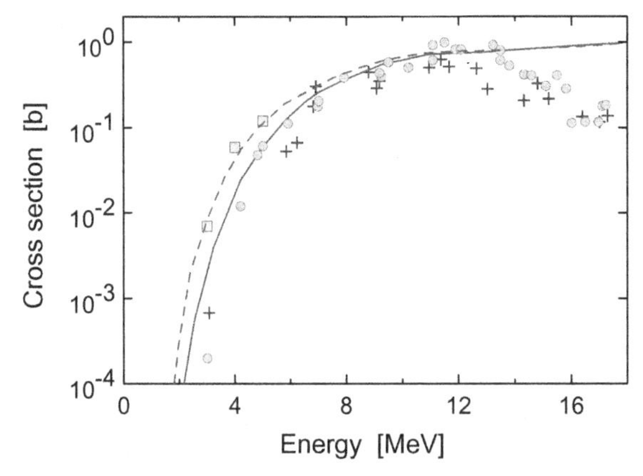

Figure 7.12. Cross-section data for two sets of ^{75}As(p,n) measurements in comparison with Hauser–Feshbach predictions [RT01]. Also shown is the experimental cross-section of ^{85}Rb(p,n) (crosses with error bars) in comparison with HF predictions, dashed line. (Adapted from [Rap06].)

Here $S_{m,o}$ is the channel separation energy, and the summation over excited states above the highest experimentally known state ν_m is changed to an integration over the level density ρ. The summation over target states μ in Eq. (7.124) has to be generalized accordingly.

The important ingredients of statistical-model calculations, as indicated in the above equations, are the particle and γ-transmission coefficients T and the level density of excited states ρ. Therefore, the reliability of such calculations is determined by the accuracy with which these components can be evaluated (often for unstable nuclei).

Figure 7.12 shows the cross-section data for two sets of $^{75}\text{As(p, n)}$ measurements in comparison with Hauser–Feshbach predictions (solid line: [RT01]). Also shown is the experimental cross-section of $^{85}\text{Rb(p, n)}$ (crosses with error bars) in comparison with HF predictions (dashed line: [RT01]). The experimental results are on average lower than the HF predictions.

7.5.5. *Beta-decay, electron capture and neutrinos*

Beta-decay, electron capture and neutrino scattering involve similar operators and nuclear matrix elements. We thus consider only the case of beta-decay. One can make an analogy of β-decay with the emission of electromagnetic radiation by the nucleus, induced by the time dependent interaction between the system that irradiates and the electromagnetic field. In the case of β-decay the weak force is the agent responsible for the decay.

It can be understood as a perturbation, i.e., it is small compared with the forces responsible for maintaining the initial and final quasi-stationary states. Expression (4.22) for the disintegration constant,

$$\Lambda = \frac{2\pi}{\hbar}|\mathcal{M}_{if}|^2 \frac{dN}{dE_T}, \tag{7.126}$$

with

$$\mathcal{M}_{if} = \int \Psi_f^* \mathcal{V} \Psi_i \, d^3r, \tag{7.127}$$

can be applied, with $\mathcal{V}$ being the operator associated to the weak force, dN/dE_T the density of available final states with the disintegration energy E_T, Ψ_i the wave function of the parent nucleus and Ψ_f the wave function of the final system, composed by the residual nucleus, the electron (or positron) and the antineutrino (or neutrino):

$$\Psi_f = \Psi_R \Psi_e \Psi_\nu. \tag{7.128}$$

Let us initially examine the integral in Eq. (7.127): the wave functions of the product in Eq. (7.128) must be normalized; if Ψ_e and Ψ_ν are the free particle wave functions — the correction due to the influence of the Coulomb field of the nucleus will be commented upon later — the normalization can be done in a cubic box of side a. Since it is irrelevant to work with traveling or standing waves, an expression of the plane wave type for Ψ_e and Ψ_ν can be employed. In this case the normalization gives $A = 1/\sqrt{V}$, where $V = a^3$ is the box volume. Thus we can write for the lepton wave functions,

$$\Psi_e = \frac{1}{\sqrt{V}} e^{i\mathbf{p}_e \cdot \mathbf{r}/\hbar}, \quad \text{and} \quad \Psi_\nu = \frac{1}{\sqrt{V}} e^{i\mathbf{p}_\nu \cdot \mathbf{r}/\hbar}, \tag{7.129}$$

where $\mathbf{p}_e$ and $\mathbf{p}_\nu$ are the electron and neutrino moments. The product that appears in Eq. (7.128) can be written in the form of a power series

$$\Psi_e \Psi_\nu = \frac{1}{V}\left[1 + \frac{i(\mathbf{p}_e + \mathbf{p}_\nu)\cdot \mathbf{r}}{\hbar} + \cdots\right]. \tag{7.130}$$

If we take into account that the wavelengths associated to the leptons are very large compared to the nuclear dimensions (a 1 MeV electron, for instance, has $\lambda = 897\,\text{fm}$), we shall see that in the proximity of the nucleus the first term in Eq. (7.130) is largely dominant; Ψ_e and Ψ_ν can be considered constant and their product equal to $1/V$. A partial wave expansion of the plane wave shows that this first term is part of the $l = 0$ component of the expansion, i.e. the lepton orbital momentum vanishes and the transitions with $l = 0$ are said to be *allowed transitions*. In some circumstances, however, the matrix element M_{if} of Eq. (7.127) vanishes, when Eq. (7.130) is reduced to just its first term. In this case the remaining terms must be taken into account in the evaluation of M_{if}. Their contribution is, as we have seen, small, and transitions where this happens are said to be *forbidden transitions*, although they are not really forbidden but instead less probable than the allowed ones. If only the first term of Eq. (7.130) vanishes then in Eq. (7.127) we have a first-forbidden transition; if the first two vanish, a second-forbidden transition and so on. These high order terms correspond to the sum of the leptons orbital momenta equal to 1, 2, etc. The higher the order of forbiddenness the lower the decay constant λ will be by a factor $pr/\hbar \cong 10^{-4}$. One example is the decay of ${}^{115}_{49}\text{In} \rightarrow {}^{115}_{50}\text{Sn}$, where the first non-zero term is the fifth one: the decay constant is very small and the half-life of this process is about 10^{14} years.

For the solution of the integral (7.127) it is necessary to know the form $\mathcal{V}$ of the weak interaction. Fermi did not take into account in his theory the spins of the particles involved in the process (the nucleons, the electron and the neutrino all have spin $\frac{1}{2}\hbar$) and, in this case, the matrix element constructed from the interaction ν has more or less a simple non-relativistic expression:

$$\mathcal{M}_{if}^F = g_F M_{if}^F \tag{7.131}$$

with

$$|M_{if}^F|^2 = \sum_{m_f} \left| \int \Psi_f^* \left(\sum_k t_\pm^k \right) \Psi_i \, d^3r \right|^2 , \tag{7.132}$$

where in Eq. (7.131) one has made explicit the factor g_F, the coupling constant for Fermi transitions, being a measure of the weak interaction intensity for this case. In this sense it has an equivalent role to the charge in the electromagnetic interaction. The matrix element M_{if}^F is now dimensionless.

The first sum of Eq. (7.132) should be taken over the values of the magnetic quantum number m of the final nucleus and the second over all the nucleons of the initial nucleus, where the operators $t_+ = t_x + it_y$ and $t_- = t_x - it_y$ are constructed from isospin operators [Ber07]. t_+ transforms a neutron into a proton and should be used in β^--decay. t_- has the opposite effect and is used in β^+-decay. In simple calculations using the shell model, the sum over k reduces to a few, or even 1, valence nucleons. In some of these cases the matrix element M_{if} can be easily obtained and this allows determination, as we shall see later, of the value of the constant g_F, whose value is around 10^{-4} MeV fm^3.

Carrying on the analysis of (7.126), we can infer how to write the density dN/dE_T. For that, let us initially evaluate λ for a given total relativistic energy E_e of the electron. Thus, dN now represents the possible number of states for the neutrino energy in the interval between E_ν and $E_\nu + dE_\nu$. Here $E_T = E_e + E_\nu$, and with fixed E_e, $dE_T = dE_\nu$. It is then possible to apply the calculation done for the number of possible energy states in a Fermi gas contained in a volume V. Thus, using the formula

$$dn(k) = \frac{1}{2}\frac{k^2 dk}{\pi^2}V, \tag{7.133}$$

and recalling that for the neutrino $k = p/\hbar = E/\hbar c$, we immediately arrive at

$$\frac{dN}{dE_T} = \frac{dN_\nu}{dE_\nu} = \frac{V}{2\pi^2(\hbar c)^3}(E_T - E_e)^2. \tag{7.134}$$

Before establishing the final form of Eq. (7.126), it is convenient to add a correction factor. This is normally introduced to take into account the nuclear Coulomb field effects over the electron wave function, which in reality could not be represented by a plane wave. This factor, that depends on the atomic number Z and on the final electron energy E_e, is referred to as the *Fermi function* $F(Z, E_e)$. We have seen that the electron wave function $\Psi_e(Z, \mathbf{r})$ is essentially constant inside the nucleus. It can be replaced by its value at the center $\Psi_e(Z, 0)$, and the Fermi function is the factor that corrects the probability to find the electron inside the nucleus, i.e., $|\Psi(Z,0)|^2 = F(Z, E_e)|\Psi(0,0)|^2$ where $\Psi(0,0)$ is the electron wave function without the Coulomb interaction. The Fermi function has a non-relativistic expression given by

$$F(Z, E_e) = \frac{2\pi\eta}{1 - e^{-2\pi\eta}}, \tag{7.135}$$

where $\eta = \pm Ze^2/\hbar v_e$, and the plus (minus) sign is valid for the electron (positron). v_e is the electron velocity corresponding to E_e.

Collecting now all the factors and omitting the index from the electron energy, expression (7.126) takes the form

$$\Lambda(E) = \frac{F(Z, E)}{V\pi\hbar^4 c^3}|\mathcal{M}_{if}|^2(E_T - E)^2, \tag{7.136}$$

where it is explicit that $\Lambda(E)$ refers to just one energy E of the emitted electron.

From Eq. (7.136) we can get the probability per unit time of the emission of an electron with energy between E and $E + dE$, being enough for this to multiply the rate $\Lambda(E)$ by the number of possible states for the electron in that energy interval. Using again (4.22), on this occasion for the electron energy, one finds that

$$\Lambda(E)dN = \frac{F(Z,E)|\mathcal{M}_{if}|^2}{2\pi^3\hbar^7c^6}E(E^2 - m^2c^4)^{1/2}(E_T - E)^2dE, \qquad (7.137)$$

where m is the electron mass.

The integral of Eq. (7.137) is

$$\Lambda = \frac{m^5 g_F^2 c^4 |M_{if}|^2}{2\pi^3\hbar^7} f(Z, E_T), \qquad (7.138)$$

where the function

$$f(Z, E_T) = \frac{1}{m^5c^{10}} \int_0^{E_T} F(Z,E)E(E^2 - m^2c^4)^{1/2}(E_T - E)^2\, dE, \qquad (7.139)$$

known as the *Fermi integral*, is dimensionless and usually presented in curves that are a function of the atomic number Z and of the electron maximum energy E_T.

Equation (7.138) allows us to examine the influence of the matrix element M_{if} in the evaluation of the decay constant Λ. Using the relation between Λ and the half-life $t_{1/2}$, Eq. (7.138) can be rewritten as

$$ft_{1/2} = \frac{1.386\pi^3\hbar^7}{g_F^2 m^5 c^4 |M_{if}|^2}. \qquad (7.140)$$

We see that the product $ft_{1/2}$ or, simply, ft, depends only on M_{if}: the larger the matrix element value the more probable the occurrence of the transition is.

In some special cases the matrix element M_{if} is easily evaluated. This occurs, for example, in a β^+ transition from ^{14}O to ^{14}N. It is a $0^+ \to 0^+$ transition with $M_{if} = \sqrt{2}$, where the measured half-live gives for ft a value near 8 hours. This allows us to use Eq. (7.140) for determining the coupling constant, resulting in the value $g_F \cong 10^{-4}$ MeV fm^3, mentioned in the last section. Transitions in which the value of M_{if} is near unity produce the lowest values of ft and are called *superallowed.*

Forbidden transitions can have ft values several orders of magnitude larger than the allowed ones. This is due to the natural difficulty in creating an electron–neutrino pair with $l > 0$. This can be shown by a simple classical calculation: suppose a β-decay with $Q = 1\,\text{MeV}$. With the assumption that the electron is emitted with the total energy, it would have near the nuclear surface a maximum angular momentum $m_e vR \cong 0.05\hbar$, for the case of a heavy nucleus with $R = 7.4\,\text{fm}$. This shows how the $l = 1$ value is improbable, and this is even more true for larger l values. From this fact results a very large range of ft values, and is common to use $\log_{10} ft$, with t given in seconds, as a measure of the decay probability of a given state by β-emission.

Gamow–Teller transitions

When a β-decay leads an initial nucleus of spin I_i to a final nucleus of spin I_f, with angular momentum and parity conservation,[2] its contribution to the decay constant can be very small. We need, in association with this, to create some selection rules for I_i, I_f and the parities of the initial and final states. To establish these rules it is necessary initially to distinguish between two possible situations: in the first one, the electron and the neutrino have opposite spins and do not contribute to the angular momentum balance. Such transitions are called *Fermi transitions* and for those it is easy to see that

$$I_i = I_f + l \quad \text{Fermi}, \tag{7.141}$$

where l is the orbital angular momentum carried by the electron and the neutrino. In turn, for *Gamow–Teller transitions*, the lepton spins are parallel and contribute with one unit to the momenta:

$$I_i = I_f + l + 1 \quad \text{Gamow–Teller}. \tag{7.142}$$

The Gamow–Teller transitions are not accounted for by Fermi theory that, as we have seen, does not take into account the spins of the particles. We can show that, with the introduction of spin, the matrix element of Eq. (7.132) is modified by the additional presence of the three components of the Pauli spin operator

$$|M_{GT}|^2 = \sum_{m_f} \sum_{x} \left| \int \Psi_f^* \left(\sum_k t_\pm^k \sigma_x^k \right) \Psi_i \, d^3r \right|^2 , \tag{7.143}$$

[2]The nuclear wave function actually has a contribution of the weak force that does not conserve parity.

where $\sum_x$ represents the sum over the Pauli matrices, σ_x, σ_y and σ_z. The index k means again that the operators t and σ act on the nucleon k of the initial nucleus, whose wavefunction is Ψ_i. If Fermi and Gamow–Teller transitions are both possible one has to write Eq. (7.131) in a complete form: $|\mathcal{M}_{if}|^2 = g_F^2|M_F|^2 + g_{GT}^2|M_{GT}|^2$. We have already seen that g_F is of the order of $10^{-4}\,\text{MeV}\cdot\text{fm}^3$. One can show through several examples that the coupling constant g_{GT} for a Gamow–Teller transition has a value a little larger than g_F.

In general we do not have a precise probe of the nuclear GT response apart from weak interactions themselves. However a good approximate probe is provided by forward-angle (p,n) scattering off nuclei. The (p,n) studies demonstrate that the GT strength tends to concentrate in a broad resonance centered at a position $\delta = E_{GT} - E_{IAS}$ relative to the IAS given by

$$\delta \sim \left(7.0 - 28.9\frac{N-Z}{A}\right) \text{MeV}. \tag{7.144}$$

Thus while the peak of the GT resonance is substantially above the IAS for $N \sim Z$ nuclei, it drops with increasing neutron excess, with $\delta \sim 0$ for Pb. A typical value for the full width at half maximum Γ is $\sim$5 MeV. The angular distribution of GT $(N, Z) + \nu_e \to (N-1, Z+1) + e^-$ reactions is $3 - \beta\cos\theta_e$, corresponding to a gentle peaking in the backward direction.

The allowed Gamow–Teller (spin-flip) and Fermi weak interaction operators are the appropriate operators when one probes the nucleus at a wavelength — that is, at a size scale — where the nucleus responds like an elementary particle. We can then characterize its response by its macroscopic quantum numbers, the spin and charge. On the other hand, the nucleus is a composite object and, therefore, if it is probed at shorter length scales, all kinds of interesting radial excitations will result, analogous to the vibrations of a drumhead. For a reaction like neutrino scattering off a nucleus, the full operator involves the additional factor, the second term in the expansion of Eq. (7.130). Thus the full charge operator includes a "first forbidden" term

$$\sum_{i=1}^{A} \mathbf{r}_i \tau_3(i) \tag{7.145}$$

and similarly for the spin operator

$$\sum_{i=1}^{A} [\mathbf{r}_i \otimes \sigma(i)]_{J=0,1,2}\tau_3(i). \tag{7.146}$$

These operators generate collective radial excitations, leading to the giant resonance excitations in nuclei. The giant resonances are typically at an excitation energy of 20–25 MeV in light nuclei. One important property is that these operators satisfy a sum rule of the form

$$\sum_f \left| \langle f | \sum_{i=1}^{A} r(i)\tau_3(i) | i \rangle \right|^2 \sim \frac{NZ}{A} \sim \frac{A}{4}, \tag{7.147}$$

where the sum extends over a complete set of final nuclear states. These first-forbidden operators tend to dominate the cross-sections for scattering the high energy supernova neutrinos (ν_μs and ν_τs), with $E_\nu \sim 25$ MeV, off light nuclei. From the sum rule above, it follows that nuclear cross-sections per target *nucleon* are roughly constant.

7.5.6. *Field theory models*

Field theories adopt a completely independent approach for nuclear physics calculations in which the concept of nuclear potentials is not used. The basic method of field theories is to start with a Lagrangian for the fields. From this Lagrangian one can "read" the Feynman diagrams and make practical calculations, although not without bypassing well-known complications such as regularization and renormalization. Quantum chromodynamics (QCD) is the proper quantum field theory for nuclear physics. But it is a very hard task to bridge the physics from QCD to the one in low-energy nuclear processes. *Effective field theory* (EFT) tries to help in this construction by making use of the concept of the separation of scales. One can form small expansion parameters from the ratios of short and long distance scales, defined by

$$\epsilon = \frac{\text{short distance scales}}{\text{long distance scales}} \tag{7.148}$$

and try to explain physical observables in terms of powers of ϵ.

In low-energy nuclear processes, the characteristic momenta are much smaller than the mass of the pion, which is the lightest hadron that mediates the strong interaction. In this regime, one often uses the pionless effective field theory, in which pions are treated as heavy particles and are integrated out of the theory [KSW96]. In this theory, the dynamical degrees of freedom are nucleons. The pion and the delta resonance degrees of freedom are hidden in the contact interactions between nucleons. The scales of the problem are the nucleon–nucleon scattering length, a (Eq. (5.34)), the binding energy, B, and the typical nucleon momentum k in the center-of-mass frame. Then, the nucleon–nucleon interactions are calculated perturbatively with the small expansion parameter

$$p = \frac{(1/a, \text{ or } B, \text{ or } k)}{\Lambda}, \tag{7.149}$$

which is the ratio of the light to the heavy scale. The heavy scale Λ is set by the pion mass ($m_\pi \sim 140$ MeV).

The pionless effective Lagrangian will only involve the nucleon field $\Psi^T = (p, n)$ and its derivatives. It must obey the symmetries observed in strong interactions at low energies, such as parity, time-reversal, and Galilean invariance. The Lagrangian can then be written as a series of local operators with increasing dimensions. In the limit where the energy goes to zero, the interactions of lowest dimension dominate. To leading order (LO), the relevant Lagrangian ($\hbar = c = 1$) is given by

$$\mathcal{L} = \Psi^\dagger \left(i\partial_t + \frac{\nabla^2}{2m} \right) \Psi - C_0(\Psi^T \mathcal{P} \Psi)(\Psi^T \mathcal{P} \Psi)^\dagger, \tag{7.150}$$

where m is the nucleon mass [CRS99]. The projection operators $\mathcal{P}$ enforce the correct spin and isospin quantum numbers in the channels under investigation. For spin–singlet interactions $\mathcal{P}_i = \sigma_2 \tau_2 \tau_i / \sqrt{8}$, while for spin–triplet interactions $\mathcal{P}_i = \sigma_2 \sigma_i \tau_2 / \sqrt{8}$.

The Feynman diagram rules can be directly "read" from the Lagrangian at hand. In the case that the scattering length a is large, i.e., $a \gg 1/\Lambda$, as it is in the nucleon–nucleon system, the full scattering amplitude T is obtained from an infinite sum of such Feynman diagrams (see Fig. 7.13), leading to a geometric series that can be written analytically as

$$T(p) = \frac{C_0}{1 - C_0 J(p)}, \quad J(p) = \int \frac{d^3q}{(2\pi)^3} \frac{1}{E - \mathbf{q}^2/m + i\epsilon}, \tag{7.151}$$

where $E = p^2/m$ is the total center-of-mass energy (see, e.g., Ref. [BK02]). The integral is linearly divergent but is finite using *dimensional regularization*. One gets

$$J = -(\mu + ip)\frac{m}{4\pi},$$

where μ is the regularization parameter. The scattering amplitude $T(p)$ has then the same structure as the s-wave partial wave amplitude (Eq. (5.39)

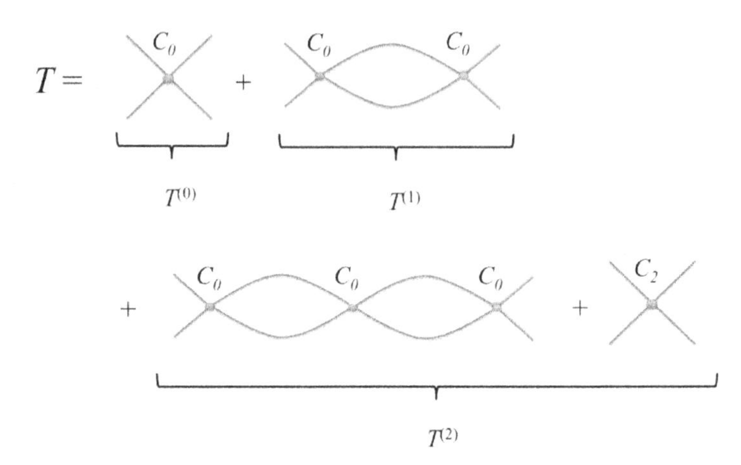

Figure 7.13. Feynman diagram series for NN-scattering in effective field theory.

with proper normalization),

$$T = -\frac{4\pi}{p\cot\delta - ip},$$

and one obtains the effective range expansion (5.37) for the phase shift δ,

$$p\cot\delta = -\frac{1}{a} + r_0\frac{p^2}{2} + \cdots$$

in the zero-momentum limit when the coupling constant takes the renormalized value

$$C_0(\mu) = \frac{4\pi}{m}\frac{1}{(1/a - \mu)}. \tag{7.152}$$

To leading order, we see that the effective range vanishes, or $r_0 = 0$. The small inverse $1/a$ scattering length is given by the difference between two large quantities. For example, in proton–neutron scattering we have $a_{pn} = -23.7\,\mathrm{fm}$ in the pn spin–singlet channel. Choosing the value $\mu = m_\pi$ for the regularization parameter, one obtains $C_0 = 3.54\,\mathrm{fm}^2$. Physical results should be independent of the exact value of the renormalization mass μ as long as $1/a < \mu \ll m_\pi$.

The theory is now ready for practical applications. For example, this procedure has been applied to obtain the electromagnetic form factor of the deuteron, the electromagnetic polarizability and the Compton scattering cross-section for the deuteron, the radiative neutron capture on protons, and the continuum structure of halo nuclei. Based on the same effective field theory, the three-nucleon system and neutron–deuteron scattering have

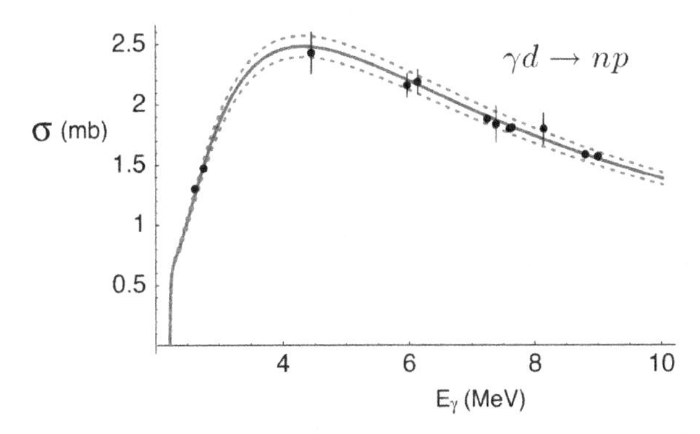

Figure 7.14. The cross-section for $d \to np$. The curves correspond to EFT calculations for cold $np \to d$ and the dashed lines denote the 3% theoretical uncertainty. Ref. [Rup00] has further reduced this uncertainty to below 1%.

been investigated [BK02]. Better agreement with data can be obtained at higher orders (next-to-leading order [NLO], next-to-next-to leading order [N^2LO], etc.). For nuclear processes involving momenta p comparable to m_π, the starting effective, pionfull, Lagrangian is more complicated. But the basic field theoretic method remains the same. The EFT unifies single-particle approaches in a model-independent framework, with the added power counting that allows for an *a priori* estimate of errors. Concepts of quantum field theory, such as regularization and renormalization, are key ingredients of the theory.

In nuclear astrophysics, this theory has been applied to $np \to d\gamma$ for Big Bang nucleosynthesis [Rup00]; νd reactions for supernovae physics [KR99] and the solar pp fusion process [But00]. EFT has also been used to deduce observables in reactions with halo nuclei and loosely bound states, with promising applications to astrophysics [BHK02]. To date, perhaps the most enlightening application of EFT for nuclear physics is the $np \to d\gamma$ cross-section, specially because there is no data at the energies of relevance for Big Bang nucleosynthesis (BBN). EFT has provided a calculation with 1% error [Rup00] in the energy range relevant to BBN. The EFT predictions also agree with a very recent measurement of the inverse process in the same energy region (see Fig. 7.14).

7.6. Exercises

1. The mass fractions of ^{1}H and ^{4}He at the time of the Sun's birth are equal to 0.71 and 0.27, respectively. Calculate the ratio of the corresponding number densities.

2. (a) What is the most probable kinetic energy of an hydrogen atom at the interior of the Sun ($T = 1.5 \times 10^7$ K)? (b) What fraction of these particles would have kinetic energy in excess of 100 keV?
3. Calculate the Gamow window for the ^{14}N(p,γ)^{15}O reaction for temperatures of 0.01 to 0.1 GK.
4. Calculate the reaction rate for ^{14}N(p,γ)^{15}O for the resonant contribution, with $E_{cm} = 0.259$ MeV, $(\omega\gamma)_r = 13$ meV (see Eq. (7.57)) and for non-resonant component, with $S = 1.61$ keV b.
5. Suppose the iron core of a supernova star has a mass of 1.4 $M_\odot$ (the Sun's mass is $M_\odot = 1.99 \times 10^{30}$ kg) and a radius of 100 km and that it collapses to a uniform sphere of neutrons of radius 10 km. Assume that the virial theorem

$$2\langle T \rangle + \langle V \rangle = 0$$

holds, where $\langle T \rangle$ is the average of the internal kinetic energy and $\langle V \rangle$ the average of the gravitational potential energy. $E = \langle T \rangle + \langle V \rangle$ is the total mechanical energy of the system. Calculate the energy consumed in neutronization and the number of electron neutrinos produced. Given that the remaining energy is radiated as neutrino–antineutrino pairs of all kinds of average energy 12 + 12 MeV, calculate the total number of neutrinos radiated.
6. Given that the supernova of Exercise 5 is at a distance of 163,000 light years, calculate the total number of neutrinos of all types arriving in each square meter at the Earth. Also estimate the number of reactions

$$\overline{\nu}_e + \mathrm{p} \longrightarrow \mathrm{n} + e^+$$

that will occur in 1000 tonnes of water. Assume that the cross-section is given by

$$\sigma = \frac{4 p_e E_e G_F^2}{\pi \hbar^4 c^3},$$

where p_e and E_e are the positron momentum and energy respectively and G_F is the Fermi coupling constant. Assume that only one-sixth of the neutrinos are electron neutrinos.
7. The rate of energy delivered by the Sun to the Earth is known as the solar constant and equals 1.4×10^6 erg cm^{-2} s^{-1}. Knowing that the distance between the Sun and the Earth is 1.5×10^8 km, give a lower estimate of the rate at which the Sun is losing mass to supply the radiated energy.

8. Calculate the energy radiated during the contraction of the primordial gas into what is now the Sun (the Sun's diameter is 1.4×10^6 km). What energy would be released if the solar diameter suddenly shrank by 10%?
9. Given that the Sun was originally composed of 71% hydrogen by weight and assuming it has generated energy at its present rate (3.86×10^{36} W) for about 5×10^9 years by converting hydrogen into helium, estimate the time it will take to burn 10% of its remaining hydrogen. Take the energy release per helium nucleus created to be 26 MeV.
10. The CNO cycle that may contribute to energy production in stars similar to the Sun begins with the reaction $\mathrm{p} + {}^{12}\mathrm{C} \longrightarrow {}^{13}\mathrm{N} + \gamma$. Assuming the temperature near the center of the Sun to be 15×10^6 K, find the peak energy and width of the reaction rate.
11. Show that the solution of the following equation

$$\nabla^2 \Phi_s(\mathbf{r}) = \lambda_D^{-2} \Phi_s(\mathbf{r}) - 4\pi Z e \delta^3(\mathbf{r})$$

is given by

$$\Phi_s(\mathbf{r}) = \frac{1}{r} Z e \exp\left[-\frac{r}{\lambda_D}\right],$$

where $\lambda_D = (kT/4\pi n_e e^2)^{1/2}$ is the Debye screening length.

12. The reaction ${}^{12}\mathrm{C}(\mathrm{p}, \gamma)^{13}\mathrm{N}$ has a peak at 424 keV center-of-mass energy, corresponding to a $J^\pi = \frac{1}{2}^+$ resonance. The resonance has a full width at half maximum $\Gamma = 40$ keV. This width is essentially the proton width, since the only other channel is Γ_γ, which is much smaller than Γ_p. What is the value of the dimensionless reduced width θ_l^2 for that state?
13. Obtain the results of Eq. (7.34) from Eq. (7.32). Follow the instructions preceding the equation.
14. Show that for non-resonant reactions in the Sun, Eq. (7.11) becomes

$$r_{12} = \frac{n_1 n_2}{1+\delta_{12}} 7.21 \cdot 10^{-19}\,\mathrm{cm}^3/\mathrm{s} \frac{1}{A Z_1 Z_2} \times \left[\frac{S(E_0)}{\mathrm{keV} \cdot \mathrm{barns}} (Z_1^2 Z_2^2 A)^{2/3} \left(\frac{42.5}{T_6^{1/3}}\right)^2 \exp\left(\frac{(Z_1^2 Z_2^2 A)^{1/3} 42.5}{T_6^{1/3}}\right)\right], \tag{7.153}$$

where $A = A_1 A_2/(A_1 + A_2)$.

This tells us that small $Z_1 Z_2$ is favored, and that rates are expected to rise as $e^{-1/T^{1/3}}$. In the above, T_6 is the temperature in units of 10^6 K. The factor δ_{12} is to prevent double counting for fusion of the same nuclear species.

15. A "network" calculation of the ppI cycle in the Sun has the main contributing reactions

$$\begin{aligned}
\mathrm{p}+\mathrm{p} &\to \mathrm{d}+e^{+}+\nu_e, & r_{pp} &\sim \lambda_{pp}\frac{n_p^2}{2},\\
\mathrm{d}+\mathrm{p} &\to {}^3\mathrm{He}+\gamma, & r_{pd} &\sim \lambda_{pd}\; n_p\; n_d,\\
{}^3\mathrm{He}+{}^3\mathrm{He} &\to {}^4\mathrm{He}+\mathrm{p}+\mathrm{p}, & r_{33} &\sim \lambda_{33}\frac{n_{3He}^2}{2}.
\end{aligned}$$

Here r represents a rate and $\lambda = \langle \sigma v \rangle$. From one calculated S-factor (for pp) and two that are measured, one can calculate the production of He once the composition and temperature is specified.

One feature of interest in this simple network is that d and ^{3}He both act as "catalysts": they are produced and then consumed in the burning. In a steady state process, this implies they must reach some equilibrium abundance where the production rate equals the destruction rate. That is, the general rate equation

$$\frac{dn_d}{dt} = \lambda_{pp}\frac{n_p^2}{2} - \lambda_{pd} n_p n_d$$

is satisfied at equilibrium by replacing the LHS by zero. Thus

$$\left(\frac{n_p}{n_d}\right)_{\text{equil}} = \frac{2\lambda_{pd}}{\lambda_{pp}}.$$

(a) Using the S-factors

$$S_{12}(0) = 2.5 \cdot 10^{-4}\,\text{kev b}, \quad S_{11}(0) = 4.07 \cdot 10^{-22}\ \text{kev b}$$

and the rate formula (7.153) find an equation for $(n_p/n_d)_{\text{equil}}$ as a function of $T_7^{1/3}$.

The result shows that this ratio is a decreasing function of T_7: the higher the temperature, the lower the equilibrium abundance of deuterium. Therefore in the region of the Sun where the ppI cycle is operating, the deuterium abundance is lowest in the Sun's center.

(b) Plugging in the solar core temperature show that

$$\left(\frac{n_d}{n_p}\right) = 3.6 \cdot 10^{-18}.$$

There isn't much deuterium about: using $n_p \sim 3 \cdot 10^{25}/\text{cm}^3$ one finds $n_d \sim 10^8/\text{cm}^3$.

(c) Using the value of $S_{11}(0)$ given above and Eq. (7.153) show that

$$r_{pp} \sim 0.6 \times 10^8/\text{cm}^3/\text{sec}$$

and that the typical lifetime of a deuterium nucleus at the core is

$$\tau_d \sim 1\,\text{sec}.$$

That is, deuterium is burned instantaneously and thus reaches equilibrium very, very quickly.

16. The result above allows us to write the analogous equation for ^3He as

$$\frac{dn_3}{dt} = \lambda_{pp}\frac{n_p^2}{2} - 2\lambda_{33}\frac{n_3^2}{2},$$

where the factor of two in the term on the right comes because the $^3\text{He} + {}^3\text{He}$ reaction destroys two ^3He nuclei. Thus at equilibrium

$$\left(\frac{n_3}{n_p}\right)_{\text{equil}} = \sqrt{\frac{\lambda_{pp}}{2\lambda_{33}}}.$$

Using

$$S_{33}(0) = 5.15 \cdot 10^3\,\text{keV b}$$

do again the rate algebra to find

$$\left(\frac{n_3}{n_p}\right)_{\text{equil}} = (1.33 \cdot 10^{-13})\exp\left(\frac{20.65}{T_7^{1/3}}\right) = \begin{pmatrix} 9.08 \cdot 10^{-6}, & T_7 = 1.5 \\ 1.24 \cdot 10^{-4}, & T_7 = 1.0 \end{pmatrix}.$$

This ratio is clearly a sharply decreasing function of T_7 and thus a sharply increasing function of r. That is, a sharp gradient in ^3He is established in the Sun.

17. The non-existence of a bound nucleus with $A = 8$ was one of the major puzzles in nuclear astrophysics. How could heavier elements than $A = 8$ be formed? Using typical values of concentration of α particles in the core of a heavy star, $n_\alpha \sim 1.5 \cdot 10^{28}/\text{cm}^3$ (corresponding to $\rho_\alpha \sim 10^5\,\text{g/cm}^3$) and $T_8 \sim 1$ one obtains

$$\frac{n(^8\text{Be})}{n(\alpha)} \sim 3.2 \times 10^{-10}.$$

Salpeter suggested that this concentration would then allow $\alpha + {}^8\mathrm{Be}$ $(\alpha + \alpha) \to {}^{12}\mathrm{C}$ to take place. Hoyle then argued that this reaction would not be fast enough to produce significant burning unless it was also resonant. Now the mass of ${}^8\mathrm{Be} + \alpha$ is 7.366 MeV, and each nucleus has $J^\pi = 0^+$. Thus s-wave capture would require a 0^+ resonance in ${}^{12}\mathrm{C}$ at ~7.4 MeV. No such state was then known, but a search by Cook, Fowler, Lauritsen, and Lauritsen revealed a 0^+ level at 7.644 MeV, with decay channels ${}^8\mathrm{Be} + \alpha$ and γ decay to the 2^+ 4.433 level in ${}^{12}\mathrm{C}$. The parameters are

$$\begin{aligned}\Gamma_\alpha &\sim 8.9\,\mathrm{eV},\\ \Gamma_\gamma &\sim 3.6 \cdot 10^{-3}\,\mathrm{eV}.\end{aligned}$$

(a) Show that

$$r_{48} = n_\alpha^3\, T_8^{-3} \exp\left(-\frac{42.9}{T_8}\right)(6.3 \cdot 10^{-54}\,\mathrm{cm}^6/\mathrm{sec}).$$

If we denote by $\omega_{3\alpha}$ the decay rate of an α in the plasma, then

$$\begin{aligned}\omega_{3\alpha} &= 3\, n_\alpha^2\, T_8^{-3} \exp\left(-\frac{42.9}{T_8}\right)(6.3 \cdot 10^{-54}\,\mathrm{cm}^6/\mathrm{sec})\\ &= \left(\frac{n_\alpha}{1.5 \cdot 10^{28}/\mathrm{cm}^3}\right)^2 (4.3 \cdot 10^3/\mathrm{sec}) T_8^{-3} \exp\left(-\frac{42.9}{T_8}\right).\end{aligned}$$

(b) Since the energy release per reaction is 7.27 MeV show that the energy produced per gram, ϵ, is

$$\epsilon = (2.5 \cdot 10^{21}\mathrm{erg/g\ sec}) \left(\frac{n_\alpha}{1.5 \cdot 10^{28}/\mathrm{cm}^3}\right)^2 T_8^{-3} \exp\left(-\frac{42.9}{T_8}\right).$$

Chapter 8

Stellar Nucleosynthesis

8.1. Introduction

The temperatures and densities in stellar cores lead to very low rates of fusion reactions. At solar core temperature ($T \sim 15\,\mathrm{MK}$) and density ($160\,\mathrm{g/cm^3}$), the energy release rate is equal to $276\,\mu$ W/cm^3, which is only a quarter of the rate at which a resting human body generates heat. In the 1930s, it was discovered that the Sun derives its energy from the fusion of hydrogen atoms, in a process called the proton–proton reaction. This thermonuclear process occurs in relatively cool stars. The proton–proton cycle involves the fusion of four hydrogen nuclei (protons) to form a single helium nucleus. Life as it exists on Earth is made possible by light and heat generated in the Sun in such thermonuclear reactions. A different kind of thermonuclear reaction, called the carbon–nitrogen cycle, creates most of the energy radiated by stars much hotter than the Sun. The carbon–nitrogen cycle involves isotopes of the elements hydrogen, carbon, nitrogen, and oxygen that are formed during a complex chain of fusion events, each releasing energy. As long as the gaseous elemental fuel for their fusion reactions remains plentiful, the stars in the cosmos continue to shine with thermonuclear brilliance.

The stellar masses and fusion processes are related by

a) $0.1 - 0.5\,M_\odot \rightarrow$ H burning but no He burning.
b) $0.5 - 8\,M_\odot \rightarrow$ H and He burning.
c) $>1.4\,M_\odot \rightarrow$ At the end of the stellar life, core will collapse — supernova (mass of the core only).
d) $<1.4\,M_\odot \rightarrow$ White dwarf (a WD is formed from the core of stars with masses $\lesssim 8\,M_\odot$ after the envelope is blown away).
e) $8 - 11\,M_\odot \rightarrow$ H, He and C burning.
f) $>11\,M_\odot \rightarrow$ All stages of thermonuclear fusion.

Table 8.1. List of main nucleosynthesis processes and nuclides created in each case.

Nucleosynthetic process	Elements created
Big Bang	^{1}H, ^{4}He, ^{2}H, ^{3}H (Li, B?)
Main sequence stars:	
Hydrogen burning	^{4}He
Helium burning	^{12}C, ^{4}He, ^{24}Mg, ^{16}O, ^{20}Ne
Carbon burning	^{24}Mg, ^{23}Na, ^{20}Ne
CNO cycle	^{4}He
x-process (spallation) and supernova (?)	Li, Be, B
α-process	^{24}Mg, ^{28}Si, ^{32}S, ^{36}Ar, ^{40}Ca
electron-capture process	^{56}Fe and other transitions
s-process	up to mass 209
r-process	up to mass 254

In the following we describe the formation of the Universe and the elements according to the so-called *Standard Model of Stellar Evolution*, which is based on the models originally developed by Bethe and Weizsäcker in the 1930's for the reactions in the Sun. A great amount of research activities in the last decades now allow us to have a very good idea of how all species of stars evolve, from quiet stars such as white dwarfs, to explosive stars, such as supernovae. Nuclear reactions at low energies are the most important physics ingredients of this research area.

Table 8.1 summarizes the nuclear processes that we will discuss in the next Sections of this chapter. We will defer the discussion on Big Bang nucleosynthesis to Chapter 9.

8.2. The Sun

What are the nuclear processes which give rise to the huge thermonuclear energy of the Sun, the latter having lasted 4.6×10^9 years (the assumed age of the Sun)? It cannot be the simple fusion of two protons, or of α-particles, or even the fusion of protons with α-particles, since none of the nuclei $^2_2$He, $^8_4$Be, and $^5_3$Li are stable. The only possibility is proton–proton fusion in the form

$$p + p \rightarrow d + e^+ + \nu_e, \tag{8.1}$$

which occurs via β-decay, i.e., due to the weak-interaction. The cross-section for this reaction for protons of energy below 1 MeV is very small, of the order of 10^{-23} b (in fact, the effective energy for this reaction in the Sun is about 20 keV). The average lifetime of protons in the Sun due to the transformation to deuterons by means of Eq. (8.1) is about 10^{10} y. This explains why the energy radiated from the Sun is approximately constant in time, and not an explosive process.

Because of the low Coulomb barrier, in the p + p reaction (E_c = 0.55 MeV), a star like the Sun would have consumed all its hydrogen quickly (note the relatively large value of $(\Delta)I_{max}$ in Table 7.1), were it not slowed down by the weakness of the weak interactions. The calculation of probability of deuteron formation consists of two separate considerations: (1) the penetration of a mutual potential barrier in a collision of two protons in a thermal bath and (2) the probability of β-decay and positron and neutrino emission. Bethe and Critchfield [Bet38] used the original Fermi theory (point interaction) for the second part, which is adequate for the low energy process.

8.2.1. *Deuterium formation*

The Hamiltonian H for the p–p interaction can be written as a sum of a nuclear term H_n and a weak-interaction term H_β. The weak interaction term is small compared to the nuclear term. Therefore, first order perturbation theory can be applied and Fermi's golden rule, Eq. (4.22), gives the differential cross-section as

$$d\sigma = \frac{\Lambda}{\text{flux}(= \text{v}_i)} = \frac{2\pi\rho(E)}{\hbar \text{v}_i}|\langle f|H_{if}|i\rangle|^2. \tag{8.2}$$

Here $\rho(E) = dN/dE$, is the density of final states in the interval dE and v_i is the relative velocity of the incoming particles.

The matrix element that appears in the differential cross-section may be written in terms of the initial state wave function Ψ_i of the two protons in the entrance channel and the final state wave function[1] Ψ_f as

$$H_{if} = \int [\Psi_d \Psi_e \Psi_\nu]^* H_\beta \Psi_i d\tau. \tag{8.3}$$

If the energy of the electron is large compared to $Z \times$ Rydberg (a *Rydberg* is $R_\infty = 2\pi^2 me^4/ch^3$), then a plane wave approximation is a good one, $\Psi_e = 1/(\sqrt{V})\exp(i\mathbf{k_e} \cdot \mathbf{r})$ where the wave function is normalized over volume V. For lower energies, typically 200 keV or less, the electron wave function could be strongly affected by nuclear charge. Apart from this, the

[1] The same arguments were used in Chapter 7 in connection with Fermi's theory for beta-decay.

final state wave function $[\Psi_d\Psi_e\Psi_\nu]$ has a deuteron part Ψ_d whose radial part rapidly vanishes outside the nuclear domain (R_0), so that the integration need not extend much beyond $r \simeq R_0$ (for example, the deuteron radius $R_d = 1.7\,\text{fm}$). Because of the Q-value of 0.42 MeV for the reaction, the kinetic energy of the electron ($K_e \leq 0.42\,\text{MeV}$) and the average energy of the neutrinos ($\bar{E}_\nu = 0.26\,\text{MeV}$) are low enough so that for both electrons and neutrino wave functions, the product $kR_0 \leq 2.2 \times 10^{-3}$ and the exponential can be approximated by the first term of the Taylor expansion, as in Eq. (7.130). Therefore, the expectation value of the Hamiltonian governed by the coupling constant g is

$$H_{if} = \int [\Psi_\mathrm{d}\Psi_\mathrm{e}\Psi_\nu]^* \mathrm{H}_\beta \Psi_\mathrm{i} d\tau = \frac{\mathrm{g}}{\mathrm{V}} \int [\Psi_\mathrm{d}]^* \Psi_\mathrm{i} d\tau. \tag{8.4}$$

For the reaction $\mathrm{p} + \mathrm{p} \to \mathrm{d} + \mathrm{e}^+ + \nu_\mathrm{e}$, the deuterium in its ground state has $J_f^\pi = 1^+$, with a predominant relative orbital angular momentum $l_f = 0$ and $S_f = 1$ (triplet S-state). For a maximum probability (superallowed transitions), there are no changes in the orbital angular momentum between the initial and final states. Hence, the initial $\mathrm{p}+\mathrm{p}$ must have $l_i = 0$. For identical particles the Pauli principle requires $S_i = 0$, so that the total wave function will be antisymmetric in space and spin coordinates. Hence, we must have $|S_i = 0, l_i = 0\rangle \to |S_f = 1, l_f = 0\rangle$, which is a pure Gamow–Teller transition.

The spin matrix element in Eq. (8.2) for σ is obtained from summing over the final states, averaging over the initial states and dividing by 2 to take into account that we have two identical particles in the initial state. The rest of the calculation yields the same result as Eq. (7.137) ($\sigma = \Lambda/\mathrm{v}_i$), which we rewrite as

$$\sigma = \frac{m^5c^4}{2\pi^3\hbar^7\nu_i} f(E) g^2 \frac{M_{space}^2 M_{spin}^2}{2}, \tag{8.5}$$

where

$$M_{spin}^2 = \frac{(2J+1)}{(2J_1+1)(2J_2+1)} = 3.$$

The space matrix element is

$$M_{space} = \int_0^\infty \chi_f(r)\chi_i(r) r^2 dr. \tag{8.6}$$

At large energies, the factor $f(E)$, given by Eq. (7.139), behaves as

$$f(E) \propto \frac{1}{30} E^5. \tag{8.7}$$

The wave functions $\chi(r)$ in Eq. (8.6) involve Coulomb wave functions for barrier penetration at (low) stellar energies. One needs only the s-wave part for the wave function of the deuteron ψ_d, as the d-wave part makes no contribution to the matrix element [Fri51], although its contribution to the normalization has to be accounted for. The wave function of the initial two-proton system ψ_p is normalized to a plane wave of unit amplitude, and only the s-wave part is needed. As explained in Section 7.5, the asymptotic form of ψ_p is given in terms of regular and irregular Coulomb functions, and related to the s-wave phase shifts in p–p scattering data. The result of a numerical integration of Eq. (8.6) is a very small cross-section of $\sigma = 10^{-47}\,\mathrm{cm}^2$ at a laboratory beam energy of $E_p = 1\,\mathrm{MeV}$, which cannot be measured experimentally.

The reaction Eq. (8.1) is a non-resonant reaction and its the rate varies smoothly with stellar temperatures. Numerically, one gets $S(0) = 3.8 \times 10^{-22}\,\mathrm{keV\ b}$ and $dS(0)/dE = 4.2 \times 10^{-24}\,\mathrm{b}$. In the center of the Sun, $T_6 = 15$, this gives $\langle \sigma v \rangle_{pp} = 1.2 \times 10^{-43}\,\mathrm{cm}^3\ \mathrm{s}^{-1}$ [Ade11]. For density in the center of the Sun $\rho = 100\,\mathrm{gm\ cm}^{-3}$ and equal mixture of hydrogen and helium ($X_H = X_{He} = 0.5$), the mean life of a hydrogen nucleus against conversion to deuterium is $\tau_H(H) = 1/N_H \langle \sigma v \rangle_{pp} \sim 10^{10}\,\mathrm{yr}$, which is comparable to the age of the old stars. The cross-section is so small because of the weak interaction and the suppression ($\sim 1/100$) due to the Coulomb barrier. The smallness of the reaction rate is the primary reason why stars consume their nuclear fuel of hydrogen so slowly.

8.2.2. *Deuterium burning*

Once deuterium is produced, the next burning step in the Sun follows through the reaction

$$\mathrm{d} + \mathrm{p} \ \rightarrow \ {}^3\mathrm{He} + \gamma. \tag{8.8}$$

This is a nonresonant direct capture reaction to the ^{3}He ground state with a Q-value of 5.497 MeV and $S(0) = 2.5 \times 10^{-3}\,\mathrm{keV}$ barn. The remaining of the (three) pp-chains in Fig. 8.1 start out with ^{3}He manufactured from $\mathrm{d} + \mathrm{p}$ reaction. The reactions $\mathrm{d(d,p)t}$, $\mathrm{d(d,n)^3He}$, $\mathrm{d(^3He,p)^4He}$, and $\mathrm{d(^3He,\gamma)^5Li}$ have a larger cross-section than the above. However, because

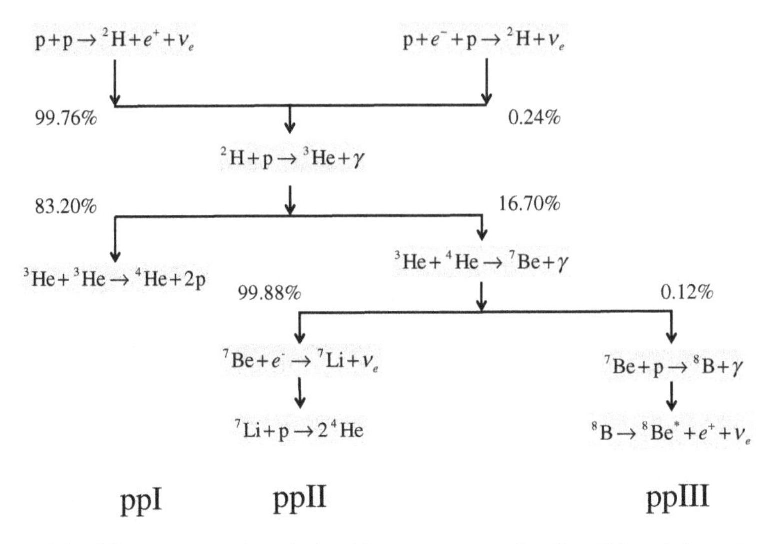

Figure 8.1. The p–p reaction chain. The percentage for the different branches are calculated in the center of the Sun [Ade11].

of the overwhelmingly large number of protons in the stellar thermonuclear reactors, the process involving protons on deuterium dominates. The overall rate of the pp-chain is not determined by this reaction because it is too fast compared to that of Eq. (8.1).

The ratio of deuterium to hydrogen has an extremely small value, that is, deuterium is readily destroyed in thermonuclear burning. The equation governing the deuterium (D) abundance is

$$\frac{d\mathrm{Y_D}}{dt} = r_{pp} - r_{pd} = \frac{\mathrm{Y_H}^2}{2}\langle\sigma\nu\rangle_{pp} - \mathrm{Y_H Y_D}\langle\sigma\nu\rangle_{pd}. \tag{8.9}$$

This rate reaches a state of quasi-equilibrium, yielding

$$\frac{\mathrm{Y_D}}{\mathrm{Y_H}} = \frac{\langle\sigma\nu\rangle_{pp}}{2\langle\sigma\nu\rangle_{pd}}, \tag{8.10}$$

The S-factor for the reaction in the numerator is about $S = 3.8\times 10^{-22}$ keV b, and for the denominator $S = 2.5\times 10^{-4}$ keV b. This yields a ratio equal to 5.6×10^{-18} at $T_6 = 5$ and 1.7×10^{-18} at $T_6 = 40$. For the solar system however, this ratio is 1.5×10^{-4} while the observed $(\mathrm{Y_D}/\mathrm{Y_H})$ ratio in the Cosmos is $\sim 10^{-5}$. The higher cosmic ratio is due to primordial nucleosynthesis in the early phase of the Universe before the stars formed. Stars destroy the deuterium in their core due to the above reaction.

The quasi-equilibrium is reached within lifetime of D in the Sun equal to

$$N_A\langle\sigma\nu\rangle_{\mathrm{p+d}} = 10^{-2}\,\mathrm{cm^3s^{-1}\ mole^{-1}} \quad\Rightarrow\quad \tau_d = \frac{1}{Y_p\rho N_A\langle\sigma\nu\rangle_{\mathrm{p+d}}} = 2\,\mathrm{s}.$$

That is, deuterons are basically instantaneously consumed in a stellar environment.

8.2.3. ^{3}He *burning*

The pp-chain-I is completed (see Fig. 8.1) through the burning of ^{3}He via the reaction

$$^3\mathrm{He} + {}^3\mathrm{He} \rightarrow \mathrm{p} + \mathrm{p} + {}^4\mathrm{He}, \tag{8.11}$$

with an S-factor $S(0) \sim 5500\,\mathrm{keV}$ barn and Q-value$=12.86\,\mathrm{MeV}$. In addition, the reaction

$$^3\mathrm{He} + \mathrm{D} \rightarrow {}^4\mathrm{He} + \mathrm{p}, \tag{8.12}$$

has an S-factor $S(0) \sim 6240\,\mathrm{keV}$ barn, but since the deuterium concentration is very small, the first reaction dominates the destruction of ^{3}He even though both reactions have comparable $S(0)$ factors [Ade11].

^{3}He can also be consumed by reactions with ^{4}He originally synthesized in the early Universe and in population III objects. These reactions proceed through direct captures and lead to the ppII and ppIII parts of the chain. The reactions $^3\mathrm{He}(\alpha,\gamma)^7\mathrm{Be}$ and $^7\mathrm{Be}(\mathrm{p},\gamma)^8\mathrm{B}$ control the production of high energy neutrinos in the Sun [Bah89].

8.2.4. *Reactions involving* ^{7}*Be*

As shown in Fig. 8.1, about 14% of the time, ^{3}He is burned with ^{4}He radiatively to ^{7}Be. Subsequent reactions involving ^{7}Be as a first step in alternate ways complete the fusion process $4\mathrm{H} \rightarrow {}^4\mathrm{He}$ in the ppII and ppIII chains.

Electron capture

The first step of the ppII chain is the atomic electron capture reaction on ^{7}Be: $^7\mathrm{Be}+\mathrm{e}^- \rightarrow {}^7\mathrm{Li} + \nu_e$. This decay goes to the ground state and to the first excited state of ^{7}Li at $E_x = 0.478\,\mathrm{keV}$, with $J^\pi = \frac{1}{2}^-$. The percentage of decays to the excited state is 10.4% and the energy released in the reaction, with a Q-value of $0.862\,\mathrm{keV}$, is carried away by monoenergetic neutrinos with energies: $E_\nu = 862$ and $384\,\mathrm{keV}$. The mean life of the decay is $\tau = 76.9\,\mathrm{d}$.

The capture rate is obtained from Fermi's golden rule, Eq. (8.2), and the fact that the wave functions of both the initial nucleus and the final one vanish rapidly outside the nuclear domain while the electron wave function within the nucleus can be approximated as its value at the origin. The neutrino wave function can be represented by a plane wave normalized to volume V, so that

$$H_{if} = \Psi_e(0) g \int \Psi^*_{^7\mathrm{Li}} \Psi_{^7\mathrm{Be}} d^3r = \Psi_e(0) g M_n,$$

where $M_n = \int \Psi^*_{^7\mathrm{Li}} \Psi_{^7\mathrm{Be}} d^3r$ is the nuclear matrix element. Therefore, we obtain for the capture rate

$$\Lambda_{EC} = \frac{1}{\tau_{EC}} = \left(\frac{g^2 M_n^2}{\pi c^3 \hbar^4} \right) E_\nu^2 |\Psi_e(0)|^2. \tag{8.13}$$

In the atomic capture process, any of the various electron shells contribute to the capture rate; however the K-shell gives the dominant contribution. At temperatures inside the Sun, for example, $T_6 = 15$, and nuclei such as ^{7}Be are largely ionized. The nuclei however are immersed in a sea of free electrons resulting from the ionized process and therefore electron capture from continuum states is possible. Since all factors in the capture of continuum electrons in the Sun are approximately the same as those in the case of atomic electron capture, except for the respective electron densities, the ^{7}Be lifetime in a star, τ_s, is related to the terrestrial lifetime τ_t by

$$\frac{\tau_s}{\tau_t} \sim \frac{2|\Psi_t(0)|^2}{|\Psi_s(0)|^2}, \tag{8.14}$$

where $|\Psi_s(0)|^2$ is the density of the free electrons in the star, $n_e = \rho/m_H$ at the nucleus, ρ being the stellar density. The factor of 2 in the denominator takes care of the two spin states in the calculation of the λ_t whereas the corresponding λ_s is calculated by averaging over these two orientations.

Taking into account the distortions of the electron wave function due to the thermally averaged Coulomb interaction with nuclei of charge Z and the contribution due to hydrogen (of mass fraction X_H) and heavier nuclei, one obtains the continuum capture rate,

$$\tau_s = \frac{2|\Psi_t(0)|^2 \tau_t}{(\rho/M_H)[(1+X_H)/2] 2\pi Z \alpha (m_e c^2/3kT)^{1/2}}, \tag{8.15}$$

with $|\Psi_e(0)|^2 \sim (Z/a_0)^3/\pi$. This equation can be used to calculate the lifetime of the ^{7}Be nucleus in the Sun [Bah69]:

$$\tau_s(^7\text{Be}) = 4.72 \times 10^8 \frac{T_6^{1/2}}{\rho(1 + X_H)} \text{ s.} \tag{8.16}$$

The temperature dependence comes from the nuclear Coulomb field corrections to the electron wave function which are thermally averaged. This yields a continuum capture rate of $\tau_s(^7\text{Be}) = 140\,\text{d}$ as compared to the terrestrial mean life of $\tau_t = 76.9\,\text{d}$ [Bah69b]. A contribution from some ^{7}Be atoms, which are only partially ionized, also has to be considered. Under solar conditions, the K-shell electrons from partially ionized atoms give another 21% increase in the total decay rate. Including this, one obtains the solar lifetime of a ^{7}Be nucleus as: $\tau_\odot(^7\text{Be}) = 120\,\text{d}$.

Formation of ^{8}B

To complete the ppIII part of the pp-chain, we notice that ^{7}Be is consumed by the proton capture reaction $^7\text{Be}(\text{p}, \alpha)^8\text{B}$. In the Sun, this reaction occurs only 0.02% of the time. The proton capture on ^{7}Be proceeds at energies away below the 640 keV resonance via the direct capture process. This reaction has a weighted average $S(0) = 0.0238\,\text{keV barn}$ [Fil83]. ^{8}B is a radioactive nucleus with a lifetime $\tau = 1.1\,\text{s}$,

$$^8\text{B} \rightarrow {}^8\text{Be} + \text{e}^+ + \nu_e. \tag{8.17}$$

Due to the selection rules, the positron decay of $^8\text{B}(J^\pi = 2^+)$ goes mainly to the $\Gamma = 1.6\,\text{MeV}$ broad excited state in ^{8}Be at excitation energy $E_x = 2.94\,\text{MeV}$ $(J^\pi = 2^+)$. This state has very short lifetime and quickly decays into two α-particles. The average energy of the neutrinos from ^{8}B reactions is $\bar{E_\nu}(^8\text{B}) = 7.3\,\text{MeV}$.

8.2.5. *The demise of the Sun*

The Sun is about halfway through its life as a star, which started 4.5 billion years ago. As its hydrogen fuel is exhausted, the Sun will contract, but then heat up even more as it next burns helium. The heat will cause it to expand and even consume the Earth. The Sun will then become a red giant (discussed below) and the surface will cool from 5500 K to 4000 K. Eventually, the light elements in the outer layers will boil off and the Sun will contract to the size of the Earth with a final mass that will be half its current mass. The Sun will cool down to become a white dwarf and then a cold black dwarf.

8.3. Massive Stars

8.3.1. *The CNO cycle*

The Sun gets most of its energy generation through the pp-chain reactions (see Fig. 8.3). However, as the central temperature (in stars more massive than the Sun) gets higher, the CNO cycle (see below for reaction sequence) comes to dominate over the pp-chain at T_6 near 20 (this changeover assumes the solar CNO abundance; the transition temperature depends upon CNO abundance in the star).

Early generations of stars (population II and population III stars) generated energy primarily through the pp-chain. These stars are still shining in globular clusters and have mass smaller than that of the Sun. Later generation stars formed from the debris of heavier stars that contained heavy elements. In second and third generation stars, slightly heavier than the Sun, higher central temperatures are possible because of higher gravity. Then, hydrogen burning can take place through a chain of reactions involving heavy elements C, N, and O which have some reasonable abundance (exceeding 1%) compared to other heavy elements like Li, Be, B which are extremely low in abundance, even though the Coulomb barriers of Li, Be, B are smaller.

The CN cycle is a part of the CNO cycle (see Fig. 8.2, cycle I) and involves the reaction chain

$$^{12}\mathrm{C}(\mathrm{p},\gamma)^{13}\mathrm{N}(\mathrm{e}^+\nu_e)^{13}\mathrm{C}(\mathrm{p},\gamma)^{14}\mathrm{N}(\mathrm{p},\gamma)^{15}\mathrm{O}(\mathrm{e}^+\nu)^{15}\mathrm{N}(\mathrm{p},\alpha)^{12}\mathrm{C}. \tag{8.18}$$

As before, one produces an α-particle out of four protons $4p \to {}^4\mathrm{He} + 2\mathrm{e}^+ + 2\nu_e$ with a $Q = 26.73$. The ^{12}C and ^{14}N act as catalysts as their nuclei remain at the end of the cycle, while the ^{12}C nuclei act as seeds that can be reused, despite their minuscule abundance compared to the hydrogen. The branch to the cycle II leads to a loss of the catalytic material from the CN cycle through the $^{15}\mathrm{N}(\mathrm{p},\gamma)^{16}\mathrm{O}$ reactions. However, the catalytic material is subsequently returned to the CN cycle by the reaction: $^{16}\mathrm{O}(\mathrm{p},\gamma)^{17}\mathrm{F}(\mathrm{e}^+\nu_e)^{17}\mathrm{O}(\mathrm{p},\alpha)^{14}\mathrm{N}$.

In the CN cycle (see Fig. 8.2), the two neutrinos involved in the beta decays (of ^{13}N ($t_{1/2} = 9.97$ min) and ^{15}O ($t_{1/2} = 122.24$ s)) are of relatively low energy. The rate of the energy production, due to the conversion of the protons into α-particles, is governed by the slowest thermonuclear reaction

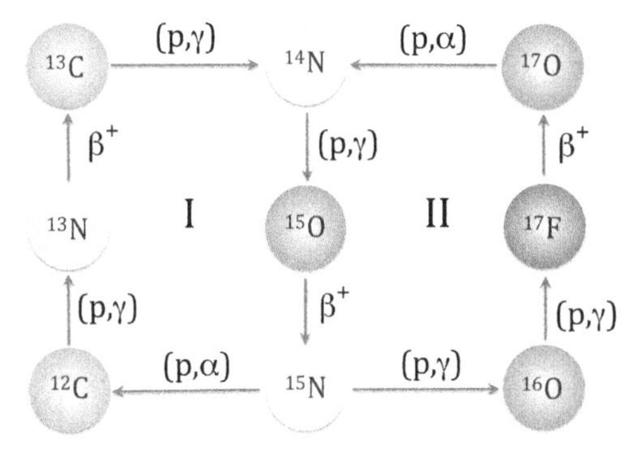

Figure 8.2. The CNO bi-cycle. The CN cycle [Bet38], marked I, produces about 1% of solar energy and significant fluxes of solar neutrinos from the Sun.

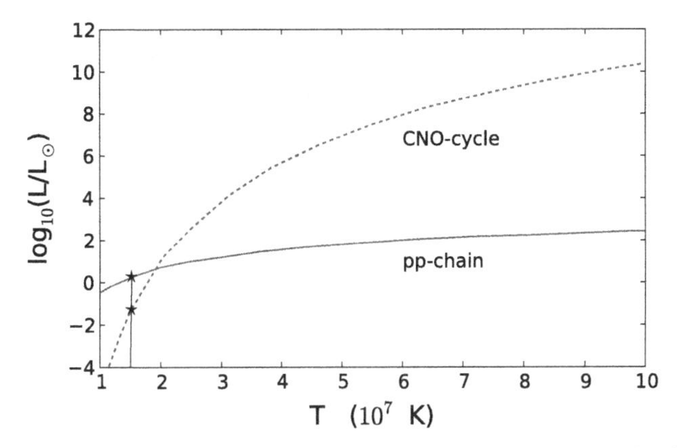

Figure 8.3. The stellar energy production as a function of temperature for the pp chain and CNO cycles, showing the dominance of the former at solar temperatures (marked by 'star' symbols).

in the cycle. The nitrogen isotopes have the highest Coulomb barriers, because of their $Z = 7$. Among them $^{14}\mathrm{N}(\mathrm{p}, \gamma)^{15}\mathrm{O}$ is the slowest because this reaction having a final state photon is governed by electromagnetic forces while that involving the other nitrogen isotope, $^{15}\mathrm{N}(\mathrm{p}, \alpha)^{12}\mathrm{C}$, is governed by strong forces and is therefore faster.

From the CN cycle, there is actually a branching off from ^{15}N by the reaction $^{15}\mathrm{N}(\mathrm{p}, \gamma)^{16}\mathrm{O}$. This leaks to cycle II in Fig. 8.2 and involves isotopes of oxygen. Sometimes it is called the ON cycle. The nitrogen is returned to the CN cycle through ^{14}N. Together the CN and the ON cycles constitutes the CNO bi-cycle. The two cycles differ considerably in their

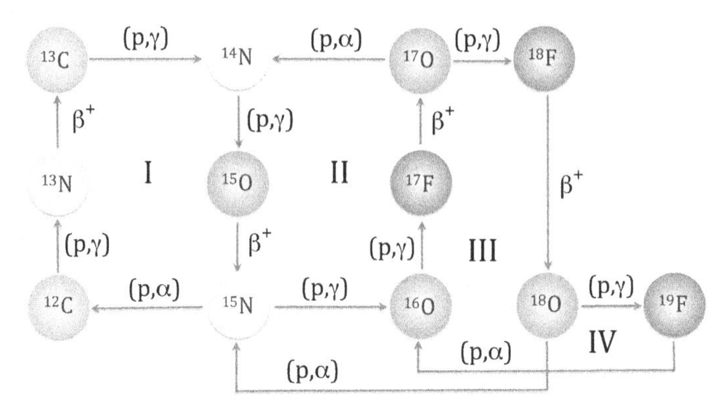

Figure 8.4. The various CNO cycles. The left part is the CN cycle where only C and N serve as catalysts for the conversion of four protons into ^{4}He. Here the slowest fusion reaction is the (p,γ) reaction on ^{14}N whereas the slower β-decay has a half-life of 9.97 m. In the CNO cycle II (middle), there is leakage from the CN cycle to the ON cycle through the branching at ^{15}N. The flow is returned to the CN cycle (which cycles 1000 times for each ON cycle) through $^{17}\mathrm{O}(\mathrm{p}, \alpha)^{14}\mathrm{N}$. The right part represents additional cycles linking into the CNO cycle through the $^{17}\mathrm{O}(\mathrm{p}, \gamma)^{18}\mathrm{F}$ reaction.

relative cycle-rates: the ON cycle operates only once for every 1000 cycles of the main CN cycle. This results from the fact that the S-factor for the reaction $^{15}\mathrm{N}(\mathrm{p}, \alpha)^{12}\mathrm{C}$ is about 1000 times larger than that for the reaction $^{15}\mathrm{N}(\mathrm{p}, \gamma)^{16}\mathrm{O}$. An extended CNO cycle is shown in Fig. 8.4.

8.3.2. *Hot CNO and rp-process*

The CNO cycle, as discussed in the previous section, is relevant for temperatures of $T_6 \geq 20$, which are found in quiescently hydrogen burning stars with solar composition, only slightly more massive than the Sun. But hydrogen burning by means of CNO cycles can also occur at much larger temperatures ($T \sim 10^8$–10^9 K) than those found in the interiors of ordinary, "main sequence", stars. As a few examples, we cite: (a) hydrogen burning at the accreting surface of a neutron star or, (b) explosive burning on the surface of a white dwarf (*novae*), or (c) burning in the outer layers of a supernova shock heated material.

Hot CNO cycles, as they are commonly known, operate on a rapid timescale (few seconds) and β-unstable nuclei like ^{13}N will live long enough to be burned by thermonuclear charged particle reactions. Unlike the normal CNO cycle, hydrogen burning in the hot CNO is limited by the β-decay lifetimes of the proton-rich nuclei such as ^{14}O and ^{15}O rather than the proton capture rate of ^{14}N. For temperatures, $T \geq 5 \times 10^8$ K, material

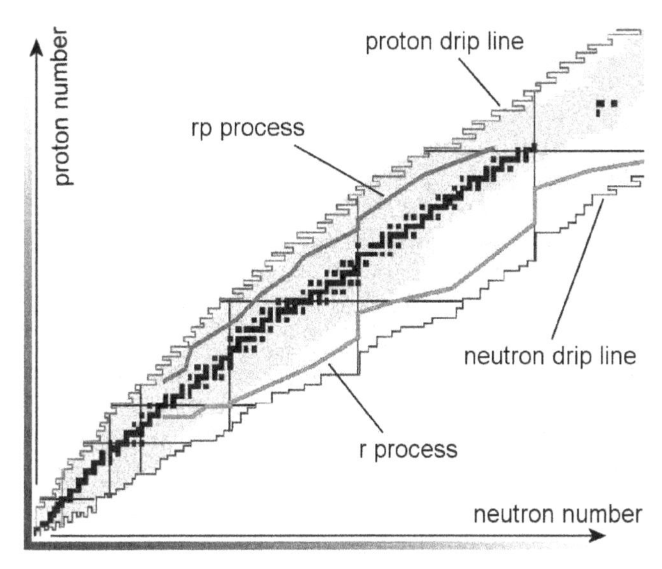

Figure 8.5. Schematic paths of r-process and rp-process in the (N,Z) plane with respect to the valley of beta-stability, the neutron drip and the proton drip lines.

can leak out of the cycles leading to the formation of heavier nuclei. This is known as *rapid proton capture*, or *rp-process.*

The path of the rp-process is analogous to the *r-process* by neutron capture, converting CNO nuclei into isotopes near the region of proton unbound nuclei (*the proton drip line*). For each neutron number, a maximum mass number A is reached where the proton capture must wait until β^+-decay occurs before the buildup of heavier nuclei can take place. In contrast to the r-process (to be discussed later), the rate of the rp-process is increasingly hindered due to the increasing Coulomb barrier of heavier and higher-Z nuclei. Hence, the rp-process does not reach the proton drip line, running closer to the beta-stability valley and through a path where the β^+-decay rate times are similar to the proton capture times. A comparison of the reaction paths of rp- and r-processes in the (N,Z) plane is shown in Fig. 8.5.

8.3.3. *Triple-α capture*

After hydrogen burning in the core of the star has exhausted its fuel, the helium core contracts slowly. Its density and temperature goes up as gravitational energy released is converted to internal kinetic energy. The contraction also heats hydrogen at the edge of the helium core, igniting the

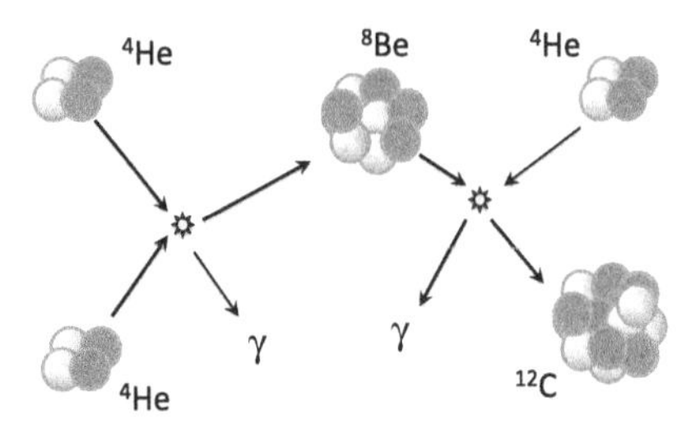

Figure 8.6. The triple-alpha process is the main source of energy production in red giants and red supergiants in which the core temperature has reached at least 100 million K. Two nuclei of helium (alpha particles) collide, fuse, and form a nucleus of beryllium. During a second collision between beryllium and helium, carbon is formed.

hydrogen to burn in a shell. The temperature in the core eventually reaches $T_6 = 100 - 200$ and the density reaches $\rho_c = 10^2 - 10^5\,\mathrm{gm\,cm^{-3}}$. At this stage, the core starts burning ^{4}He steadily. The fusion of three α particles to produce a ^{12}C nucleus is the most probable path to heavier elements [Opi51,Sal52,Sal57]. This occurs in two steps: (a) first by the fusion of two α particles producing a ^{8}Be nucleus, which is unstable to α-breakup, (b) followed fusion of ^{8}Be with another α-particle to produce ^{12}C (see Fig. 8.6).

^{8}Be has a lifetime of only 10^{-16} s (corresponding to a width $\Gamma = 6.8\,\mathrm{eV}$). This is long compared to the time 10^{-19} s that two α-particles scatter past each other with kinetic energies comparable to the Q-value of the reaction, namely, $Q = -92.1\,\mathrm{keV}$. Therefore, it is possible that the reaction $\alpha +{}^8\mathrm{Be} \to {}^{12}\mathrm{C}$ occurs before ^{8}Be decays, if a sufficiently large density of ^{8}Be exists. One can calculate the equilibrium concentration of ^{8}Be by means of the Saha Equation, Eq. (4.97),

$$N_{12} = \frac{N_1 N_2}{2}\left(\frac{2\pi}{\mu kT}\right)^{3/2}\hbar^3 \frac{(2J+1)}{(2J_1+1)(2J_2+1)}\exp\left(-\frac{E}{kT}\right), \qquad (8.19)$$

at the relevant temperature $T_6 = 11$ and density $\rho = 10^5\,\mathrm{gm\ cm^{-3}}$. One obtains

$$\frac{\mathrm{N}(^8\mathrm{Be})}{\mathrm{N}(^4\mathrm{He})} = 5.2 \times 10^{-10}. \qquad (8.20)$$

The amount of ^{12}C produced in this way, for the conditions inside a star at the tip of the red-giant branch, is insufficient to explain the observed abundance of ^{12}C, unless the reaction proceeds through a resonance [Hoy53,Hoy54]. An s-wave ($l = 0$) resonance in ^{12}C near the threshold of ^{8}Be + α reaction greatly speeds up the capture rate. Since ^{8}Be and

^{4}He both have $J^\pi = 0^+$, an s-wave resonance implies that the resonant state has to be 0^+ in ^{12}C. Hoyle suggested that the excitation energy in ^{12}C was $E_x \sim 7.68\,\text{MeV}$ [Hoy53,Hoy54] and this state was found experimentally a few years later [Coo57]. The *Hoyle state* has a total width of $\Gamma = 8.9 \pm 1.08\,\text{eV}$ [RR88], mostly due to α-decay, as γ-decay from this excited state to the ground state of ^{12}C cannot occur, since all participating states have $J^\pi = 0^+$ and $0^+ \to 0^+$ γ-decays are forbidden. Thus, the partial width due to γ-decay is several thousand times smaller than that due to α-decay, and $\Gamma = \Gamma_\alpha + \Gamma_{rad} \sim \Gamma_\alpha$. Experimentally, one has found that $\Gamma_{rad} = \Gamma_\gamma + \Gamma_{e^+e^-} = 3.67 \pm 0.50\,\text{meV}$, which is dominated by the photon de-excitation width, $\Gamma_\gamma = 3.58 \pm 0.46\,\text{meV}$.

The reaction rate for the formation of ^{12}C is given by

$$r_{3\alpha} = N_{^8\text{Be}} N_\alpha \langle \sigma \nu \rangle_{^8\text{Be}+\alpha}, \tag{8.21}$$

where $N_{^8\text{Be}}$ and N_α are the number densities of ^{8}Be and ^{4}He. The average over the Maxwellian distribution is obtained from Eq. (7.56), with $\Gamma_p = \Gamma_\alpha$ dominating over Γ_γ and $(\Gamma_\alpha \Gamma_\gamma / \Gamma) \sim \Gamma_\gamma$. This limit usually holds for resonances of energy sufficiently high so that the incident particle width (Γ_γ) dominates the natural width of the state (Γ_γ). Moreover, we can use the Saha equilibrium condition, Eq. (8.19),

$$N(^8\text{Be}) = N_\alpha^2 \omega f \frac{h^3}{(2\pi\mu kT)^{3/2}} \exp\left(-\frac{E}{kT}\right), \tag{8.22}$$

where we have introduced the screening factor f, according to the discussion in Section 7.4.1.

The triple-alpha reaction rate is obtained by calculating the equilibrium concentration of the resonant state of ^{12}C and multiplying it by the gamma-decay rate $\Gamma_\gamma/\hbar$ which leads to the ground state of ^{12}C. One obtains

$$r_{3\alpha} = N_{^8\text{Be}} N_\alpha \hbar^2 \left(\frac{2\pi}{\mu kT}\right)^{3/2} \omega f \Gamma_\gamma \exp\left(-\frac{E'}{kT}\right), \tag{8.23}$$

where μ is the reduced mass of ^{8}Be and α. According to the arguments given above, this can be rewritten as

$$r_{3\alpha \to ^{12}C} = \frac{N_\alpha^3}{2} 3^{3/2} \left(\frac{2\pi\hbar^2}{M_\alpha kT}\right)^3 f \frac{\Gamma_\alpha \Gamma_\gamma}{\Gamma \hbar} \exp\left(-\frac{Q}{kT}\right). \tag{8.24}$$

The Q-value of the reaction is the sum of $E'(^8\text{Be}+\alpha) = E_R = 287\,\text{keV}$ and $E(\alpha+\alpha) = |Q| = 92\,\text{keV}$, namely, $Q_{3\alpha} = (M_{^{12}\text{C}^*} - 3M_\alpha)c^2 = 379.38 \pm 0.20\,\text{keV}$.

The energy generation rate for the triple-alpha reaction is

$$\epsilon_{3\alpha} = \frac{r_{3\alpha}Q_{3\alpha}}{\rho} = 3.9\times 10^{11}\,\frac{\rho^2 X_\alpha^3}{T_8^3} f \exp\left(-\frac{42.94}{T_8}\right) \text{ erg gm}^{-1}\text{ s}^{-1}, \quad (8.25)$$

yielding for a temperature $T_8 \sim 1$,

$$\epsilon_{3\alpha} \sim \left(584\,\frac{\text{ergs}}{\text{g sec}}\right)\left(\frac{N_\alpha}{1.5\times 10^{28}/\text{cm}^3}\right)^2. \quad (8.26)$$

This agrees reasonably with numerical calculations, yielding an energy production in red giants of about 100 ergs/g sec.

Near a temperature value of T_0, the triple alpha reaction has a very strong temperature dependence. This temperature sensitivity can be obtained by a Taylor series expansion around T_0. One obtains,

$$\epsilon_{3\alpha}(T) \sim T^{-3}e^{-42.9/T} \sim \epsilon_{3\alpha}(T_0)\left(\frac{T}{T_0}\right)^{(42.9-3T_0)/T_0}. \quad (8.27)$$

That is,

$$\epsilon_{3\alpha}(T) \sim \left(\frac{T}{T_0}\right)^{40} N_\alpha^2. \quad (8.28)$$

Hence, a small temperature increase gives rise to greatly accelerated reaction rate and energy production. At a sufficiently high temperature and density, the helium gas is highly explosive. When helium thermonuclear burning is ignited in the stellar core under degenerate conditions, an unstable and sometimes explosive condition develops, known as *helium flash*. At maximum, the luminosity from this fusion is $\gg 10^{11}L_\odot$, i.e., like a supernovae! However, almost none of this energy reaches the surface; it is all absorbed in the expansion of the non-degenerate outer layers. As the flash proceeds, the degeneracy in the core is removed, and the core expands. Models of the helium flash are complicated and uncertain. It is likely that the actual ignition occurs off-center, due to plasma neutrino losses that occur in the densest regions of the star. (This process is similar to pair-production: in dense environments at temperature $T > 10^8$ K, electrons may interact with other electrons to emit neutrino–antineutrino pairs, instead of photons. Since the neutrino energy is lost, this process can cool the regions closest to the center of the star.) The details of the flash (i.e., chemical mixing in the core, and the overall symmetry and duration of the event) depend on the detailed physics.

8.3.4. *Red giants*

By an inspection of the HR diagram, Fig. 1.2, one has figured out that when a star exhausts the supply of hydrogen by nuclear fusion processes in its core, the core contracts and its temperature increases, causing the outer layers of the star to expand and cool. The star's luminosity increases greatly, and it becomes a *red giant*, following a track leading into the upper-right hand corner of the HR diagram. Eventually, once the temperature in the core has reached approximately 3×10^8 K, helium burning begins. The onset of helium burning in the core halts the star's cooling and increases its luminosity. The star moves back towards the left hand side of the HR diagram. After the completion of helium burning in the core, the star again moves to the right and upwards on the diagram. Its path is almost aligned with its previous red giant track, hence the name *asymptotic giant branch* (AGB).

After its formation, ^{12}C is burned into ^{16}O by the α-capture reaction

$$^{12}\mathrm{C} + \alpha \rightarrow {}^{16}\mathrm{O} + \gamma. \tag{8.29}$$

If this reaction proceeds too efficiently, then all the carbon will be burned up to oxygen. Carbon is however the most abundant element in the Universe after hydrogen, helium and oxygen, and the cosmic C/O ratio is about 0.6. In fact, the O and C burning reactions and the conversion of He into C and O take place in similar stellar core temperature and density conditions. Major ashes of He burning in red giant stars are C and O. Red giants are the source of the galactic supply of ^{12}C and ^{16}O.

The $^{12}\mathrm{C}(\alpha,\gamma)^{16}\mathrm{O}$ reaction is very complicated, as displayed in Figure 7.11. It contains non-resonant capture and interfering subthreshold resonances. The S-factor has been estimated to be about 0.3 MeV b [LK83]. Following our usual techniques, the reaction rate for it is

$$\omega_{12C} = \left(\frac{N_\alpha}{7.5\cdot 10^{26}/\mathrm{cm}^3}\right)(2.2\cdot 10^{13}/\mathrm{s})e^{-69.3/T_8^{1/3}}T_8^{-2/3}. \tag{8.30}$$

Note that $N_\alpha = 7.5\cdot 10^{26}/\mathrm{cm}^3$ corresponds to a red giant density of $10^4\,\mathrm{g/cm}^3$ and an α fraction of 0.5. Numerically, one gets

$$\omega_{12C} = \left(\frac{N_\alpha}{7.5\cdot 10^{26}/\mathrm{cm}^3}\right)\begin{bmatrix}1.76\cdot 10^{-17}/\mathrm{s}, & T_8 = 1\\ 1.79\cdot 10^{-11}/\mathrm{s}, & T_8 = 2\end{bmatrix}, \tag{8.31}$$

which yield ^{12}C lifetimes of $1.8\cdot 10^9$ y and $1.8\cdot 10^3$ y, respectively.

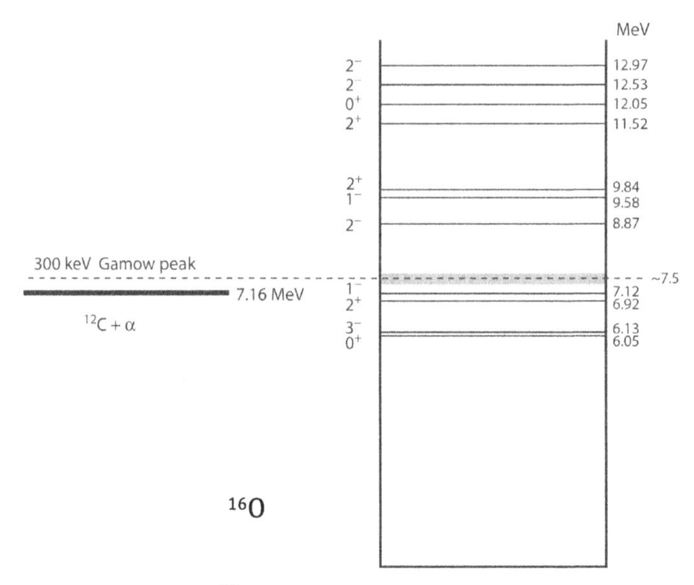

Figure 8.7. Energy levels of ^{16}O nucleus near and above the alpha-particle threshold of capture on ^{12}C. The reaction rate is influenced mainly by the high energy tails of two subthreshold resonances in ^{16}O at $E_R = -45\,\mathrm{keV}$ and $E_R = -245\,\mathrm{keV}$, plus the low energy tail of another high-lying broad resonance at 9580 keV.

In Fig. 8.7 the threshold for the $^{12}\mathrm{C} + {}^4\mathrm{He}$ reaction, with respect to the ground state of the ^{16}O nucleus, is shown on the left of the ^{16}O energy levels. The Gamow energy for temperatures $T_9 = 0.1$ and above indicates that for the expected central temperatures, the effective stellar (center of mass) energy region is near $E_0 = 0.3\,\mathrm{MeV}$. This energy region is reached by the low energy tail of a broad resonance centered at $E_{CM} = 2.42\,\mathrm{MeV}$ above the threshold (the $J^\pi = 1^-$ state at 9.58 MeV above the ground state of ^{16}O) with a (relatively large) resonance width of 400 keV. But there are also two subthreshold resonances in ^{16}O (at $E_X = 7.12\,\mathrm{MeV}$ and $E_X = 6.92\,\mathrm{MeV}$), that is, at $-45\,\mathrm{keV}$ ($J^\pi = 1^-$) and $-245\,\mathrm{keV}$ ($J^\pi = 2^+$) below the α-particle threshold that contribute to reaction rate by their high energy tails. However, electric dipole (E1) γ-decay of the 7.12 MeV state is inhibited by isospin selection rules. The contribution to the S-factors from the resonances and subthreshold states yield the total S-factor of about 0.3 MeV barn. This is low enough that ^{12}C is not burned entirely to ^{16}O, namely, the ratio C/O ~ 0.1 is obtained. The complexity of these interfering levels is difficult to explain microscopically. Theoretically, one often resorts to the $\mathcal{R}$-matrix formalism (Chapter 7) to fit the experimental data and extrapolate the results to the $E_0 = 300\,\text{keV}$ region, as shown in Fig. 7.11.

The ^{16}O formed by α-capture on ^{12}C do not react by further α-capture as in

$$^{16}\text{O} + {}^{4}\text{He} \to {}^{20}\text{Ne} + \gamma. \tag{8.32}$$

This happens because spin–parity selection rules disallow many possible transitions for neighboring resonances and subthreshold states in ^{20}Ne. It seems that only direct capture reactions are effective, which for (α, γ) reactions yields very small cross-sections. Hence, the destruction of ^{16}O via $^{16}\text{O}(\alpha, \gamma)^{20}\text{Ne}$ reaction proceeds at a very slow rate during the stage of helium burning in red giant stars and the major ashes in such stars are carbon and oxygen. In fact, these elements have their galactic origin in the red giants.

The synthesis of two important elements for the evolution of life as we know on the Earth have depended upon fortuitous circumstances of nuclear properties and selection rules for nuclear reactions. These are: (1) the mass of the unstable lowest (ground) state of ^{8}Be being close to the combined mass of two α-particles; (2) there is a resonance in ^{12}C at 7.65 MeV which enhances the alpha addition reaction (the second step); and (3) parity conservation has protected ^{16}O from being destroyed in the $^{16}\text{O}(\alpha, \gamma)^{20}\text{Ne}$ reactions by making the 4.97 MeV excited state in ^{20}Ne of unnatural parity.

8.4. Advanced Burning Stages

As the helium burning progresses, the stellar core is increasingly made up of C and O. At the end of helium burning, all hydrogen and helium is converted into a mixture of C and O. Since H and He are the most abundant elements in the original gas from which the star formed, the amount of C and O are far more than the traces of heavy elements in the gas cloud [Woo86]. The C + O rich core is surrounded by He burning shells and a helium rich layer, and surrounded by a hydrogen burning shell and a non-burning hydrogen-rich envelope. When the helium burning ceases to provide sufficient power, the star begins to contract again under its own gravity. According to the virial theorem (see Section 6.2.3 and Eq. (6.31)) the temperature of the helium–exhausted core rises. The contraction continues until the next nuclear fuel begins to burn at rapid enough rate or until electron degeneracy pressure stops the infall.

8.4.1. *Carbon burning*

In stars more massive than $M \geq 8 - 10\, M_\odot$, the contracting C + O core remains non-degenerate until C starts burning at $T \sim 5 \times 10^8 K$ and $\rho = 3 \times 10^6\, \mathrm{g\, cm^{-3}}$. Enough power is then generated, the contraction stops, and quiescent C-burning proceeds. Two $^{12}\mathrm{C}$ have a combined mass at an excitation energy of 14 MeV in the compound nucleus of $^{24}\mathrm{Mg}$. At this energy there are many compound nuclear states, and the Gamow window at the relevant temperature is about 1 MeV. A large number of resonant states can contribute to the decay of the compound nucleus. Many reaction channels are available, including [Cla84]

$$\begin{aligned} {}^{12}\mathrm{C} + {}^{12}\mathrm{C} &\to {}^{20}\mathrm{Ne} + {}^{4}\mathrm{He}\ (Q = 4.62\,\mathrm{MeV}), \\ &\to {}^{23}\mathrm{Na} + \mathrm{p}\ (Q = 2.24\,\mathrm{MeV}), \\ &\to {}^{23}\mathrm{Mg} + \mathrm{n}\ (Q = -2.62\,\mathrm{MeV}). \end{aligned} \tag{8.33}$$

The rate for this reaction per pair of $^{12}\mathrm{C}$ nuclei is [Ree59]

$$\log(\lambda_{12,12}) = \log(f_{12,12}) + 4.3 - \frac{36.55(1 + 0.1T_9)^{1/3}}{T_9^{1/3}} - \frac{2}{3}\log(T_9), \tag{8.34}$$

where $f_{12,12}$ is a screening factor and T_9 is the temperature in units of a billion degrees K. The produced protons and alpha particles are quickly consumed through the reaction chain

$${}^{12}\mathrm{C}(\mathrm{p},\gamma){}^{13}\mathrm{N}(\mathrm{e}^+\nu_e){}^{13}\mathrm{C}(\alpha,\mathrm{n}){}^{16}\mathrm{O}.$$

Hence, the net effect is that the free proton is converted into a free neutron and the α-particle is consumed with $^{12}\mathrm{C}$ into $^{16}\mathrm{O}$. The α-particles are also captured by other alpha-particles.

At the end of carbon burning, nuclei like $^{16}\mathrm{O}$, $^{20}\mathrm{Ne}$, $^{24}\mathrm{Mg}$ and $^{28}\mathrm{Si}$ are formed. These reactions increase the energy output by the initial carbon burning and each pair of $^{12}\mathrm{C}$ release about 13 MeV [Ree59] of energy. At the end of carbon burning phase other reactions such as $^{12}\mathrm{C} + {}^{16}\mathrm{O}$ and $^{12}\mathrm{C} + {}^{20}\mathrm{Ne}$ will also take place. But their reaction rates are smaller than for $^{12}\mathrm{C} + {}^{12}\mathrm{C}$ due to their increased Coulomb barriers.

During the carbon-burning and subsequent stages, the dominant energy loss from the star is due to the released neutrinos. The neutrino luminosity is a sensitive function of core temperature and quickly becomes larger than the surface photon luminosity of the star. The energy production timescale of the star, due to the neutrino emission, becomes very short and the core evolves very rapidly compared to the gravitational cooling timescale $t_{th} = GM_s^2/RL_s$ (see Section 6.2.4 and Eq. (6.38)). There is not enough time to pass this information to the surface, since this happens by slow photon diffusion. Thus, the surface temperature does not change appreciably as the core evolves beyond the carbon burning stage. Hence, it may not be possible only by means of astronomical observations to know whether the core is close to a supernova stage or has many thousands of years of hydrostatic thermonuclear burning still left.

8.4.2. *Neon burning*

Carbon burning produces mainly neon, sodium and magnesium. Aluminum and silicon are also produced in small quantities by the capture of α, p and n released during carbon burning. When carbon fuel is exhausted, the core contracts and its temperature T_c increases. At $T_9 \sim 1$, energetic photons from the high energy tail of the Planck distribution start disintegrating ^{20}Ne by means of the reaction $^{20}\mathrm{Ne} + \gamma \rightarrow {}^{16}\mathrm{O} + {}^{4}\mathrm{He}$. Alpha particles are released from a nucleus at approximately the same energy as a nucleon. For example, the alpha separation energy in ^{20}Ne is 4.73 MeV, compared to the neutron and proton separation energies of 8 MeV. Thus, the major photonuclear reactions are $(\gamma, \mathrm{n}), (\gamma, \mathrm{p})$ and (γ, α).

The photodisintegration decay rate proceeding through an excited state E_x is given by

$$\lambda(\gamma, \alpha) = \left[\exp\left(-\frac{E_x}{kT}\right)\frac{2J_R + 1}{2J_0 + 1}\frac{\Gamma_\gamma}{\Gamma}\right]\frac{\Gamma_\alpha}{\hbar}, \tag{8.35}$$

where the first factor in square brackets on the right hand side is the probability of finding the nucleus in the excited state E_x and spin J_R (with J_0 being the ground state spin), while the second factor $\Gamma_\alpha/\hbar$ is the decay rate of the excited state with an alpha particle emission. Since $E_x = E_R + Q$, we have

$$\lambda(\gamma, \alpha) = \frac{\exp(-Q/kT)}{\hbar(2J_0 + 1)}(2J_R + 1)\frac{\Gamma_\alpha \Gamma_\gamma}{\Gamma}\exp(-E_R/kT). \tag{8.36}$$

At $T_9 \geq 1$, the photodisintegration is dominated by the 5.63 MeV level in ^{20}Ne. At approximately $T_9 \sim 1.5$, the photodissociation rate becomes greater than the rate for alpha capture on ^{16}O to produce ^{20}Ne (that is, the reverse reaction). Therefore, a net dissociation of ^{20}Ne prevails over its production. The released ^{4}He further reacts with ^{20}Ne and leads to

$^4\text{He} + {}^{20}\text{Ne} \rightarrow {}^{24}\text{Mg} + \gamma$. The net result of the photodissociation of two ^{20}Ne nuclei is $2 \times^{20} \text{Ne} \rightarrow {}^{16}\text{O} + {}^{24}\text{Mg}$ with a Q-value of 4.58 MeV. The brief neon burning phase concludes at T_9 close to $\sim$1.

8.4.3. *Oxygen burning*

At the end of the neon burning the core is left with a mixture of alpha particle nuclei ^{16}O and ^{24}Mg. After this, the core contracts and temperature increases up to $T_9 \sim 2$, at which point ^{16}O begins to react another ^{16}O by means of

$$^{16}\text{O} + {}^{16}\text{O} \rightarrow {}^{28}\text{Si} + {}^{4}\text{He} \tag{8.37}$$

$$\rightarrow {}^{32}\text{S} + \gamma. \tag{8.38}$$

The first reaction takes place approximately 50% of the time with a Q-value of 9.59 MeV. In addition to Si and S, the oxygen burning phase also produces Ar, Ca and trace amounts of Cl, K, ..., up to Sc. At $T_9 \sim 3$, the produced ^{28}Si begins to burn in what is known as the Si burning phase.

8.4.4. *Silicon burning*

Most stages of stellar burning involve thermonuclear fusion of nuclei to produce higher Z and A nuclei. But, as discussed above, neon burning is an exception because photons are sufficiently energetic to photodissociate neon, before the temperature rises sufficiently to ignite reactions among oxygen nuclei. Neon burning takes place with the addition of helium nuclei to the undissociated neon rather than overcoming the Coulomb barrier of two neon nuclei. This trend continues in the silicon burning phase. In general, a photodisintegration channel becomes important when the temperature rises to the point that the Q-value, Eq. (7.16), is smaller than approximately $30\,kT$.

Typical Q-values for reactions among stable nuclei above silicon are 8–12 MeV. Hence, photodisintegration of the nuclear products of neon and oxygen burning are important when the temperature exceeds $T_9 \geq 3$. Nuclei with smaller binding energies are destroyed by photodissociation and many nuclear reactions involving α-particles, protons and neutrons interacting with nuclei with mass $A = 28 - 65$ take place. These nuclear reactions are primarily of a rearrangement type, in which a particle is photo-ejected from one nucleus and captured by another. A large number of reaction chains and cycles occur and more and more stable nuclei are formed as the rearrangement proceeds. For example, due to the large number of free neutrons, many (n,γ)-reactions (*radiative neutron capture*) occur. The reactions lead to a large abundance of elements in the iron mass region, which have the largest binding energy per nucleon. Since ^{56}Fe has the maximum binding energy per nucleon, as shown in Fig. 3.2, the reactions stop around the iron-group nuclei.

In the mass range $A = 28 - 65$, the levels in the compound nuclei that form in the reactions during silicon burning are so dense that they overlap. Moreover, at the high temperatures that are involved ($T_9 = 3-5$), a quasi-equilibrium of backward and forward reactions exist between groups of nuclei, which are connected by a few, slow rate reactions known as *bottlenecks*. However, as more reactions consume the nuclear fuel and thermal energy decreases due to escaping neutrinos, various nuclear reactions may no longer occur leading to a *freeze-out* of the thermonuclear process.

Weak interaction processes such as electron capture and beta decay of nuclei are important, by influencing the Y_e and thereby the reaction flow. These ultimately affect both the stellar core density and entropy structures, and it is important to track and include the changing Y_e (the number of electrons per nucleon) of the core material not only in the silicon burning phase, but also from earlier oxygen burning phases.

For temperatures above 3×10^9 K more photo-nuclear processes appear. These yield more nuclei to be burned and heavier nuclei are produced

$$\gamma + {}^{28}_{14}\mathrm{Si} \rightarrow {}^{24}_{12}\mathrm{Mg} + {}^{4}_{2}\mathrm{He}, \quad {}^{4}_{2}\mathrm{He} + {}^{28}_{14}\mathrm{O} \rightarrow {}^{32}_{16}\mathrm{S} + \gamma, \text{ etc.} \tag{8.39}$$

For $A > 100$ the distribution of nuclei cannot be explained in terms of fusion reactions with charged particles. They are formed by the successive capture of slow neutrons and of β^--decay. The maxima of the element distribution of $N = 50$, 82, 126 are due to the small capture cross-sections corresponding to the magic numbers. This yields an ash of isotopes at the observed element distribution.

8.5. Synthesis of Heavy Elements

So far we have been dealing primarily with charged particle reactions and photodisintegration which lead to the production of lighter elements ($1 \leq A \leq 40$) and the recombination reactions for the production of elements $40 \leq A \leq 65$. However, the heavier elements ($A \geq 65$), because of their high charge and relatively weak stability, cannot be produced by these two processes. It was therefore natural to investigate the hypothesis of neutron induced reactions on the elements that are formed already in the various thermonuclear burning stages, and in particular on the iron-group elements.

The study of the nuclear reaction chains in stellar evolution shows that during certain phases large neutron fluxes are released in the core of a star. On the other hand, the analysis of the relative abundance of elements shows certain patterns which can be explained in terms of the neutron absorption cross-sections of these elements. If the heavier elements above the iron peak were to be synthesized during for example charged particle thermonuclear reactions during silicon burning, their abundance would drop very much more steeply with increasing mass (larger and larger Coulomb barriers) than the observed behavior of abundance curves, which shows a much smaller than expected decrease [Sue56]. It is thus natural to believe that heavy elements are made instead by thermal neutron capture [Bur57].

Two distinct neutron processes are required to make the heavier elements. The *slow neutron capture* process (s-process) has a lifetime for β-decay τ_β shorter than the competing neutron capture time τ_n (i.e. $\tau_\beta \leq \tau_n$). This makes the s-process nucleosynthesis run through the valley of β-stability. The *rapid neutron capture* r-process on the other hand requires $\tau_n \ll \tau_\beta$. This process takes place in extremely neutron-rich environments, because the neutron capture timescale is inversely proportional to the ambient neutron density. The r-process, in contrast to the s-process, goes through very neutron rich and unstable nuclei that are far off the valley of stability. The relevant properties of such nuclei are most often not known experimentally, and are usually estimated theoretically. Some of the key parameters are the half-lives of the β-unstable nuclei along the s-process path. But the nuclear half-life in stellar environments can change due to transitions from not just the ground state of the parent nucleus, but also because its excited states are thermally populated.

8.5.1. *The s-process*

In the s-process, when a neutron is captured, the resulting nucleus has time to β decay to a more stable nucleus. Hence, the reaction is proportional to the neutron capture rate, which controls the path to heavier nuclei along the *valley of stability*. Along this path, a number of stable nuclei are avoided because nuclei (N, Z) and $(N+2, Z-2)$, when N and Z are even, are stable to β decay, while odd–odd nuclei $(N+1, Z-1)$ are unstable (see Fig. 3.5). The odd–odd nucleus has an unpaired proton and an unpaired neutron, accounting for its unfavorable ground state energy. The ($N + 2$, $Z - 2$) nucleus can be "shielded" from production in the s-process. Thus the existence of such isotopes with significant abundances indicates that a second process, other than the s-process, must also occur.

The general equation describing the path along the s-process is

$$\frac{dN_A(t)}{dt} = N_n(t)N_{A-1}(t)\langle\sigma\nu\rangle_{A-1} - N_n(t)N_A(t)\langle\sigma\nu\rangle_A - \lambda_\beta(t)N_A(t), \tag{8.40}$$

where $N_n(t)$ is the neutron density at time t and λ_β is the β decay rate. This is a set of complicated equations to solve because the initial conditions would have to be fully specified, and the time evolution of N_n given. The equation allows for destruction of mass number A by either neutron capture or β decay. If we define an average cross-section $\langle\sigma\nu\rangle = \sigma_A\langle\nu\rangle$, and assume that neutron capture is faster than β-decay, then

$$\frac{dN_A(t)}{dt} = \langle\nu\rangle N_n(t)(\sigma_{A-1}N_{A-1} - \sigma_A N_A). \tag{8.41}$$

If equilibrium is reached, $dN/dt = 0$, and

$$\frac{N_A}{N_{A-1}} = \frac{\sigma_{A-1}}{\sigma_A}. \tag{8.42}$$

That is, the abundance achieved is inversely proportional to the neutron cross-section: if the capture rate is slow, then mass piles up at that target number. The same argument applies if β decay is faster than neutron capture. The low neutron capture cross-sections at the closed shells should result in mass peaks, as observation shows. It also follows that equilibrium will set in most quickly in the broad plateaus between the mass peaks: mass must pile up at the closed shells before the closed shell is surpassed.

Several sites have been suggested for the s-process, but one well accepted site is in the helium-burning shell of a red giant, where temperatures are sufficiently high to liberate neutrons by the reaction $^{22}\text{Ne}(\alpha, n)^{25}\text{Mg}$. The ^{22}Ne is produced from helium burning on the elements the CNO cycle left after hydrogen burning. The s-process cannot proceed beyond ^{209}Bi because neutron capture on this isotope leads to a decay chain that ends with α emission. This is a gap the s-process cannot cross. We conclude that the *transuranic elements* must have some other origin.

8.5.2. *The r-process*

In the r-process, the β-decay properties of the nuclei regulate the reaction flow. The r-process lasts for a few seconds, in a dense neutron environment, $n_n \sim 10^{20} - 10^{25}$ cm^{-3}. The neutron densities in the s-process are much smaller, $n_n \sim 10^8\,\text{cm}^{-3}$. Nuclei above the iron group up to about $A = 90$ are produced in massive stars mainly by the s-process. Above $A = 100$ the s-process does very little in massive stars and most of its contribution in this mass range is believed to occur in *Asymptotic Giant Branch* (AGB) stars.

In the r-process the neutron capture is fast compared to β decay rates and thus the equilibrium maintained in $(n, \gamma) \leftrightarrow (\gamma, n)$. Neutron capture fills up the available bound levels in the nucleus until this equilibrium sets in: the new Fermi level depends on the temperature and the relative n/γ abundance. Each β^- capture converting n $\rightarrow$ p opens up a hole in the neutron Fermi sea allowing another neutron to be captured. The nucleosynthesis path is along highly unstable, exotic, and neutron-rich nuclei (see Fig. 8.8).

The reaction

$$A + n \rightarrow (A + 1) + \gamma$$

is *exothermic*. Assuming a resonant reaction (the level density is high in heavy nuclei), and neglecting spins, Eq. (7.56) yields

$$\langle \sigma \nu \rangle_{(n,\gamma)} = \left(\frac{2\pi}{\mu k T} \right)^{3/2} \frac{\Gamma_n \Gamma_\gamma}{\Gamma} e^{-E_R/KT}, \tag{8.43}$$

where $E_R \sim 0$ is the resonance energy. Thus the rate is

$$r_{(n,\gamma)} \sim N_n N_A \left(\frac{2\pi}{\mu k T} \right)^{3/2} \frac{\Gamma_n \Gamma_\gamma}{\Gamma}. \tag{8.44}$$

The inverse, (γ, n), reaction requires knowledge of the photon number density given by the Bose–Einstein distribution of photons, Eq. (4.67). The high-energy tail of the normalized distribution can thus be written $N(E_\gamma) \sim (E_\gamma^2/N_\gamma \pi^2)e^{-E_\gamma/kT}$, where the approximate expression for the total photon density (by integrating the distribution over all energies) is $N_\gamma \sim (\pi/13)(kT)^3$, and where we used $\hbar = c = 1$.

The resonant cross-section for photons is obtained from the Breit–Wigner formula, Eq. (5.84). Inserting the asymptotic form of the Maxwell–Boltzmann distribution in the reaction rate integral and pulling out the

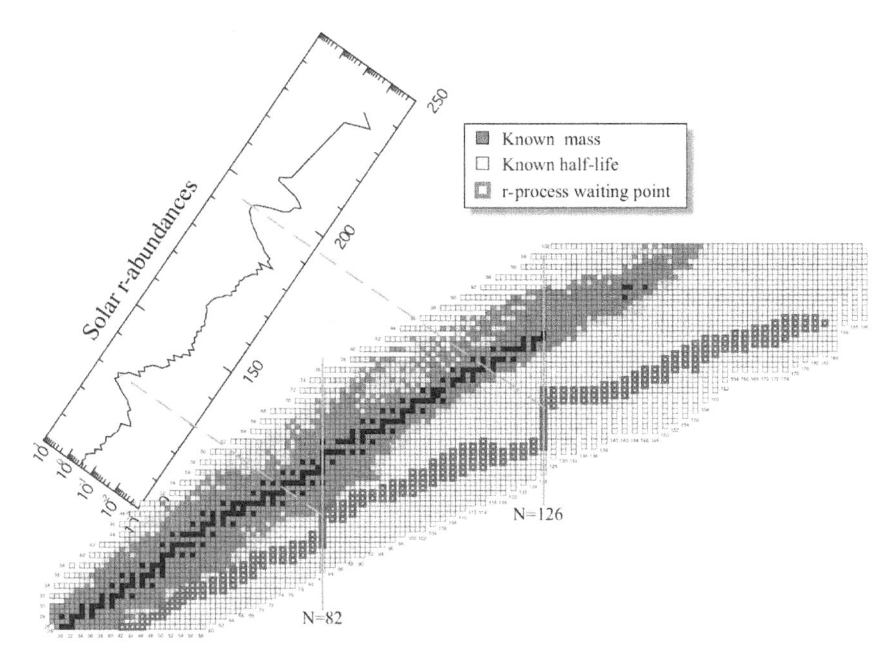

Figure 8.8. The r-process path defined by their waiting point nuclei. After decay to stability the abundances of the r-process progenitors produce the observed solar r-process abundance distribution. The r-process paths run generally through neutron-rich nuclei with experimentally unknown masses and half-lives. (Image courtesy of Hendrik Schatz).

slowly varying parts of the integrand (as we did in Section 7.5.5), the rate becomes

$$r_{(\gamma,n)} \sim 2N_{A+1}\frac{\Gamma_\gamma\Gamma_n}{\Gamma}e^{-E_R/kT}. \tag{8.45}$$

Equating the (n,γ) and (γ,n) rates and taking $N_A \sim N_{A-1}$ yields

$$N_n \sim \frac{2}{(\hbar c)^3}\left(\frac{\mu c^2 kT}{2\pi}\right)^{3/2} e^{-E_R/kT},$$

where the $\hbar$s and cs have been properly inserted to give the right dimensions. Now E_R is essentially the binding energy. An estimate of it can be made by assuming that $N_n \sim 3\times 10^{23}/\text{cm}^3$ and $T_9 \sim 1$. We obtain that the binding energy is ~2.4 MeV. Hence, under such conditions, the neutrons are bound by about 30 times kT, a value that is small compared to a typical binding of 8 MeV for a normal nucleus.

In Fig. 8.9 we show the relative distribution of elements in our galaxy. It has two distinct regions: in the region $A < 100$ it decreases

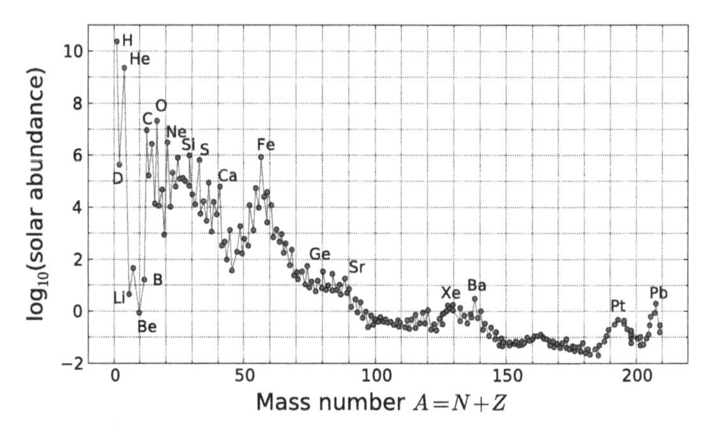

Figure 8.9. The abundances of isotopes in the solar system as a function of atomic mass. The abundances are normalized so that the total abundance of silicon is 10^6.

with A approximately like an exponential, whereas for $A > 100$ it is approximately constant, except for the peaks in the region of the magic numbers $Z = 50$ and $N = 50, 82, 126$.

When a gap at the shell closures $N = 82{,}126$ is reached in the r-process, the neutron number of the nucleus remains fixed until the nucleus can change sufficiently to overcome the gap: N remains fixed while successive β decays occur. The (N, Z) path is along increasing Z with fixed N: every β decay is followed by a (n, γ) reaction to fill the open neutron hole, but no further neutrons can be captured until the gap is overcome. The closed neutron shells are called the *waiting points*, because it takes a long time for the successive β decays to occur to allow progression through higher N nuclei. The β decays are slow at the shell closures. Just as in the s-process, the abundance of a given isotope is inversely proportional to the β decay lifetime. Thus mass builds up at the waiting points, forming the large abundance peaks seen in Fig. 8.9.

After completion of the r-process, the nuclei decay back to the valley of stability by β decay. Neutron spallation (fragmenting the nuclei) might occur, shifting the mass number A to a lower value. The conversion of neutrons into protons during beta-decay also shifts the r-process peaks at $N \sim 82, 126$ to a lower N. Such effects are clearly seen in the abundance distribution: the r-process peaks are shifted to lower N relative to the s-process peaks. It might also be that the r-process can proceed to very heavy nuclei ($A \sim 270$) where it is finally ended by β-delayed and neutron-induced fission, which feeds matter back into the process at an $A \sim A_{max}/2$.

The conditions during the r-process, $\rho(n) \sim 10^{20}$ cm^{-3}, $T \sim 10^9$K, and $t \sim 1$ s, are explosive conditions. One believes that possible sites for the r-process are (a) neutronized atmosphere above proto-neutron star in a Type II supernova, (b) neutron-rich jets from supernovae or neutron star mergers, (c) inhomogeneous Big Bang, (d) He/C zones in Type II supernovae, (e) red giant He flash, and (f) ν spallation neutrons in He zone. In the sites (d–f), the neutron density ρ(n) are expected to be lower than those in (a–c).

8.5.3. *Pycnonuclear reactions*

Inside compact astrophysical objects, fusion reactions can be divided into thermonuclear and *pycnonuclear* processes (see Section 7.4.1). Thermonuclear fusion takes place in hot and dilute plasmas inside stars with only the high-energetic components of the velocity distribution being important, whereas pycnonuclear fusion happens at rather high densities where mostly low-energy nuclei contribute to the fusion process. In terms of the *plasma parameter* $\Gamma = Ze^2/akT$, where a is the inter-particle distance, thermonuclear reactions occur when

$$\Gamma \ll 1 \ll E_0/kT, \tag{8.46}$$

where E_0 is the Gamow peak energy, Eq. (7.33). As Γ becomes comparable to unity (high densities), charge screening by the electrons and ion correlation become important (see Section 7.4.1).

For $\Gamma \gtrsim 1$, the screened field observed by a nucleus is no longer approximated by Eq. (7.70); in addition, ion correlation becomes important. The accurate treatment requires advanced statistical plasma theory [IOT94]. Phenomenologically, one finds that we can approximate electron screening by Eq. (7.70) as long as one replaces the Debye length with the interparticle distance a. For $E_0 > kT$ this yields a correction to the reaction rate $\langle\sigma\nu\rangle$ due to electron screening, of the order of $\exp(\Gamma_e)$, where $\Gamma_e = V_C(a_e)/kT$ is the plasma parameter at the average interelectron distance a_e (V_C is the Coulomb potential). Analogously, a correction due to ion screening, with ion correlations, results in increased reaction rates by a factor $\exp(\Gamma_i)$, where Γ_i is the plasma parameter at the average inter-ion distance. The mentioned corrections are large for plasmas at low temperature, since T (keV) $= 0.02 Z_1 Z_2 [\rho(\mathrm{g/cm^3})]^{1/3}/\Gamma_i$. Therefore they only affect reaction rates, which are too small to be of any practical interest to astrophysics.

Pycnonuclear reactions in neutron stars are considered as a possible energy source for γ-ray bursts. Neutron-rich nuclei are arranged on a lattice, the spacing of which is shrinking under the gravitational pressure of accreting material. At a critical distance nuclear fusion of lattice nuclei sets in. Detailed knowledge of the nuclear matter distribution is crucial for the determination of the pyconuclear fusion rates and their dependence on the densities inside the neutron star's crust.

Pycnonuclear reactions practically do not depend on the temperature and occur even at $T = 0$ [Ga38,Wil40,SvH69]. When $\Gamma_i > 170$, each ion is frozen in a crystalline lattice and oscillates with frequency ω around its equilibrium position. The ground state of this set of quantum-mechanical oscillators is given by $E_B = (4\pi n_i i Z^2 e^2/3m_i)^{1/2}$, where Z, m_i, and n_i are the ion charge, mass, and number density, respectively. When E_B is much larger than the ion thermal energy (that is, when $E_B/kT \gtrsim 10$), each nucleus performs small amplitude oscillations proportional to the interparticle distance $a \sim \rho^{-1/3}$, and can only interact with the nuclei in the neighboring lattice position. According to Eq. (7.30), the dominating exponential factor in the penetration factor scales as $R_C^{1/2} \sim a^{1/2} \sim \rho^{-1/6}$. The penetration factor is then proportional to $\exp(-\rho^{1/6})$, without any dependence on temperature. Accurate computations of the carbon–carbon reaction rate at $\Gamma_i > 170$ and $E_B/kT > 20$ confirm this behavior, giving the result $\langle \sigma v \rangle = 10^7 \rho_8^{-0.6} \exp(-258\rho_8^{-1/6})$ cm^3/s, where ρ_8 is the density in units of 10^{-8} g/cm^3 [SvH69,IOT94]. This pycnonuclear regime is thought to be responsible for sudden power release by very dense and relatively cold white dwarfs with a carbon core. When this core is compressed to density about 10^9 g/cm^3 the reaction rate increases rapidly approaching the value it would take at very high temperature and low density, thus igniting carbon combustion.

In accreting neutron stars (NS), the heat released due to the infall of matter and thermonuclear reactions in the surface layers is radiated away by photons from the surface and cannot warm up the NS interiors. The accreted matter sinks into the NS crust under the weight of newly accreted material. The density gradually increases in a matter element, and the nuclei undergo transformations — beta captures, absorption and emission of neutrons, and pycnonuclear reactions. The nuclei evolve then into highly exotic atomic nuclei which are unstable in laboratory conditions but stable in dense matter ($\rho \gtrsim 10^9$ g cm^{-3}). The main energy release occurs at higher densities, $\rho \sim 10^{12}$–10^{13} g cm^{-3}, several hundred meters under the NS surface, in pycnonuclear reactions. For iron matter at $\rho \sim 10^9$ g cm^{-3}, reactions such as ^{34}Ne + ^{34}Ne → ^{68}Ca, ^{36}Ne + ^{36}Ne → ^{72}Ca, and ^{48}Mg + ^{48}Mg → ^{96}Cr can occur. The total energy release is about 1.45 MeV per accreted nucleon. The total NS heating power is determined by the *mass accretion rate*, dM/dt,

$$L = 1.45\,\text{MeV}\left(\frac{dM/dt}{m_N}\right) \approx 8.74 \times 10^{33}\left(\frac{dM/dt}{10^{-10}M_\odot\ \text{y}^{-1}}\right)\ \text{erg s}^{-1}, \tag{8.47}$$

where m_N is the nucleon mass. It produces deep crustal heating of the star, spreading by thermal conductivity over the entire NS and warming it up.

8.6. Cataclysmic Events

8.6.1. *Novae*

Occasionally, a new star appears in the sky, increases its brightness to a maximum value, and decays afterwards until its visual disappearance. Such stars are known as *novae*. A nova is a binary star system which undergoes a sudden, spectacular brightening, by a factor of up to a million or so, before dimming to its pre-nova state. Observation suggests that more the half the stars are part of binary systems. Novae consist of a white dwarf primary in close orbit around an orange/red dwarf or (in some cases) giant secondary, the fuel for the outbursts being gas plundered from the larger star by the white dwarf. In contrast with supernovae, which are one-time events accompanied by the total destruction of a star, novae leave the host stars essentially intact and capable of repeating the event.

The Milky Way experiences a few dozen novae per year. Spectroscopic observation of nova ejecta nebulae has shown that they are enriched in elements such as He, C, N, O, Ne, and Mg. The contribution of novae to the interstellar medium is not great; novae supply only 1/50 as much material to the galaxy as supernovae, and only 1/200 as much as red giant and supergiant stars.

If the distance between two stars which form a binary system is close enough and the radius of one of the stars becomes significantly large during its evolution, mass may flow from the evolving star to another. The *Roche lobe* is the region of space around one of the binary stars within which orbiting material is gravitationally bound to that star. If the material of the star extends beyond its Roche lobe, it can escape the gravitational pull of the star, and it will fall in through the inner *Lagrangian point* towards

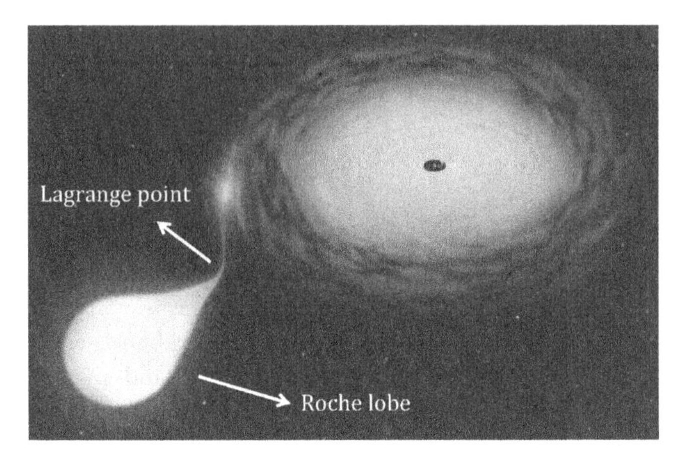

Figure 8.10. Close to each star, the potential is dominated by the gravity of that star, and the equipotential surface is a sphere around the center of that star. Further out, the equipotential surfaces are deformed. They assume a pear-like shape. For a critical value of the potential the equipotential surfaces of the two stars touch, at the inner Lagrangian point.

the other star (see Fig. 8.10). The Roche lobe is a tear-drop shaped region bounded by a gravitational equipotential, with the apex of the tear-drop pointing towards the other star. It is different from the Roche limit which is the distance at which an object held together only by gravity begins to break up due to tidal forces of another body. It is useful to approximate the Roche lobe as a sphere of the same volume. Then, an approximate formula for the radius of this sphere can be obtained as [Pac71]

$$\frac{r_1}{A} = 0.38 + 0.2 \log\left(\frac{M_1}{M_2}\right), \quad \text{for } 0.3 < \frac{M_1}{M_2} < 20, \tag{8.48}$$

and

$$\frac{r_1}{A} = 0.46224 \left(\frac{M_1}{M_1 + M_2}\right)^{1/3}, \quad \text{for } \frac{M_1}{M_2} < 0.3, \tag{8.49}$$

where A is the semimajor axis of the system and r_1 is the radius of the Roche lobe around mass M_1.

The transfer of matter from the companion star to the white dwarf is a continuous process. Fresh matter arrives at the outer edge of the accretion disk from the companion star, spirals through the disk, and accretes onto the white dwarf. The matter accreting from the binary companion builds up on the surface of the white dwarf, creating a layer of hydrogen.

This layer sits on top of the rest of the white dwarf, which is composed mainly of carbon and oxygen. As the hydrogen builds up, the pressure and temperature at the white dwarf's surface grows and grows. Eventually, it may become hot enough to fuse the hydrogen into helium. When about 1/100,000 of a solar mass of hydrogen-rich matter has been accreted, the temperature and density at the base of the accreted matter become so large that a nuclear explosion is triggered and the white dwarf's surface layer is ejected at speeds of about 700 km/second or greater (3 million kilometers per hour or greater).

The explosion and ejection are accompanied by an intense brightening, hence the name nova, meaning "new" (i.e., the star becomes visible across interstellar distances). Maximum brightness lasts only a few days. The brightness then diminishes and, in the course of several months, returns to the pre-outburst level. Only the built up surface gets blown away. The white dwarf is left intact underneath. As the blown off material floats away into space to become part of the interstellar medium, the accreting white dwarf is left behind to start the cycle all over again. Mass transfer and accretion then resume until another nova outburst occurs. Typical time intervals between outbursts are several thousand years. The white dwarf will experience outbursts as long as the companion star is able to furnish fresh hydrogen-rich matter.

8.6.2. *X-ray bursts*

Accretion in binary systems leads to different astrophysical phenomena such as novae, type Ia supernovae, and *X-ray bursts*. In X-ray binary stars, the compact object can be either a neutron star or a black hole. When a star in a binary system fills its Roche lobe, it begins to lose matter, which streams towards the neutron star. The partner star may undergo mass loss by exceeding its *Eddington luminosity*, Eq. (6.145). Some of this material may become gravitationally attracted to the neutron star. In the circumstance of a short orbital period and a massive partner star, both of these processes may contribute to the transfer of material from the companion to the neutron star.

The accreted material originates from the surface layers of the partner star and it is rich in hydrogen and helium (Fig. 8.11). Because compact stars have high gravitational fields, the material falls with a high velocity towards them, usually colliding with other accreted material, forming an accretion disk. Because an enormous amount of energy is released in a short

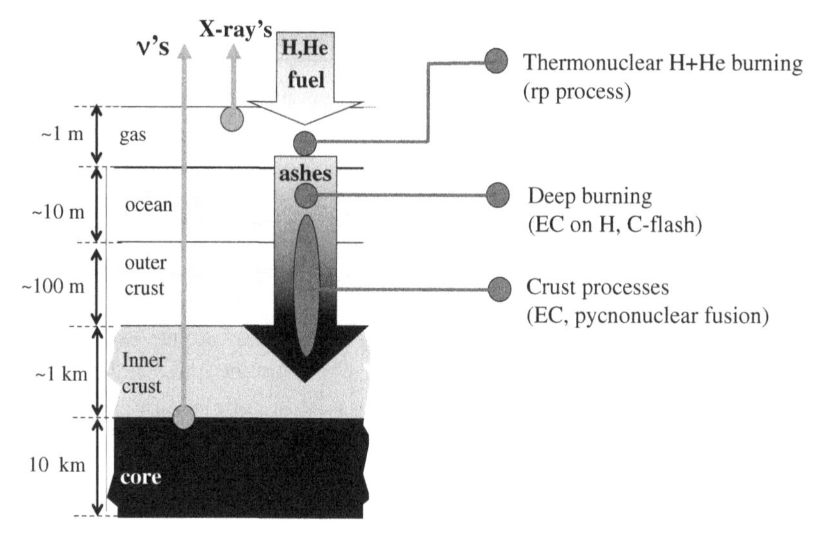

Figure 8.11. The accreted material on a neutron star gives rise to X-ray bursts due to reactions involving capture of H, He, C, electron capture (EC) and pycnonuclear reactions in the deep crust of the star. (Image courtesy of Hendrik Schatz.)

period of time, much of the energy is released as high energy photons in accordance with the theory of black body radiation, in this case X-rays. In an X-ray burst, this material accretes onto the surface of the compact star as a dense layer of electron degenerate gas, another result of the extremely high gravitational field. Degenerate matter does not follow the ideal gas law, and so changes in temperature do not lead to notable changes in pressure. After enough of this material accumulates on the surface of the compact star, thermal instabilities set off exothermic nuclear fusion reactions, which causes an increase in temperature (greater than 10^9 K), eventually giving rise to a runaway thermonuclear explosion. This explosive stellar nucleosynthesis begins with the hot CNO cycle which quickly yields to the rp-process.

X-ray bursts cannot be observed on Earth's surface because our atmosphere is opaque to X-rays. Most X-ray bursting stars exhibit recurrent bursts because the bursts are not powerful enough to disrupt the stability or orbit of either star, and the whole process may begin again. Most X-ray bursters have irregular periods, which can be on the order of a few hours to many months, depending on factors such as the masses of the stars, the distance between the two stars, the rate of accretion, and the exact composition of the accreted material.

8.6.3. *Supernovae*

Classification

Among novae, some stars present an exceptional variation in their brightness and are called *supernovae*. *Type I Supernovae* (SNI) are thermonuclear explosions of massive carbon–oxygen white dwarfs in binary stellar systems. They are important tools ("*standard candles*") of modern physics and cosmology.

Explosion of a SNI involves hundreds of nuclear species and thousands of nuclear reactions occurring at densities up to $\rho \sim 10^{10}\,\mathrm{g/cm^3}$ and temperatures up to $T \sim 10^{10}\,\mathrm{K}$. The released nuclear energy causes complex hydrodynamic phenomena such as turbulent deflagration, acceleration of burned matter by pressure forces, and an overall expansion of a SNI. Hydrodynamics and nucleosynthesis in SNI are intimately related: nucleosynthesis in SNI cannot be predicted correctly without detailed 3D hydrodynamic modeling, and the modeling cannot be correct without correct nuclear physics input. Nuclear physics and SNI simulations are intimately connected.

There are two types of supernovae:

1. *Type II supernovae* (SNII) show hydrogen in their spectra. Their light curves are rather diverse, and their peak luminosities are around $10^{42}\,\mathrm{erg\,s^{-1}}$.
2. Type I supernovae do not show hydrogen in their spectra. They are subclassified based on other considerations. If they have certain silicon lines, they are called Type Ia. Otherwise, if they have helium they are Type Ib and if they do not they are Type Ic. The Type Ia supernovae reach peak luminosities of about $2 \times 10^{43}\,\mathrm{erg\,s^{-1}}$.

Type II supernova

Type-II supernovae are defined as those showing H-lines in their spectra. It is likely that most, if not all, of the exploding massive stars still have some H-envelope left, and thus exhibit such a feature. In contrast, Type-I supernovae lack H in their ejecta.

Type II supernovae begin with a massive star, in excess of 10 solar masses, burning the hydrogen in its core under the conditions of hydrostatic equilibrium. When the hydrogen is exhausted, the core contracts until the density and temperature are reached where $3\alpha \rightarrow {}^{12}\mathrm{C}$ can take place. The He is then burned to exhaustion. This pattern — fuel exhaustion,

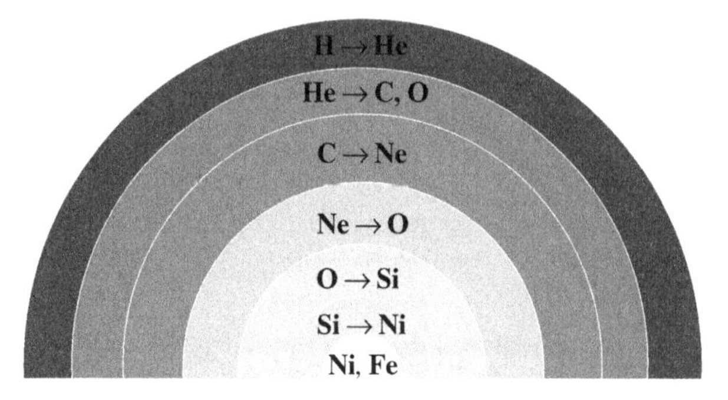

Figure 8.12. Schematic view of the "onion" structure of a $20M_\odot$ star just before a supernova explosion.

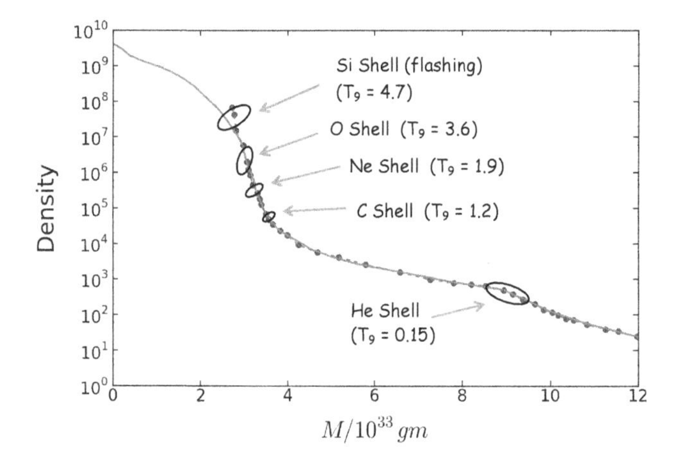

Figure 8.13. Density versus mass relation of a supernova star.

contraction, and ignition of the ashes of the previous burning cycle — repeats several times, leading finally to the explosive burning of ^{28}Si to Fe. For a heavy star, the evolution is fast: the star has to work harder to maintain itself against its own gravity, and therefore consumes its fuel faster. A 25 solar mass star would go through all of these cycles in about 7 My, with the final explosion Si burning stages taking a few days. The resulting "onion skin" structure of the precollapse star is shown in Figs. 8.12 and 8.13. One can read off the nuclear history of the star by looking from the surface inward. Starting from the center of the star, we first find a core of iron, the remnant of silicon burning. After that we pass by successive

Table 8.2. Nuclear binding energy in a scale where the ^{12}C mass is taken as zero, i.e., Δ is the nuclear binding energy relative to C.

Nucleus	Δ/nucleon (MeV)
^{12}C	0.000
^{16}O	−0.296
^{28}Si	−0.768
^{40}Ca	−0.871
^{56}Fe	−1.082
^{72}Ge	−1.008
^{98}Mo	−0.899

regions where ^{28}Si, ^{16}O, ^{12}C, ^{4}He, and ^{1}H form the dominant fraction. In the interfaces, the nuclear burning continues to happen.

The silicon burning exhausts the nuclear fuel. As we mentioned previously, the gravitational collapse of the iron core cannot be held by means of pressure heat from nuclear reactions. The collapse occurs inevitably for a massive star, since the silicon burning adds more and more material to the stellar core.

The source of energy for the supernova evolution is the nuclear binding energy. A look at the nuclear binding energy as a function of nuclear mass shows that the minimum is achieved at Fe. In a scale where the ^{12}C mass is taken as zero, the nuclear binding is shown in Table 8.2, where Δ is the nuclear binding energy relative to C. Hence, after Si burns to produce Fe, there is no further source of nuclear energy adequate to support the star. The core becomes largely supported by degeneracy pressure, with the energy generation rate being less than the stellar luminosity. The core density is about $2 \times 10^9\,\mathrm{g/cm^{-3}}$ and the temperature is $kT \sim 0.5\,\mathrm{MeV}$, or $T \sim 10^{10}\,\mathrm{K}$. The core is made of ^{56}Fe and of electrons. There are two possibilities, both accelerating the collapse:

1. At conditions present in the collapse the strong reactions and the electromagnetic reactions between the nuclei are in inverse equilibrium, that is,

$$\gamma + {}^{56}_{26}\mathrm{Fe} \leftrightarrow 13(^4\mathrm{He}) + 4\mathrm{n} - 124\,\mathrm{MeV}. \tag{8.50}$$

For example, with $\rho = 3 \times 10^9\,\mathrm{g/cm^3}$ and $T = 11 \times 10^9$ K, half of ^{56}Fe is dissociated. This dissociation takes energy from the core and causes pressure loss. The collapse is thus accelerated.

2. If the mass of the core exceeds M_{Ch}, electrons are captured by the nuclei to avoid violation of the Pauli principle

$$\mathrm{e}^- + (Z, A) \rightarrow (Z-1, A) + \nu_e. \tag{8.51}$$

The neutrinos can escape the core, taking away energy. This is again accompanied by a pressure loss due to the decrease of the free electrons (this also decreases M_{Ch}). The collapse is again accelerated.

Note that the chemical equilibrium condition arising for the reaction $e^- + p \rightarrow \nu_e + n$ is

$$\mu_e + \mu_p = \mu_n + \langle E_\nu \rangle,$$

where μ_i are the chemical potentials of the particle i. Thus the fact that neutrinos are not trapped, plus the rise in the electron Fermi surface as the density increases, leads to increased neutronization of the matter. The escaping neutrino carries off energy and lepton number. Both the electron capture and the nuclear excitation and disassociation takes energy out of the electron gas, which is the star's only source of support. This means that the collapse is very fast. Numerical simulations find that the iron core of the star ($\sim$1.2–1.5 $M_\odot$.) collapses at about 0.6 of the free fall velocity.

Supernova dynamical and diffusion scales

In terms of the density, the dynamical timescale, Eq. (6.12), is

$$t_{dyn} = \left(\frac{3}{8\pi G\rho}\right)^{1/2}.$$

At the edge of the iron core the density is $\sim 10^8\,\mathrm{g\,cm^{-3}}$ when core collapse sets in, and at the center it is $\sim 3 \times 10^9\,\mathrm{g\,cm^{-3}}$. Therefore,

$$t_{dyn} = 0.13\,\rho_8^{-1/2}\,\mathrm{s}. \tag{8.52}$$

Using techniques analogous to those used in Section 7.5.5, one can show that for neutrino energies much less than $m_N c^2 \sim 1$ GeV the cross-section for the nucleon scattering is

$$\sigma_\nu = \frac{1}{4}\sigma_0 \left(\frac{E_\nu}{m_e c^2}\right)^2, \tag{8.53}$$

where

$$\sigma_0 = \frac{4\pi G_F^2 m_e^2}{\hbar^4} = 1.76 \times 10^{-44}\,\mathrm{cm}^2. \tag{8.54}$$

The mean free path for scattering is $\lambda_\nu = 1/\langle n\sigma_\nu\rangle$, which is an average over the cross-sections for these processes. An approximate expression for the mean free path is given by

$$\lambda_\nu \sim 2 \times 10^5 \left(\frac{E_\nu}{10\,\mathrm{MeV}}\right)^{-2} \rho_{12}^{-1}\,\mathrm{cm}. \tag{8.55}$$

The typical neutrino energy is $\sim$20 MeV, so the mean free path is only $\lambda_\nu \sim 0.5\rho_{12}^{-1}$ km.

Scattering is a diffusion process, and from the diffusion equation in spherical geometry one finds that the time for a neutrino to diffuse a radial distance R is

$$t_{diff} = \frac{R^2}{3\lambda_\nu c}. \tag{8.56}$$

If we assume a uniform density sphere of mass $1.4\,M_\odot$ and estimate the neutrino energy as the Fermi energy we get $E \sim E_F = 36.8\rho_{12}^{1/3}$ MeV, and

$$t_{diff} = 5.2 \times 10^{-2}\rho_{12}\ \mathrm{s}.$$

The diffusion time scale should be compared to the dynamical timescale, t_{dyn}, from Eq. (8.52),

$$\frac{t_{diff}}{t_{dyn}} = 40\ \rho_{12}^{3/2}\ \mathrm{s}. \tag{8.57}$$

Therefore, we find that above a density of $\sim 10^{11}$ g cm^{-3} the neutrinos become completely trapped in the core. This has the consequence that the lepton number will be conserved in the core. If neutrino trapping did not set in, the lepton number would have decreased to a very low level because of beta decay and inverse beta decay (K-capture). This now instead happens after the collapse in the explosion phase when the density is low enough.

Neutrinos

The newly formed nucleus in Eq. (8.51) can also undergo beta decay. As a result, the nucleus is restored to its original state, and an electron–electron antineutrino ($\bar{\nu}_e$) pair is created. The $\bar{\nu}_e$ similarly escapes the core.

The nucleus can now endlessly repeat this sequence whereby escaping neutrinos drain the core of energy. For a nucleus containing N neutrons and Z protons, written as (N, Z), the two-step process is represented by the following reactions:

$$
\begin{aligned}
(N, Z) + e^- &\rightarrow (N+1, Z-1) + \nu_e \text{ (nuclear electron capture)}, \\
(N+1, Z-1) &\rightarrow (N, Z) + e^- + \bar{\nu}_e \text{ (nuclear beta decay)}.
\end{aligned}
\tag{8.58}
$$

During a conference in Urca, Brazil, Gamow and Schoenberg noted that the local casino appeared to drain money from gamblers much in the way these reactions drained energy from a star. They promptly dubbed this set of reactions the *Urca process.*

In the early stages of the infall the ν_es readily escape. But neutrinos are trapped when a density of $\sim 10^{12}\,\text{g/cm}^3$ is reached. At this point the neutrinos begin to scatter off the matter through both charged current and coherent neutral current processes. The neutral current neutrino scattering off nuclei is particularly important, as the scattering cross-section is off the total nuclear weak charge, which is approximately proportional to N^2, where N is the neutron number. This process transfers very little energy because the mass energy of the nucleus is so much greater than the typical energy of the neutrinos. But momentum is exchanged. Thus the neutrino "random walks" out of the star.

An important change in the physics of the collapse occurs when the density reaches $\rho_{\text{trap}} \simeq 4 \times 10^{11}\,\text{g/cm}^3$. The neutrinos become essentially confined to the core, since their diffusion time in the core is larger than the collapse time. After this point, the energy released by further gravitational collapse and the star's remaining lepton number are trapped within the star. Also, all reactions are in equilibrium, including the capture process (8.51). If we take a neutron star of 1.4 solar masses and a radius of 10 km, a rough estimate of its binding energy is

$$
\frac{GM^2}{2R} \sim 2.5 \times 10^{53}\,\text{ergs}. \tag{8.59}
$$

Thus this is roughly the trapped energy that will later be radiated in neutrinos.

The duration of the burst is set by the diffusion time scale of the neutrinos as the proto-neutron star is deleptonized and is cooling down. The mean free path from Eq. (8.55) is therefore $\sim 10^6 \rho_{14}^{-1} (E_\nu/\text{MeV})^{-2}\,\text{cm}$. Using a constant density for the proto-neutron star with mass $\sim 1.4\ M_\odot$, we have $\rho \sim 2.5\times 10^{13}(R/30\text{ km})^{-3}$. Using these expressions in the equation for the diffusion time, Eq. (8.56), we get

$$t_{diff} \sim 0.2 \left(\frac{R}{30\,\text{km}}\right)^{-1} \left(\frac{E_\nu}{100\,\text{MeV}}\right)^{2} \text{s}. \tag{8.60}$$

Typically, the neutrino energies are of the order of 100–200 MeV in the inner core. In reality, the density in the center is higher than the mean density used above, and the neutrino energies also vary by a large factor, so this number should only be taken as indicative. The fact that it is much larger than the dynamical timescale, however, shows the importance of the neutrino diffusion. More accurate calculations show that the neutrinos diffuse out on a time scale of ~2 s.

Shock waves

The degeneracy of the neutrino Fermi gas avoids a complete neutronization, directing the reaction (8.51) to the left. As a consequence, Y_e remains large during the collapse ($Y_e \approx 0.3 - 0.4$ [Be79]). To equilibrate the charge, the number of protons must also be large. To reach $Z/A = Y_e \approx 0.3 - 0.4$, the protons must be inside heavy nuclei which will therefore survive the collapse.

Two consequences follow:

1. The velocity of sound in matter rises with increasing density. The inner homologous core, with a mass $M_{HC} \sim 0.6 - 0.9$ solar masses, is that part of the iron core where the sound velocity exceeds the infall velocity. This allows any pressure variations that may develop in the homologous core during infall to even out before the collapse is completed. As a result, the homologous core collapses as a unit, retaining its density profile. That is, if nothing were to happen to prevent it, the homologous core would collapse to a point.

   The pressure is given by the degenerate electron gas that controls the whole collapse; the collapse is thus adiabatic, with the important consequence that the collapse of the most internal part of the core is *homologous*, i.e., the position $r(t)$ and the velocity $\nu(t)$ of a given element of mass of the core are related by

$$r(t) = \alpha(t) r_0; \quad \nu(t) = \frac{\dot{\alpha}}{\alpha} r(t), \tag{8.61}$$

   where r_0 is the initial position.

2. Since the nuclei remain in the core of the star, the collapse has a reasonably large order and the entropy remains small during the collapse ($S \approx 1.5\,k$ per nucleon, where k is the Boltzmann constant).

The collapse continues homologously until nuclear densities of the order of $\rho_N \approx 10^{14}\,\text{g/cm}^3$ are reached, when the matter can be considered as approximately a degenerate Fermi gas of nucleons. Since the nuclear matter has a finite compressibility, the homologous core decelerates and starts to increase again in response to the increase of nuclear matter. This eventually leads to a *shock wave* which propagates to the external core (i.e., the iron core outside the homologous core) which, during the collapse time, continues to contract reaching the supersonic velocity. The collapse break followed by the shock wave is the mechanism which creates the supernova explosion. Nonetheless, several ingredients of this scenario are still unknown, including the equation of state of the nuclear matter. The compressibility influences the available energy for the shock wave, which must be of the order of 10^{51} erg.

Initially the shock wave may carry an order of magnitude more energy than is needed to eject the mantle of the star (less than 10^{51} ergs). But as the shock wave travels through the outer iron core, it heats and melts the iron that crosses the shock front, at a loss of ~8 MeV/nucleon. The enhanced electron capture that occurs off the free protons left in the wake of the shock, coupled with the sudden reduction of the neutrino opacity of the matter ($\sigma_{coherent} \sim N^2$), greatly accelerates neutrino emission. This is another energy loss. The summed losses from shock wave heating and neutrino emission are comparable to the initial energy carried by the shock wave.

The exact mechanism for the explosion of a supernova is still controversial.

1. In the *direct mechanism*, the shock wave is not only strong enough to stop the collapse, but also to explode the exterior stellar shells. However, it seems to be the consensus among the supernovae community that realistic nuclear equations of state do not give rise to prompt explosion, or the direct mechanism
2. If the energy in the shock wave is insufficient for a direct explosion, the wave will deposit its energy in the exterior of the core, e.g., by excitation of the nuclei, being frequently followed by electronic capture and emission of neutrinos (*neutrino eruption*). Additionally, neutrinos of the all three species are generated by the production of pairs in the

hot environment. A new shock wave can be generated by the outward diffusion of neutrinos, indeed carrying the biggest part of the energy liberated in the gravitational collapse of the core ($\approx 10^{53}$ erg). If about 1% of the energy of the neutrinos is converted into kinetic energy due to the coherent neutrino–nucleus scattering, a new shock wave arises. This will be strong enough to explode the star. This process is known as the *retarded mechanism* for supernova explosion.

The delayed mechanism begins with a failed hydrodynamic explosion; after about 0.01 s the shock wave stalls at a radius of 200 – 300 km. After neutrinos diffuse out of the proto-neutron star, the majority of them starts free-streaming; however, some of them get absorbed in the vicinity of the proto-neutron star and the resultant heating of the material there increases the pressure behind the stalled shock, thereby pushing it to move again. After perhaps 0.5 s, the shock wave is revived due to neutrino heating of the nucleon "soup" left in the wake of the shock. This heating comes primarily from charged current reactions off the nucleons in that nucleon gas; quasi-elastic scattering also increases the energy transfer. This heated gas may reach 2 MeV in temperature; it has a very high entropy. Thus the energy is in the radiation, not in the matter. The pressure exerted by this gas helps to push the shock outward. It is important to note that there are limits to how effective this neutrino energy transfer can be: if matter is too far from the core, the coupling to neutrinos is too weak to deposit significant energy. If too close, the matter may be at a temperature (or soon reach a temperature) where neutrino emission cools the matter as fast as or faster than neutrino absorption heats it. In the parlance of the field, one hears the work "gain radius" to describe the region where useful heating is done.

To know which of the above mechanisms is responsible for the supernova explosion one needs to know the rate of electron capture, the nuclear compressibility, and the way neutrinos are transported. The iron core, the remnant of the explosion (the homologous core and part of the external core) will not explode and will become either a neutron star, and possibly later a *pulsar* (rotating neutron star), or a *black hole*, as in the case of more massive stars, with $M \geq 25 - 35\, M_{\odot}$.

Supernova radioactivity and light curve

The explosive nucleosynthesis in the silicon core results in several radioactive isotopes, the most important being ^{56}Ni, ^{57}Ni and ^{44}Ti. All of these have comparatively short half-lives. The decays of these elements can therefore be directly observed, and are in fact crucial for the observability of the supernova. The decays are characterized by either the half-life, $t_{1/2}$, or the exponential decay time scale, $\tau = t_{1/2}/\ln 2$.

^{56}Ni decays on a time scale of $\tau = 8.8$ days by electron capture as

$$^{56}\mathrm{Ni} \rightarrow {}^{56}\mathrm{Co} + \gamma. \tag{8.62}$$

In this process it emits gamma rays with energies 0.16–0.81 MeV. The ^{56}Co isotope resulting from this decay is, however, not stable either, but decays by electron capture or by positron decay according to

$$^{56}\mathrm{Co} \rightarrow {}^{56}\mathrm{Fe} + \gamma \ (\text{or } {}^{56}\mathrm{Fe} + e^{+}). \tag{8.63}$$

The first decay occurs in 80% of the cases and the second in the remaining 20%. In terms of energy going into gamma rays and positrons these numbers are 96.4% and 3.6%, respectively. The strongest gamma ray lines are at 0.85 MeV and 1.24 MeV. The average positron energy is 0.66 MeV. Similarly, ^{57}Ni decays by electron capture as

$$^{57}\mathrm{Ni} \rightarrow {}^{57}\mathrm{Co} + \gamma, \tag{8.64}$$

with a very short decay time $\tau = 52$ h. The more interesting decay is

$$^{57}\mathrm{Co} \rightarrow {}^{57}\mathrm{Fe} + \gamma, \tag{8.65}$$

with $\tau = 390$ days. Finally, ^{44}Ti decays first to ^{44}Sc on a time scale of $\sim$89 years,

$$^{44}\mathrm{Ti} \rightarrow {}^{44}\mathrm{Sc} + \gamma, \tag{8.66}$$

and then rapidly ($\tau = 5.4$ h) to

$$^{44}\mathrm{Sc} \rightarrow {}^{44}\mathrm{Ca} + \gamma \ (\text{or } {}^{44}\mathrm{Ca} + e^{+}). \tag{8.67}$$

The result of these radioactive decays are either gamma rays or positrons. The gamma rays are scattered by the electrons in the ejecta through Compton scattering. In each scattering they lose roughly half of their energy to the electrons. Because the energy of the gamma-rays are initially in the MeV range, much higher than the binding energies of the bound electrons in the atoms, both free and bound electrons contribute to the scattering.

This down-scattering of the gamma rays continues until the cross-section for photoelectric absorption is larger than the Compton cross-section, which occurs at an energy of ~10–100 keV, depending on the composition. The most important element for the photoelectric absorption is iron. The total gamma ray luminosity from the various decays is given by

$$\begin{aligned} L_\gamma &= 1.3 \times 10^{42} \left[\frac{M(^{56}\mathrm{Ni})}{0.1 M_\odot}\right] \exp\left(-\frac{t}{111[\mathrm{d}]}\right) \\ &+ 6.9 \times 10^{38} \left[\frac{M(^{57}\mathrm{Co})}{0.5 \times M_\odot}\right] \exp\left(-\frac{t}{390[\mathrm{d}]}\right) \\ &+ 4.1 \times 10^{36} \left[\frac{M(^{44}\mathrm{Ti})}{10^{-4} M_\odot}\right] \exp\left(-\frac{t}{89[\mathrm{y}]}\right) \mathrm{erg}\ s^{-1}. \end{aligned} \tag{8.68}$$

After shock breakout the radiation will leak out on a diffusion time scale. We have already estimated this in Eq. (8.56), which we write as

$$t_{diff} = \frac{3R^2\rho\kappa}{\pi^2 c}, \tag{8.69}$$

where κ is now the radiation opacity. This should be compared to the expansion time scale $t_{exp} = R/v$. Taking the opacity to be that of Thompson scattering, $\kappa = 0.4\,\mathrm{cm^2 g^{-1}}$, and assuming a uniform density for the envelope we get

$$\frac{t_{diff}}{t_{exp}} = 1.9 \left(\frac{M}{M_\odot}\right) \left(\frac{v}{10^4\ \mathrm{km.s^{-1}}}\right) \left(\frac{R}{10^{15}\,\mathrm{cm}}\right)^{-2}. \tag{8.70}$$

For a typical mass of $10\,M_\odot$ we find that not until the supernova has expanded to $R_{peak} \sim 4 \times 10^{15}$ cm, after $t_{peak} = R/v \sim 40$ days, can the radiation leak out faster than the ejecta expand.

The burst of neutrinos is the first evidence that a core-collapse supernova has occurred. But a few hours later the outgoing shock wave releases electromagnetic radiation initially as a ultra-violet flash. As it expands, the supernova becomes visible at optical wavelengths, with the initial rise in the light curve the result of the increasing surface area of the star combined with a relatively slow temperature decrease. The peak in the light curve occurs as the temperature of the outer layers starts to decrease. In some cases (supernova type II-P) a plateau in the falling spectrum appears corresponding to a change in opacity in the outer layer of the exploded star produced by the shock wave, which heats the star's outer envelope to over 100,000 K ionizing all the hydrogen. The ionized hydrogen has a high opacity, so that the radiation from the inner parts of the star

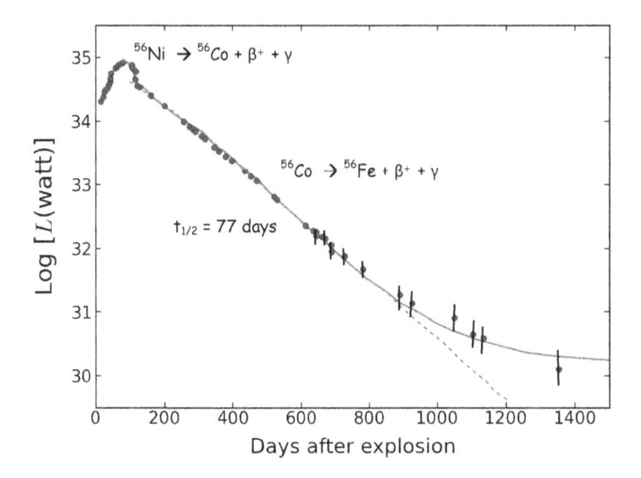

Figure 8.14. Luminosity from supernova SN1987 A as a function of time.

cannot escape, and we can only observe photons from the outermost parts of the star.

After a few weeks, the outer parts of the star have cooled sufficiently that the ionized hydrogen recombines to neutral hydrogen.This temperature is about 4,000 and 6,000 K. Neutral hydrogen is transparent at most wavelengths, and this recombination front where the opacity changes is known as the photosphere of the star. Once the hydrogen starts to recombine, photons from the hotter, inner regions of the hydrogen envelope are able to escape.

After the recombination front has passed through the entire hydrogen envelope, the plateau phase ends, and the light curves are dimmer and powered by a radioactive tail due to the conversion of ^{56}Co into ^{56}Fe. It has the same shape for all core-collapse supernovae. The ^{56}Ni decays according to (EC = electron capture) ^{56}Ni (EC, 6.1 d) ^{56}Co (EC, 77.3 d) ^{56}Fe (stable: 91.7%). The light from the SN1987 A (Fig. 8.14) supernova can be related to 0.85 and 1.24 MeV γ-lines from the decay of ^{56}Co. One can actually calculate that in the first moments after the explosion about 0.1 $M_\odot$ of ^{56}Ni was formed. Most of the Fe in the Universe is likely produced by this kind of process.

The mass of heavy elements ejected out into space in supernova explosions enriches the abundance of iron for the formation of later generation of stars. The transition elements, which are only formed in such stellar explosions, are a necessity for the existence of life on Earth and maybe elsewhere in the Universe.

8.6.4. *Other cataclysmic events*

Gamma ray bursts

First discovered by military satellites looking for the telltale signature of clandestine nuclear explosions, gamma ray bursts (GRB) are sudden, intense flashes of gamma rays, possibly due to extremely energetic explosions in distant galaxies. GRBs are detected roughly once per day from wholly random directions of the sky. Gamma rays are absorbed in the atmosphere, so GRBs must be observed by satellites. They can last from milliseconds to about an hour, although a typical burst lasts a few seconds only. The initial burst is usually followed by an "afterglow" emitting at longer wavelengths (X-ray, ultraviolet, optical, infrared and radio).

Their origin may be due to a narrow beam of intense radiation released during a supernova event, as a rapidly rotating, high-mass star collapses to form a black hole. Some of the short burst GRBs also appear to originate from a merger of binary neutron stars. Other GRBs may be due to two ultra-dense neutron stars smashing into each other.

So far, the observed GRBs are billions of light years away from Earth, although a small fraction may be magnetic eruptions on neutron stars in very nearby galaxies. Their origin at large distance from us implies that the explosions are extremely energetic, releasing as much energy in a few seconds as the Sun will in its entire 10 billion year lifetime. It also implies that they are extremely rare, at a few per galaxy per million years [Pod04]. It has been hypothesized that a gamma ray burst in the Milky Way could cause a mass extinction on Earth.

Quasars

Quasi-stellar objects, or *quasars*, are sub-arc-sec objects with tremendously strong radio signals and strange optical spectra. Such objects can outshine galaxies. They are among the most distant and oldest objects in the Universe and appear to be massive black holes in the centers of galaxies in early times, eating stars and other matter and emitting massive amounts of energy.

Active Galactic Nuclei (AGN)

Active galactic nuclei are a category of exotic objects that includes: luminous quasars, Seyfert galaxies (galaxies with nuclei that produce spectral line emission from highly ionized gas), and *blazars*. Its likely that the core of an AGN contains a supermassive black hole surrounded by an accretion disk. As matter spirals into the black hole, electromagnetic radiation and plasma jets spew outward from the poles. Blazars are AGN with jets spewing relativistic energies toward the Earth.

8.7. Exercises

1. Find the classical distance of closest approach for two protons with an energy of approach equal to 2 keV. Estimate the probability that the proton penetrates the Coulomb barrier tending to keep them apart. Compare this probability with the corresponding probability for two ^{4}He nuclei with same energy of approach.
2. Estimate the lifetime of our Sun if the p + p reaction were based on the strong interaction with an S-factor of $S = 5 \times 10^{-5}$ MeV barn.
3. The flux of energetic neutrinos from ^{8}B decay in branch III of the proton–proton chain is very dependent on the central temperature of the Sun. Confirm this by showing that the rate of the reaction producing ^{8}B,

$$p + {}^7\text{Be} \rightarrow {}^8\text{B} + \gamma,$$

   is approximately proportional to T^{14} , when the temperature is near to 1.5×10^7 K. (Hint: $\langle \sigma\nu \rangle \sim T^\beta$ and $\beta = Q = 3E_0/kT$ for $Q \gg 1$.)
4. Assume that the mass fractions of ^{1}H and ^{4}He at the time of the Sun's birth are equal to 0.71 and 0.27, respectively. Calculate the ratio of the corresponding number densities.
5. Before it explodes as a SN1a, a white dwarf has a mass of $1.4\,M_\odot$. During the explosion, all of the matter in the white dwarf is converted from helium to iron; this process converts mass to energy with an efficiency of approximately 0.1%. Calculate how much energy is released during this explosion. You will need to know that $M_\odot = 2 \times 10^{30}$ kg, and that $c = 3.0 \times 10^8$ m/s.
6. The r-process takes place in a plasma with $T_9 = 0.7$ and neutron density $n = 10^{22}/\text{cm}^3$. Following the Saha Equation, find the neutron separation energy of the nuclei taking part in the r-process, under these conditions. Here you can consider a typical nuclear mass number of $A \sim 200$, equilibrium conditions (considering only $n \leftrightarrow \gamma$) that are maintaining neighboring abundances $N_A \sim N_{A-1}$, and $g_A = g_{A+1}$.
7. A supernova explodes in the Large Magellanic Cloud (LMC), 50 kpc from Earth. The burst of neutrinos turns on suddenly at $t = 0$ and then declines in time as e^{-t/t_0}, where $t_0 \sim 2$ seconds. The distribution

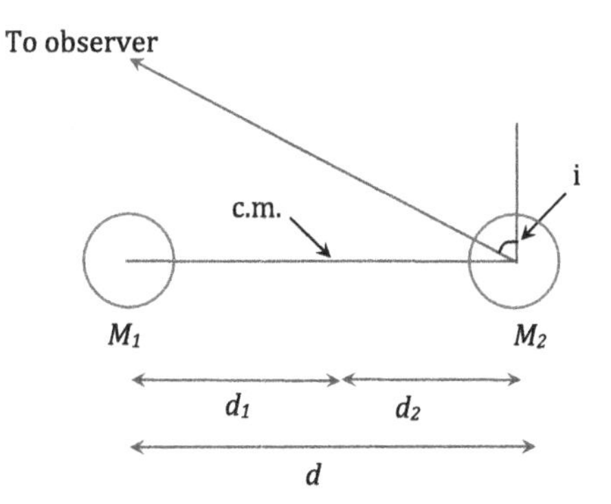

Figure 8.15. Schematic representation of an X-ray binary.

in energy can be taken as a Fermi–Dirac distribution from characterized by a temperature $T \sim 5\,\text{MeV}$. The neutrinos are observed on Earth in a detector, and the spread in the arrival times of the neutrinos recorded. Estimate what neutrino mass would cause the duration of the observed events to have approximately doubled, due to spreading of the signal during transit from the LMC to Earth. Note this is about the kinematic effects of the neutrino mass, not about oscillations.

8. Consider an X-ray binary, composing of a main sequence star of mass M_1 and a neutron star of mass M_2 which emits periodic X-ray pulses. They orbit (assuming circular orbit for simplicity) about their centre of mass with period T, as shown in Fig. 8.15, viewed in the orbital plane. Their separation is d and their distances from the centre of mass are d_1 and d_2 respectively. The angle i is the inclination of the orbital plane to the line of sight.

   a) Write down the relation between M_1, M_2, d and d_1.
   b) The light radiated from M_1 will be shifted according to the Doppler effect. Such variation of shifting is related to the projection of the orbital velocity ν along the line of sight. Find ν in terms of d_1, i and T.
   c) Kepler's third law states that in this system

$$T^2 = \frac{4\pi^2}{G(M_1 + M_2)} d^3.$$

Use Kepler's law together with the above results to show that the mass function, defined by

$$f(M_1, M_2, i) = \frac{(M_2 \sin i)^3}{(M_1 + M_2)^2}$$

can be related to the period T and the orbital velocity ν by

$$f = \frac{T\nu^3}{2\pi G}.$$

Suppose we know the "mass functions" of both stars, i.e.

$$f_1 = \frac{(M_2 \sin i)^3}{(M_1 + M_2)^2}, \quad f_2 = \frac{(M_1 \sin i)^3}{(M_1 + M_2)^2},$$

which can be determined the mass ratio $q = M_2/M_1$. Write down an expression of M_2 in terms of the f_2, q and i.

9. Calculate carefully the energy deposited in the Sun by each of the ppI, ppII, and ppIII cycles, assuming that all of the energy is deposited except for that carried off in neutrinos. For neutrinos, neglect the $\mathrm{p} + \mathrm{p} + e^-$ reaction and assume, in beta decays, that the neutrino carries, on average, half of the kinetic energy available to the outgoing neutrino and positron. Use the value of the solar constant to derive a model-independent constraint on the sum of the $\mathrm{pp} + {}^7\mathrm{Be} + {}^8\mathrm{B}$ neutrino fluxes, assuming only that the Sun burns in a steady state (so that today's neutrino fluxes and photon luminosity are coupled).
10. The observed spectrum of a given star contains extremely broad hydrogen emission lines centered approximately at their natural frequency and relatively sharp hydrogen absorption lines in the short wavelength wing of the emission lines. Explain how the observed spectrum is produced, assuming that the star is a nova surrounded by an expanding shell of material which is mostly hydrogen at 10^4 K. If the absorption lines are shifted from their natural wavelength by about ten times their width, estimate the velocity of the expanding shell.
11. The Crab Pulsar is a rotating neutron star formed by a supernova in 1054 AD. At present it has an angular velocity and an angular acceleration given by $\omega = 190\ s^{-1}$ and $d\omega/dt = -2.4 \times 10^{-9}\ s^{-2}$. It is supposed that the neutron star losses its rotation energy via magnetic dipole radiation radiating energy at a rate $dE/dt = -\alpha\omega^4$, where α is a constant.

    a) Use this model to derive the differential equation satisfied by ω.
    b) Find the time dependence of (you may assume that the initial angular velocity of the star is infinite)
    c) Estimate the age of the pulsar according to the magnetic dipole model.

d) If the age of the pulsar calculated according to the magnetic dipole model is not concordant with the real age of pulsar (945 years) could you suggest another explanation?

12. Assume that a particle exists with rest mass-energy 1200 MeV. How dense would the stellar interior have to be, if made of pure neutrons (rest mass 939 MeV), so that the extra particle is stable?
13. The CNO cycle establishes an equilibrium between the reactants within a few years. Calculate the CNO equilibrium abundances $dY(CNO)/dt = 0$ for the typical conditions of stellar hydrogen burning in massive stars, $\rho = 100\,\text{g/cm}^3$ and $T = 0.03\,\text{GK}$.
14. A neutron star accretes 10^{21} grams of hydrogen-rich matter from a binary companion star during the 5 h interval since its last burst. The accreted matter forms a thin spherical shell on the surface of the neutron star. At some instant there is runaway thermonuclear burning of the accreted matter; all the fusion reactions are completed in much less than 1 s. Assume that the fraction of the rest mass of the accreted matter that is converted to energy is equal to 0.003. As a result of the energy released, the surface of the neutron star becomes very hot and it radiates away this excess energy in the form of black body radiation. Take the radius of the neutron star to be 10 km.

    a) How much energy is released in the burst?
    b) If the burst lasts for about 10 s, what is the average luminosity of the burst? Compare this to the luminosity of the Sun.
    c) Assume that during the burst the surface of the neutron star reaches a temperature T which remains nearly constant over an interval of ~10 s. Find T.
    d) Use the Wien displacement law, Eq. 4.72, to demonstrate why these events are called X-ray bursts.

15. The r-process takes place in a plasma with $T_9 = 0.7$ and neutron density $\rho_n = 10^{22}/\text{cm}^3$. What is the neutron separation energy of the nuclei taking part in the r-process? Here you can consider a typical nuclear mass number of $A \sim 200$, equilibrium conditions (considering only $n \leftrightarrow \gamma$) that are maintaining neighboring abundances $N_A \sim N_{A-1}$, and $g_A = g_{A+1}$.
16. (a) Why is the iron peak not a single spike at the most bound isotope but has instead a significant width (in terms of mass number, for example)? (b) Why does the iron peak in the solar abundances

not peak at the most bound isotope in nature (^{62}Ni) but at ^{56}Fe instead?

17. A white dwarf accretes matter until its mass exceeds the Chandrasekhar limiting mass of $1.4\,M_\odot$, after which it collapses to become a neutron star that is 10 km in radius. Estimate the energy (including all forms, e.g., photons, neutrinos, kinetic energy of any ejected matter, etc.) that is released in the ensuing supernova explosion.
18. Calculate the total energy release in erg
    (a) from a core-collapse supernova. Assume at the end of Si burning a 1.4 solar mass iron core with radius 10,000 km, and as the final compact remnant a typical neutron star with 10 km radius.
    (b) from a thermonuclear supernova. Assume a white dwarf reaches the Chandrasekhar mass limit with a composition of 50% ^{16}O and 50% ^{12}C (by mass), and assume the final product of the explosion is mainly ^{56}Ni.
19. Calculate the energy the shock in a core-collapse supernova loses by completely dissociating into protons and neutrons a typical outer core of 0.4 solar masses of ^{54}Fe. Compare with a typical shock energy of about 1% of the explosion energy from the previous exercise (a) and discuss the consequences.
20. Consider a supernova that initially creates N_0 ^{56}Ni nuclei and no ^{56}Co nuclei. The rate of decay of nickel to cobalt is λ_{Ni}, and the rate of decay of cobalt into iron is λ_{Co}.
    (a) Suppose that at a given time there are N_{Ni} nickel-56 nuclei present in the ejected material. Write the equation for dN_{Ni}/dt, the time rate of change of the number of cobalt nuclei, and solve the equation to find N_{Ni} as a function of the time t since the explosion.
    (b) Suppose there are N_{Co} cobalt-56 present at any given time. Write down an equation for dN_{Co}/dt, the time rate of change of the number of cobalt nuclei, accounting for both creation of cobalt nuclei by decay of nickel and the destruction of cobalt nuclei by decay into iron.
    (c) Solve the differential equation you obtained in part (b) to find the number of cobalt nuclei at time t after the explosion. (*Hint:* search for solutions of the form $N_{Co} = ae^{-\lambda_{Co}t} + be^{-\lambda_{Ni}t}$, where a and b are constants, and use the fact that the explosion creates no cobalt initially.)

(d) If each nickel decay releases an energy E_{Ni} and each cobalt decay releases an energy E_{Co}, compute the total rate at which energy is released by the decay of both cobalt and nickel at time t.

(e) Verify that, for large times t, your answer to part (d) approaches the rate of energy release one would have found if the supernova had simply created N_0 cobalt nuclei at time 0. (*Hint:* use the approximation that $\lambda_{Ni} \gg \lambda_{Co}$, i.e. the decay rate of nickel-56 is much larger than the decay rate of cobalt-56.)

21. A galaxy's age can be estimated by radioactive decay of uranium. Uranium is supposedly produced as an r-process element in supernovae, and on this basis the initial abundances of ^{235}U and ^{238}U are expected to be

$$\left.\frac{^{235}\mathrm{U}}{^{238}\mathrm{U}}\right|_{initial} = 1.65.$$

The decay rates are

$$\Lambda(^{235}\mathrm{U}) = 0.9 \times 10^{-9}\ \mathrm{yr}^{-1},$$
$$\Lambda(^{238}\mathrm{U}) = 0.17 \times 10^{-9}\ \mathrm{yr}^{-1}.$$

Use the decay law $U(t) = U(0)\exp(-\Lambda t)$ to estimate the age of the galaxy. Assuming the galaxy took a minimum of an additional 10^9 yr to form in the first place, obtain an upper limit on the value of the Hubble parameter h assuming a flat matter-dominated Universe (see Chapter). (Recall that $H_0 = 100\,h\,\mathrm{km/s/Mpc}$).

Chapter 9

The Big Bang

9.1. Introduction

The Big Bang model is a broadly accepted cosmological theory for the origin and evolution of our Universe. It is based on scientific evidence and observations. According to this model, 14 billion years ago, the portion of the Universe we can observe at the present moment was, say, just a few millimeters across. The Universe has since expanded from this hot dense state into the vast and much cooler cosmos we currently inhabit. We can see remnants of this hot dense matter as the now very cold *cosmic microwave background radiation* (CMB) which is also called "relic radiation". This radiation can be "seen" by microwave detectors as a uniform glow across the entire sky. Its spectrum (i.e., the amount of radiation measured at each wavelength) is found to match that of thermal radiation from a black body, giving fairly strong evidence that the Big Bang scenario must have occurred. The Big Bang model relies on the theory of *general relativity* and on simplifying assumptions, such as *homogeneity* and *isotropy* of space. These assumptions are known as the *cosmological principle.*

The distances to far away galaxies are proportional to their redshift, which indicates that all very distant galaxies and clusters have an apparent velocity directly away from our observation point: the farther away, the higher the apparent velocity (*Hubble's law*). Thus, everything must have been closer together in the past, with extreme densities and temperatures. *Big Bang nucleosynthesis* explains the observed abundances of the light elements throughout the cosmos by the formation of light elements from nuclear processes in the rapidly expanding and cooling first minutes of the Universe. Fred Hoyle is credited with coining the term Big Bang during a 1949 radio broadcast.

Some data has been agreed upon based on observations. These are:

1. The Universe is ~13.7 billion years old.
2. After about 100–200 million years the first stars and galaxies appeared.
3. The composition of the Universe: (a) 73% dark energy, (b) 23% dark matter, (c) 4% matter: stars, planets, interstellar dust, 200 billion galaxies in total, 200 billion stars per galaxy.

9.2. In the Beginning

In 1964 it was discovered that low energy microwave radiation (at about 7.35 cm) reaches us from all directions in space (about 400 photons cm^{-3}). This is referred to as the *cosmic background radiation* (CMB) whose wavelength corresponds to radiation from a black body of temperature 2.7 K (about 0.0003 eV). The CMB is the most-precisely measured black body spectrum in nature (see Fig. 9.1). Thermodynamic calculations show that this is the temperature reached after adiabatic expansion of a very hot cloud for some 10 billion years. The Big Bang hypothesis was first discussed by Georges Lemaitre [Lem31], and subsequently advocated and developed by George Gamow, who introduced Big Bang Nucleosynthesis (BBN) and whose associates, Ralph Alpher and Robert Herman, predicted the cosmic microwave background radiation (CMB) [AH48]. Measurements have also shown that the intensity of this radiation varies slightly ("ripples") in different directions in space.

The Big Bang hypothesis suggests an instantaneous beginning of our Universe at a point at which all energy is concentrated. Ordinary nuclear

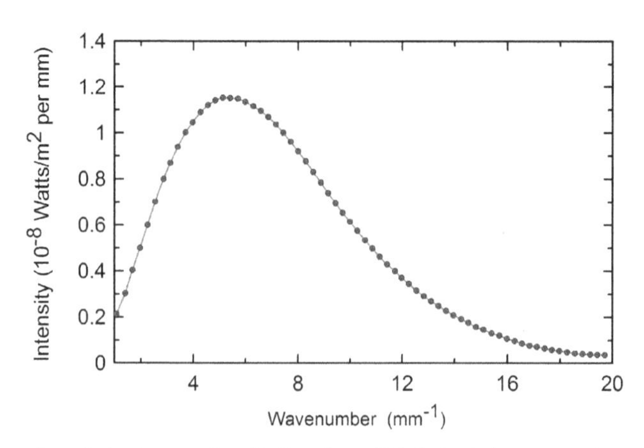

Figure 9.1. Cosmic background radiation data (filled circles) compared with a Planck distribution (solid curve) with temperature of 2.7 K.

reactions cannot model this beginning and we must turn to particle physics. The Big Bang hypothesis for the formation of the Universe was originally suggested in 1948 by Gamow, Alpher and Herman [ABG48], and later developed by the B^2FH-group [Bu57] and others.

Allowing the expansion of the Universe going backwards in time and using general relativity yields an infinite density and temperature at a finite time in the past. This singularity cannot be understood solely on the basis of general relativity. In fact, the earliest phases of the Big Bang are subject to much speculation. At the beginning, the Universe is assumed to be filled homogeneously and isotropically with an incredibly high energy density, huge temperatures and pressures, and was very rapidly expanding and cooling. The temperature must have been $\gtrsim 10^{13}$ K, but no light was emitted, because the enormous gravitational force pulled the photons back. The system was supposed to be in a unique state with no repulsion forces. However, just as a bottle of supercritical (overheated) water can explode by a phase transition, so did the Universe, and time began. The Universe expanded violently in all directions, and as age and size grew, density and temperature fell. The phase transition, which occurred at about 10^{-37} s, caused a *cosmic inflation*, during which the Universe grew exponentially [Gut81,Gut98]. After inflation stopped, the Universe consisted of an immensely dense, hot spot of photons, quarks and leptons, and their antiparticles, in thermal equilibrium, particles being created by photons and photons created by annihilation of particles. At some point a little known process called *baryogenesis* violated the conservation of baryon number, leading to a very small excess of quarks and leptons over antiquarks and antileptons, of the order of one part in 30 million. This resulted in the predominance of matter over antimatter in the present Universe.

9.2.1. *From elementary particles to nuclei*

At about 10^{-6} s, quarks and gluons combined to form baryons such as protons and neutrons. The small excess of quarks over antiquarks led to a small excess of baryons over antibaryons. The temperature was no longer high enough to create new proton–anti-proton or neutrons–antineutrons pairs, and a mass annihilation followed, leaving just one in 10^{10} of the original protons and neutrons, and none of their antiparticles. At one-hundredth of a second later all the quarks were gone, and the Universe consisted of an approximately equal number of electrons, positrons, neutrinos and photons, and a small amount of protons and neutrons; the ratio of protons to photons is assumed to have been about 10^{-9}. The temperature was about 10^{11} K and the density so high, about $4 \times 10^{9}\,\mathrm{g\,cm^{-3}}$, that even the unreactive neutrinos were hindered in escaping.

The numbers mentioned above for the conditions at the early Universe can be partly understood by considering the relations

$$E(\mathrm{MeV}) = mc^2 = 931.5\,\Delta M, \tag{9.1}$$

which gives the energy required to create a particle of mass ΔM (E in MeV, ΔM in nuclear mass units, u), and

$$E(\mathrm{MeV}) = kT = 8.61 \times 10^{-11}\,T, \tag{9.2}$$

which gives the average kinetic energy of a particle at temperature T (K, Kelvin). Note that in a thermal gas a particle has a kinetic energy of $E \sim 3kT/2$, where $k =$ Boltzmann's constant $= 0.862 \times 10^{-10}$ MeV/K (see Chapter 4). So 1 MeV $\sim kT \Rightarrow T = 1.16 \times 10^{10}$ K. That is, a MeV is the typical thermal energy of a particle in a heat bath of $T = 10$ billion degrees K.

As the Universe expands, the initial quark–gluon plasma condenses into hadrons including nucleons. Also, as the photon energy of E (eV) corresponds to the wavelength λ (m) according to

$$E(eV) = h\nu = hc/\lambda = 1.240 \times 10^{-6}/\lambda(\mathrm{m}) \tag{9.3}$$

one can estimate that the creation of a proton or a neutron (rest mass 940 MeV) out of radiation requires a temperature of 1.1×10^{13} K, corresponding to a photon wavelength of about 10^{-15} m, i.e., the size of a nucleon. At these temperatures nucleons are formed out of radiation, but are also disrupted by photons, leading to an equilibrium with about an equal number of protons and neutrons. At temperatures below the threshold formation energy, no nucleons are formed. However, it should be remembered that particles and radiation are distributed over a range of energies, according to the Boltzmann and Planck distribution laws. Thus some formation (and disruption) of nucleons occurs even at lower temperatures.

When the Universe cooled below about $kT \sim 1$ MeV ($\sim 10^{10}$ K), i.e., when e^+–e^- annihilation occurred, the particle soup would have consisted of the familiar stable particles such as p, n, e^-, and γ. The density was about that of air and the basic idea of *Big Bang Nucleosynthesis* (BBN) is that a nuclear reaction network begins with $\mathrm{n} + \mathrm{p} \rightarrow \mathrm{d} + \gamma$ and ends with light nuclear elements. This happens because the neutron lifetime is long eneough, $\tau_{1/2}(\mathrm{n}) \sim 10$ minutes. In fact, neutrons exist in our present day world only because they bind in nuclei. Free neutrons have enough energy to decay to protons via beta decay. Bound neutrons do not because their binding energy makes this decay energetically impossible. At this stage of the Big Bang, most protons remained uncombined as hydrogen nuclei. As the Universe cooled, the rest mass energy density of matter came to gravitationally dominate that of the photon radiation. After about 400,000 years the electrons and nuclei combined into atoms (mostly hydrogen); hence the radiation decoupled from matter and continued through space largely unimpeded. This is the relic radiation, known as the cosmic microwave background radiation.

Today's observed elemental mass abundance is 75% H (1 mass unit) and 25% ^{4}He (4 mass units), i.e., for every ^{4}He there must be about 12 protons. Thus there are 2 neutrons and 14 protons in the sum (see Fig. 9.2) of all masses. Therefore,

$$\frac{\mathrm{n}}{\mathrm{p}} \sim \frac{1}{7} \tag{9.4}$$

in our Universe. A few minutes after the Big Bang, starting with only protons and neutrons, nuclei up to lithium and beryllium were formed in small amounts. Some boron may also have been formed, but the BBN stopped before significant carbon could be formed. Carbon requires far higher helium density and longer time than were available during BBN. Thus, BBN essentially stopped due to drops in temperature and density

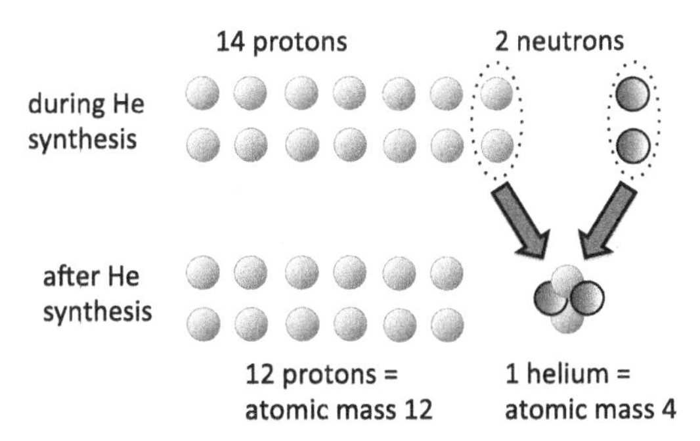

Figure 9.2. The Big Bang theory predicts that when the Universe was still very hot ($T \sim 10^9$ K, and ~1 minute old), protons outnumbered neutrons 7:1. When ^{2}H and He nuclei formed, most of the neutrons formed He nuclei, yielding about 1 He nucleus for every 12 H nuclei, equivalent to 75% of the Universe's mass being hydrogen nuclei and 25% being helium nuclei. This is indeed the abundance of He and H we observe today.

as the Universe continued to expand. Before we discuss BBN in detail, we summarize elements of cosmology required to understand the Big Bang epoch and the BBN.

9.2.2. *Space, particles, and their relation to time*

We start with the *Friedmann–Robertson–Walker (FRW) metric*, Eq. (D.104), where

$$\begin{aligned} k &= +1 \leftrightarrow \text{curved space, finite,} \\ k &= 0 \leftrightarrow \text{flat space, infinite,} \\ k &= -1 \leftrightarrow \text{curved space, infinite.} \end{aligned} \tag{9.5}$$

In the FRW metric, the coordinate r is dimensionless: it is not the usual r. The dimension of length is carried by $a(t)$. The evolution of $a(t)$ is determined by the Friedmann equation, Eq. (2.18). When the Universe is dominated by relativistic or radiation-like particles the density is $\rho \propto 1/a^4$, as we have shown after Eq. (2.52). Then

$$\frac{d\rho/dt}{\rho} = -4\frac{da/dt}{a}. \tag{9.6}$$

Plugging this result into Friedmann equation we get

$$\rho(t) = \frac{3}{32\pi G(t+\beta)^2}, \tag{9.7}$$

where β is an integration constant.

For a system with temperature T, the density of massive particles increases with T^3 because the momentum states (k_x, k_y, k_z) will fill up roughly to (T, T, T) (in units such that $k = 1$). If we calculate the energy density of relativistic particles, an extra power of T comes from the fact that each particle has an energy $\sim T$. Thus the energy density, which in the early Universe is dominated by light, relativistic particles (the photon, electron and positron, and the three neutrinos) is given by $\rho \sim T^4$.

We can quantify this result by starting with the energy density of photons in the background calculated from $\rho_\gamma = \int E_\gamma dn_\gamma$ where the density of states is given by

$$dn_\gamma = \frac{g_\gamma}{2\pi^2} \frac{k_\gamma^2}{\exp(E_\gamma/T) - 1} dk_\gamma \tag{9.8}$$

and $g_\gamma = 2$ is the number of spin polarizations for the photon while $E_\gamma = k_\gamma$ is the photon energy (momentum). ($\hbar = c = k = 1$). Performing the integration gives

$$\rho_\gamma = \frac{\pi^2}{15} T^4, \tag{9.9}$$

which is the familiar blackbody result.

In general, at very early times, at very high temperatures, other particle degrees of freedom join the radiation background when $T \sim m_i$ for each particle type i, if that type is brought into thermal equilibrium through interactions. In equilibrium the energy density of a particle type i is given by $\rho_i = \int E_i dn_i$ and

$$dn_i = \frac{g_i}{2\pi^2} \frac{q^2}{\exp[(E_i - \mu_i)/T] \pm 1} dq, \tag{9.10}$$

where again g_i counts the total number of degrees of freedom for type i, $E_i = (m_i^2 + q_i^2)^{1/2}$. μ_i is the chemical potential if present and $\pm$ corresponds to either Fermi or Bose statistics. Note that if $E_i \gg m_i$, the energy is dominated by the momentum term, just like for photons. However, the majority of particles in either the Bose or Fermi distribution have energy $E_i \sim T$. Hence, if $T \gg m_i$, the mass of the particles are irrelevant: they behave like radiation, and this is true for both bosons and fermions (assuming that the chemical potential is much less than T as well.)

In this relativistic limit, when we also assume $\mu_i \ll m_i$ (non-degenerate), we find

$$\rho = g_i \frac{\pi^2}{15} T^4 \times \begin{cases} 7/8 & \text{(fermions)} \\ 1 & \text{(bosons)} \end{cases}$$
$$n = g_i \frac{\zeta(3)}{\pi^2} T^3 \times \begin{cases} 3/4 & \text{(fermions)} \\ 1 & \text{(bosons)} \end{cases}$$
$$p = \frac{\rho}{3}, \tag{9.11}$$

where $\zeta(s) = \sum_1^\infty 1/n^s$ is the Riemann zeta function, and the last line gives the pressure. The ratio of the energy density to the number density gives the average energy per particle,

$$\langle E \rangle = \frac{\rho}{n} \simeq \begin{cases} 3.15\,T & \text{(fermions)} \\ 2.70\,T & \text{(bosons)} \end{cases}, \tag{9.12}$$

which we will very often just take to be $\langle E \rangle \simeq 3kT$.

In the limit that $T \gg m_i$ the total energy density can be conveniently expressed by[1]

$$\rho = g_*(T)\frac{\pi^2}{30}T^4 = \left(\sum_B g_B + \frac{7}{8}\sum_F g_F\right)\frac{\pi^2}{30}T^4 \equiv \frac{\pi^2}{30}N(T)T^4, \qquad (9.13)$$

where $g_*(T)$ are the *effective degrees of freedom*, $g_{B(F)}$ are the total number of boson (fermion) degrees of freedom and the sum runs over all boson (fermion) states with $m \ll T$. The factor of 7/8 is due to the difference between the Fermi and Bose integrals. Equation (9.13) defines $N(T)$ by taking into account new particle degrees of freedom as the temperature is raised.

A relationship between the age of the Universe and its temperature is obtained by using Eqs. (9.13) and (9.7), which gives (with $\beta = 0$)

$$t = \left(\frac{90}{32\pi^3 GN(T)}\right)^{1/2} T^{-2}. \qquad (9.14)$$

Put into a more convenient form

$$tT_{MeV}^2 = 2.4[N(T)]^{-1/2}, \qquad (9.15)$$

where t is measured in seconds and T_{MeV} is in units of MeV.

The value of $N(T)$ at any given temperature depends on the particle physics model. In the standard model, we can specify $N(T)$ up to temperatures of 100 GeV. The change in N can be seen in Table 9.1, where

Table 9.1. Effective numbers of degrees of freedom in the standard model.

Temperature	New Particles	$4N(T)$
$T < m_e$	γ's + ν's	29
$m_e < T < m_\mu$	$e^\pm$	43
$m_\mu < T < m_\pi$	$\mu^\pm$	57
$m_\pi < T < Tc$	π's	69
$T_c < T < m_{strange}$	$-\pi$'s + $u, \bar{u}, d, \bar{d}$ + gluons	205
$m_{\mathrm{strange}} < T < m_{charm}$	$s, \bar{s}$	247
$m_{charm} < T < m_\tau$	$c, \bar{c}$	289
$m_\tau < T < m_{bottom}$	$\tau^\pm$	303
$m_{bottom} < T < m_{W,Z}$	$b, \bar{b}$	345
$m_{W,Z} < T < m_{top}$	$W^\pm, Z$	381
$m_{top} < T < m_{Higgs}$	$t, \bar{t}$	423
$m_{Higgs} < T$	H^o	427

[1]Again, for convenience $\hbar = c = k = 1$.

Table 9.2. Properties of the early Universe: particle energy, temperature, radius and elapsed time.

	T(K)	a/a_0	t(s)
~10 MeV	10^{11}	1.9×10^{-11}	0.0108
~1 MeV	10^{10}	1.9×10^{-10}	1.103
~100 keV	10^{9}	2.6×10^{-9}	182
~10 keV	10^{8}	2.7×10^{-8}	19200

T_c corresponds to the confinement–deconfinement transition between quarks and hadrons. It has been assumed that $m_{Higgs} > m_{top}$.

When the Universe is dominated by relativistic, or radiation-like, particles $\rho \propto 1/a^4$ and the above expressions yield

$$a \sim T^{-1} \sim \sqrt{t}, \tag{9.16}$$

and more careful calculations allows one to calculate the results in Table 9.2.

9.3. Big Bang Timeline

The distinct epochs of the Universe, from the time of the Big Bang, are:

1. *Planck era* — All four known forces are unified.
2. *Grand Unified Theory (GUT) era* — Gravity "freezes out" and becomes distinct.
3. *Electroweak era* — The nuclear strong force "freezes out" and becomes distinct.
4. *Particle era* — Particles begin to form.
5. *Nucleosynthesis era* — Nuclear fusion creates helium, and tiny amount of heavier elements.
6. *Era of nuclei* — Electrons are not yet bound to nuclei.
7. *Era of atoms* — Electrons recombine to form neutral atoms, and the first stars are born.
8. *Era of galaxies* — Galaxies begin to form, leading up to the present.

The earliest eras were very short lasting, and very high energy. The first few eras are when the laws of physics were considerably different from what they are now, but we can still predict some of the behavior. Table 9.3 summarizes the time and temperature of each phase. In the next sections we investigate some of these epochs more closely.

Table 9.3. Temperature and elapsed time of the Universe since its beginning.

Time after Big Bang	Temperature (K)	Event
5.39×10^{-44} s	—	Appearance of space, time and energy
10^{-43} s	10^{31}	Gravity decouples
10^{-35} s	10^{28}	Strong force and electro-weak force decouple
10^{-33} to 10^{-32} s	10^{27}	Inflation
1×10^{-10} s	10^{15}	Electromagnetic and weak force decouple
3×10^{-10} to 5×10^{-6} s	$\sim 10^{13}$	Equilibrium of quarks, antiquarks
6×10^{-6}	1.4×10^{12}	Formation of protons and neutrons
10 s	3.9×10^{9}	Equilibrium of electrons and positrons
3.8 m	9×10^{8}	Formation of ^{2}H, ^{3}He, and ^{4}He nuclei
700,000 y	3000	Electrons captured by nuclei

9.3.1. *Time $t = 0$*

Given the composition of the Universe in terms of matter, radiation, and vacuum energy, and given the isotropy and homogeneity of space, general relativity predicts the evolution of the Universe. Except in the case where the Universe is dominated by a cosmological constant early on in its history, we see that the prediction is that the Universe started expanding from zero size. How this expansion started, and where the matter and other sources of energy density came from, the model says nothing about.

Strictly, the time $t = 0$ is not a part of the model, since the density of the Universe goes to infinity at this point, and then general relativity breaks down. In this sense, the Big Bang model is really a model for how the Universe evolved once the expansion had started. We want to know how that happened, and, similarly, we are also interested in extending our understanding of the Universe all the way back to the beginning.

9.3.2. *Inflation*

Inflation [Gut81] occurs when the vacuum energy contribution dominates the ordinary density and curvature terms in Eq. (2.29). Assuming these are negligible and substituting $\Lambda =$ constant results in

$$R \propto \exp(t) \quad \text{(inflationary Universe).} \tag{9.17}$$

The acceleration is positive, corresponding to an accelerating expanding Universe called an *inflationary Universe.*

We have derived the Friedmann equation without use of *General Relativity.* But, as we shall see in the next sections, the derivation of this equation from General Relativity brings new insights in our understanding of the Universe.

Before $t = 10^{-3}$ s, the Universe was dominated by quantum mechanics. A measure for this point in time is the Compton wavelength of the Universe

$$R_c = \frac{\hbar}{Mc}. \tag{9.18}$$

Equating the Compton wavelength and the Schwartzschild radius (Eq. (2.106)) leads to the so called *Planck mass*

$$M_{Pl} = \left(\frac{\hbar c}{G}\right)^{1/2} = 1.2 \times 10^{19}\,\mathrm{GeV}/c^2, \tag{9.19}$$

which is equivalent to $T = 10^{32}$ K. Via the uncertainty relation one can get a Planck time and length, i.e., $t_{Pl} = 5 \times 10^{-44}$ s and $l_{Pl} = 3 \times 10^{-33}$ cm. At these lengths and times, time itself becomes undefined, as we already mentioned in Chapter 2.

Since the Universe is the ultimate black body, Wien's law, Eq. (4.72) applies, i.e., $T\lambda(= R) = \text{const}$ and therefore $T(t) \sim 1/R(t)$. Then,

$$\rho \sim \left(\frac{1}{R}\right)^n, \quad n = 3, 4. \tag{9.20}$$

This reinserted into Eq. (2.18) results in

$$\frac{dR}{dt} = \kappa R^{1-n/2}, \tag{9.21}$$

which integrated becomes

$$\int_{t_1}^{t_2} \frac{dR}{R^{1+n/2}} = t_1 - t_2 = \frac{2}{n}\,\frac{1}{\kappa R^{-n/2}}\bigg|_{t_1}^{t_2}. \tag{9.22}$$

Replacing κ by Hubble's constant and setting $t_1 = 0$, and $t_2 = t_{Universe} = t_0$, we get

$$t_0 = \frac{2}{n}\frac{1}{H(t)}, \tag{9.23}$$

which relates the age of the Universe to the presently observed Hubble constant. As the Universe is by now matter dominated, one has

$$t_0 = \frac{2}{3H}. \tag{9.24}$$

As we discussed in Chapter 2, the above derivation is valid if $\Lambda = 0$. For a non-zero cosmological constant, we obtained in Eq. (2.56) that the age of the Universe is larger than it would have been obtained were the cosmological constant not present. And there is strong experimental evidence that indicates that the expansion of the Universe is accelerating, so that there is a non-zero cosmological constant.

9.3.3. $t \sim 0.01$ s

At this time, $T \sim 10^{11}$ K, and $kT \sim 10\,\text{MeV} \gg 2m_ec^2$. Neutrinos, electrons and positrons are kept in chemical equilibrium by neutral- and charged-current interactions. The neutral current reaction (Z_0 exchange) has the same strength for all three neutrino flavors (ν_e, ν_μ, ν_τ). The charged-current reaction involves only electron-family particles. The analogous reactions involving muon-family particles do not occur because of the temperature. The mass of two muons exceeds 200 MeV, and thus cannot be produced at this time.

There are similar charge-changing reactions coupling n $\leftrightarrow$ p:

$$\begin{aligned}
&\mathrm{p} + e^- \leftrightarrow \mathrm{n} + \nu_e, \\
&\mathrm{n} + e^+ \leftrightarrow \mathrm{p} + \bar{\nu}_e, \\
&\mathrm{p} \leftrightarrow \mathrm{n} + e^+ + \nu_e, \\
&\mathrm{n} \leftrightarrow \mathrm{p} + e^- + \bar{\nu}_e.
\end{aligned} \tag{9.25}$$

The relative numbers of protons and neutrons follow the Boltzmann distribution, i.e.,

$$\frac{\mathrm{n}}{\mathrm{p}} = \frac{e^{-m_n/kT}}{e^{-m_p/kT}} = e^{-\Delta m/kT}, \tag{9.26}$$

where $m(n) = 939.566\,\text{MeV}$, $m(p) = 938.272\,\text{MeV}$, and $\Delta m = 1.294\,\text{MeV}$. At $T = 10^{11}$ K, $kT = 8.62\,\text{MeV}$. Thus at this temperature, n/p $= 0.86$. Although this temperature is far above the temperature of nucleosynthesis, the n/p ratio has already begun to drop.

9.3.4. $t \sim 0.1$ s

In about 0.1 s the temperature is assumed to have decreased to 3×10^{10} K (corresponding to 2.6 MeV). Before this temperature is reached the ν_μ and ν_τ have fallen out of equilibrium. This occurs because the rate for interactions with electrons is too slow to keep up with the rate of expansion of the Universe. The muon and tauon flavor reactions off electrons and positrons are only about 1/7 as strong as for electron neutrinos. That is why these "heavy flavors" decouple first. At around 3 MeV the ν_e also decouple.

Now, the equilibrium between protons, neutrons, electrons and neutrinos can be written

$$\begin{aligned} \mathrm{p} + \overline{\nu} &= \mathrm{n} + e^+ (Q = -1.80\,\mathrm{MeV}) \quad \text{and} \\ \mathrm{n} + \nu &= \mathrm{p} + e^- (Q = 0.78\,\mathrm{MeV}). \end{aligned} \tag{9.27}$$

The mass of the neutron exceeds that of the proton by a small margin of 0.00013885 u, corresponding to 1.29 MeV; thus the reaction on the left of Eq. (9.27) requires energy, while the reaction on the right of it releases energy. The formation of protons is therefore favored over neutrons, leading to 38% neutrons and 62% protons.

Now we need to know:

- What is the time scale for n $\leftrightarrow$ p reactions?
- What is the time scale describing the expansion of the Universe that forms the comparison scale?

The time scale for the n $\leftrightarrow$ p reactions can be posed as that for a neutron in our thermal bath to convert to a proton via n + $\nu_e \leftrightarrow$ p + e$^-$. That rate is

$$r(T) \sim \langle \sigma v \rangle \eta_\nu(t), \tag{9.28}$$

where η is the electron neutrino number density and $v \sim c$ is the relative velocity of the neutrino and neutron. The cross-section is roughly

$$\sigma \sim G_F^2 E_\nu^2 \sim G_F^2 (kT)^2. \tag{9.29}$$

A typical cross-section for the interaction of a neutrino with any other particle via the weak nuclear force is

$$\sigma_\nu \sim 10^{-47}\,\mathrm{m}^2 \left(\frac{kT}{1\ \mathrm{MeV}}\right)^2. \tag{9.30}$$

(Compare this to the Thomson cross-section for the interaction of electrons via the electromagnetic force: $\sigma_T = 6.65 \times 10^{-29}\,\mathrm{m}^2$.)

The number of states/unit volume for relativistic particles should go like $(kT)^3$. Thus,

$$r(T) \sim (kT)^5 G_F^2. \tag{9.31}$$

G_F has the dimensions of $1/\text{mass}^2$, as it represents the exchange of a W boson: so there is a coupling constant on each end and a propagator that goes like $1/M_W^2$ (since all momenta of the scattering particles are much, much less than M_W). A much easier way to remember is

$$G_F \sim \frac{10^{-5}}{M_N^2}, \tag{9.32}$$

where M_N is the nucleon mass. Thus

$$r(T) \sim (kT)^5 \frac{10^{-10}}{M_N^4}. \tag{9.33}$$

Had we done things more carefully, and written in proper units, the result is [Wein08]

$$r(T) \sim \left(\frac{0.76}{\text{s}}\right) \left(\frac{kT}{\text{MeV}}\right)^5. \tag{9.34}$$

This gives a neutron lifetime at $3 \times 10^{10}\,\text{K}$ of about 0.011 s.

We see that weak rates evolve very rapidly in the early Universe, dropping as T^5. At some point one reaches a T where weak rates are so slow that they cannot keep up with the other changes in the Universe. Furthermore, the transition (range of T) over which this "freezeout" occurs should be rather narrow, since the functional dependence on T is steep.

A timescale for comparison is the age of the Universe at this epoch. One can obtain this estimate from the Hubble parameter, which has dimensions of 1/s and describes the rate of change of the Universe at any given time. According to the Friedmann Equation (2.18), it is given by $\sqrt{8\pi G\rho(t)/3}$ and Eq. (9.13) tells us that $\rho(T) = (N\pi^2/30)(kT)^4 \sim 3.5(kT)^4$. Thus,

$$\text{Hubble rate} \sim 5.4(kT)^2\sqrt{G}. \tag{9.35}$$

The value of G in MeV units is $6.7 \times 10^{-45}\,\text{MeV}^{-2}$. Thus, inserting the needed $\hbar$ we have

$$\text{Hubble rate} \sim 4.4 \times 10^{-22} \left(\frac{kT}{\text{MeV}}\right)^2 \frac{\text{MeV}}{\hbar} \sim \frac{0.67}{\text{s}} \left(\frac{kT}{\text{MeV}}\right)^2, \tag{9.36}$$

or about 0.1 s. This is ~10 times longer than our neutron lifetime meaning that the weak rates are easily keeping the chemical equilibrium between neutrons and protons at this time.

9.3.5. *Decoupling*

The temperature characterizing the epoch of decoupling occurs when the neutron lifetime and the Hubble rate are comparable, which yields

$$\frac{0.67}{\mathrm{s}}\left(\frac{kT}{\mathrm{MeV}}\right)^2 = \frac{0.76}{\mathrm{s}}\left(\frac{kT}{\mathrm{MeV}}\right)^5, \tag{9.37}$$

or $kT \sim 1\,\mathrm{MeV}$.

As temperature and density further decreases the neutrinos begin to behave like free particles, and below $10^{10}\,\mathrm{K}$ they cease to play any active role in the formation sequence (matter became transparent to the neutrinos). The temperature corresponds then to $\sim 1\,\mathrm{MeV}$, i.e., about the threshold energy for formation of positron/electron pairs. Consequently they begin to annihilate each other, leaving, for some reason, a small excess of electrons. Though neutrons and protons may react at this temperature, the thermal energies are still high enough to destroy any heavier nuclides eventually formed.

Because the neutrinos are decoupled, the energy released by $e^{\pm}$ annihilation heats up the photon background relative to the neutrinos. The neutrino entropy must be treated separately from the entropy of interacting particles. If we denote $T_>$, the temperature of photons and $e^{\pm}$ before annihilation, we also have $T_\nu = T_>$ as well. The entropy density of the interacting particles at $T = T_>$ is just

$$s_> = \frac{4}{3}\frac{\rho_>}{T_>} = \frac{4}{3}\left(2 + \frac{7}{2}\right)\left(\frac{\pi^2}{30}\right)T_>^3, \tag{9.38}$$

while at $T = T_<$, the temperature of the photons just after $e^{\pm}$ annihilation, the entropy density is

$$s_< = \frac{4}{3}\frac{\rho_<}{T_<} = \frac{4}{3}(2)\left(\frac{\pi^2}{30}\right)T_<^3, \tag{9.39}$$

and assuming that the number of degrees of freedom remains constant (conservation of entropy) $s_< = s_>$ and $(T_</T_>)^3 = 11/4$. Thus, the photon background is at higher temperature than the neutrinos because the $e^{\pm}$ annihilation energy could not be shared among the neutrinos, and

$$T_\nu = (4/11)^{1/3} T_\gamma \simeq 1.9\,\mathrm{K}. \tag{9.40}$$

Now that we have the temperature at which the n $\leftrightarrow$ p system breaks out of chemical equilibrium, we can evaluate the corresponding n/p ratio

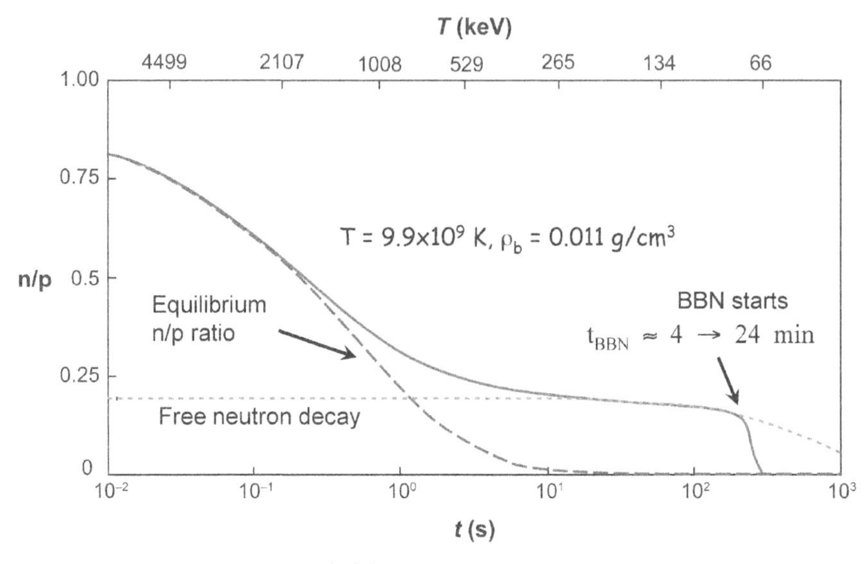

Figure 9.3. Neutron-to-proton (n/p) ratio as a function of time and temperature. The dashed curve is the equilibrium n/p ratio given by $\exp(-\Delta m/kT)$. The dotted curve is due to the free-neutron decay, $\exp(-t/\tau_n)$. The solid curve indicates the resulting n/p variation. The steep decline at a few hundred seconds is the result of the onset of BBN [Stei07].

at this point:

$$\frac{\mathrm{n}}{\mathrm{p}} \sim e^{-\Delta m/kT} \sim 0.25. \tag{9.41}$$

Note this is not 1/7, but it is also not too far from it. And we expect it too continue to drop. As the temperature drops, the ν-induced reactions continue to push things toward the proton-rich side, as shown in Fig. 9.3. In the next 10 s (1 MeV is about 1 s after the Big Bang) the n/p ratio will drop to about $0.17 \sim 1/6$ [Wein08]. And after that there continues to be a slow decrease in the neutron percentage because of neutron β decay — but this has the timescale of 10 minutes.

Nuclei must form in less than 10 minutes, since one cannot have much β decay and still get 1/7 for the n/p ratio. The nucleon gas in the early Universe is relatively dilute, with the consequence that nuclei must be made by two-body reactions. The only two-nucleon bound state is the deuteron, with a binding energy of just 2.24 MeV.

For reasons we do not fully understand, when the very early Universe cooled, it left over a net baryon number. That is, we have neutrons and protons in our Universe, not antineutrons and antiprotons. Presumably at very early times there were approximately equal numbers of quarks and antiquarks, but not precisely equal numbers. As the Universe cooled down, quarks and antiquarks (nucleons and antinucleons) annihilated each other. At the end — for reasons connected with baryon number violation, CP violation, and non-equilibrium physics — a small residual baryon number excess remained. The resulting baryon/photon ratio is

$$\frac{\rho_N}{\rho_\gamma} = \eta \sim 10^{-9}. \tag{9.42}$$

So consider the two possible states of $\mathrm{n}+\mathrm{p}$

$$\mathrm{n}+\mathrm{p} \leftrightarrow \mathrm{d}+\gamma. \tag{9.43}$$

This reaction is electromagnetic and therefore fast, so we can reasonably assume it is in equilibrium in the early Universe. The equilibrium condition is

$$\begin{aligned}&(\text{number of n} + \text{p}) \times (\textit{destruction rate})\\ &\quad\sim (\text{number of d} + \gamma) \times (\textit{destruction rate}).\end{aligned} \tag{9.44}$$

This then yields

$$\frac{\#(\mathrm{n}+\mathrm{p})}{\#(\mathrm{d}+\gamma)} \sim \frac{\rho_d \rho_\gamma e^{-\Delta m/kT}}{\rho_p \rho_n}. \tag{9.45}$$

Now $\rho_p/\rho_\gamma \sim \eta$. The time when deuterium forms can be defined as the time when $\rho_n \sim \rho_d$, that is, when half of the neutrons are in deuterium nuclei. It follows that the epoch of nucleosynthesis is given by

$$e^{-2.24\,\mathrm{MeV}/kT} \sim \eta \sim 10^{-9} \tag{9.46}$$

and thus

$$T_{\text{deuterium}} \sim 1.25 \times 10^9\,\mathrm{K} \quad \text{or} \quad t_{\text{deuterium}} \sim 100\,\mathrm{s}. \tag{9.47}$$

That is, nucleosynthesis starts at about 100 s after the Big Bang. If we followed the n/p from the weak freeze-out (~1 s) to 100 s by a numerical integration we would find

$$\frac{\mathrm{n}}{\mathrm{p}} \sim 0.15 \sim \frac{1}{7}. \tag{9.48}$$

That is, we understand the n/p ratio as a "fingerprint" of the thermal physics of the early Universe.

Modern determinations of the primordial ^{4}He abundance, which is essentially the same as the n/p ratio is, are quite precise. A precise calculation should be able to relate a fundamental and unknown cosmological parameter η to the primordial ^{4}He abundance! (That is how the value of 10^{-9} arises.) Furthermore the larger η is, the higher the temperature T characterizing nucleosynthesis. That is, the smaller the baryon number of the Universe and the later the epoch of nucleosynthesis.

The discussion above shows that the temperature for nucleosynthesis is about 100 keV, when the naive guess might have placed it at about 2 MeV, the binding energy of deuterium. The explanation is the very small value of η. The BBN temperature, ~100 keV, corresponds according to Eqs. (9.14) and (9.15) to timescales less than about 200 s. The typical cross-section and reaction rate for the first link in the nucleosynthetic chain yields

$$\sigma v(\mathrm{p} + \mathrm{n} \rightarrow \mathrm{d} + \gamma) \simeq 5 \times 10^{-20}\,\mathrm{cm}^3/\mathrm{s}. \tag{9.49}$$

This implies that it was necessary to achieve a density $\rho \sim 1/\sigma v t \sim 10^{17}\,\mathrm{cm}^{-3}$.

The density of baryons today is known approximately from the density of visible matter to be $\rho_0 \sim 10^{-7}\,\mathrm{cm}^{-3}$ and since we know that that the density n scales as $R^{-3} \sim T^3$, the temperature today must be $T_0 = (\rho_0/\rho)^{1/3} T_{\mathrm{BBN}} \sim 10\,\mathrm{K}$, which is a very good estimate.

9.3.6. $t \gtrsim 3$ *min*

After 14 s the temperature had decreased to 3×10^9 K (0.27 MeV) and 3 min later to about 10^9 K (<0.1 MeV). Now, with the number of electrons, protons and neutrons about equal (though the Universe mostly consisted of photons and neutrinos), some protons and neutrons reacted to form stable nuclides like deuterium and helium. However, for the deuteron to be stable, the temperature must decrease below the Q-value for its formation, i.e. to about 10^{10} K and, in reality, to the much lower value of about 10^9 K, because of the high photon flux which may dissociate the deuteron into a proton and a neutron. Two deuterium atoms then fuse, probably in several steps as discussed below, to form He. He is an extremely stable nucleus, not easily destroyed, as compared to nuclides with masses >4, whose binding energies (per nucleon) are only a few MeV (Fig. 3.2).

As the Universe expanded, the probability for particle collisions decreased, while the kinetic energy available for fusion reactions was reduced. Therefore, the nucleon build-up in practice stopped with ^{4}He, leading to an average universal composition of 73% hydrogen and 27% helium. A very small amount of deuterium atoms was still left, as well as a minute fraction of heavier atoms, formed by the effects of the "Boltzmann tail" and "quantum tunneling". The remaining free neutrons (half-life 10.4 min) now decayed to protons. The situation about 35 minutes after time zero was then the following: the temperature was 3×10^8 K, density about 10^{-4} kg m^{-3}. The Universe consisted of 69% photons, 31% neutrinos, and a fraction of 10^{-9} of particles consisting of 72–78% hydrogen and of 28–22% helium, and a corresponding number of free electrons, all rapidly expanding in all directions of space.

9.3.7. *$t \gtrsim$ years*

It was still too hot for the electrons to join the hydrogen and helium ions to form neutral atoms. This did not occur until about 500,000 years later, when temperature had dropped to a few 1,000 K. Photons were no longer energetic enough to keep electrons from binding to protons. This marked the time of *recombination*: ionized plasma of electrons and protons (with some Helium nuclei) combined into mostly neutral hydrogen atoms (with some Helium atoms).

The disappearance of free electrons broke the thermal contact between radiation and matter, and radiation continued to expand freely. Photons interact less with bound electrons than free electrons, so radiation began to stream freely after recombination. This marked the event of decoupling of matter and radiation. In the adiabatic expansion of the fireball the radiation cooled further to the cosmic microwave background (CMB) level of 2.7 K measured today. After decoupling, radiation has been freely streaming through Universe. The CMB is the redshifted remnant of radiation that was last in contact with matter at $z \sim 1000$.

Matter also began to evolve freely at that time; when it finally became cool enough, galaxies began to form. The recent observation that the cosmic background radiation shows ripples in intensity in various directions of space indicates a slightly uneven ejection of matter into space, allowing gravitational forces to act, condensing the denser cloud parts into even more dense regions, or "islands", which with time separated from each other, leaving seemingly empty space in between. Within these clouds, or proto-galaxies, local higher densities lead to the formation of stars.

9.4. BBN Nucleosynthesis

A detailed modeling of nucleosynthesis uses a system of differential mass fractions equations involving all of the abundances and reaction rates. The rates of nuclear reactions are parameterizations of experimental data. An example of predicted mass fractions versus temperature and time is shown in Fig. 9.5.

We have discussed above the reaction n + p $\rightarrow$ d + γ, which lead us to the conclusion that the neutron to proton ratio is n/p $\sim$ 1/7. After the

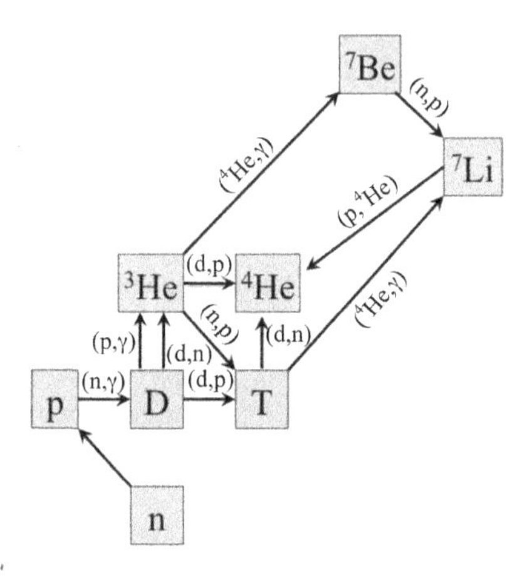

Figure 9.4. The main nuclear reaction chains for Big Bang nucleosynthesis.

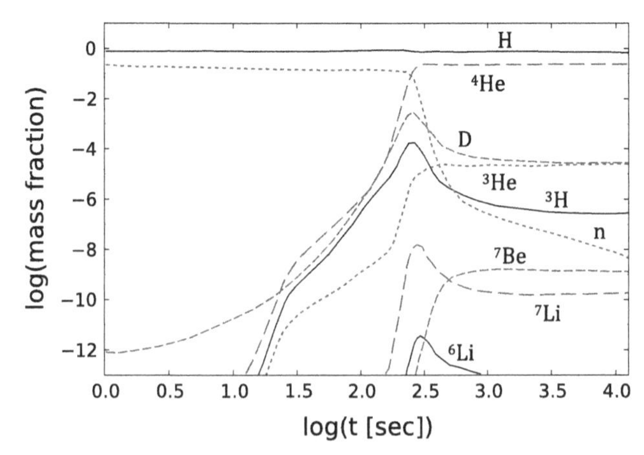

Figure 9.5. Evolution of the mass and number fractions of primordial elements. ^{4}He is given as a mass fraction while the other elements are presented as number fractions.

deuteron is formed, the reactions

$$\begin{gathered} d(n,\gamma)^3H, \quad d(d,p)^3H, \\ d(p,\gamma)^3He, \quad d(d,n)^3He, \\ {}^3He(n,p)^3H \end{gathered} \tag{9.50}$$

follow. At long times, ^{3}H β-decays to ^{3}He.

What follows next are the reactions

$$\begin{gathered} {}^3H(p,\gamma)^4He, \quad {}^3H(d,n)^4He, \\ {}^3He(n,\gamma)^4He, \quad {}^3He(d,p)^4He, \quad {}^3He({}^3He,2p)^4He. \end{gathered} \tag{9.51}$$

These occur at center of mass energies around $\sim$100 keV and are strongly suppressed by Coulomb repulsion.

Trace amounts of ^{7}Li and ^{7}Be are then generated through the reactions

$${}^4He({}^3H,\gamma)^7Li, \quad {}^4He({}^3He,\gamma)^7Be. \tag{9.52}$$

^{7}Be later decays to ^{7}Li, but only when it becomes an atom. That is, after recombination.

The usual explanation for the termination of the reaction network with the species mentioned above is that there are no stable nuclei with $A = 5$ and $A = 8$. Thus the obvious potential reactions for going further, ^{4}He + ^{4}He and ^{4}He + p, are ineffective. Actually, this common explanation is not quite true. If one cheats and makes up stable isotopes at these mass numbers with modest binding energies, the chain still largely terminates as above. The reason is that the Coulomb barriers at these low temperatures (100 keV) become increasingly hard to penetrate as Z increases.

The repulsive Coulomb barrier near a nucleus is $\alpha Z_1 Z_2/r \sim 1.44 Z_1 Z_2$ (1 fm)$/r$. Hence, if $Z_1 = Z_2 = 2$ and $r \sim 2$ fm, this is about 3 MeV. So it is very hard for charged particles with kinetic energies of $\sim$100 keV to tunnel through the Coulomb barrier to the regions where the strong nuclear force can bind the reacting particles to form a new nucleus.

Once the bottleneck to forming deuterium is broken, the rest of the network described above proceeds quickly to produce ^{4}He.

A few more comments are noteworthy:

1. The larger η is, the larger $T_{nucleosynthesis}$, and thus the larger the n/p ratio. Therefore larger η leads to larger ^{4}He abundance.
2. One can think of d, ^{3}He as "catalysts" in the network: they are produced and then consumed, and thus reach an equilibrium value

that depends on the competition between production and consumption. What happens if $T_{nucleosynthesis}$ is increased? The production channel is effectively $\mathrm{n} + \mathrm{p}$, where there is no Coulomb barrier. Destruction channels include reactions like $\mathrm{d} + \mathrm{d}$, $\mathrm{d} + \mathrm{p}$, etc, which are Coulomb inhibited. So increasing T affects the destruction channels more, since Coulomb barrier penetration is exponential, enhancing the destruction. The conclusion is that higher $T_{nucleosynthesis}$ should produce lower d, ^{3}He. ^{3}He is produced (10^{-5}/H) (see Fig. 9.6) in low mass stars and should slowly enrich over the history of the Universe.

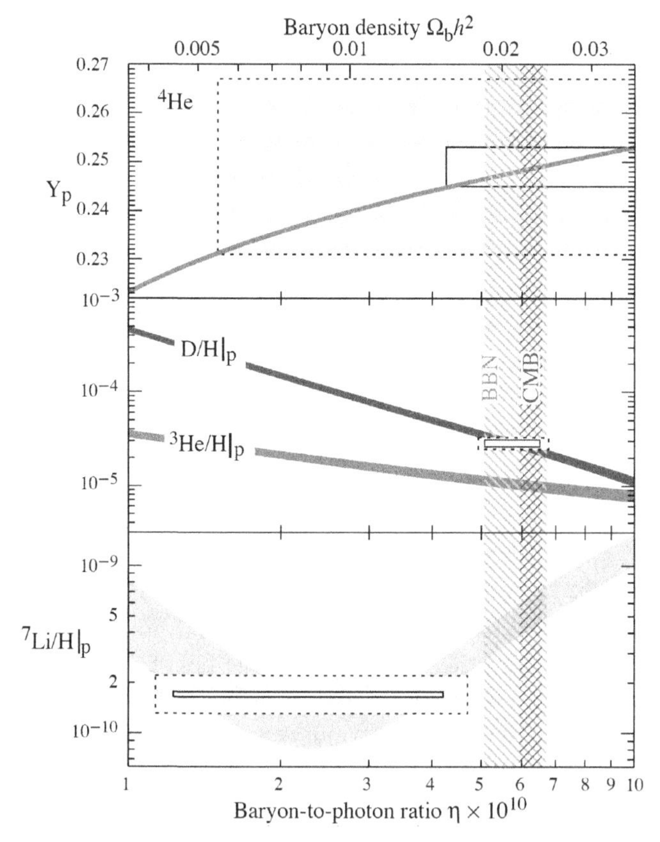

Figure 9.6. BBN abundance yields vs. baryon density for a homogeneous Universe. The bands show the 95% confidence level (CL) range. Boxes indicate the observed light element abundances (smaller boxes: $\pm 2\sigma$ statistical errors; larger boxes: $\pm 2\sigma$ statistical and systematic errors). The narrow vertical band indicates the CMB measure of the cosmic baryon density, while the wider band indicates the concordance range of direct measurements of the light element abundances [SA09].

3. For low η (and therefore low T) ^{7}Li is made and destroyed by:

$$^4\text{He}(^3\text{H},\gamma)^7\text{Li} \quad \text{and} \quad ^7\text{Li}(\text{p},\alpha)^4\text{He}. \tag{9.53}$$

The second reaction has the more effective Coulomb barriers (affecting both initial and final states). Thus a lower T will more effectively turn off the destruction of ^{7}Li than its production, and therefore a lower T (with low η) means more ^{7}Li.

But it turns out that for high η, the primary way to make ^{7}Li changes. High η means more ^{3}He, as we have noted, so

$$^4\text{He}(^3\text{He},\gamma)^7\text{Be} \tag{9.54}$$

becomes important. This production clearly benefits from high T because of the Coulomb barrier. Furthermore there are very few neutrons around to kill this isotope by (n,α). Thus, ^{7}Li produced as ^{7}Be begins to turn up again at high T and high η.

According to BBN calculations (Fig. 9.6), ^{7}Li/H is required to be $\simeq 5\times10^{-10}$. Observation in low metallicity population II halo stars shows about 1.5×10^{-10}. The missing factor can be explained by modeling of stellar atmospheres. Outside the solar system and the local interstellar medium (which are all chemically evolved), lithium has been observed in the absorption spectra of very old, very metal-poor stars in the halo of the galaxy or in similarly metal-poor galactic globular cluster stars. These metal-poor targets are ideal for probing the primordial abundance of lithium. Observational data suggest that although lithium may have been destroyed within many stars, the overall trend is that its galactic abundance has increased with time. Therefore, to probe the BBN yield of ^{7}Li, the key data are from the oldest, most metal-poor halo or GGC stars in the galaxy.

Unlike ^{7}Li, ^{6}Li is predicted to be formed at a very low level in Big Bang nucleosynthesis, $^6\text{Li/H} = 10^{-14}$. Whereas most elements are produced by stellar nucleosynthesis, lithium is mainly destroyed in stellar interiors by thermonuclear reactions with protons. In fact, ^{6}Li is rapidly consumed at stellar temperatures higher than 2×10^6 K. The major source of ^{6}Li has been thought for decades to be the interaction of galactic cosmic rays with the interstellar medium. The low energy capture reaction $^6\text{Li}(\text{p},\gamma)^7\text{Be}$ plays an important role in the consumption of ^{6}Li and formation of ^{7}Be.

For detailed calculations of BBN one needs a nuclear reaction network tracing the density and temperature dependence of the Big Bang. Some reactions in such a network are shown in Fig. 9.4. Some of the nuclear reactions involved will be discussed for hydrogen burning of stars (pp chain), as they are identical. However, nuclear reactions in the Big Bang happen at higher temperatures than in stars. Only p, d, ^{3}He, ^{4}He and ^{7}Li are observed with some certainty as primordial. The primordial concentration of these isotopes puts restrictions on the baryon density of the Universe. The baryon density can be varied, and it is somewhat Hubble constant dependent. The result is the plot shown in Fig. 9.6.

9.5. Cosmological Issues

9.5.1. *Dark matter and dark energy*

The need for dark matter was originally pointed out in 1933 [Zw33] when it was realized that the velocities of individual galaxies located within the Coma cluster were quite large, and that this cluster would be gravitationally bound only if its total mass substantially exceeded the sum of the masses of its component galaxies. For clusters which have relaxed to dynamical equilibrium the mean kinetic and potential energies are related by the virial theorem $K + U/2 = 0$ where $U = -GM^2/R$ is the potential energy of a cluster of radius R, $K \simeq 3M\langle v_r^2\rangle/2$ is the kinetic energy and $\langle v^2\rangle^{1/2}$ is the dispersion in the line-of-sight velocity of cluster galaxies. (Clusters in the Abell catalogue typically have $R = 1.5h^{-1}$ Mpc.) This relation allows us to infer the mean gravitational potential energy if the kinetic energy is accurately known. The mass-to-light ratio in clusters can be as large as $M/L \simeq 300M_\odot/L_\odot$. However since most of the mass in clusters is in the form of hot, X-ray emitting intracluster gas, the extent of dark matter in these objects is estimated to be $M/M_{lum} \simeq 20$, where M_{lum} is the total mass of luminous matter including stars and gas.

In individual galaxies the presence of dark matter has been convincingly established through the use of Kepler's third law $v = \sqrt{GM(r)/r}$ to determine the rotation curve $v(r)$ at a given radial distance from the galactic center. Observations of galaxies taken at distances large enough for there to be no luminous galactic component indicate that, instead of declining at the expected rate $v \simeq r^{-1/2}$ if $M \simeq$ constant, the velocity curves flattened out to $v \simeq$ constant implying $M(r) \propto r$ (see Fig. 9.7). This observation suggests that the mass of galaxies continues to grow even when there is no luminous component to account for this increase. Velocity curves

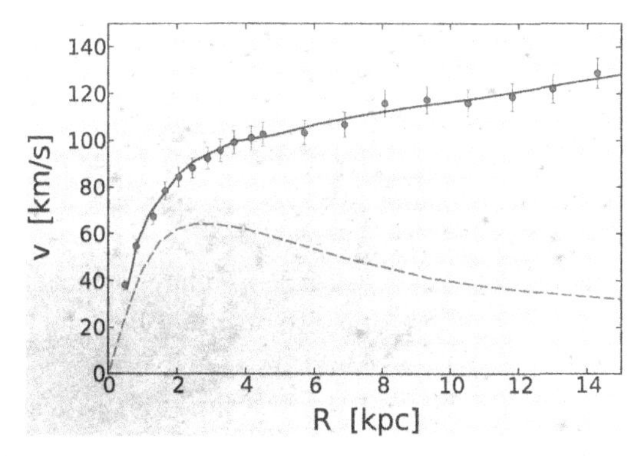

Figure 9.7. The observed rotation curve of the dwarf spiral galaxy M33 extends considerably beyond its optical image (shown superimposed).

have been compiled for over 1000 spiral galaxies usually by measuring the 21 cm emission line from neutral hydrogen (HI). The results indicate that $M/L = (10 - 20)M_\odot/L_\odot$ in spiral galaxies and in ellipticals, while this ratio can increase to $M/L \simeq (200 - 600)M_\odot/L_\odot$ in low surface brightness galaxies and in dwarfs.

Present astronomical observations seem to give the abundance D/H $\simeq 3 \times 10^{-5}$. This implies that $\Omega_b \simeq 0.02\,h^{-2}$, where h is the Hubble constant in terms of $100\,\mathrm{km\,s^{-1}\,Mpc^{-1}}$ (Fig. 9.6). So with the current value of $h \simeq 0.7$ the baryonic density is about 3% of the needed density to close the Universe. Thus, there is not enough baryonic matter to account for the dynamical observation of gravitational effects in galaxies, clusters of galaxies and for results from cosmic background radiation. Therefore *dark non-baryonic matter* has been postulated to exist. Non-baryonic dark matter can interact only by the weak (not necessary, but possible) and gravitational interactions. To produce the gravity effects observed in galaxies and clusters, matter bound to these systems is needed (cold dark matter). In addition, the Universe may be filled homogeneously with light particles with non-zero mass (hot dark matter). Non-baryonic hot dark matter (particles are assumed to have decoupled from the rest of matter/radiation when they were relativistic and so have a very large velocity dispersion — hence called "hot"). Cold dark matter particles, on the other hand, have a very small velocity dispersion and decouple from the rest of matter/radiation when they are non-relativistic.

MACHOs are not possible candidates for dark matter due to the constraint on baryon number from cosmic nucleosynthesis. But they contribute to a part of the invisible Universe (see discussion at the end of

Section 6.5). An example is brown dwarfs, which are objects whose mass is between twice that of Jupiter and the lower mass limit for nuclear reactions (8% of the mass of our Sun). Brown dwarfs are failed stars with insufficient density to start nuclear fusion. A better possibility to explain dark matter are much smaller particles, called *Weakly Interacting Massive Particles* (WIMPs). WIMPs are elementary particles with very tiny interactions with ordinary matter (likely only the weak force). They do not emit or absorb photons. With hardly any interactions, they would be very hard to detect. But they have gravity. Neutrinos could in principle fit the bill, but their mass is too low. The WIMPS needed are a kind of particle that has not been discovered yet.

Observations suggest that the expansion of the Universe is speeding up rather than slowing down [Rie98,Per99]. The case for an accelerating Universe also receives independent support from CMB and large scale structure studies. All three data sets can be simultaneously satisfied if one postulates that the dominant component of the Universe is relatively smooth and has a large negative pressure, the dark energy. The simplest example of dark energy is a cosmological constant, introduced in Chapter 2. As we discussed before, the Friedmann Equation (2.23), can be written as

$$\ddot{a} = -\frac{4\pi G}{3} a\rho_m + \frac{\Lambda a}{3}, \tag{9.55}$$

which can also be rewritten in the form of a force law (with $M = 4\pi a^3 \rho_m/3$),

$$F_{\text{rep}} = -\frac{GM}{a^2} + \frac{\Lambda a}{3}, \tag{9.56}$$

which demonstrates that the cosmological constant gives rise to a repulsive force whose value increases with distance. The repulsive nature of Λ could be responsible for the acceleration of the Universe.

Although introduced into physics in 1917, the physical basis for a cosmological constant remained a bit of a mystery until the 1960s, when it was realized that zero-point vacuum fluctuations must respect Lorenz invariance and therefore have the form $\langle T_{ik} \rangle = \Lambda g_{ik}$ [Zel68]. As it turns out, the vacuum expectation value of the energy momentum is divergent for both bosonic and fermionic fields, and this gives rise to what is known as the cosmological constant problem. Indeed the effective cosmological constant generated by vacuum fluctuations is

$$\frac{\Lambda}{8\pi G} = \langle T_{00} \rangle_{vac} \propto \int_0^\infty \sqrt{k^2 + m^2} k^4 dk. \tag{9.57}$$

Since the integral diverges as k^4 one gets an infinite value for the vacuum energy. Even if one chooses to "regularize" the above integral by imposing an ultraviolet cutoff at the Planck scale, one is still left with an enormously large value for the vacuum energy $\langle T_{00}\rangle_{vac} \simeq c^5/G^2\hbar \sim 10^{76}\,\text{GeV}^4$ which is 123 orders of magnitude larger than the currently observed $\rho_\Lambda \simeq 10^{-47}\,\text{GeV}^4$. This is considered the worst prediction in theoretical physics!

9.5.2. *Inhomogeneous Big Bang*

Basic assumptions of the standard Big Bang model are that the Universe is isotropic and homogeneous. What happens if one gives up on these conditions? The idea is based on the results of quantum chromodynamics which predicts a phase transition from a *quark–gluon plasma* to normal matter around $T_{crit} = 170\,\text{MeV}$ (see Fig. 5.9). Following this, to overcome some of the difficulties of BNN and to produce more baryonic matter, *inhomogeneous Big Bang models* have been proposed. In these models the density is not homogeneous, but lumpy. This happens in three steps.

(a) Hadronic (H) and quark–gluon plasma (QGP) phase exist in equilibrium,
(b) When the temperature of the Universe, due to its expansion, drops below T_{crit}, the QGP regions transform to hadronic bubbles (see Fig. 9.8). The energy released by this transition keeps the Universe at T_{crit}. At some point the released energy is insufficient to balance the temperature reduction due to the expansion. The two phases (QGP and H regions) are then no longer in equilibrium of pressure and chemical potential. The phases decouple.
(c) With decreasing temperatures, the remaining QGP regions transform to hadronic matter, however at higher densities. During equilibrium the baryon number density was much larger in the QGP regions than in the H regions as it is statistically easier to make nearly massless quarks than heavy baryons; former QGP regions have high baryon densities while former H regions have low baryon densities. The distance between different bubbles of high density might have been 1–10 m at T_{crit}; at the beginning of BBN these distances have expanded to $10^3 - 10^4\,\text{m}$ ($T \sim 100\,\text{keV}$). As neutrons are chargeless, they move "freely" in the plasma and have a much larger mean-free path than protons.

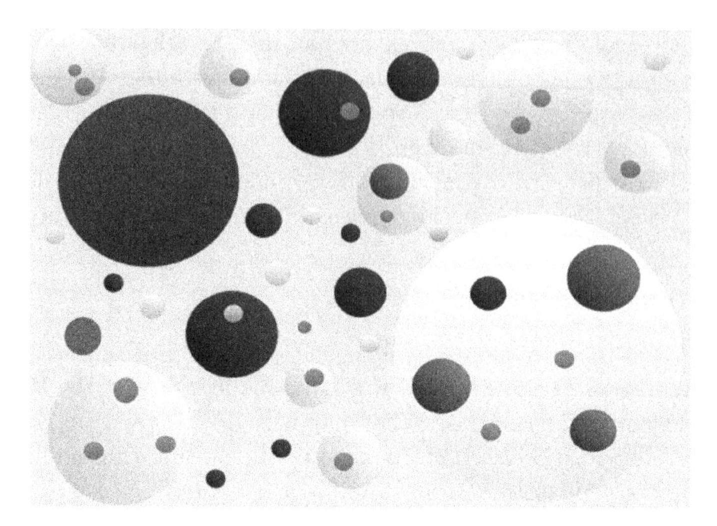

Figure 9.8. In the inhomogeneous Big Bang model, when the temperature of the Universe drops below T_{crit} (due to its expansion), the QGP regions transform to hadronic bubbles. The energy released by this transition keeps the Universe at T_{crit}.

(d) When BBN starts, the neutron number density is homogeneous in the entire Universe, while protons are mainly confined to the high-density regions. p/n is different in high-density and low-density regions. This leads to a different BBN chain.

Nucleosynthesis in inhomogeneous Big Bang models are considerably dependent on neutron capture reactions on light nuclei. Such reactions are also of crucial relevance for s-process nucleosynthesis in red giant stars. To determine the reaction rates for such different temperature conditions, the neutron capture cross-sections need to be known for a wide energy range. A possible signature of baryon-number-inhomogeneous Big Bang is the presence of a high primordial lithium abundance, or a high abundance of beryllium and boron isotopes. As previously mentioned, inhomogeneous Big Bang models involve chain reactions such as [App88] ^{1}H(n, γ)^{2}H(n, γ)^{3}H(d,n)^{4}He(t, γ)^{7}Li(n, γ)^{8}Li, and ^{8}Li(α, n)^{11}B (n, γ)^{12}B(β^-)^{12}C(n, γ)^{13}C, etc., paving the way to heavier nuclei.

9.5.3. *The horizon and the flatness problem*

Consider a galaxy ten billion light years away from us in one direction, and another in the opposite direction. The total distance between them is twenty billion light years. Light from the first has not yet reached the second, because the Universe is only 13.7 billion years old. In a more general sense, there are portions of the Universe that are visible to us, but invisible to each other, outside each other's respective particle horizons. The two galaxies in question cannot have shared any sort of information; they are not in "causal contact". One would expect, then, that their physical properties would be different, and more generally, that the Universe as a whole would have varying properties in different areas.

Contrary to this expectation, the Universe is in fact extremely homogeneous. For instance, the CMB, which fills the Universe, is almost precisely the same temperature everywhere in the sky, about 2.725 K. This presents a serious problem if the Universe had started with even slightly different temperatures in different areas, then there would simply be no way it could have evened itself out to a common temperature by this point in time. Quantum physics demands that this initial temperature difference should have actually existed at the Big Bang due to the uncertainty principle, such that there is no way that the Universe could have formed with precisely the same properties everywhere.

According to the Big Bang model, as the density of the Universe dropped while it expanded, it eventually reached a point where photons decoupled from the plasma and spread out into the Universe as a burst of light. This is thought to have occurred about 500,000 years after the Big Bang. The volume of any possible information exchange at that time was 900,000 light years across, using the speed of light and the rate of expansion of space in the early Universe. Instead, the entire sky has the same temperature, a volume 1000 times larger.

Inflationary theory, discussed in Chapter 2, allows for a solution to the problem by positing a short 10^{-32} s period of exponential expansion within the first minute or so of the history of the Universe (see Fig. 9.9). During inflation, the Universe would have increased in size by an enormous factor, a factor of 10^{26}. Assume that very early the cosmological constant Λ is not zero, but at a rather high value, such that $\rho_{vac} \gg \rho_{rad}$. This density or stored energy may turn itself rapidly into space so that

$$R(t) = R(t_0)\exp[H_*(t - t_0)], \quad T(t) = T(t_0)\exp[-H_*(t - t_0)], \qquad (9.58)$$

where

$$H_* = \left(\frac{8\pi G}{3}\right)^{1/2} \rho_{vac}^{1/2}. \qquad (9.59)$$

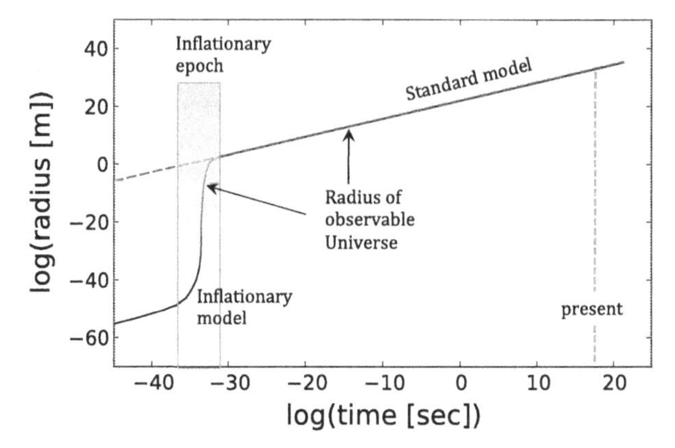

Figure 9.9. The expansion of Universe during the brief "inflation era" was much greater than the present expansion of the Universe.

If it happened at Grand Unification Theory (GUT) time (arbitrary choice!) one would need about a factor 10^{26} to connect the horizon and the size of the Universe ($\rightarrow H_*\Delta t \simeq 60$). As we've seen in Chapter 2, for a radiation dominated Universe, Eq. (2.57) yields

$$\frac{R(t_1)}{R(t_2)} = \left(\frac{t_1}{t_2}\right)^{1/2}. \tag{9.60}$$

As you increase energy the four different interactions (gravity, weak, electromagnetic and strong) start to look more and more equal, i.e., the energy dependent coupling constants intersect. This has been shown experimentally for the electromagnetic (a unified theory in itself) and the weak interaction. For these two interactions this happens at a bit more than 100 GeV. The mediator of the electroweak interaction are γ, Z_0, and $W_\pm$, all of which are, of course, vector bosons. Similarly it is suspected that at about 10^{15} GeV the electroweak and the strong coupling constant agree. That is the Grand Unified Theory (GUT). Again, mediating vector bosons "X_i" (i at least equal to 8) would exist. Such particles could change leptons (only electroweak) into baryons (strong, electroweak), and vice versa. This would allow decays like $\text{p} \rightarrow e^+ + \pi_0$. Above 10^{15} GeV in temperature these particles X_i would form the main mass of the Universe.

It can be speculated that the decay of these particles violates charge–parity (CP) symmetry. There is indeed one particle/antiparticle decay known where CP violation has been observed, i.e., the neutral kaon system.

$$K^0 \rightarrow e^+ + \pi^- \nu_e, \quad \bar{K}^0 \rightarrow e^- + \pi^+ \bar{\nu}_e. \tag{9.61}$$

Present observations have revealed the *flatness problem*: why is the density of the Universe so close to the critical density ρ_{crit}? Or why is $\Omega = \rho/\rho_{crit}$ so close to 1? The scaling law for Ω leads to

$$\Omega(t_2) = 1 + [\Omega(t_1) - 1]\frac{t_2}{t_1}. \tag{9.62}$$

If one scales between the GUT time (t_1) and now ($\simeq 10^{10}\,\mathrm{y} = t_2$), $\Omega(t_1)$ has to be tuned to 10^{-53} digits, unless $\Omega(t_1) = \Omega(t_2) = 1$ for reasons of principle. The Universe has been found to be very homogeneous. However, using Eq. (9.60) reveals, e.g., for the GUT time $R(t_1) = 10\,\mathrm{cm}$, while the causality length, i.e., the length by which domains of space can communicate with each other is $L_c = 3 \times 10^{10} \times 10^{-35}\,\mathrm{cm} = 3 \times 10^{-25}\,\mathrm{cm}$. That means space at that time is fractured into $N = (10/3 \times 10^{-25})^3 = 3.7 \times 10^{76}$ regions not connected by causality. Why is it then so homogeneous?

Combining Eqs. (9.60) and (9.62) results in $\Omega(t_2) = 1 \pm 10^{-52}$ independent of the initial matter and radiation density of the Universe. The effect of inflation is therefore: an enormous growth size, an enormous cooling down, $\Omega = 1$, and the creation of lots of radiation density which will partially turn into matter density.

If correct, inflation solves the horizon problem by suggesting that prior to the inflationary period the entire Universe was causally connected, and it was during this period that the physical properties evened out. Inflation then expanded it rapidly, freezing in these properties all over the sky; at this point the Universe would be forced to be almost perfectly homogeneous, as the information needed to change it from that state was no longer causally connected. In the modern era distant areas in the sky appear to be unconnected causally, but in fact were much closer together in the past. Inflation also solves the flatness problem: An inflated curved space has to look flat! As a model for inflation a two potential model (like for a supercooled fluid) may be chosen, where for Λ_1 one minimum is present and for Λ_2 another one, as shown in Fig. 9.10. It must be assumed then that either for unknown reasons the Universe settled itself first in the higher potential or that Λ itself is time dependent.

9.5.4. *Matter–antimatter asymmetry in the universe*

When Universe cooled, nucleons/antinucleons and electrons/positrons should have completely annihilated. From where comes the matter excess? Several observational facts on antimatter are well known: (a) Antimatter

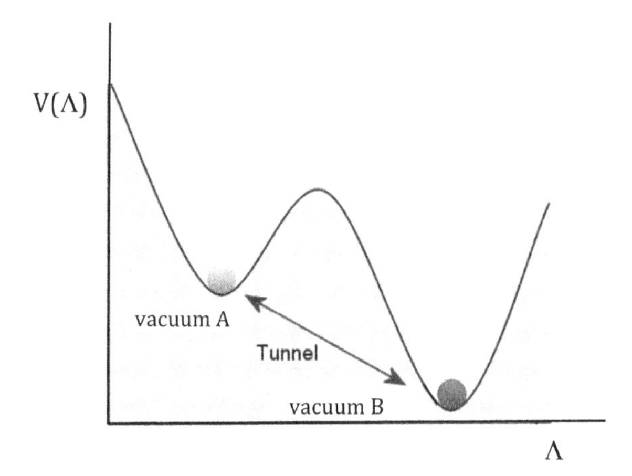

Figure 9.10. Two potential model of inflation with quantum tunneling between the potential minima.

is rare on Earth; anti-atoms were first produced in a laboratory (CERN) in 1996 [Ba96]. (b) Antiprotons are seen in cosmic rays at 10^{-4} of protons; they are secondary products from cosmic ray collisions with the interstellar medium. Cosmic rays are proof for baryon–antibaryon asymmetry on galactic scales. (c) If matter and antimatter galaxies existed in the same galaxy cluster there should be strong γ-ray emission from N$\bar{\text{N}}$ annihilation. The absence of such γ-ray flux is evidence that such clusters of galaxies $(10^{13} - 10^{14}\,M_\odot)$ are either all baryons or all antibaryons. Note that in a baryon-symmetric Universe (not ours!) n and $\bar{\text{n}}$ would annihilate and at $T \sim 22\,\text{MeV}$ the baryon-to-photon ratio would be about 10^{-19}, over 10^9 less than observed. The evidence against primary forms of antimatter in the Universe is quite strong. The observed number of baryons in the Universe, over the number of photons, is $\eta = n_b/n_\gamma \sim 10^{-10}$.

The most reasonable assumption is that the very early Universe $T > 38\,\text{MeV}$ possessed an asymmetry between n_b and $n_{\hat{b}}$ which prevented the annihilation catastrophe. There are three basic ingredients necessary to generate a non-zero baryon number from an initially baryon-symmetric state: (a) baryon (and lepton) number violation, (b) $\mathcal{C}$ and $\mathcal{CP}$ violation (charge conjugation $\mathcal{C}$, parity $\mathcal{P}$) and (c) non-equilibrium conditions.

Baryon number violation — There must obviously be a violation of baryon number. If baryon number is to be conserved in all interactions, the present asymmetry can only come from initial conditions. Grand Unified Theories (GUT) predict violation of baryon and lepton number conservation. For example GUT predicts that protons decay; but the predicted lifetimes are already ruled out by experiment.

$\mathcal{CP}$ violation — Even in the presence of non-baryon number conserving interactions, no baryon asymmetry would develop if charge conjugation $\mathcal{C}$ and $\mathcal{CP}$ ($\mathcal{P}$ = parity conjugation are conserved: in the absence of a preference for matter or antimatter, even baryon number non-conserving reactions would produce matter and antimatter at the same rate; no asymmetry would develop [Sak67]. $\mathcal{P}$ is maximally violated in weak interactions; $\mathcal{CP}$ violation is observed in neutral kaons.

Conventionally e^-, π^-, and ν_e in Eq. (9.61) are considered matter. The first decay (K_0) happens 0.7% more often. The neutral kaon exists in a strange state, mixing itself with its antiparticle. Thus, matter–antimatter symmetry ($\mathcal{CP}$) is broken. As $\mathcal{CPT}$ symmetry must be conserved, time reversal symmetry is also broken. $\mathcal{CP}$ violations in the K_0 decay is, however, part of the standard model of elementary particle physics and not sufficient to explain matter-antimatter symmetry. If, however, the X particle in GUT would have only a 0.3% matter–antimatter asymmetry in decay, that would be sufficient and would also explain the baryon/photon ratio of 10^{-9} as the final outcome of this asymmetry.

Non-equilibrium conditions — In chemical equilibrium the entropy is maximal when the chemical potentials associated with all non-conserved quantum numbers vanish. Further, particle and antiparticle masses are guaranteed to be equal by $\mathcal{CPT}$ invariance. Thus, in thermal equilibrium the phase space density of baryons and antibaryons, given by $[1 + \exp (p^2 + m^2)^{1/2}/kT]^{-1}$, are necessarily identical, implying $n_B = n_{\bar{B}}$. The necessary non-equilibrium conditions are provided by the expansion of the Universe, as departures from equilibrium can occur if the expansion rate is faster than the particle interaction rates.

9.5.5. *The demise of the universe*

Here we discuss some possibilities for the ultimate fate of the Universe [AL97,KS00].

The big rip

If dark energy prevails and increases in time, the rate of acceleration of the Universe will increase and all matter, from galaxies to atoms, will eventually disintegrate into unbound particles and radiation. In other words, they will be *ripped apart* by the repulsive energy force. In this scenario, the Universe will end up in a singularity, as the dark energy density becomes infinite.

The big freeze

All stars will eventually burn out. They will either become black dwarfs, neutron stars, or black holes. White dwarfs eventually become black dwarfs when they cool — if they have mass less than 1.4 $M_\odot$. If they are heavier, they collapse and become neutron stars if their mass is between 1.4 and 2 $M_\odot$, and they become black holes if they are more massive. The black holes will swallow some of the other stars they encounter. Most of the stars, as well as interstellar gas and dust, will eventually be hurled into intergalactic space whenever it accidentally reaches escape velocity through its random encounters with other stars.

White dwarfs cool to black dwarfs with a temperature of at most 5 K in about 10^{17} years, and the galaxies will boil away stars in about 10^{19} years. In 10^{23} years the dead stars will actually boil off from the galactic clusters, so that the clusters will disintegrate. In 10^{23} years the cosmic background radiation will have cooled down to about 10^{-13} K. As the Universe expands, mass spread out to the point where each particle, or chunk of matter, is completely alone in the vastness of space.

All matter (and atoms) tend to "ionize" gradually losing their atoms (or electrons) and protons, despite the low temperature. Just give them enough time. If the Universe is expanding exponentially thanks to a nonzero cosmological constant, the density of matter goes to zero. But the temperature does not go to zero. It approaches a particular non-zero value. So all forms of matter made from protons, neutrons and electrons will ionize at some point.

In an accelerated expanding universe, each pair of inertial observers will lose contact, because they get redshifted out of sight. This is called a "cosmological horizon" and, like the horizon of a black hole, the cosmological horizon emits thermal radiation at a specific temperature: the Hawking radiation. Its temperature depends on the value of the cosmological constant. Thus, given a low enough density of matter, the Hawking temperature is enough to ionize all forms of matter, given enough time! A solar mass black hole will evaporate in about 10^{67} years due to the Hawking radiation (discussed in Chapter 2), much larger than the age of the Universe. As black holes evaporate, they will emit photons and other particles, and for a while there will be some radiation around. This process will eventually cease. Neutron stars will at some point quantum-tunnel and become black holes, which then Hawking-radiate away.

Thus, in a far future everything consists of isolated stable particles: electrons, neutrinos, and protons (unless protons decay). The density of these particles will go to zero, and each one will lose contact from all the rest by a cosmological horizon, making them unable to interact. The photons will come into thermal equilibrium forming blackbody radiation at the temperature of the cosmological horizon — about 10^{-30} K.

The big crunch and big bounce

This theory assumes that the average density of the universe is enough to stop its expansion and begin contracting. The Universe would then collapse into a dimensionless singularity. Thus, the Big Bang could have occurred immediately after the big crunch of a preceding Universe. This model leads to an oscillatory Universe scenario: big crunch followed by Big Bang, and so on.

Multiverses and parallel Universes

In this theory, our Universe is just one of several among other parallel Universes. The vacuum in one Universe can tunnel to a lower vacuum state in another Universe. The end result is that there will many still-expanding Universes where their vacuum has not yet decayed, while some others might end up in a big freeze or big rip. New Universes can be produced in this way, so the Multiverse (or system of parallel Universes) as a whole never ends.

9.6. Exercises

1. Why is the epoch of recombination also called the decoupling epoch?
2. Calculate the number of degrees of freedom relevant to the very early high temperature universe, assuming the relativistic particles are the muons, electrons, photons, and three types of neutrinos, in units where the photon yields a weight of 2.
3. The entropy per degree of freedom of a particle species at temperature T is given by

$$s = \frac{2k\pi^2}{45} g\alpha(kT)^3, \tag{9.63}$$

where k is Boltzmann's constant, $\alpha = 1$ for bosons and $\alpha = 7/8$ for fermions. Photons have $g = 2$ degrees of freedom (two possible polarizations) and so do electrons and positrons (two spin states for each).
For high enough temperatures, the photons are in thermal equilibrium with a plasma of electrons and positrons. When the Universe cools the electrons and positrons annihilate into photons. Assuming this process happens at constant entropy, so that $s(g_i, T_i) = s(g_f, T_f)$ where i and f correspond to before (initial) and after (final) the electrons and positrons annihilate, how much are the photons heated compared to the neutrinos (which are not heated, as they had already decoupled from the photons as discussed in this chapter)? Given that the cosmic microwave background temperature is 2.728 K, what temperature do you predict the cosmic neutrino background to be at?

4. Explain how the anisotropy of the microwave background yields information about the physical conditions in the Universe at the moment of decoupling.
5. Suppose the neutron decay time were $\tau_n = 89\,\mathrm{s}$ instead of $\tau_n = 890\,\mathrm{s}$, with all other physical parameters unchanged. Estimate Y_{max}, the maximum possible mass fraction in ^{4}He, assuming that all available neutrons are incorporated into ^{4}He nuclei.
6. Suppose the difference in rest energy of the neutron and proton were $\Delta_{np} = (m_n - m_p)c^2 = 0.129\,\mathrm{MeV}$ instead of $\Delta_{np} = 1.29\,\mathrm{MeV}$, with all other physical parameters unchanged. Estimate Y_{max}, the maximum possible mass fraction in ^{4}He, assuming that all available neutrons are incorporated into ^{4}He nuclei.
7. Show that in the early Universe, when photons, electrons, neutrinos and their antiparticles coexisted in equilibrium, time and temperature were related by $tT^2 \simeq 0.74\,\mathrm{s\,MeV}^2$.
8. Use the Saha equation to calculate the temperature of recombination, which you can define as the time when there is an equal amount of neutral hydrogen and free protons/electrons. You can simplify the hydrogen atom to be a single state at $-13.6\,\mathrm{eV}$.
9. Generalize the Saha equation for the case of two protons combining with two neutrons directly — not through deuterium — to form ^{4}He. Taking $\eta = 10^{-9}$, find the temperature where half of the neutrons are bound in ^{4}He. How does this temperature compare to the deuteron bottleneck temperature?

10. Show that we can write the baryon-to-photon ratio $\eta = n_B/n_\gamma$ as

$$\eta = x\Omega_B h^2 \left(\frac{T}{T_0}\right)^\alpha,$$

with $T_0 = 2.73\,\text{K}$ and find the constant of proportionality x and the exponent α.

11. About 3 s after the onset of the Big Bang, the neutron–proton ratio became frozen when the temperature was still as high as $10^{10}\,\text{K}$ ($kT \simeq 0.8\,\text{MeV}$). About 250 s later, fusion reactions took place converting neutrons and protons into ^{4}He nuclei. Show that the resulting ratio of the masses of hydrogen and helium in the Universe was close to 3. The neutron half-life = 10.24 min and the neutron–proton mass difference is 1.29 MeV.

12. The total luminosity of the stars in our galaxy is $L \sim 10^{10}\,L_\odot$. Suppose that the luminosity of our galaxy has been constant for the past 10 Gyr. How much energy has our galaxy emitted in the form of starlight during that time? Most stars are powered by the fusion of H into ^{4}He, with the release of 28.4 MeV for every helium nucleus formed. How many helium nuclei have been created within stars in our galaxy over the course of the past 10 Gyr, assuming that the fusion of H into ^{4}He is the only significant energy source? If the baryonic mass of our galaxy is $M \sim 10^{11}\,M_\odot$, by what amount has the helium fraction Y of our galaxy been increased over its primordial value $Y_4 = 0.24$?

13. Consider the temperature T_d for deuterium formation. Consider a two state problem of d + d vs. ^{4}He in equilibrium. Looking up masses, find the temperature where half of the deuterons would have combined to form He. Is this temperature lower or higher than T_d? Why, in the Big Bang, don't two neutrons and two protons combine directly to form He, before the time of deuterium formation?

14. (a) Derive an approximate relation for the redshift of matter–radiation equality, by extrapolating backwards from the matter and radiation densities at the present epoch. (b) If a significant fraction of the primordial mass density is comprised of particles that decay into relativistic species at early times, how does this affect the epoch of matter–radiation equality? (Hint: remember that the relevant density of "radiation" for computing the epoch of matter–radiation equality includes the energy density in particles that are relativistic at that time.)

15. Suppose that a black hole is in a field of dark matter particles. These particles interact only gravitationally, and in particular not electromagnetically. Would an accretion disk form?
16. The measurement of the total width of the Z resonance at particle accelerators led to a precise value of the number of neutrino species, $N_\nu = 3$. Explain how this number was inferred from the Z-width.
17. Could the dark mass in the center of our galaxy be $3.5 \times 10^6\, M_\odot$ of dark matter, if the dark matter is composed of fairly light elementary particles (say, roughly the mass of a proton)?
18. Suppose that a neutron star was born with a 10^{-3} s period and a 10^{15} G magnetic field. How long would it take to double its period, and how much luminosity would it average in that time? This is a model being considered for some types of gamma ray bursts.
19. (a) Above what temperature is the mean energy of a photon in a thermal distribution sufficient to ionize hydrogen atoms? At what redshift did the radiation in the Universe have this temperature? Was the Universe matter- or radiation-dominated at this time?

    (b) Above what temperature is the mean energy of a photon in a thermal distribution sufficient to disassociate a helium nucleus (no need to calculate the exact energy required)? At what redshift did the radiation in the Universe have this temperature? Was the Universe matter- or radiation-dominated at this time?
20. What effects can you think of resulting from a rapid rotation of the supernova core just prior to collapse?
21. Suppose that black holes of mass $10^{-8}\, M_\odot$ made up all the dark matter in the halo of our galaxy. How far away would you expect the nearest such black hole to be? How frequently would you expect such a black hole to pass within 1 AU of the Sun? (An order-of-magnitude estimate is sufficient.)

    Suppose that MACHOs of mass $10^{-3}\, M_\odot$ (about the mass of Jupiter) made up all the dark matter in the halo of our galaxy. How far away would you expect the nearest MACHO to be? How frequently would such a MACHO pass within 1 AU of the Sun? (Again, an order-of-magnitude estimate will suffice.)
22. A neutrino with mass 10 eV is a dark matter candidate. At what temperature would the thermal energy of such a neutrino equal its mass energy?

    Before that epoch the neutrinos would be relativistic. Making the appropriate assumption as to whether the Universe is matter or radiation dominated (see Chapter 2), estimate how far the neutrinos would be able to travel up to that epoch. Assume they travel at the speed of light while relativistic. Since that time, the neutrinos have been non-relativistic and so have had negligible velocity, but the Universe has still been expanding. What is the factor of expansion, and what is the distance in the Universe as measured today that the neutrinos would have been able to travel? How does this scale compare to the scale of a typical structure in the Universe?

Chapter 10

Rays

10.1. Introduction

The terrestrial surface is bombarded by a constant flux of charged particles coming from the cosmos. A measurement of the intensity of this radiation shows that it decreases until a certain altitude, but it grows quickly again for larger altitudes. This invalidates the hypothesis that the radiation is coming from the radioactive decay of elements in the terrestrial ground and it is indeed proven that this activity originates from particles accelerated by astrophysical sources (*primary radiation*) and of collisions of these particles with the interstellar gas or with atoms of the upper part of the atmosphere (*secondary radiation*). These particles, known as *cosmic rays*, are studied in several places of the Earth — from the deepness of mines to the heights of satellites — and in the regions of the solar system explored by the man. Figure 10.1 shows the cascade process due to the collision of a cosmic ray in the terrestrial atmosphere.

Cosmic rays can be broadly defined as the massive particles, photons (γ rays, X-rays, ultraviolet and infrared radiation, ...), neutrinos, and exotics (WIMPS, axions,...) striking the Earth. Cosmic rays can be of either galactic (including solar) or extragalactic origin. The intensity of the primary radiation is approximately 2–4 particles per second and per square centimeter. This radiation consists essentially of nuclei (98% in number) and of electrons (2%). Among the nuclei, the predominance is of protons (87% of the mass) and of α-particles (12% of the mass). The remaining 1% includes all nuclei, with smaller abundance the larger the mass of the nucleus is.

In this chapter we discuss the main characteristics of cosmic rays, their composition, energy distribution and how they relate to the material studied in the previous chapters.

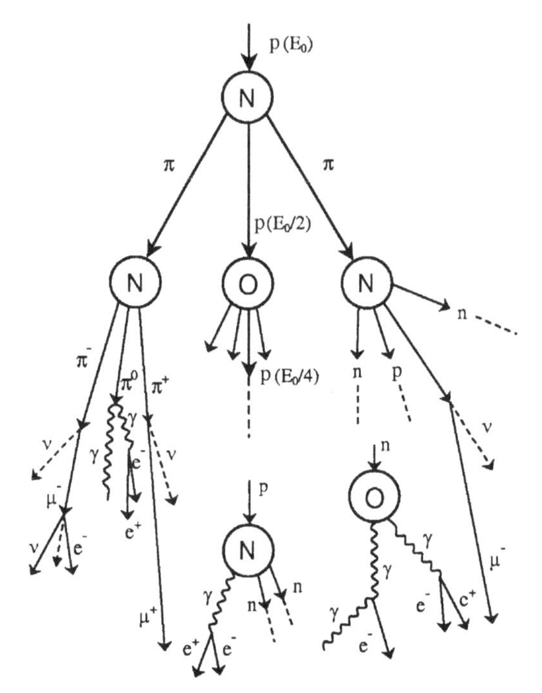

Figure 10.1. Schematic diagram of the cascade process due to the collision of a cosmic ray in the terrestrial atmosphere.

10.2. Composition and Energy Distribution

10.2.1. *Composition*

The chemical composition and kinetic energy distribution of (primary) cosmic rays is shown in Fig. 10.2. This distribution is approximately independent of energy, at least over the dominant energy range of 10 MeV/nucleon through several GeV/nucleon. One recognizes that the main component of cosmic rays are protons, with additionally around 10% of helium and an even smaller admixture of heavier elements.

The figure also shows the chemical distribution of the elements in our solar system, which differs from that of the cosmic rays in some remarkable ways. The most dramatic of these is an enormous enrichment in the cosmic rays of the elements Li/Be/B. Note also that there is enrichment is even Z elements relative to odd Z. Finally the cosmic rays are relatively enriched in heavy elements relative to H and He.

Although not shown in the figure, many elements heavier than the iron group have been measured with typical abundances of 10^{-5} of iron. Much

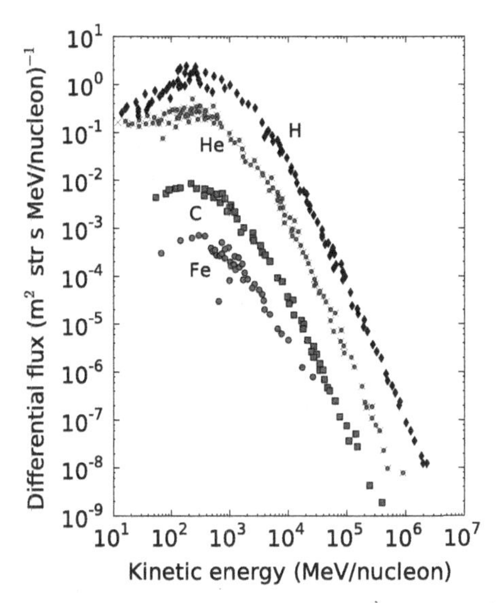

Figure 10.2. Main elements of the composition of cosmic rays (data collected from Ref. [Yao06]).

of this information was gained from satellite and spacecraft measurements over the last decade. Some of the conclusions:

(1) Abundances of even Z elements with $30 \lesssim Z \lesssim 60$ are in reasonable agreement with solar system abundances.
(2) In the region $62 \lesssim Z \lesssim 80$, which includes the platinum–lead region, abundances are enhanced relative to solar by about a factor of two. This suggests an enhancement in r-process elements, which dominate this mass region.

There are obvious connections between other astrophysics we have discussed (e.g., if the r-process site is core-collapse supernovae, then one would expect enrichment in r-process nuclei as supernovae are also believed to be the primary acceleration mechanism for lower energy cosmic rays) and possible deviations from solar abundances in the cosmic rays.

For some nuclei, like Li, Be and B, the percent of cosmic rays, albeit small, is several orders of magnitude larger than the average of these elements in the Universe. That is due to the fact that these elements are not nucleosynthesis products in stars, and they should be part of the secondary radiation. The proportion of several isotopes of an element in the cosmic rays can also be different from the universal average. The *isotopic ratio*

$^3He/^4He$, for example, is 200 times larger in cosmic rays than the same average isotopic ratio in all the cosmos.

In the electron component the positron percentage is about 10%. On the other hand, the anti-proton/proton ratio is of the order of 10^{-4}. These antiparticles belong, probably, to secondary radiation. We can calculate the minimal energy of a proton able to produce anti-protons scattering on another proton at rest. Because of baryon number conservation, anti-proton production is first possible in $pp \to ppp\bar{p}$ with $s = 2m_p^2 + 2E_p m_p \geq 16m_p^2$ or $E_p \geq 7m_p$. Furthermore, the cross-section of this reaction is small close to the threshold. Hence the anti-proton flux at small energies should be strongly suppressed, if anti-protons are only produced as secondaries in cosmic ray interactions.

10.2.2. *Energy distribution*

The energy spectrum of the primary cosmic rays has an average of 10 GeV/nucleon. For the nuclei the significant contribution begins at 10 MeV/nucleon and it grows until reaching a maximum at 300 MeV/nucleon. At this maximum, the proton flux is 2 protons/(m^2·s·MeV·steradian). The intensity of cosmic rays of energy 1 Gev/nucleon or greater is about 1/cm^2s. The energy density corresponding to this is thus about 1 ev/cm^3. This can be compared to the energy density of stellar light of 0.3 eV/cm^3. If we confine ourselves to the particle constituents (protons, nuclei, leptons), their motion in the galaxy has been roughly randomized by the *galactic magnetic field.* Thus they provide very little information about the direction of the source.

The primary cosmic rays are approximately isotropic and are also constant over a very long time scale ($\sim 10^9$ y).

The energy distribution of cosmic rays from about 10^{10} eV to about 10^{15} eV has a power-law distribution

$$N(E) \propto E^{-(1.6-1.7)}.$$

However there is a break or "*knee*" in the curve at about 10^{15} eV. The slope sharpens above this knee (see Fig. 10.3), falling as

$$N(E) \propto E^{-(2.0-2.2)},$$

eventually steepening to an exponent of above -2.7. The knee and *ankle* are more visible in Fig. 10.4 where the spectrum is divided by $E^{-2.5}$. The knee is generally attributed to the fact that supernovae acceleration of cosmic

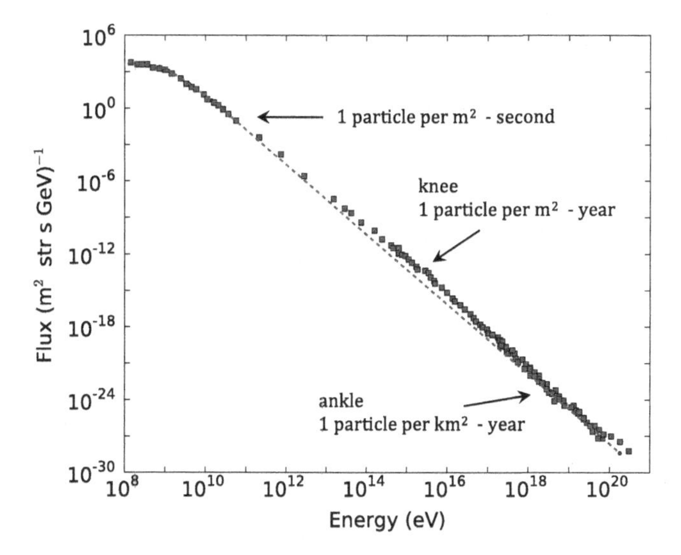

Figure 10.3. The energy distribution of cosmic rays.

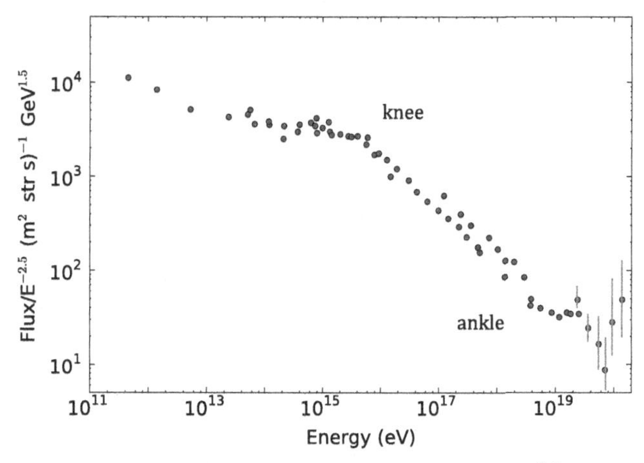

Figure 10.4. The energy spectrum of cosmic rays (divided by $E_9^{-2.5}$) showing a "*kink*", or "*knee*" and an "*ankle*". The energy E_9 is given in GeV.

rays is limited to about this energy. This would argue that the cosmic rays above this energy either have a different origin, or were further accelerated after production. However the sharpness of the knee has troubled many experts: it is very difficult to find natural models producing such a defined break.

10.2.3. *Propagation and origin*

The solar wind impedes and slows the incoming cosmic rays, reducing their energy and preventing the lowest energy ones from reaching the Earth. This effect is known as *solar modulation*. The Sun has an 11-year activity cycle which is reflected in the ability of the solar wind to modulate cosmic rays. As a result, the cosmic ray intensity at Earth is anti-correlated with the level of solar activity, i.e., when solar activity is high and there are lots of sunspots, the cosmic ray intensity at Earth is low, and vice versa.

The properties listed in the previous sections do not allow us to obtain a definitive answer on the origin of the cosmic rays. The most accepted hypothesis is that supernovae and neutron stars in our galaxy are the responsible agents for this radiation and it is related to the fact that in our galaxy a supernova explodes every 30 years on the average. Each one of these explosions liberates about $10^{44} - 10^{45.5}$ J of energy. To explain the energy spectrum of the cosmic rays it is also necessary to suppose that it is influenced by the re-acceleration of the particles by galactic magnetic fields.

The most common assumption is that the cosmic rays are confined within the galactic disk, where the mass density is high, but with some gradual leaking out of the disk. This model does a good job of explaining the energy-dependence of cosmic rays. Other models are also popular, including closed models where cosmic rays are fully confined, then explaining lifetimes through devices such as a combination of a few nearby and many distant cosmic ray sources. The relatively large presence of light and heavy nuclei would be a consequence of fragmentation processes, that is, of the interaction of protons and of α-particles with remnant nuclei in gases in the sidereal space. In modeling the origin of cosmic rays, the first conclusion, given their richness in metals, is that they must come from highly evolved stars such as those that undergo supernovae. The abundance of r-process nuclei can be taken as an evidence of supernova dominance, for those who accept that supernovae are the r-process site. But studies of the isotopic composition as the knee is approached shows that the composition changes: the spectrum of protons becomes noticeably steeper in energy, while the iron group elements do not show such a dramatic change. Above the knee — at energies above 10^{16} eV the galactic magnetic field is too weak to appreciably trap particles. Thus it is probable that at these high energies the character of the cosmic rays changes from primarily galactic to primarily extragalactic. This also suggests a natural explanation for the relative enrichment of heavy nuclei in the vicinity of the knee: the upper energy of confined particles should vary as Z, since the cyclotron frequency for particles of the same velocity varies as Z × B. Above 10^{18} eV/n the cosmic rays freely stream through galaxies.

Assuming a density of particles in space of $1/\text{cm}^3$, the mean free path of a cosmic ray is about 3×10^{22} m, or a mean time path of 3×10^6 y. This last figure is taken as the lifetime of a particle present in the cosmic rays in our galaxy. During this time, several other particles at different points in the Universe are produced, with a wide energy spectrum, and emitted in several directions. Therefore, the large lifetime would explain the isotropy and the temporal constancy of the cosmic rays, since fluctuations in the place of emission, in the production time, and in the intensity of the source are smoothed out.

The conventional explanation for the most dramatic isotopic anomaly in the cosmic ray, the enrichment in Li/Be/B by about six orders of magnitude, is that they are produced in the interstellar medium when accelerated protons collide with C, N, and O. The enrichment of odd-A nuclei (these also tend to be relatively rare in their solar distribution since stellar processes tend to favor production of more stable even A nuclei) is usually also attributed to spallation reactions off more abundant even-A nuclei. These associations immediately lead to some interesting physics conclusions because, from the known density of cosmic rays (at least in the Earth's vicinity) and from known spallation cross-sections, one can estimate the amount of material through which a typical cosmic ray propagates. Although the estimates are model dependent, typical values are $4-6\,\text{g/cm}^2$ for the effective thickness. Now the mass density within intergalactic space is about 1 protron/cm^3, or about 1.7×10^{-24} g/cm^3. Thus taking a velocity of c, we can crudely estimate the cosmic ray lifetime

$$1.7 \times 10^{-24}\text{g/cm}^3 \times (3 \times 10^{10}\text{cm/s}) \times t = (4-6)\text{g/cm}^2.$$

So this gives $t \sim 3 \times 10^6$ y, in accordance with the estimate mentioned above.

This calculation assumes an average galactic mass density that is not known by direct measurement. Thus it is nice that a more direct estimate of the galactic cosmic ray lifetime is provided by cosmic ray radioactive isotopes. The best chronometer is one that has a lifetime in the ballpark of the estimate above. ^{10}Be, with a lifetime of 1.51×10^6 y, is thus quite suitable. It is a cosmic ray spallation product: this guarantees that it is born as a cosmic ray. Its abundance can be normalized to those of the other, stable Li/Be/B isotopes: the spallation cross-sections are known. Thus the absence of ^{10}Be in the cosmic ray spectrum would indicate that the typical cosmic ray lifetime is much larger than 1.51×10^6 y. The survival probability should also depend on the ^{10}Be energy, due to time dilation effects. One observes a reduction in ^{10}Be to about $(0.2-0.3)$ of its expected instantaneous production, relative to other Li/Be/B isotopes. From this one concludes $t \sim (2-3) \times 10^7$ years. This suggests that the mass density estimate used above (in our first calculation) may have been too high by a factor of $5-10$.

During certain short periods, a small part of the cosmic rays that reach the Earth is coming from the Sun. The emission of α-particles and of energetic protons by the Sun happens in the period of maximum solar activity, during which up to 10 eruptions in the chromosphere are observed in a period of one year. The energy of the emitted particles is of the order of 400 MeV, or smaller. However, this radiation of short period can reach 10^8 particles/cm^2/s. The acceleration of these particles is possibly due to the magnetic fields originated by the large currents in the solar plasma.

The magnetic field of the Earth has a large effect on the cosmic rays during their entrance into the atmosphere. In the first place, this field prevents particles with momentum smaller than a certain value $p_{\rm min}$ from penetrating the atmosphere. $p_{\rm min}$ is related to the geomagnetic latitude at the arrival point of the cosmic ray. At the equator, this value is of the order of 15 GeV/c, while at the poles it is of order of zero.

In the second place, due to the electromagnetic Lorentz force, the magnetic field of the Earth impedes the entrance of cosmic rays in the atmosphere inside a given solid angle of incidence with respect to the horizon. For positively charged particles coming from the east, there is a totally forbidden angular area above the horizon. Thus, this leads to an east–west asymmetry in the direction of incidence of the cosmic rays, which can be easily observed with simple detectors.

Finally, the magnetic field of the Earth creates a trap around the Earth for particles with certain values of energy and momentum. In these areas, the accumulation of particles is responsible for the effect of belts of particles around the Earth (*van Allen belts*). Another consequence of this effect is the *aurora borealis*, or *australis*, a light that appears in the sky close to the Earth's poles due to the scattering of sunlight on these particles. Figure 10.5 shows the intensity of the several components of the secondary cosmic rays as a function of the distance crossed in the atmosphere.

The cosmic rays that penetrate the terrestrial atmosphere collide with the atoms of oxygen and nitrogen, originating an effect of cascade collisions (Fig. 10.1). The first collisions yield secondary protons, as well as neutrons and pions. In small amounts kaons, α-particles and secondary nuclei are also

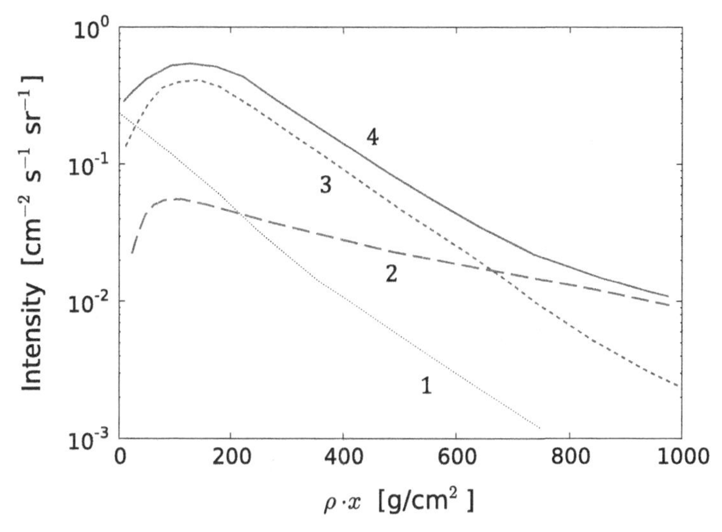

Figure 10.5. Intensity of the several components of the secondary cosmic rays as a function of the distance crossed in the atmosphere: 1 – nuclear, 2 – electron-positron, 3 – muons, 4 – total.

produced. Due to the beta-decay of unstable nuclei, muons, neutrinos and gamma rays are created. The muons are particles of large penetrability in matter and they constitute the so-called "hard component" of the cosmic rays. The photons give rise to electrons and free positrons ("soft" component). In Fig. 10.5 the relative intensity of the three components are presented as a function of the mass per unit of area of the atmosphere penetrated by them (at sea altitude this corresponds to $1000\,\text{g/cm}^2$). Neither the primary cosmic rays, nor the secondary nuclear component, can reach the terrestrial surface with an appreciable probability. Here only electrons, photons and muons are registered. Most of the energy of the primary cosmic rays is transformed into ionization energy of the molecules of the atmosphere.

10.2.4. *Scattering by cosmic background photons*

There are three main energy loss processes for protons propagating over cosmological distances: adiabatic energy losses due to the expansion of the Universe, $-(dE/dt)/E = H_0$, and e^+e^- pair- or pion-production on photons of the cosmic microwave background (CMB). The relative energy loss per unit time of a particle (due to interactions with the CMB) can be estimated as

$$\frac{1}{E}\frac{dE}{dt} = \langle y\sigma n_\gamma \rangle, \tag{10.1}$$

where $y = (E - E')/E$ is the energy fraction lost per interaction, $n_\gamma \approx 410/\text{cm}^3$ is the density of CMB photons with temperature $T \approx 2.7\,\text{K}$ and the brackets $\langle \ldots \rangle$ remind us that we should perform an average of the differential cross-section with the momentum distribution of photons. Their typical energy is about $10^{-3}\,\text{eV}$.

Since the produced e^+e^- pair in the process $p + \gamma \to p + e^+ + e^-$ is light, this energy loss process has a low threshold energy but leads in turn only to a small energy loss per interaction, $y = 2m_e/m_p \approx 10^{-3}$. The threshold energy of CMB photons follows from $(k_p + k_\gamma)^2 \geq (m_p + 2m_e)^2$ as $E_p \geq m_e m_p / E_\gamma \sim 2 \times 10^{18}\,\text{eV}$. The cross-section of the reaction is a factor $\sim 2 \times 10^{-4}$ smaller than Compton scattering.

Let $p_\mu = (\epsilon, \mathbf{p})$ be the photon four-momentum. Then $|\mathbf{p}| = \epsilon$. Let $P_\mu = (\omega, \mathbf{P})$ be the proton four-momentum. The center-of-mass energy is

$$(P + p)^\mu (P + p)_\mu = (\omega + \epsilon)^2 = \epsilon_{CM}^2.$$

If ϵ_{CM} exceeds $m_\pi + M_N$ then the reaction $\gamma + N \to \pi + N$ is possible, degrading the nucleon energy.

The center-of-mass energy is a Lorentz invariant quantity, so it can be evaluated in the laboratory frame

$$\epsilon_{CM}^2 = M_N^2 + \omega\epsilon - \mathbf{P} \cdot \mathbf{p}.$$

As the cosmic background photons are moving in all directions, we are free to maximize the expression by taking $\cos\theta \sim -1$. As the incident nucleon is highly relativistic, $(\epsilon_{CM}^{max})^2 \sim M_N^2 + 2\omega\epsilon$. Thus the requirement for photoproduction is $\epsilon_{CM}^{max} \gtrsim m_\pi + M_N$, or

$$\omega \gtrsim \frac{m_\pi^2 + 2M_N m_\pi}{2\epsilon} \sim 1.4 \times 10^{20}\,\text{eV}. \tag{10.2}$$

This results in a mean free path for protons of about 10^8 light years for protons at $\sim 10^{20}\,\text{eV}$ or higher energies, a distance substantially smaller than the horizon. Thus if the origin of such cosmic rays is all of extragalactic space, there should be a very sharp cutoff in the cosmic ray flux at about this energy. This is called the *Greisen–Zatsepin–Kuzmin cutoff* (GZK cutoff for short). Yet evidence for such a cutoff has not been found experimentally.

For a more detailed estimate of the above result, we consider the cross-section for pion production using a Breit–Wigner formula (see Chapter 4) with the lowest lying nucleon resonance Δ^+ as an intermediate state,

$$\sigma_{\mathrm{BW}}(E) = \frac{(2J+1)}{(2s_1+1)(2s_2+1)} \frac{\pi}{p_{\mathrm{cms}}^2} \frac{b_{\mathrm{in}} b_{\mathrm{out}} \Gamma^2}{(E-M_R)^2+\Gamma^2/4}, \tag{10.3}$$

where $p_{\mathrm{cms}}^2 = (s - m_N^2)^2/(4s)$, $M_R = m_\Delta = 1.230\,\mathrm{GeV}$, $J = 3/2$, $\Gamma_{\mathrm{tot}} = 0.118\,\mathrm{GeV}$, and $b_{\mathrm{in}} \simeq b_{\mathrm{out}} = 0.55\%$. At resonance, we obtain $\sigma_{\mathrm{BW}} \sim 0.4$ mbarn in good agreement with experimental data. We estimate the energy loss length well above threshold $E_{\mathrm{th}} \sim 4 \times 10^{19}\,\mathrm{eV}$ with $\sigma \sim 0.1$ mbarn and $y = 0.5$ as

$$l_{\mathrm{GZK}}^{-1} = \frac{1}{E}\frac{dE}{dt} \approx 0.5 \times 400/\mathrm{cm}^3 \times 10^{-28}\,\mathrm{cm}^2 \approx 2 \times 10^{-26}\,\mathrm{cm}^{-1} \tag{10.4}$$

or $l_{\mathrm{GZK}} \sim 17\,\mathrm{Mpc}$. Thus the energy loss length of a proton with $E \gtrsim 10^{20}\,\mathrm{eV}$ is comparable to the distance of the closest galaxy clusters and we should see only local sources at these energies.

The cutoff for nuclei is more severe. At a considerably lower energy nuclei can absorb (in their rest frame) a photon of energy $\sim$10 MeV, resulting in photodistintegration. This leads to a GZK cutoff for iron nuclei of $\sim 10^{19}\,\mathrm{eV}$. In fact, the dominant loss process for nuclei of energy $E \gtrsim 10^{19}\,\mathrm{eV}$ is photodisintegration $A + \gamma \to (A-1) + N$ in the CMB and the infrared background due to the giant dipole resonance. The threshold for this reaction follows from the binding energy per nucleon, $\sim$10 MeV. Photodisintegration leads to a suppression of the flux of nuclei above an energy that varies between $3 \times 10^{19}\,\mathrm{eV}$ for He and $8 \times 10^{19}\,\mathrm{eV}$ for Fe.

10.3. Neutrinos

10.3.1. *Neutrinos from the sun*

The Sun serves as a very important test case for a variety of problems related to stellar structure and evolution, as well as to fundamental physics. The nuclear reactions in the Sun, in particular, the pp-*chains*, has been discussed in Chapter 9. The central temperature $T_\odot$ of the Sun is a nice example of a physical quantity which can be determined by means of solar neutrino detection, provided that the relevant nuclear physics is known (and neutrino properties are also known). Surprisingly enough for a star that has all the reasons to be considered as one of the dullest astrophysical objects, the Sun has been for years at the center of various controversies. One of them was the *solar neutrino problem*, referring to the fact that the pioneering ^{37}Cl neutrino-capture experiments carried out over the years in the Homestake gold mine observed a neutrino flux that was substantially smaller than the one predicted by the solar models [Bah98]. That puzzle led to a flurry of theoretical activities, and to the development of new detectors. These activities have transformed the original solar neutrino problem into other problems. The relative levels of 'responsibility' of particle physics, nuclear physics or astrophysics in these discrepancies have been debated ever since.

Several experiments support the idea of "oscillations" between different neutrinos types [Bah98, Fu98]. In the neutrino oscillation picture, the electron-neutrino, detected by the chlorine experiment, can transform into a muon-neutrino on its way to the Earth from the center of the Sun. This would explain the smaller number of electron-neutrinos observed at the Earth.

SSM calculations [Bah98] predict $T_\odot$ with an accuracy of 1% or even better. In order to appreciate such a result, let us remind that the central temperature of Earth is known with an accuracy of about 20%. However, let us remind that this is a theoretical prediction which, as any result in physics, demands observational evidence.

The fluxes of ^{8}B and ^{7}Be neutrinos (see Fig. 10.6) are given by:

$$\Phi(B) = c_B S_{17} \frac{S_{34}}{\sqrt{S_{33}}} T_\odot^{20}, \quad \Phi(Be) = c_{Be} \frac{S_{34}}{\sqrt{S_{33}}} T_\odot^{10}, \tag{10.5}$$

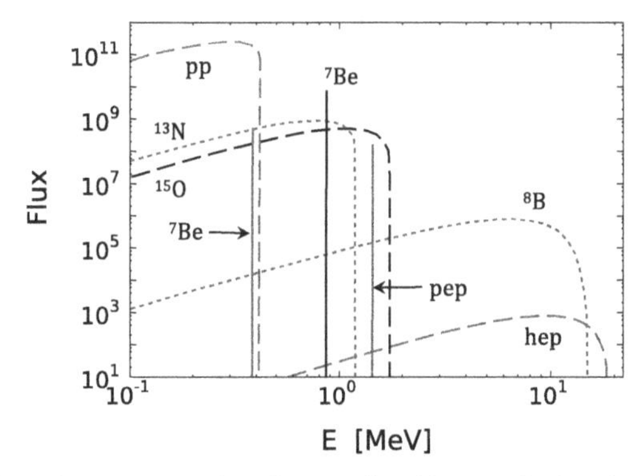

Figure 10.6. The solar neutrino flux (in cm^{-2} s^{-1}), as calculated with the Standard Solar Model (SSM) [Bah98].

where S_{ij} are the low energy astrophysical factors for nuclear reactions between nuclei with atomic mass numbers i and j, while c_B and c_{Be} are well determined constants.

The high powers of $T_\odot$ in the above equations imply that the measured neutrino fluxes are strongly sensitive to $T_\odot$, i.e. ^{7}Be and ^{8}B neutrinos in principle are good thermometers for the innermost part of the Sun. On the other hand, the relevant nuclear physics has to be known, which justifies the present theoretical and experimental efforts for better determinations of S_{ij}.

10.3.2. *Neutrino oscillations*

One odd feature of particle physics is that neutrinos, which are not required by any symmetry to be massless, must nevertheless be much lighter than any of the other known fermions. For instance, the current limit on the $\overline{\nu}_e$ mass is $\lesssim$2.2 eV. The standard model of particle physics requires neutrinos to be massless, but the reasons are not fundamental. Dirac mass terms m_D, analogous to the mass terms for other fermions, cannot be constructed because the model contains no right-handed neutrino fields. If a neutrino has a mass m, we mean that as it propagates through free space, its energy and momentum are related in the usual way for this mass. Thus if we have two neutrinos, we can label those neutrinos according to the eigenstates of the free Hamiltonian, that is, as mass eigenstates.

But neutrinos are produced by the weak interaction. In this case, we have another set of eigenstates, the *flavor eigenstates*. We can define a ν_e as the neutrino that accompanies the positron in β decay. Likewise we label by ν_μ the neutrino produced in muon decay.

The question is: are the eigenstates of the free Hamiltonian and of the weak interaction Hamiltonian identical? Most likely the answer is no: we know this is the case with the quarks, since the different families (the analog of the mass eigenstates) do interact through the weak interaction. That is, the up quark decays not only to the down quark, but also occasionally to the strange quark. Thus we suspect that the weak interaction and mass eigenstates, while spanning the same two-neutrino space, are not coincident: the mass eigenstates $|\nu_1\rangle$ and $|\nu_2\rangle$ (with masses m_1 and m_2) are related to the weak interaction eigenstates by [Mad01]

$$\begin{aligned}|\nu_e\rangle &= \cos\theta_v|\nu_1\rangle + \sin\theta_v|\nu_2\rangle,\\ |\nu_\mu\rangle &= -\sin\theta_v|\nu_1\rangle + \cos\theta_v|\nu_2\rangle,\end{aligned} \tag{10.6}$$

where θ_v is the (vacuum) mixing angle.

An immediate consequence is that a state produced as a $|\nu_e\rangle$ or a $|\nu_\mu\rangle$ at some time t — for example, a neutrino produced in β decay — does not remain a pure flavor eigenstate as it propagates away from the source. The different mass eigenstates comprising the neutrino will accumulate different phases as they propagate downstream, a phenomenon known as *vacuum oscillations* (vacuum because the experiment is done in free space). To see the effect, suppose the neutrino produced in a β decay is a momentum eigenstate. At time $t = 0$

$$|\nu(t=0)\rangle = |\nu_e\rangle = \cos\theta_v|\nu_1\rangle + \sin\theta_v|\nu_2\rangle. \tag{10.7}$$

Each eigenstate subsequently propagates with a phase

$$\exp[i(\mathbf{k}\cdot\mathbf{x} - \omega t)] = \exp\left[i\left(\mathbf{k}\cdot\mathbf{x} - \sqrt{m_i^2 + k^2 t}\right)\right]. \tag{10.8}$$

But if the neutrino mass is small compared to the neutrino momentum/energy, one can write

$$\sqrt{m_i^2 + k^2} \sim k\left(1 + \frac{m_i^2}{2k^2}\right). \tag{10.9}$$

Thus we conclude

$$\begin{aligned} |\nu(t)\rangle &= \exp[i(\mathbf{k}\cdot\mathbf{x} - kt - (m_1^2 + m_2^2)t/4k)] \\ &\quad \times[\cos\theta_v|\nu_1\rangle e^{i\delta m^2 t/4k} + \sin\theta_v|\nu_2\rangle e^{-i\delta m^2 t/4k}]. \end{aligned} \tag{10.10}$$

We see there is a common average phase (which has no physical consequence) as well as a beat phase that depends on

$$\delta m^2 = m_2^2 - m_1^2. \tag{10.11}$$

Now it is a simple matter to calculate the probability that our neutrino state remains a $|\nu_e\rangle$ at time t

$$P_{\nu_e}(t) = |\langle\nu_e|\nu(t)\rangle|^2 = 1 - \sin^2 2\theta_v \sin^2\left(\frac{\delta m^2 t}{4k}\right) \to 1 - \frac{1}{2}\sin^2 2\theta_v \tag{10.12}$$

where the limit on the right is appropriate for large t as the average of rapid harmonic oscillations for large t yields $\langle\sin^2(\alpha t)\rangle = 1/2$. Now $E \sim k$, where E is the neutrino energy, by our assumption that the neutrino masses are small compared to k. We can reinsert the implicit constants to write the probability in terms of the distance x of the neutrino from its source,

$$P_\nu(x) = 1 - \sin^2 2\theta_v \sin^2\left(\frac{\delta m^2 c^4 x}{4\hbar c E}\right). \tag{10.13}$$

If the oscillation length

$$L_o = \frac{4\pi\hbar c E}{\delta m^2 c^4} \tag{10.14}$$

is comparable to or shorter than one astronomical unit, a reduction in the solar ν_e flux would be expected in terrestrial neutrino observations.

We now present the results above in a slightly more general way. The analog of Eq. (10.10) for an initial muon neutrino ($|\nu(t=0)\rangle = |\nu_\mu\rangle$) is

$$\begin{aligned}|\nu(t)\rangle &= e^{i(\mathbf{k}\cdot\mathbf{x} - kt - (m_1^2+m_2^2)t/4k)}[-\sin\theta_v|\nu_1\rangle e^{i\delta m^2 t/4k} \\ &\quad + \cos\theta_v|\nu_2\rangle e^{-i\delta m^2 t/4k}].\end{aligned} \tag{10.15}$$

Now if we compare Eqs. (10.10) and (10.15) we see that they are special cases of a more general problem. Suppose we write our initial neutrino wave function as

$$|\nu(t=0)\rangle = a_e(t=0)|\nu_e\rangle + a_\mu(t=0)|\nu_\mu\rangle. \tag{10.16}$$

Then Eqs. (10.10) and (10.15) tell us that the subsequent propagation is described by changes in $a_e(x)$ and $a_\mu(x)$ according to

$$i\frac{d}{dx}\begin{pmatrix} a_e \\ a_\mu \end{pmatrix} = \frac{1}{4E}\begin{pmatrix} -\delta m^2\cos 2\theta_v & \delta m^2 \sin 2\theta_v \\ \delta m^2 \sin 2\theta_v & \delta m^2 \cos 2\theta_v \end{pmatrix}\begin{pmatrix} a_e \\ a_\mu \end{pmatrix}. \tag{10.17}$$

Note that the common phase has been ignored: it can be absorbed into the overall phase of the coefficients a_e and a_μ, and thus has no consequence. Also, we have equated $x = t$, that is, set $c = 1$.

Neutrino oscillation change due to the density dependence of the neutrino effective mass which greatly enhances oscillation probabilities: a ν_e is adiabatically transformed into a ν_μ as it traverses a critical density within the Sun [Wol78,MS85]. It became clear that the Sun was not only an excellent neutrino source, but also a natural regenerator for cleverly enhancing the effects of flavor mixing

$$i\frac{d}{dx}\begin{pmatrix} a_e \\ a_\mu \end{pmatrix} = \frac{1}{4E}\begin{pmatrix} 2E\sqrt{2}G_F\rho(x) - \delta m^2 \cos 2\theta_v & \delta m^2 \sin 2\theta_v \\ \delta m^2 \sin 2\theta_v & -2E\sqrt{2}G_F\rho(x) + \delta m^2 \cos 2\theta_v \end{pmatrix}\begin{pmatrix} a_e \\ a_\mu \end{pmatrix}, \tag{10.18}$$

where G_F is the weak coupling constant and $\rho(x)$ the solar electron density. If $\rho(x) = 0$, this is exactly our previous result and can be trivially integrated to give the vacuum oscillation solutions given above. The new contribution to the diagonal elements, $2E\sqrt{2}G_F\rho(x)$, represents the effective contribution to the M_ν^2 matrix that arises from neutrino–electron scattering. The indices of refraction of electron and muon neutrinos differ because the former scatter by charged and neutral currents, while the latter have only neutral current interactions. The difference in the forward scattering amplitudes determines the density-dependent splitting of the diagonal elements of the matter equation, the generalization of Eq. (10.17).

It is helpful to rewrite this equation in a basis consisting of the light and heavy local mass eigenstates (i.e., the states that diagonalize the right-hand side of Eq. (10.18)),

$$\begin{aligned} |\nu_L(x)\rangle &= \cos\theta(x)|\nu_e\rangle - \sin\theta(x)|\nu_\mu\rangle, \\ |\nu_H(x)\rangle &= \sin\theta(x)|\nu_e\rangle + \cos\theta(x)|\nu_\mu\rangle. \end{aligned} \tag{10.19}$$

The local mixing angle is defined by

$$\begin{aligned} \sin 2\theta(x) &= \frac{\sin 2\theta_v}{\sqrt{X^2(x) + \sin^2 2\theta_v}}, \\ \cos 2\theta(x) &= \frac{-X(x)}{\sqrt{X^2(x) + \sin^2 2\theta_v}}, \end{aligned} \tag{10.20}$$

where $X(x) = 2\sqrt{2}G_F\rho(x)E/\delta m^2 - \cos 2\theta_v$. Thus $\theta(x)$ ranges from θ_v to $\pi/2$ as the density $\rho(x)$ goes from 0 to ∞.

If we define

$$|\nu(x)\rangle = a_H(x)|\nu_H(x)\rangle + a_L(x)|\nu_L(x)\rangle, \tag{10.21}$$

the neutrino propagation can be rewritten in terms of the local mass eigenstates

$$i\frac{d}{dx}\begin{pmatrix} a_H \\ a_L \end{pmatrix} = \begin{pmatrix} \lambda(x) & i\alpha(x) \\ -i\alpha(x) & -\lambda(x) \end{pmatrix}\begin{pmatrix} a_H \\ a_L \end{pmatrix}, \tag{10.22}$$

with the splitting of the local mass eigenstates determined by

$$2\lambda(x) = \frac{\delta m^2}{2E}\sqrt{X^2(x) + \sin^2 2\theta_v} \tag{10.23}$$

and with mixing of these eigenstates governed by the density gradient

$$\alpha(x) = \left(\frac{E}{\delta m^2}\right)\frac{\sqrt{2}G_F \frac{d}{dx}\rho(x)\sin 2\theta_v}{X^2(x) + \sin^2 2\theta_v}. \tag{10.24}$$

The results above are quite interesting: the local mass eigenstates diagonalize the matrix if the density is constant. In such a limit, the problem is no more complicated than our original vacuum oscillation case, although our mixing angle is changed because of the matter effects. But if the density is not constant, the mass eigenstates evolve as the density changes. This is the *Mikheyev–Smirnov–Wolfenstein (MSW) effect.* Note that the splitting achieves its minimum value, $\frac{\delta m^2}{2E}\sin 2\theta_v$, at a critical density $\rho_c = \rho(x_c)$

$$2\sqrt{2}EG_F\rho_c = \delta m^2 \cos 2\theta_v \tag{10.25}$$

that defines the point where the diagonal elements of the original flavor matrix cross.

Our local-mass-eigenstate form of the propagation equation can be trivially integrated if the splitting of the diagonal elements is large compared to the off-diagonal elements (see Eq. (10.22)),

$$\gamma(x) = \left|\frac{\lambda(x)}{\alpha(x)}\right| = \frac{\sin^2 2\theta_v}{\cos 2\theta_v}\frac{\delta m^2}{2E}\frac{1}{\left|\frac{1}{\rho_c}\frac{d\rho(x)}{dx}\right|}\frac{[X(x)^2 + \sin^2 2\theta_v]^{3/2}}{\sin^3 2\theta_v} \gg 1, \tag{10.26}$$

a condition that becomes particularly stringent near the crossing point,

$$\gamma_c = \gamma(x_c) = \frac{\sin^2 2\theta_v}{\cos 2\theta_v}\frac{\delta m^2}{2E}\frac{1}{\left|\frac{1}{\rho_c}\frac{d\rho(x)}{dx}\Big|_{x=x_c}\right|} \gg 1. \tag{10.27}$$

That is, adiabaticity depends on the density scale height at the crossing point. The resulting adiabatic electron neutrino survival probability [Bet86],

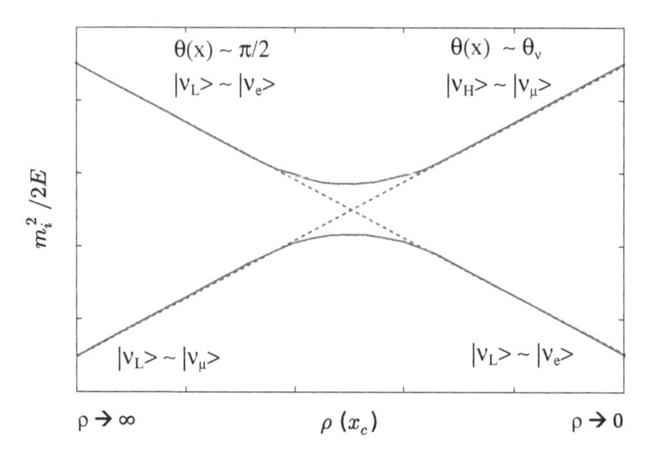

Figure 10.7. Schematic illustration of the MSW crossing. The dashed lines correspond to the electron–electron and muon–muon diagonal elements of the M_ν^2 matrix in the flavor basis. Their intersection defines the level-crossing density ρ_c. The solid lines are the trajectories of the light and heavy local mass eigenstates. If the electron neutrino is produced at high density and propagates adiabatically, it will follow the heavy-mass trajectory, emerging from the Sun as a ν_μ.

valid when $\gamma_c \gg 1$, is

$$P_{\nu_e}^{\rm adiab} = \frac{1}{2} + \frac{1}{2}\cos 2\theta_v \cos 2\theta_i, \tag{10.28}$$

where $\theta_i = \theta(x_i)$ is the local mixing angle at the density where the neutrino was produced.

The physical picture behind this derivation is illustrated in Fig. 10.7. One makes the usual assumption that, in vacuum, the ν_e is almost identical to the light mass eigenstate, $\nu_L(0)$, i.e., $m_1 < m_2$ and $\cos\theta_v \sim 1$. But as the density increases, the matter effects make the ν_e heavier than the ν_μ, with $\nu_e \to \nu_H(x)$ as $\rho(x)$ becomes large. The special property of the Sun is that it produces ν_es at high density that then propagate to the vacuum where they are measured. The adiabatic approximation tells us that if initially $\nu_e \sim \nu_H(x)$, the neutrino will remain on the heavy mass trajectory provided the density changes slowly. That is, if the solar density gradient is sufficiently gentle, the neutrino will emerge from the Sun as the heavy vacuum eigenstate, $\sim \nu_\mu$. This guarantees nearly complete conversion of ν_es into ν_μs.

The MSW mechanism provides a natural explanation for the pattern of observed solar neutrino fluxes. While it requires profound new physics, both massive neutrinos and neutrino mixing are expected in extended models.

10.3.3. *Atmospheric neutrinos*

We have seen that the hadronic cosmic rays impinging on the Earth have an energy distribution that peaks at around 1 GeV. These cosmic rays then interact in the top of the atmosphere, producing secondary showers of hadrons, leptons, and neutrinos. The atmosphere acts, in effect, as a beam stop for the incident cosmic rays. Experiments done with terrestrial accelerators tells us the products of these reactions are pions and kaons. Consider the reaction

$$p + p \to p + n + \pi^+.$$

The produced pion decays by weak interactions

$$\pi^+ \to e^+ + \nu_e \quad \text{or} \quad \mu^+ + \nu_\mu.$$

Now recall that only left-handed particles and right handed antiparticles couple in the standard model. Thus envision the decay in the rest frame of the pion. Momentum conservation has the two decay particles going out back-to-back. Thus the orbital angular momentum is out of the scattering plane. If the positron were massless, its spin would be along its direction of motion (right-handed), while the spin of the neutrino is anti-aligned (left-handed). Thus there is unbalanced angular momentum pointing in the direction of the positron.

Therefore the reaction would be forbidden were it not for the finite lepton mass. The transition probability is proportional to $1 - (\nu_e/c)^2 \sim (2m_e/m_\mu)^2 \sim 10^{-4}$ while in the case of decay into a muon, the muon is nonrelativistic and thus not suppressed by a similar kinematic factor. As a result, the $\mu^+ + \nu_\mu$ decay accounts for about 99.99% of the decays of the π^+. The muon then decays by

$$\mu^+ \to e^+ + \nu_e + \bar{\nu}_\mu.$$

The net effect is a neutrino flavor ratio $(\nu_\mu + \bar{\nu}_\mu)/\nu_e = 2$. The same arguments follow for kaons or pions of either sign (with an interchange $\nu \leftrightarrow \bar{\nu}$ in the case of a negative charge).

These neutrinos will penetrate the Earth and generate interactions in underground detectors, producing in charged-current interactions detectable muons and electrons. The observed ratio $(\mu/e)_{\text{data}}$ is closer to one than to the expected two. This anomaly is reminiscent of the solar neutrino puzzle and prompted explanations in terms of neutrino oscillation. The path-length is limited by the Earth's diameter and the neutrino energy is on the order of 1 GeV. As the vacuum oscillation probability is given by

$$P_{\nu_\mu \to \nu_\mu} = 1 - \sin^2(2\theta)\sin^2(\pi L/L_o), \tag{10.29}$$

where L_o is the oscillation length, we see that the largest effects will occur for $L = L_o/2$, which cannot exceed $R_\odot \sim 10^4$ km. But from Eq. (10.13)

$$\left(\frac{L_o}{\text{km}}\right) = \frac{2.48(E/\text{GeV})}{\delta m^2 c^4(\text{eV}^2)}. \tag{10.30}$$

Thus if we take $L_o \sim 2 \cdot 10^4$ km and $E \sim$1 GeV, we find atmospheric neutrinos are sensitive to

$$\delta m^2 \gtrsim 10^{-4} eV^2. \tag{10.31}$$

10.3.4. *Supernova neutrinos*

A supernova (SN) radiates almost all of its binding energy in the form of neutrinos, most of which have energies in the range 10 – 30 MeV. These neutrinos come in all flavors, and are emitted over a timescale of several tens of seconds. The neutrino luminosity of a gravitational collapse-driven supernova is typically 100 times its optical luminosity (see Section 8.6.3). Neutrinos emerge from the core of a supernova promptly after core collapse, whereas the photon signal may take hours or days to emerge from the stellar envelope. The neutrino signal can therefore give information about the very early stages of core collapse, which is inaccessible to other kinds of astronomy. The generation and propagation of neutrinos on their way out of the supernova explosion was discussed in Section 8.6.3.

Neutrinos from supernovae can be detected with water Ćerenkov detectors, usually through the absorption of electron anti-neutrinos on protons:

$$\bar{\nu_e} + \text{p} \to \text{n} + e^+. \tag{10.32}$$

The positron retains most of the energy of the incoming neutrino, and is detected from its Ćerenkov light. *Ćerenkov radiation* is electromagnetic radiation emitted when a charged particle (such as an electron) passes through a dielectric medium at a speed greater than the phase velocity of light in that medium.

Another detection method is the neutrino–electron elastic scattering events (neutral current and charged current),

$$\nu_x + e \to \nu_x + e, \tag{10.33}$$

where ν_x is a neutrino of any flavor. Although their cross-sections are two orders of magnitude smaller than for the inverse beta decay reaction, the elastic scattering reactions are of particular significance for water Ćerenkov detectors, because unlike the products of the inverse beta decay reaction, the electrons scattered by neutrinos retain some of the direction information of the incoming neutrinos. Information about the direction of motion of the scattered electron is available from the Ćerenkov light cone, so a Ćerenkov detector observing this reaction has some potential for reconstructing the direction to the supernova source. The famous supernova SN1987A, a gravitational collapse event in the Large Magellanic Cloud outside our galaxy, was the first to have its neutrino signal detected. It was the first supernova visible to the naked eye since early in the seventeenth century [Arn89].

The neutrinos emitted during the explosions of core-collapse SN pass through a very large density gradient and undergo the MSW resonant flavor conversion. The flavor conversion can give information on neutrino mass hierarchy for ν_e, ν_μ and ν_τ. A crucial feature in the study of SN neutrinos comes from the collective neutrino–neutrino interaction at very high densities of the core and this may change the emitted flux of different flavors substantially.

10.4. Gamma Rays

Some discussion on gamma rays has been presented in Section 8.6.4. The properties of the Earth's atmosphere divides gamma ray astronomy into two halves. From the ultraviolet to gamma rays of energy $\sim 20\,\mathrm{GeV}$ the atmosphere is opaque. Thus observations in this energy range must be done with instruments mounted in satellites, carried by balloons, or by space-bound detectors. The ease of detecting the radiation generally goes up with energy, but the strengths of typical astrophysical sources go down. For gamma rays above $20\,\mathrm{GeV}$, interactions in the atmosphere produce showers that can be observed either by the Ćerenkov light produced by the secondaries or, at high altitude, by direct detection of the secondaries.

Nuclear gamma rays originate from nuclear decays. An example is ^{26}Al, which has a 720,000 y lifetime for decay to ^{26}Mg. The decay of the 5^+ ground state populates the first two excited states of ^{26}Mg, which are 2^+ states with energies of 2.938 and 1.809 MeV. The latter state is populated 97% of the time, so the primary signature is a 1.809 MeV γ. The 2.938 MeV state decays to the 1.809 MeV level, so a small number of 1.129 MeV γ's are also produced. The primary site for producing ^{26}Al is thought to be Type II and IIb supernovae. In a similar way, ^{44}Ti decay proceeds with a ~60 year half-life to the ground state of ^{44}Sc, which in turn electron captures to ^{44}Ca. The order of the states in Sc is 2^+ (gs), 1^- (68 keV), and 0^- (146 keV). The decay feeds the second excited state 98% of the time, which then decays through the 1^- state to the ground state, producing γ's of 78 and 68 keV. The subsequent decay to ^{44}Ca has a 4 hour lifetime and produces a 1.157 MeV γ, as the 2^+ first excited state of ^{44}Ca is populated 99% of the time. Various types of supernovae are thought to produce ^{44}Ti, including both Type I and II. As in the case of the ^{26}Al line, the galaxy is effectively transparent to the produced γ ray, so the detection provides a measure of the very recent supernova rate free from worries about obscuration. Typical productions of ^{44}Ti from supernovae are, according to modelers, on the order of 10^{-4} $M_\odot$ per event.

Blazars are highly variable and are bright radio sources. The radio structure consists of knotty jets moving outward at high velocities. The luminosity in gamma rays can exceed that from other wavelengths by up to two orders of magnitude. The variability of the emission can be fast, less than a week. The density of high energy gammas at the source are sufficient that photon–photon pair production would keep them trapped, unless the gammas are highly beamed, as in a relativistic jet. Blazars have been observed producing very high energy gamma rays, up to TeV scales. But most blazars emit low-energy radiation that does not extend into the X-ray. The interpretation is that blazars accelerate electrons to high energies, which then radiate soft synchrotron radiation and hard gammas by inverse Compton scattering.

The plasma in the jet is moving towards us, boosting the energy of the emitted gammas, relative to the jet rest frame. In terms of the redshift z, the Doppler factor is

$$D = \frac{\sqrt{1 - v^2/c^2}}{(1+z)(1 - v\cos\theta/c)}, \tag{10.34}$$

where θ is the observation angle relative to the jet axis. This is the standard relativistic Doppler shift corrected by the redshift. The observed energy of the gamma rays is $E_{max} \sim D\gamma m_e c^2$, where $\gamma m_e c^2$ is the maximum energy of the electrons in the jet rest frame. The Doppler factor also appears in the calculation of the photon density in the blob, and thus of the blob's opacity to high energy γ's. The argument goes as follows: if Δt_{obs} is the fastest observed TeV gamma ray flare variability, then the radius of the blob emitting the photons must be less than $\sim cD\Delta t_{obs}$. Thus a larger D means a lower photon density. Other interesting sources of high energy γ's are isolated and binary pulsars. Gamma ray emission can be very strong, representing 10% or more of the spin-down energy of some pulsars.

10.5. Exercises

1. At what speed, $\beta = v/c$, is a particle's kinetic energy equal to its rest mass energy?
2. (a) What is the speed, $\beta = v/c$, of an electron ($m_e = 0.511\,\mathrm{MeV/c^2}$) with a momentum of 10 MeV/c?
   (b) What is the speed, $\beta = v/c$, of a proton ($m_p = 940\,\mathrm{MeV/c^2}$) with a momentum of 10 MeV/c?
3. A supernova explodes in the Large Magellanic Cloud (LMC), 50 kpc from Earth. The burst of neutrinos turns on suddenly at $t = 0$ and then declines in time as $\exp(-t/t_0)$, where $t_0 \sim 2\,\mathrm{s}$. The distribution in energy can be taken as Fermi–Dirac characterized by a temperature $T \sim 5\,\mathrm{MeV}$. The neutrinos are observed on Earth in a detector, and the spread in the arrival times of the neutrinos recorded. Estimate what neutrino mass would cause the duration of the observed events to approximately double, due to spreading of the signal during transit from the LMC to Earth. Note this is about the kinematic effects of the neutrino mass, not about oscillations.
4. A proton with a kinetic energy of 10^{10} eV collides with a proton at rest. Find (a) the velocity of the center of mass, (b) the total momentum and total energy in the laboratory frame, and (c) the kinetic energy of the particles in center of mass frame.
5. Consider a muon ($m_0 = 0.106\,\mathrm{GeV/c^2}$) with 1 GeV energy in the lab frame. In the muon's rest frame it only lives (on average) $t = 2\,\mathrm{ms}$.

   (a) On average how long does this particle live in the laboratory?
   (b) On average how far does this particle travel in the laboratory before decaying?

6. Consider an anti-proton with 10 GeV/c of momentum in the lab frame.

   (a) What is the energy of the anti-proton in the lab frame?
   (b) How fast is the anti-proton moving in the lab frame?

7. What is the approximate spectrum of cosmic rays about 1 GeV? What are the features at 10^{15} eV and 10^{19} eV? What causes the near-isotropy at low energy? What composes cosmic rays?
8. Find the minimal energy E_{th} of a proton scattering on a photon with the typical energy of the cosmic microwave background ($T \simeq 2.7\,\mathrm{K}$) for the process $p + \gamma \to p + \pi_0$.
9. Consider the 2-particle decay $\pi_0 \to 2\gamma$. What are the minimal and maximal photon energies, if the pion moves with velocity v? What is the shape of the photon spectrum dN/dE?
10. How much energy is available in the CM for production of particles from a 10 GeV/c anti-proton colliding with a proton at rest?
11. Assuming fixed target proton–proton collisions is to be used to create two anti-protons, what energy proton beam (E_b) is necessary?
12. An unstable particle is at rest and suddenly breaks up into two fragments. No external forces act on the particle or its fragments. One of the fragments has a velocity of +0.800c and a mass of 1.67×10^{-24} kg, and the other has a mass of 5.01×10^{-27} kg. What is the velocity of the less massive fragment?
13. Can a single photon in free space split into an electron and positron? Demonstrate.
14. What are the main features of cosmic rays below about 10^{15} eV that suggest a supernova shock origin?
15. Assume that the mean radius of dust grains in intergalactic space with a uniform number density n grains/cm^3 is given by R. (a) Show that the mean free path d_0 of a photon in interstellar dust is given by $d_0 = 1/(n\pi R^2)$. (b) Starlight traveling toward an Earth observer a distance d from a star has intensity $I = I_0 \exp(d/d_0)$. In the vicinity of the Sun measurement of I yields $d_0 = 3000\,\mathrm{ly}$. If $R = 10^{25}$ cm, calculate n. (c) The average mass density of solid material in the galaxy is $2\,\mathrm{g/cm^3}$ and in the disk the density of stars is about $1\mathrm{M}_\odot/300\,(\mathrm{ly})^3$. Calculate the ratio of the mass of dust to $1\mathrm{M}_\odot$ in $300\,(\mathrm{ly})^3$.
16. Microlensing occurs when an object of stellar mass or less causes a temporary brightening in a background star. How could you tell a lens that is a $10\,M_\odot$ black hole from a $10\,M_\odot$ main sequence star?
17. Low-energy cosmic rays. The radius of curvature R (in centimeters) of a particle of energy E (in ergs) and nuclear charge Ze (where $e = 4.8 \times 10^{-10}$ esu) in a magnetic field of strength B (in Gauss) is $R = E/(ZeB)$. Consider a supernova that occurs a distance $d = 1000\,\mathrm{pc}$ from us, and

assume that the average strength of the interstellar magnetic field is $B = 3 \times 10^{-6}$ Gauss.

(a) What is the energy in GeV needed for an iron nucleus so that $R = d$? At this energy and above, the nucleus will basically come directly at us.

(b) If $R \ll d$, the tangled nature of the interstellar magnetic field means that the nucleus will basically undergo a random walk. Note, though, that the nucleus will not lose significant energy from the magnetic deflections. In this random walk, the nucleus will effectively take $(d/R)^2$ steps of length R each to go a net distance d. Use this, plus special relativistic principles, to compute the energy E_{min} (in GeV) below which most ^{10}Be nuclei (with rest-frame half-life of 3.9×10^6 yr) will decay in transit from the distance $d = 1000\,\text{pc}$.

18. Assume that the neutrinos are relativistic, and distributed as a black-body. Their background has an expected temperature of $T = 1.9\,\text{K}$. Assuming for simplicity that all the neutrinos are at the thermal peak $E = 2.7kT$, compute to within a factor of 5 the volume of water (at $1\,\text{g}\,\text{cm}^{-3}$) needed so that we would expect 100 scatterings in one year (so that we have good statistics and can derive interesting quantities). Assume that all electrons in the water can potentially scatter the neutrinos, and that any scattering will be detected.

(a) Find a body of water with approximately this volume.

(b) Give a simple argument that, in reality, your answer is a tremendous underestimate of the volume actually needed to detect electron scattering of these neutrinos.

19. Consider MSW neutrino oscillations within a supernova involving the electron neutrino, for an atmospheric neutrino δm^2 of $3 \times 10^{-3}\,\text{eV}^2$.

(a) Calculate the critical density at which the level crossing will occur. Using typical supernova shell densities locate where in the star's mantle this crossing would occur.

(b) Calculate the matter contribution to the electron neutrino effective m^2 for a 10 MeV neutrino just after it has left the neutrinosphere, e.g., at a matter density of $10^{11}\,\text{g/cm}^3$. You can assume the matter has an equal number of protons and neutrons.

20. In the 1940s Enrico Fermi proposed that cosmic ray acceleration is like a tennis game between giants. Molecular clouds drift this randomly. These clouds have magnetic fields, so if cosmic rays hit them, they curve around and eventually move out in the opposite direction from where they came. If the cloud was initially moving towards the cosmic ray, this results in a gain of energy, but if the cloud was initially moving away from the cosmic ray this results in the loss of energy.

    (a) Derive the average rate of energy gain for cosmic rays in a simplified situation. Here we assume that

    (1) the cosmic rays are all moving relativistically,

    (2) as seen in the Galactic rest frame, the molecular clouds move with equal probability left or right on a straight line (you can call it the x-axis),

    (3) the cosmic rays themselves only move left or right on the x-axis, both before and after their scattering,

    (4) all clouds move with a speed of 10^{-4} c in the Galactic rest frame, and

    (5) cosmic rays interact with the clouds every 1000 years as seen in the Galactic reference frame, and the clouds themselves are 1 pc in size. We assume that the cosmic rays bounce straight back from the clouds, like a mirror, with no energy loss as seen in the cloud rest frame. With this setup, if a cosmic ray starts with energy $E \gg m_p c^2$, how long will it take to reach an energy $2E$?

    (b) Discuss whether this is a viable mechanism for producing cosmic rays up to 10^{15} eV if the initial energy is $\ll 10^{15}$ eV.

21. The typical energy of a neutrino in the Cosmic Neutrino Background is $E_\nu \sim kT_\nu \sim 10^{-4}$ eV. What is the approximate interaction cross-section, Eq. (9.30), for one of these cosmic neutrinos? Suppose you had a large lump of ^{56}Fe (with density $\rho = 7900\,\mathrm{kg\,m^{-3}}$). What is the number density of protons, neutrons, and electrons within the lump of iron? How far, on average, would a cosmic neutrino travel through the iron before interacting with a proton, neutron, or electron? (Assume that the cross-section for interaction is simply σ_ν, Eq. (9.30), regardless of the type of particle the neutrino interacts with.)

Appendix A

Units

A.1. Microscopic Units

In the scattering of microscopic systems, typical values of lengths, masses, charges, energies etc. are extremely small when expressed in the scales of the Gaussian unit system or any other macroscopic system of units. Therefore, it is convenient to introduce appropriate microscopic units. We discuss this problem below, in two situations where different scales are used: (a) the scattering of nuclei at non-relativistic energies, and (b) the scattering of atoms, ions and molecules.

A.1.1. *Nuclear collisions*

In the scattering of nuclear systems, the typical size is the nuclear radius: $R \sim 10^{-13} - 10^{-12}\,\text{cm}$. One then uses as unit of length the Fermi[1] (fm), defined by the relation: $1\,\text{fm} = 10^{-13}\,\text{cm}$. This leads naturally to the cross-section unit of fm^2; $1\,\text{fm}^2 = 10^{-26}\,\text{cm}^2$. However, the cross-sections are usually expressed in barns; $1\,\text{b} = 10^{-24}\,\text{cm}$; 1 mb (*millibarns*) $= 10^{-3}$ b; 1 μb (*microbarns*) $= 10^{-6}$ b; 1 nb (*nanobarns*) $= 10^{-9}$ b; 1 pb (*picobarns*) $= 10^{-12}$ b.

In nuclear physics, electromagnetic forces play a major role. In such systems, the natural unit of charge is the absolute value of the electronic charge e, which has the value $e = 4.803\,\text{esu} = 1.6 \times 10^{-19}\,\text{C}$. Usually, a scattering experiment uses a beam of charged projectiles accelerated by electromagnetic forces. In an electrostatic accelerator the kinetic energy of a projectile of charge q subjected to a potential ΔU is $E_{lab} = q\,\Delta U$.

[1]This unit is also known as femtometer. For this reason, one uses the notation 'fm' instead of 'Fm'.

This leads to the use of the electron Volt, defined as the kinetic energy acquired by an electron through the acceleration by a potential of 1 Volt. It can be easily checked that $1\,\text{eV} = 1.602 \times 10^{-12}$ erg. For investigations of the internal structure of nuclei, the projectile's energy is usually high enough to surmount the potential barrier (resulting from the interplay between short range attractive forces and Coulomb repulsion) and excite the intrinsic states of interest. This amounts to energies of the order of MeV (10^6 eV) or GeV (10^9 eV). Nowadays, the beam energy is frequently expressed by the ratio between the collision energy in the laboratory frame to the number of nucleons in the projectile, denoted by ϵ. That is $E_{lab} = \epsilon \cdot A_p$. It can be readily checked that $E_{cm} = \epsilon \cdot A_p A_T/(A_p + A_T)$.

Particle masses are frequently multiplied by c^2 and given in MeV. Some examples are:

$$\begin{aligned} m_p\, c^2 &= 938.271\,\text{MeV}, \\ m_n\, c^2 &= 939.565\,\text{MeV}, \\ m_e\, c^2 &= 510.998\,\text{keV}, \\ m_N\, c^2 &= 931.494\,\text{MeV}, \\ M(^{12}\text{C})\, c^2 &= 12 \times m_N\, c^2, \end{aligned}$$

where the labels p, n, e stand respectively for proton, neutron and electron, m_N is the average mass of a bound nucleon inside the nucleus and $M(^{12}\text{C})$ is the mass of the $A = 12$ carbon isotope.

The Coulomb potential energy in a collision between nuclei with atomic numbers Z_P and Z_T can be written (for the Gaussian system of units)

$$V(r) = \frac{Z_p Z_T\, e^2}{r}.$$

In order to express it in MeV and r in fm, we must evaluate e^2 accordingly. Writing

$$e^2 \equiv \frac{e^2}{\hbar c}\, \hbar c,$$

and using the explicit value of the fine structure constant

$$\alpha = \frac{e^2}{\hbar c} = \frac{1}{137.036}$$

and the constant

$$\hbar c = 197.327\,\text{MeV fm},$$

we get

$$e^2 = 1.44\ \text{MeV fm}.$$

A.1.2. *Collisions of atoms, ion or molecules*

Atoms and molecules are about 10^5 times larger than nuclei. Their radii can be measured in Angstroms, where $1\,\text{Å} = 10^{-8} cm = 10^5\,\text{fm}$. On the other hand their appropriate energy scale is much smaller. Typical excitation energies are of the order of the eV or a fraction of it. The Coulomb energy can be expressed in terms of these units using the relations

$$\hbar c = 1973.29\ \text{eV Å},$$
$$e^2 = 14.4\ \text{eV Å}.$$

However, in practical applications Hartree's atomic units (au) are used. The basic quantities in this system of units are discussed below.

In Hartree's atomic unit system, the unit of length is the radius of the classical orbit of the electron in a hydrogen atom. According to Bohr's atomic model, this radius is

$$1au = a_0 = \frac{\hbar^2}{m_e e^2} = 5.29 \times 10^{-9}\,\text{cm} = 0.529\,\text{Å}.$$

An important remark is that, despite the use of atomic units for lengths, cross-sections in atomic and molecular physics are usually given in cm^2. The units of mass and charge are mass and the charge of the electron, that is m_e and e. To complement the unit system, one adopts $\hbar$ as the unit of action and, as in the Gaussian system, sets $\varepsilon_0 = 1/4\pi$, $\mu_0 = 4\pi/c^2$. In this way, in au we have: $m_e = e = a_0 = \hbar = 4\pi\varepsilon_0 = \mu_0 c^2/4\pi = 1$. The atomic units for other physical quantities are defined in terms of these basic units. We give below a few examples. (a) The unit of velocity is that of an electron in Bohr's orbit, v_0. It can be readily checked that $1\,au = v_0 = \alpha c = 2.18769 \times 10^8\,\text{cm/s}$, so that $c = 137.036\,au$. The au of energy, the Hartree (H) is twice the ionization energy of the hydrogen atom. That is: $1\,\text{H} = e^2/a_0 = 27.2116\,\text{eV}$. (b) The unit of time is $1\,au = a_0/v_0 = 2.41889 \times 10^{-17}\,\text{s}$.

In the case of electromagnetic radiation, the photon energy can also be specified by the frequency or by the wavelength. The spectroscopic tradition uses MHz (MegaHertz) for photon frequencies in the infrared and microwave regions and wave numbers in the visible and ultraviolet regions. With this in mind, the energy of 1 *au* corresponds to a frequency of $\nu = 6.57968 \times 10^9$ MHz, wavelength $\lambda = 455.633$ Å and wave number $k = 219475\,\text{cm}^{-1}$. This energy can also be expressed as a temperature (in units of the Boltzmann constant, $k = 8.617 \times 10^{-5}$ eV/K), 1 $au = 3.15777 \times 10^5$ K.

A.2. Constants and Conversion Factors

The fundamental constants presented here are recommended by the CODATA (Committee on Data for Science and Technology), USA. The values are periodically revised and can be accessed electronically at the NIST (National Institute of Standards and Technology) website. The values inside parenthesis indicate the standard deviation. For example, in the first line the error in the electron charge must be understood as $0.000000063 \times 10^{-19}$ C.

A.2.1. *Constants*

Electric charge	$e = 1.602176462(63) \times 10^{-19}\text{ C} = 1.200\sqrt{\text{MeV}\cdot\text{fm}}$ $e^2 = 1.440\text{ MeV}\cdot\text{fm}$
Planck constant	$h = 6.62606876(52) \times 10^{-27}\text{ erg}\cdot\text{s}$ $= 4.13566727(16) \times 10^{-21}\text{ MeV}\cdot\text{s}$ $\hbar = h/2\pi = 1.054571596(82) \times 10^{-27}\text{ erg}\cdot\text{s}$ $= 6.58211889(26) \times 10^{-22}\text{ MeV}\cdot\text{s}$
Speed of light	$c = 299792458\text{ m/s}$ $\hbar c = 1.9733 \times 10^{-11}\text{ Mev.cm} = 197.327\text{ MeV}\cdot\text{fm}$
Gravitational constant	$G = 6.673(10) \times 10^{-11}\text{ m}^3\cdot\text{kg}^{-1}\cdot\text{s}^{-2}$
Boltzmann constant	$k = 1.3806503(24) \times 10^{-16}\text{ erg/K}$
Avogadro number	$N_A = 6.02214199(47) \times 10^{23}\text{ mol}^{-1}$
Molar volume	$V_m = 22.413996(39)$ l/mol (273.15 K; 101325 Pa)
Faraday constant	$F = 96485.3415(39)$ C/mol
Compton wavelengths	$\bar{\lambda}_e = \hbar/m_e c = 386.1592642(28)$ fm (electron) $\bar{\lambda}_p = \hbar/m_p c = 02103089089(16)$ fm (proton)
Nuclear magneton	$\mu_N = 3.152451238(24) \times 10^{-14}$ MeV/T
Bohr magneton	$\mu_B = 5.788381749(43) \times 10^{-11}$ MeV/T
Fine structure constant	$\alpha = e^2/\hbar c = 1/137.03599976(50)$
Electron classical radius	$r_e = e^2/m_e c^2 = 2.817940285(31)$ fm
Bohr radius	$a_0 = \hbar^2/m_e e^2 = 0.5291772083(19) \times 10^{-8}$ cm

A.2.2. *Masses*

Electron	$m_e = 9.10938188(72) \times 10^{-28}$ g $= 5.485799110(12) \times 10^{-4}$ u
	$= 0.510998902(21)$ MeV$/c^2$
Muon	$m_\mu = 0.1134289168(34)$ u $= 105.6583568(52)$ MeV$/c^2$
Pions	$m_{\pi^0} = 134.9764(6)$ MeV$/c^2$
	$m_{\pi^\pm} = 139.56995(35)$ MeV$/c^2$
Proton	$m_p = 1.67262158(13) \times 10^{-24}$ g $= 1.00727646688(13)$ u
	$= 938.271998(38)$ MeV$/c^2$
Neutron	$m_n = 1.00866491578(55)$ u $= 939.565330(38)$ MeV$/c^2$
Hydrogen atom	$m_H = 1.007825036(11)$ u $= 938.791$ MeV$/c^2$

A.2.3. *Conversion factors*

Length	1 fermi = 1 fm = 10^{-15} m = 10^{-13} cm
Area	1 barn = 1 b = 10^{-24} cm^2 = 10^2 fm^2
Mass	1 unit of atomic mass = 1 u = (1/12) m($^{12}_{6}$C)
	$= 1.66053873(13) \times 10^{-24}$ g = 931.494013(37) MeV$/c^2$
	$= 1822.872\, m_e$
Energy	1 eV = $1.602176462(63) \times 10^{-12}$ erg
	$= 1.073544206(43) \times 10^{-9}$ u $\cdot$ c^2
	1 erg = 10^{-7} J
Temperature ($k = 1$)	1 MeV = 1.16×10^{10} K = 1.78×10^{-30} kg

A.3. Astronomical Units

A.3.1. *Distances*

1. The typical separation between stars in the disk of the Milky Way (our own galaxy) is around a parsec

$$1\,\text{pc} = 3.1 \times 10^{16}\text{m}$$

and this is a convenient unit to use. For comparison, the distance from the Earth to the Sun, which defines the *Astronomical Unit*, is

$$1\,\text{AU} = 1.5 \times 10^{11}\text{m} = 0.000005\,\text{pc}.$$

2. Most of the stars in our galaxy — including the Sun — are contained in a disc with a radius of about

$$10\,\text{kpc} = 10^4\,\text{pc}$$

and thickness of 300 pc; these are the so-called "Population I" stars. There is also a nuclear bulge and a spheroidal distribution of "Population II" stars; many of the latter are contained in globular clusters, each of which has a radius of about 10 pc and contains around 106 stars (see Fig. 1.1). These features are common to all spiral galaxies.
3. The typical separation between large galaxies like our own is around

$$1\,\text{Mpc} = 10^6\,\text{pc},$$

but many galaxies are assembled into groups or clusters where the separation may be much smaller. For example, the Milky Way is part of a Local Group and this also comprises the Large Magellanic Cloud (LMC) at a distance of 55 kpc, the Small Magellanic Cloud (SMC) at 67 kpc, M31 (Andromeda) at 710 kpc, M33 at 850 kpc, and several dozen dwarf galaxies. Other nearby groups are M81 at a distance of 2.9 Mpc and M101 at 6.8 Mpc.
4. 1 pc $= 2.06 \times 10^5$ AU $= 3.1 \times 10^{18}$ cm was originally defined as the distance to a star with a parallax equal to one second of arc. A parallax is an angle at which the radius of Earth's orbit around the Sun appears at the distance of the star. Note: 2.06×10^5 is the number of seconds of arc in 1 radian.
5. $d_H = c/H_0 \simeq 1.4 \times 10^{28}$ cm $\simeq 4$ Gpc $=$ Hubble distance, where $H_0 \simeq 70\,\text{km s}^{-1}\,\text{Mpc}^{-1}$ is Hubble constant; $c = 3 \times 10^{10}\,\text{cm s}^{-1}$ is the speed of light. Hubble distance is approximately the radius of observable Universe, with us at the "center".
6. $1\ R_\odot \simeq 7 \times 10^{10}$ cm $=$ solar radius. Most stars have radii between 10^{-2} $R_\odot$ (white dwarfs) and 10^3 $R_\odot$ (red supergiants); neutron stars have radii of about $10^6\,\text{cm} = 10\,\text{km}$.

A.3.2. *Time*

1. 1 year $= 3 \times 10^7$ s.
2. $H_0^{-1} = d_H c^{-1} \simeq 1.4 \times 10^{10}$ years $=$ Hubble time, approximately the age of the Universe known to us.

A.3.3. *Masses*

1. The mass of the Sun

$$1\,M_\odot = 2 \times 10^{30}\ \text{kg}$$

is a convenient unit to use for this purpose. All stars have a mass in the range $0.1\,\text{M}_\odot$ to $300\,\text{M}_\odot$, smaller objects being too small to ignite their nuclear fuel and larger ones being unstable to pulsations.
2. Globular clusters have a mass of around $10^6\,\text{M}_\odot$, while spiral galaxies like our own all have a mass of order $10^{11}\,\text{M}_\odot$. There are other sorts of galaxies and these span a much larger mass range — from $10^8\,\text{M}_\odot$ to $10^{12}\,\text{M}_\odot$, but spirals are most numerous.
3. Clusters have a mass of around $10^{13}\,\text{M}_\odot$, superclusters around $10^{15}\,\text{M}_\odot$, and filaments around $10^{17}\,\text{M}_\odot$. The visible Universe itself has a total mass of around $10^{22}\,\text{M}_\odot$, so the number of galaxies in the Universe is comparable to the number of stars in the galaxy.

A.3.4. *Luminosities, magnitudes*

1. $L_\odot = 4 \times 10^{33}\,\text{erg s}^{-1}$ = solar luminosity.
2. Known stars have luminosity in the range $10^{-5} - 10^6\,L_\odot$.
3. $M_{bol} = 4.8 - 2.5\log(L/L_\odot)$ = absolute bolometric magnitude of a star with a luminosity L. "Bolometric" means integrated over all stellar spectrum: infrared, optical, ultraviolet. $M_{bol,\odot} = +4.74$.
4. $M_V = M_{bol} - BC$ = absolute visual magnitude of a star; BC is a bolometric correction, and V indicates that we are referring to that part of stellar radiation that is emitted in the "visual" part of the spectrum, i.e. at about $5 \times 10^{-5}\,\text{cm}$. BC depends on stellar temperature. $BC_\odot = -0.08$.
5. M_B = absolute blue magnitude of a star; B indicates that we are referring to that part of stellar radiation that is emitted in the "blue" part of the spectrum, i.e. at about $4 \times 10^{-5}\,\text{cm}$.
6. $m_{bol} = M_{bol} + 5\log(d/10\,\text{pc})$ = apparent bolometric magnitude of a star at a distance d.
7. $V = M_V + 5\log(d/10\,\text{pc})$ = apparent visual magnitude of a star as seen in the sky.
8. $B = M_B + 5\log(d/10\,\text{pc})$ = apparent blue magnitude of a star as seen in the sky.

9. $B - V = M_B - M_V$ = difference between visual and blue magnitudes; it is called a color index, and it is a measure of a color. i.e. of a shape of stellar spectrum between 4×10^{-5} and 5×10^{-5} cm. Very hot stars are blue, and may have $B - V = -0.3$, whereas very cold stars are red and may have $B - V = +1.5$. In general, color index is a good indicator of the temperature of the stellar "surface", or photosphere.

A.3.5. *Temperatures and luminosities*

1. Temperature is measured in Kelvins (K). A unit area of a black body radiates a "flux" of energy given as

$$\Phi = \sigma T^4,$$

where $\sigma = 5.67 \times 10^{-5}$ erg s^{-1} cm^{-2} deg^{-4} is the Stefan–Boltzman constant. The flux of energy is measured in [erg s^{-1} cm^{-2}].

2. A star with a radius R and luminosity L has an "effective" temperature T_{eff} defined with the relation

$$L = 4\sigma R^2 T_{eff}^4.$$

The sun has $T_{eff} = 5.8 \times 10^3$ K.
The coolest hydrogen-burning stars have $T_{eff} \simeq 2 \times 10^3$ K.
The hottest main sequence stars have $T_{eff} \simeq 5 \times 10^4$ K.
The hottest white dwarfs have $T_{eff} \simeq 3 \times 10^5$ K.
The hottest neutron stars have $T_{eff} \simeq 3 \times 10^7$ K.

Appendix B

Space and Time

B.1. Introduction

Special relativity is founded on two basic postulates:

1. *Galilean invariance: The laws of nature are independent of any uniform, translational motion of the reference frame.*
2. *The speed c of light in empty space is independent of the motion of its source.*

The second postulate implies that c takes the same constant value in all inertial frames [Ein05]. Transformations between inertial frames follow, which have far reaching physical consequences. The speed of light in vacuum is

$$c = 299,792,458\,\mathrm{m/s}. \tag{B.1}$$

The units for time and length are not independent, as the speed of light is a universal constant. This allows us to define natural units, which are frequently used in nuclear, particle and astrophysics. In these units, $c = 1$ is taken as a dimensionless constant, and $1\ [s] = 299,792,458\,\mathrm{m}$ holds. Then c disappears in calculations. Appropriate factors have to be recovered by dimensional analysis to convert back to conventional units, e.g., $x = t$ in natural units and in conventional units $x = ct$ with x in meters and c is given by Eq. (B.1).

B.2. Transformation between Moving Frames

B.2.1. *Lorentz invariance*

Consider two inertial frames with uniform relative motion $\mathbf{v}$: F with coordinates $(t, \mathbf{x})$ and F' with coordinates $(t', \mathbf{x}')$, as shown in Fig. B.1.

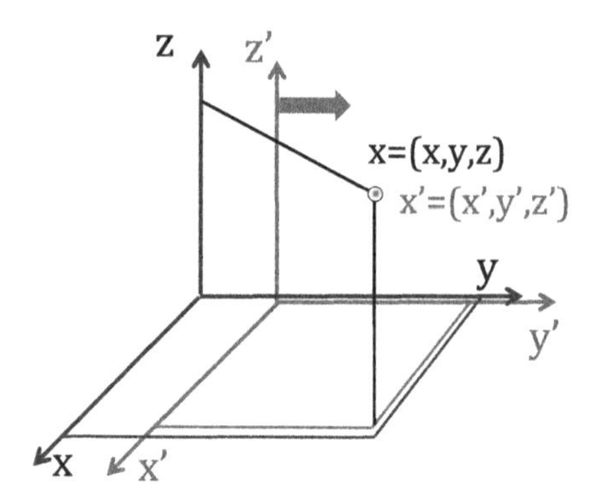

Figure B.1. Coordinates in different reference frames.

At time $t = t' = 0$ their two origins coincide. Imagine a spherical shell of radiation originating at time $t = 0$ from $\mathbf{x} = \mathbf{x}' = 0$. A light wavefront is described by

$$c^2t^2 - x^2 - y^2 - z^2 = 0 \text{ in } F, \tag{B.2}$$

and by

$$c^2t'^2 - x'^2 - y'^2 - z'^2 = 0 \text{ in } F'. \tag{B.3}$$

Thus, the fact that the speed of light is constant in both systems introduces an invariant measure of distances in the space-time which we can exploit with a simplified notation for space-time variables, which we discuss next.

We define *4-vectors* ($\alpha = 0, 1, 2, 3$) by

$$(x^\alpha) = (ct,\ \mathbf{x}) \text{ and } (x_\alpha) = (ct,\ -\mathbf{x}). \tag{B.4}$$

Due to a more general notation, to be explained in section B.3, the components x^α are called *contravariant* and the components x_α *covariant*. In matrix notation the contravariant 4-vector (x^α) is represented by a column and the covariant 4-vector (x_α) as a row.

The *summation convention* is defined by

$$x_\alpha x^\alpha = \sum_{\alpha=0}^{3} x_\alpha x^\alpha, \tag{B.5}$$

and will be employed from here on. Equations (B.2) and (B.3) read then

$$x_\alpha x^\alpha = x'_\alpha x'^\alpha = 0. \tag{B.6}$$

Lorentz transformations are defined as the group of transformations which leave the distance $s^2 = x_\alpha x^\alpha$ invariant:

$$x_\alpha x^\alpha = x'_\alpha x'^\alpha = s^2, \tag{B.7}$$

where $\mathbf{x}$ and t can have any value. Equation (B.7) implies (B.6), but the reverse is not true. The scale transformation $x'^\alpha = \lambda\, x^\alpha$ leaves (B.6) invariant, but not (B.7). But, if the initial condition $\mathbf{x}' = \mathbf{x} = 0$ for $t' = t = 0$ is removed, the equation $(x_\alpha - y_\alpha)(x^\alpha - y^\alpha) = (x'_\alpha - y'_\alpha)(x'^\alpha - y'^\alpha)$ still holds.

Poincaré transformations are defined as the group of transformations which leave

$$s^2 = (x_\alpha - y_\alpha)(x^\alpha - y^\alpha) \text{ invariant.} \tag{B.8}$$

In contrast to the Lorentz transformations the Poincaré transformations include invariance under *translations*

$$x^\alpha \to x^\alpha + a^\alpha \quad \text{and} \quad y^\alpha \to y^\alpha + a^\alpha \tag{B.9}$$

where a^α is a constant vector.

B.2.2. *Minkowski space*

Equation (B.8) gives the invariant metric of the *Minkowski space.* The difference from the norm of 4-dimensional Euclidean space is the relative minus sign between time and space components. The *light cone* of a 4-vector x_0^α is defined as the set of vectors x^α which satisfy

$$(x - x_0)^2 = (x_\alpha - x_{0\alpha})\,(x^\alpha - x_0^\alpha) = 0.$$

The light cone separates events which are *timelike* and *spacelike* with respect to x_0^α. That is,

$$(x - x_0)^2 > 0 \quad \text{for timelike}$$

and

$$(x - x_0)^2 < 0 \quad \text{for spacelike.}$$

The time ordering of spacelike points is distinct in different inertial frames, whereas it is the same for timelike points. For the choice $x_0^\alpha = 0$ this Minkowski space situation is depicted in Fig. B.2. On the abscissa we have the projection of the three dimensional Euclidean space on $r = |\mathbf{x}|$. The regions *future* and *past* of this figure are the timelike points of $x_0 = 0$, whereas *elsewhere* are the spacelike points.

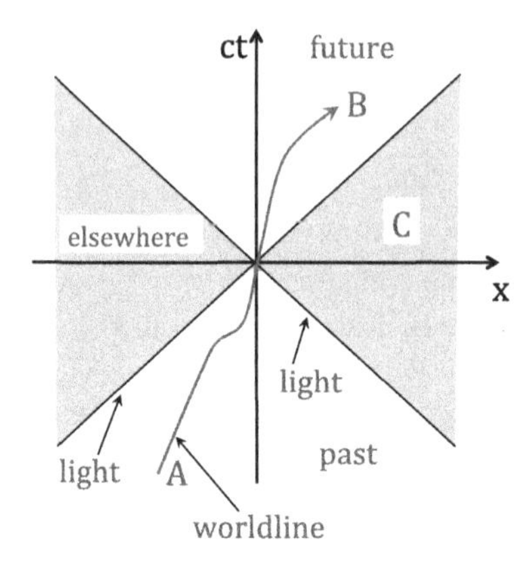

Figure B.2. As seen from the spacetime point at the origin, the spacetime points in the forward light cone are in the future, those in the backward light cone are in the past and the spacelike points are "elsewhere", because their time-ordering depends on the inertial frame chosen.

B.2.3. *Addition of velocities*

Assuming **v** in x-direction and considering a one-dimension case,

$$c^2t^2 - (x^1)^2 = c^2t'^2 - (x'^{\,1})^2 \tag{B.10}$$

We define $\beta = v/c$ and look for a linear transformation

$$\begin{pmatrix} x'^{\,0} \\ x'^{\,1} \end{pmatrix} = \begin{pmatrix} a & b \\ d & e \end{pmatrix} \begin{pmatrix} x^0 \\ x^1 \end{pmatrix} \tag{B.11}$$

which fulfills (B.10) for all $x^0 = ct$, x^1. Choosing $\begin{pmatrix} x^0 \\ x^1 \end{pmatrix} = \begin{pmatrix} 1 \\ 0 \end{pmatrix}$ gives

$$a^2 - d^2 = 1 \Rightarrow a = \cosh\xi, \quad d = \pm\sinh\xi \tag{B.12}$$

and choosing $\begin{pmatrix} x^0 \\ x^1 \end{pmatrix} = \begin{pmatrix} 0 \\ 1 \end{pmatrix}$ gives

$$b^2 - e^2 = -1 \Rightarrow e = \cosh\eta, \quad b = \pm\sinh\eta. \tag{B.13}$$

Using $\begin{pmatrix} x^0 \\ x^1 \end{pmatrix} = \begin{pmatrix} 1 \\ 1 \end{pmatrix}$ leads to

$$[\cosh\xi + \sinh\eta]^2 - [\sinh\xi + \cosh\eta]^2 = 0 \;\Rightarrow\; \xi = \eta.$$

In equation (B.12) we use the convention $d = -\sinh\xi$. We obtain

$$\begin{pmatrix} a & b \\ d & e \end{pmatrix} = \begin{pmatrix} \cosh\xi & -\sinh\xi \\ -\sinh\xi & \cosh\xi \end{pmatrix}, \tag{B.14}$$

where ξ is called *rapidity*. It behaves as an angle in a hyperbolic geometry.

In components (B.14) reads

$$x'^0 = +x^0 \cosh\xi - x^1 \sinh\xi,$$
$$x'^1 = -x^0 \sinh\xi + x^1 \cosh\xi. \tag{B.15}$$

For spacelike points such as $x^1 > x^0 > 0$, a value ξ_0 for the rapidity exists, so that

$$0 = +x^0 \cosh\xi_0 - x^1 \sinh\xi_0.$$

Therefore,

$$\text{sign}(x'^0) = -\text{sign}(x^0)$$

for $\xi > \xi_0$, with reverts the time-ordering. For timelike points such a reversal of the time-ordering is impossible as it would mean $|x^0| > |x^1|$. In Figure B.2 we call this spacelike (with respect to $x_0 = 0$) region *elsewhere* in contrast to *future* and *past*.

From the point of view of frame F, the origin $x'^1 = 0$ of F moves with constant velocity v. In F this corresponds to the equation

$$0 = -x^0 \sinh\xi + x^1 \cosh\xi.$$

Therefore, the rapidity is related to the velocity between the frames by

$$\beta = \frac{v}{c} = \frac{x^1}{x^0} = \frac{\sinh\xi}{\cosh\xi} = \tanh\xi. \tag{B.16}$$

The *Lorentz gamma-factor* is defined by

$$\gamma = \cosh\xi = \frac{1}{\sqrt{1-\beta^2}} \quad \text{and} \quad \gamma\,\beta = \sinh\xi. \tag{B.17}$$

The transformation (B.11) can also be written as

$$x'^0 = \gamma\,(x^0 - \beta\,x^1)\,,$$
$$x'^1 = \gamma\,(x^1 - \beta\,x^0)\,, \tag{B.18}$$

or

$$ct' = \gamma\,(ct - \beta\,x),$$
$$x' = \gamma\,(x - \beta\,ct). \tag{B.19}$$

These equations are often called *Lorentz transformations.*

Two subsequent Lorentz transformations with rapidity ξ_1 and ξ_2 combine as

$$\begin{pmatrix} +\cosh\xi_2 & -\sinh\xi_2 \\ -\sinh\xi_2 & +\cosh\xi_2 \end{pmatrix} \begin{pmatrix} +\cosh\xi_1 & -\sinh\xi_1 \\ -\sinh\xi_1 & +\cosh\xi_1 \end{pmatrix}$$
$$= \begin{pmatrix} +\cosh(\xi_2+\xi_1) & -\sinh(\xi_2+\xi_1) \\ -\sinh(\xi_2+\xi_1) & +\cosh(\xi_2+\xi_1) \end{pmatrix}. \tag{B.20}$$

The rapidities add up as

$$\xi = \xi_1 + \xi_2 \tag{B.21}$$

in the same way as velocities do under Galilei transformations or angles for rotations about the same axis. The inverse to the transformation with rapidity ξ_1 is obtained for $\xi_2 = -\xi_1$.

The relativistic addition of velocities can be obtained from (B.21). For $\beta_1 = \tanh\xi_1$ and $\beta_2 = \tanh\xi_2$,

$$\beta = \tanh(\xi_1 + \xi_2) = \frac{\beta_1 + \beta_2}{1 + \beta_1\beta_2}. \tag{B.22}$$

As expected, the velocities cannot add to values greater than the speed of light. The difference between Galilean and relativistic velocity addition is shown schematically in Fig. B.3.

Lorentz transformations imply the existence of a *time dilatation*, because a moving clock ticks slower. In F the position of the origin of

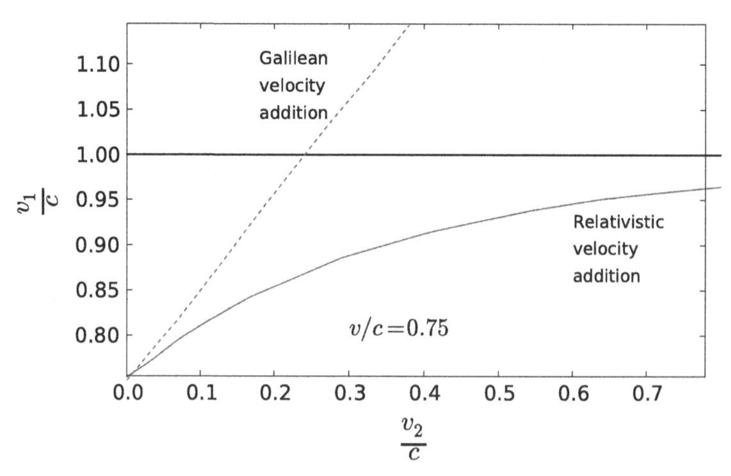

Figure B.3. Difference between Galilean, $v_2 = v_1 + v$, and relativistic velocity addition, $v_2 = (v_1 + v)/(1 + vv_1/c^2)$.

F' is given by

$$x^1 = \frac{v\,x^0}{c} = x^0 \tanh \xi$$

and the Lorentz transformation (B.15) gives

$$x^{'0} = x^0 \cosh \xi - x^0 \sinh \xi \tanh \xi = \frac{\cosh^2 \xi - \sinh^2 \xi}{\cosh \xi} x^0 = \frac{x^0}{\cosh \xi} < x^0. \tag{B.23}$$

Now, in F' the position of the origin of F is given by

$$x^{'1} = -x^{'0} \tanh \xi$$

and with this relation between $x^{'1}$ and $x^{'0}$ the inverse Lorentz transformation gives

$$x^0 = \frac{x^{'0}}{\cosh \xi}.$$

This is not inconsistent, because equal times at separate points in one frame are not equal in another. Note that the definition of time in one frame relies on the constant speed of light (see Fig. B.4). In particle physics this effect is observed for the lifetimes of unstable particles.

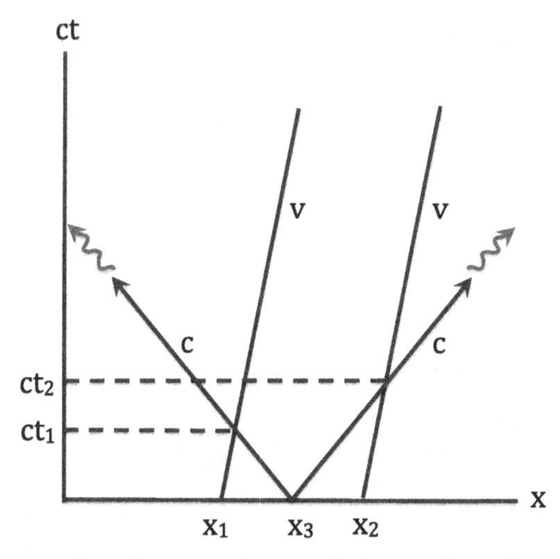

Figure B.4. Observers in a frame moving at velocity v will see the event happening at $x = x_3$ and $t = 0$ at different times.

To test time dilatation, we have to send a particle on a roundtrip. For this, an infinitesimal form of equation (B.23) is needed. Since $x^0 = x^1 = 0$ does not have to coincide with $x'^0 = x'^1 = 0$, we can use Poincaré transformations. We assume that a light beam originates in F at (x_0^0, x_0^1) and in F' at $(x_0'^0, x_0'^1)$. This generalizes equation (B.10) to

$$(x'^0 - x_0'^0)^2 - (x'^1 - x_0'^1)^2 = (x^0 - x_0^0)^2 - (x^1 - x_0^1)^2,$$

and the Lorentz transformations become

$$\begin{aligned}(x'^0 - x_0'^0) &= \gamma\,[(x^0 - x_0^0) - \beta\,(x^1 - x_0^1)],\\ (x'^1 - x_0'^1) &= \gamma\,[(x^1 - x_0^1) - \beta\,(x^0 - x_0^0)],\end{aligned} \tag{B.24}$$

or

$$\begin{pmatrix} x'^0 - x_0'^0 \\ x'^1 - x_0'^1 \end{pmatrix} = \begin{pmatrix} \cosh\xi & -\sinh\xi \\ -\sinh\xi & \cosh\xi \end{pmatrix} \begin{pmatrix} x^0 - x_0^0 \\ x^1 - x_0^1 \end{pmatrix}. \tag{B.25}$$

In addition we have invariance under translations (B.9).

The Minkowski space allows us to describe the motion of particles by means of *world lines.* For a particle traveling along its world line, we can introduce the concept of *proper time.* For a particle moving with velocity $v(t)$, $dx^1 = \beta\, dx^0$, and the infinitesimal invariant along its 2-dimensional world line is

$$(ds)^2 = (dx^0)^2 - (dx^1)^2 = (c\,dt)^2\,(1 - \beta^2). \tag{B.26}$$

Each instantaneous rest frame of the particle is an inertial frame. Increasing the time by $d\tau$ in such a frame yields a Lorentz invariant quantity

$$d\tau \equiv ds = dt\,\sqrt{1 - \beta^2} = \frac{dt}{\gamma} = \frac{dt}{\cosh\xi}, \tag{B.27}$$

where τ is called *proper time.* Note that in the rest frame of the particle its position does not change, and that is why $ds = d\tau$.

As $\gamma(\tau) \geq 1$ time dilatation follows

$$t_2 - t_1 = \int_{\tau_1}^{\tau_2} \gamma(\tau)\, d\tau = \int_{\tau_1}^{\tau_2} \cosh\xi(\tau)\, d\tau \geq \tau_2 - \tau_1, \tag{B.28}$$

which proves that a moving clock runs more slowly than a stationary clock. Equation (B.28) applies to general paths of a clock, including those in acceleration.

B.2.4. *Acceleration*

Considering an acceleration a in the instantaneous rest frame and defining $\alpha = a/c$, one has

$$d\beta = d\xi = \alpha\, d\tau \tag{B.29}$$

in the instantaneous rest frame. Note that in another frame, $d\beta = d\xi/\gamma^2$, but in the instantaneous rest frame, $\beta = 0$, or $\gamma = 1$. From the addition theorem of velocities (B.22) one obtains

$$d\beta = \frac{\alpha d\tau + \beta}{1 + \alpha d\tau\, \beta} - \beta = \alpha\,(1 - \beta^2)\, d\tau. \tag{B.30}$$

For the rapidity, in the particle's instantaneous rest frame,

$$d\xi = (\alpha\, d\tau + \xi) - \xi = \alpha\, d\tau. \tag{B.31}$$

In terms of the proper time, the change of rapidity is analogous to the change of velocity in non-relativistic mechanics,

$$\xi - \xi_0 = \int_{\tau_0}^{\tau} \alpha(\tau)\, d\tau. \tag{B.32}$$

In the special case that α is constant, the integration yields

$$\xi(\tau) = \alpha\tau + \xi_0 \quad \text{for } \alpha \text{ constant.} \tag{B.33}$$

B.3. Tensors

B.3.1. *Definition*

A general transformation $x \to x'$ is denoted by

$$x'^{\,\alpha} = x'^{\,\alpha}(x) = x'^{\,\alpha}\,(x^0, x^1, x^2, x^3), \;\; \alpha = 0, 1, 2, 3, \tag{B.34}$$

which means that $x'^{\,\alpha}$ is a function of four variables and differentiable with respect to each of its arguments.

A *scalar* means a quantity whose value is not changed under the transformation between frames, such as the proper time. We denote a *four-vector* $A^\alpha, (\alpha = 0, 1, 2, 3)$ as *contravariant* if its components transform according to

$$A'^{\alpha} = \frac{\partial x'^{\alpha}}{\partial x^{\beta}} A^{\beta}. \tag{B.35}$$

An example of contravariant vector is $A^\alpha = dx^\alpha$, where (B.35) reduces to the well-known rule for the differential of a function of several variables $(f^\alpha(x) = x'^\alpha(x))$:

$$dx'^\alpha = \frac{\partial x'^\alpha}{\partial x^\beta} dx^\beta.$$

Although dx^α is contravariant, the vector x^α is not always contravariant.

For a linear transformation

$$x'^\alpha = a^\alpha{}_\beta \, x^\beta$$

with space-time independent coefficients $a^\alpha{}_\beta$, one finds

$$\frac{\partial x'^\alpha}{\partial x^\beta} = a^\alpha{}_\beta .$$

A four-vector is *covariant* if it transforms like

$$B'_\alpha = \frac{\partial x^\beta}{\partial x'^\alpha} B_\beta. \tag{B.36}$$

An example is

$$B_\alpha = \partial_\alpha = \frac{\partial}{\partial x^\alpha}, \tag{B.37}$$

since

$$\frac{\partial}{\partial x'^\alpha} = \frac{\partial x^\beta}{\partial x'^\alpha} \frac{\partial}{\partial x^\beta}.$$

For two vectors, we define the *inner* or *scalar product* as a product of the components of their covariant and contravariant parts,

$$B \cdot A = B_\alpha A^\alpha. \tag{B.38}$$

Using Eqs. (B.35) and (B.36) we see that the scalar product is an invariant under the transformation (B.34),

$$B' \cdot A' = \frac{\partial x^\beta}{\partial x'^\alpha} \frac{\partial x'^\alpha}{\partial x^\gamma} B_\beta A^\gamma = \frac{\partial x^\beta}{\partial x^\gamma} B_\beta A^\gamma = \delta^\beta{}_\gamma B_\beta A^\gamma = B \cdot A,$$

where we have introduced the *Kronecker delta*,

$$\delta^\alpha{}_\beta = \delta_\alpha{}^\beta = \begin{cases} 1, & \text{for } \alpha = \beta, \\ 0, & \text{for } \alpha \neq \beta. \end{cases} \tag{B.39}$$

Tensors of rank k are quantities with k indices, such as

$$T^{\alpha_1 \alpha_2 \cdots}{}_{\cdots \alpha_i \cdots \alpha_k}.$$

For instance, vectors are *rank* one tensors. For tensors, upper indices transform contravariant and the lower transform covariant. In this way, a contravariant tensor of rank two $F^{\alpha\beta}$ consists of 16 quantities that transform according to

$$F'^{\alpha\beta} = \frac{\partial x'^{\alpha}}{\partial x^{\gamma}} \frac{\partial x'^{\beta}}{\partial x^{\delta}} F^{\gamma\delta}.$$

A covariant tensor of rank two $G_{\alpha\beta}$ transforms as

$$G'_{\alpha\beta} = \frac{\partial x^{\gamma}}{\partial x'^{\alpha}} \frac{\partial x^{\delta}}{\partial x'^{\beta}} G_{\gamma\delta}.$$

In analogy with (B.38), the *inner product* or *contraction* with respect to a pair of indices, either on the same tensor or between different tensors, is defined so that one index is contravariant and the other covariant.

A *symmetric tensor* in α and β obeys the relation

$$S^{\cdots\alpha\cdots\beta\cdots} = S^{\cdots\beta\cdots\alpha\cdots}.$$

For an *antisymmetric tensor* in α and β,

$$A^{\cdots\alpha\cdots\beta\cdots} = -A^{\cdots\beta\cdots\alpha\cdots}.$$

One can easily show that if $S^{\cdots\alpha\cdots\beta}$ is a symmetric and $A^{\cdots\alpha\cdots\beta}$ an antisymmetric tensor, then

$$S^{\cdots\alpha\cdots\beta\cdots} A_{\cdots\alpha\cdots\beta\cdots} = 0. \tag{B.40}$$

Also, every tensor can be written as a sum of its symmetric and antisymmetric parts in two of its indices.

B.3.2. *Metric tensor*

In special relativity, the infinitesimal interval ds defines the proper time $c\,d\tau = ds$,

$$(ds)^2 = (dx^0)^2 - (dx^1)^2 - (dx^2)^2 - (dx^3)^2, \tag{B.41}$$

where we have used the convention that dx^{α} is a contravariant vector. Now, we define a *metric tensor* $g_{\alpha\beta}$ so that Eq. (B.41) becomes

$$(ds)^2 = g_{\alpha\beta}\, dx^{\alpha} dx^{\beta}. \tag{B.42}$$

Comparing (B.41) and (B.42) we see that $g_{\alpha\beta}$ is diagonal:

$$g_{00} = 1,\ g_{11} = g_{22} = g_{33} = -1 \quad \text{and} \quad g_{\alpha\beta} = 0 \text{ for } \alpha \neq \beta. \tag{B.43}$$

This is the *Minkowski metric.*

Next we compare (B.42) with the invariant scalar product (B.38), and we conclude that

$$x_\alpha = g_{\alpha\beta}\, x^\beta .$$

Thus, the covariant metric tensor lowers the indices, i.e., it transforms a contravariant into a covariant vector. Now, we define the contravariant metric tensor $g^{\alpha\beta}$ so that it raises indices:

$$x^\alpha = g^{\alpha\beta}\, x_\beta ,$$

which implies that

$$g_{\alpha\gamma}\, g^{\gamma\beta} = \delta_\alpha^{\ \beta} . \tag{B.44}$$

Solving this equation for $g^{\alpha\beta}$ yields, using the diagonal matrix (B.43),

$$g^{\alpha\beta} = g_{\alpha\beta} . \tag{B.45}$$

The following equations hold:

$$A^\alpha = \begin{pmatrix} A^0 \\ \mathbf{A} \end{pmatrix}, \qquad A_\alpha = (A^0, -\mathbf{A})$$

and, because of Eq. (B.37),

$$(\partial_\alpha) = \left(\frac{\partial}{c\partial t}, \boldsymbol{\nabla} \right), \qquad (\partial^\alpha) = \begin{pmatrix} \dfrac{\partial}{c\partial t} \\ -\boldsymbol{\nabla} \end{pmatrix} . \tag{B.46}$$

The *four-divergence* of a four-vector

$$\partial^\alpha A_\alpha = \partial_\alpha A^\alpha = \frac{\partial A^0}{\partial x^0} + \nabla \cdot \mathbf{A}$$

and the *d'Alembertian* (four-dimensional Laplace) operator

$$\Box = \partial_\alpha \partial^\alpha = \left(\frac{\partial}{\partial x^0} \right)^2 - \nabla^2$$

are also invariants. The notation $\triangle = \nabla^2$ is often used for the (3-dimensional) Laplace operator.

B.4. Lorentz Group

A group of linear transformations

$$x'^{\alpha} = a^{\alpha}{}_{\beta}\, x^{\beta}, \quad \left(\text{so that } \frac{\partial x'^{\alpha}}{\partial x^{\beta}} = a^{\alpha}{}_{\beta}\right) \tag{B.47}$$

stays invariant if

$$x'_{\alpha}\, x'^{\alpha} = a_{\alpha}{}^{\beta} x_{\beta}\, a^{\alpha}{}_{\gamma} x^{\gamma} = x_{\alpha}\, x^{\alpha} = \delta^{\beta}{}_{\gamma} x_{\beta} x^{\gamma}.$$

As the $x_{\beta}x^{\gamma}$ are independent, one has

$$a_{\alpha}{}^{\beta}\, a^{\alpha}{}_{\gamma} = \delta^{\beta}{}_{\gamma}, \quad \text{or} \quad a_{\alpha\beta}\, a^{\alpha}{}_{\gamma} = g_{\beta\gamma}, \quad \text{or} \quad a^{\delta}{}_{\beta}\, g_{\delta\alpha}\, a^{\alpha}{}_{\gamma} = g_{\beta\gamma}.$$

In matrix notation

$$\tilde{A} g A = g, \tag{B.48}$$

where $g = (g_{\beta\alpha})$,

$$A = (a^{\beta}{}_{\alpha}) = \begin{pmatrix} a^0{}_0 & a^0{}_1 & a^0{}_2 & a^0{}_3 \\ a^1{}_0 & a^1{}_1 & a^1{}_2 & a^1{}_3 \\ a^2{}_0 & a^2{}_1 & a^2{}_2 & a^2{}_3 \\ a^3{}_0 & a^3{}_1 & a^3{}_2 & a^3{}_3 \end{pmatrix}, \tag{B.49}$$

and $\tilde{A} = (\tilde{a}_{\beta}{}^{\alpha})$, where $\tilde{a}_{\beta}{}^{\alpha} = a^{\alpha}{}_{\beta}$ is the transpose of the matrix $A = (a^{\beta}{}_{\alpha})$,

$$\tilde{A} = (\tilde{a}_{\beta}{}^{\alpha}) = \begin{pmatrix} \tilde{a}_0{}^0 & \tilde{a}_0{}^1 & \tilde{a}_0{}^2 & \tilde{a}_0{}^3 \\ \tilde{a}_1{}^0 & \tilde{a}_1{}^1 & \tilde{a}_1{}^2 & \tilde{a}_1{}^3 \\ \tilde{a}_2{}^0 & \tilde{a}_2{}^1 & \tilde{a}_2{}^2 & \tilde{a}_2{}^3 \\ \tilde{a}_3{}^0 & \tilde{a}_3{}^1 & \tilde{a}_3{}^2 & \tilde{a}_3{}^3 \end{pmatrix} = \begin{pmatrix} a^0{}_0 & a^1{}_0 & a^2{}_0 & a^3{}_0 \\ a^0{}_1 & a^1{}_1 & a^2{}_1 & a^3{}_1 \\ a^0{}_2 & a^1{}_2 & a^2{}_2 & a^3{}_2 \\ a^0{}_3 & a^1{}_3 & a^2{}_3 & a^3{}_3 \end{pmatrix}. \tag{B.50}$$

In the transpose matrix the row indices are contravariant and the column indices are covariant, opposite to the definition (B.4) for vectors and ordinary matrices.

Taking the determinant of both sides of Eq. (B.48) yields $\det(\tilde{A} g A) = \det(g)\det(A)^2 = \det(g)$. Because $\det(g) = -1$, this yields

$$\det(A) = \pm 1. \tag{B.51}$$

Proper Lorentz transformations are continuously connected through the identity transformation $A = \mathbf{1}$. All other Lorentz transformations are *improper*. Proper transformations have necessarily $\det(A) = 1$. For improper Lorentz transformations it is sufficient, but not necessary, to have

$\det(A) = -1$. Simultaneous space and time inversion, such that $A = -\mathbf{1}$, is an improper Lorentz transformation but with $\det(A) = +1$.

Since A and g are 4×4 matrices, one has 16 equations for $4^2 = 16$ elements of A. But they are not all independent because of symmetry under transposition. The off-diagonal equations are identical in pairs. Therefore, there are $4 + 6 = 10$ linearly independent equations for the 16 elements of A. We conclude that the Lorentz group has only *six free parameters*.

B.4.1. *Lorentz boost*

To construct A explicitly, for proper Lorentz transformations, one assumes that

$$A = e^L = \sum_{n=0}^{\infty} \frac{L^n}{n!},$$

where L is a 4×4 matrix. Then

$$\det(A) = \det(e^L) = e^{\mathrm{Tr}(L)}, \tag{B.52}$$

and $\det(A) = +1$ implies that L is traceless. Equation (B.48) can be written

$$g\tilde{A}g = A^{-1}. \tag{B.53}$$

Since $g^2 = \mathbf{1}$, $(g\tilde{L}g)^n = g\tilde{L}^n g$, and $\mathbf{1} = \left(\sum_{n=0}^{\infty} L^n/n!\right)\left(\sum_{n=0}^{\infty}(-L)^n/n!\right)$, we get

$$\tilde{A} = e^{\tilde{L}}, \quad g\tilde{A}g = e^{g\tilde{L}g} \quad \text{and} \quad A^{-1} = e^{-L}.$$

Hence, (B.53) is equivalent to

$$g\tilde{L}g = -L \text{ or } \widetilde{(gL)} = -gL.$$

The matrix gL is thus antisymmetric and the general form of L is

$$L = \begin{pmatrix} 0 & l^0{}_1 & l^0{}_2 & l^0{}_3 \\ l^0{}_1 & 0 & l^1{}_2 & l^1{}_3 \\ l^0{}_2 & -l^1{}_2 & 0 & l^2{}_3 \\ l^0{}_3 & -l^1{}_3 & -l^2{}_3 & 0 \end{pmatrix}. \tag{B.54}$$

We can also expand L in terms of six *generators*

$$L = -\sum_{i=1}^{3}(\omega_i S_i + \xi_i K_i) \text{ and } A = e^{-\sum_{i=1}^{3}(\omega_i S_i + \xi_i K_i)}, \tag{B.55}$$

such that

$$S_1 = \begin{pmatrix} 0 & 0 & 0 & 0 \\ 0 & 0 & 0 & 0 \\ 0 & 0 & 0 & -1 \\ 0 & 0 & 1 & 0 \end{pmatrix}, \quad S_2 = \begin{pmatrix} 0 & 0 & 0 & 0 \\ 0 & 0 & 0 & 1 \\ 0 & 0 & 0 & 0 \\ 0 & -1 & 0 & 0 \end{pmatrix}, \quad S_3 = \begin{pmatrix} 0 & 0 & 0 & 0 \\ 0 & 0 & -1 & 0 \\ 0 & 1 & 0 & 0 \\ 0 & 0 & 0 & 0 \end{pmatrix}, \tag{B.56}$$

and

$$K_1 = \begin{pmatrix} 0 & 1 & 0 & 0 \\ 1 & 0 & 0 & 0 \\ 0 & 0 & 0 & 0 \\ 0 & 0 & 0 & 0 \end{pmatrix}, \quad K_2 = \begin{pmatrix} 0 & 0 & 1 & 0 \\ 0 & 0 & 0 & 0 \\ 1 & 0 & 0 & 0 \\ 0 & 0 & 0 & 0 \end{pmatrix}, \quad K_3 = \begin{pmatrix} 0 & 0 & 0 & 1 \\ 0 & 0 & 0 & 0 \\ 0 & 0 & 0 & 0 \\ 1 & 0 & 0 & 0 \end{pmatrix}. \tag{B.57}$$

These operators satisfy the *Lie algebra* commutation relations

$$[S_i, S_j] = \sum_{k=1}^{3} \epsilon^{ijk} S_k, \quad [S_i, K_j] = \sum_{k=1}^{3} \epsilon^{ijk} K_k, \quad [K_i, K_j] = -\sum_{k=1}^{3} \epsilon^{ijk} S_k,$$

where the commutator of two matrices is defined by $[A, B] = AB - BA$. In the Equation above, ϵ_{ijk} is the *Levi-Civita tensor*, defined so that

$$\epsilon^{i_1 i_2 \dots i_n} = \begin{cases} +1, & \text{for } (i_1, i_2, \dots, i_n) \\ & \quad \text{being an even permutation of } (1, 2, \dots, n), \\ -1, & \text{for } (i_1, i_2, \dots, i_n) \\ & \quad \text{being an odd permutation of } (1, 2, \dots, n), \\ 0, & \text{otherwise.} \end{cases} \tag{B.58}$$

As an example, let us take $\xi = \omega_1 = \omega_2 = 0$ and $\omega_3 = \omega$. Then,

$$A = e^{-\omega S_3} = \begin{pmatrix} 1 & 0 & 0 & 0 \\ 0 & \cos\omega & \sin\omega & 0 \\ 0 & -\sin\omega & \cos\omega & 0 \\ 0 & 0 & 0 & 1 \end{pmatrix}, \tag{B.59}$$

which describes a rotation by the *angle* ω (in the clockwise sense) around the $\hat{e}_3$ axis. On the other hand, if we choose $\omega = \xi_2 = \xi_3 = 0$ and $\xi_1 = \xi$, then

$$A = e^{-\xi K_1} = \begin{pmatrix} \cosh\xi & -\sinh\xi & 0 & 0 \\ -\sinh\xi & \cosh\xi & 0 & 0 \\ 0 & 0 & 1 & 0 \\ 0 & 0 & 0 & 1 \end{pmatrix} \tag{B.60}$$

is obtained, where ξ is our previously defined *boost parameter* or *rapidity.* The parameters ω_i, ξ_i, $(i = 1, 2, 3)$ are real, because Equation (B.47) implies that the elements of A have to be real.

B.4.2. *Relativistic kinematics*

From Eq. (B.60) one gets the Lorentz boost transformation,

$$\begin{aligned} x'^{\,0} &= x^0 \cosh\xi - x^1 \sinh\xi, \\ x'^{\,1} &= -x^0 \sinh\xi + x^1 \cosh\xi, \\ x'^{\,i} &= x^i, \ (i = 2, 3). \end{aligned} \tag{B.61}$$

For an arbitrary vector $\mathbf{A}$, it is convenient to decompose $\mathbf{A}$ into components parallel and perpendicular to $\boldsymbol{\beta} = \mathbf{v}/c$. Denoting $\hat{\boldsymbol{\beta}}$ the unit vector in $\boldsymbol{\beta}$ direction,

$$\mathbf{A} = A^{\|}\hat{\boldsymbol{\beta}} + \mathbf{A}^{\perp} \text{ with } A^{\|} = \hat{\boldsymbol{\beta}} \cdot \mathbf{A}.$$

The Lorentz transformation law is

$$\begin{aligned} A'^{\,0} &= A^0 \cosh\xi - A^{\|} \sinh\xi = \gamma(A^0 - \beta A^{\|}), \\ A'^{\,\|} &= -A^0 \sinh\xi + A^{\|} \cosh\xi = \gamma(-\beta A^0 + A^{\|}), \\ \mathbf{A}'^{\perp} &= \mathbf{A}^{\perp}. \end{aligned} \tag{B.62}$$

We can connect the subscript notation $A_{\|}$ and $\mathbf{A}_{\perp}$ to covariant vectors: $A_{\|} = -A^{\|}$ and $\mathbf{A}_{\perp} = -\mathbf{A}^{\perp}$.

If a particle moves with respect to F' with velocity $\mathbf{u}'$, then

$$x'^{\,i} = c^{-1} u'^{\,i} x'^{\,0}.$$

Equations (B.18) imply that

$$\gamma\,(x^1 - \beta x^0) = c^{-1} u'^{\,1} \gamma\,(x^0 - \beta x^1),$$

or

$$\gamma\,\left(1 + \frac{u'^{\,1} v}{c^2}\right)\,x^1 = c^{-1}\gamma\,(u'^1 + v)\,x^0.$$

Using the definition of the velocity in F, $\mathbf{x} = c^{-1}\mathbf{u}\,x^0$, gives

$$u^1 = c\,\frac{x^1}{x^0} = \frac{u'^1 + v}{1 + \frac{u'^1 v}{c^2}}. \tag{B.63}$$

Analogously, for the two other components, we get

$$u^i = \frac{u'^i}{\gamma\left(1 + \frac{u'^1 v}{c^2}\right)}, \quad (i = 2, 3). \tag{B.64}$$

In these equations, $\mathbf{v}$ was chosen to be along the x^1-axis. For general $\mathbf{v}$ we can decompose $\mathbf{u}$ into its components parallel and perpendicular to the $\mathbf{v}$

$$\mathbf{u} = u^{\|}\hat{\mathbf{v}} + \mathbf{u}^{\perp},$$

where $\hat{\mathbf{v}}$ is the unit vector in $\mathbf{v}$ direction, and one obtains

$$u^{\|} = \frac{u'^{\|} + v}{1 + \frac{u'^{\|} v}{c^2}} \quad \text{and} \quad \mathbf{u}^{\perp} = \frac{\mathbf{u}'^{\perp}}{\gamma\left(1 + \frac{u'^{\|} v}{c^2}\right)}. \tag{B.65}$$

It is then obvious that the velocity itself is not part of a 4-vector. The relativistic generalization is given in subsection (B.5).

The concepts of *world lines* in Minkowski space and *proper time* are generalized immediately to four dimensions. Assume the particle moves with velocity $\mathbf{v}(t)$, then $d\mathbf{x} = \beta dx^0$, and the infinitesimal invariant along its world line is

$$(ds)^2 = (dx^0)^2 - (d\mathbf{x})^2 = (c\,dt)^2\,(1 - \beta^2) \tag{B.66}$$

and the relations (B.27) and (B.28) follow as in the two dimensional case.

B.4.3. *Waves*

A spacetime picture of a wave is shown in Fig. B.5. The wave fronts are represented by the dashed lines in this figure. The phase speed of the wave divided by the speed of light is λ/cT, which is also the inverse of the slope of the wave fronts. The wavelength, λ, is the interval between successive wave fronts in the space direction, while c times the period, T, is the interval in the time direction. Since "space direction" and "time direction" are not relativistically invariant concepts, it is clear that the period and wavelength of a wave will be different in different coordinate systems.

Let us choose coordinates with respect to an inertial frame F. In complex notation a plane wave is defined by the equation

$$W(x) = W(x^0, \mathbf{x}) = W_0 \exp[\,i\,(k^0\,x^0 - \mathbf{k}\cdot\mathbf{x})], \tag{B.67}$$

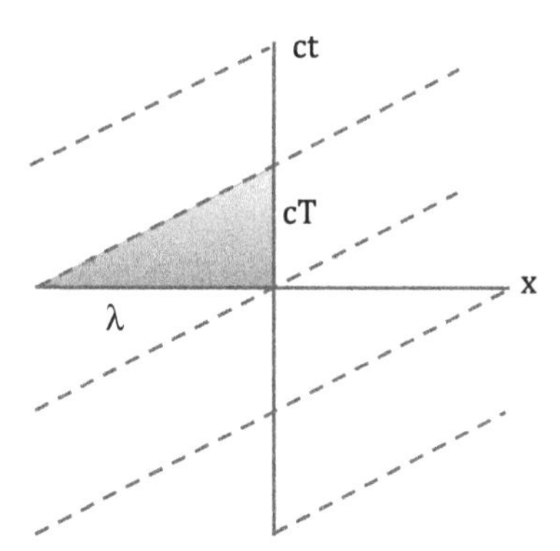

Figure B.5. Wave fronts in spacetime are represented by the dashed lines. The wavelength λ and the period T are shown for the illustrated wave.

where $W_0 = U_0 + i\, V_0$ is a complex *amplitude.* The vector $\mathbf{k}$ is called *wave vector.* It becomes a 4-vector (k^α) by identifying

$$k^0 = \frac{\omega}{c} \tag{B.68}$$

as its zero-component, where ω is the *angular frequency* of the wave.

Waves of the form (B.67) may either propagate in a medium (water, air, shock waves, etc.) or in vacuum (light waves, particle waves in quantum mechanics). We are interested in the latter case, as the other defines a preferred inertial frame, namely the one where the medium is at rest. The *phase* of the wave is defined by

$$\Phi(x) = \Phi(x^0, \mathbf{x}) = k^0\, x^0 - \mathbf{k}\cdot\mathbf{x} = \omega\, t - \mathbf{k}\cdot\mathbf{x}. \tag{B.69}$$

When (k^α) is a 4-vector, it follows that the phase is a scalar, *invariant* under Lorentz transformations

$$\Phi'(x') = k'_\alpha\, x'^\alpha = k_\alpha\, x^\alpha = \Phi(x). \tag{B.70}$$

For an observer at a fixed position $\mathbf{x}$ (note that the term $\mathbf{k}\cdot\mathbf{x}$ is then constant) the wave performs a periodic motion with *period*

$$T = \frac{2\pi}{\omega} = \frac{1}{\nu}, \tag{B.71}$$

where ν is the *frequency.* In particular, the phase (and hence the wave) takes identical values on the two-dimensional planes perpendicular to $\mathbf{k}$. Namely, let $\hat{\mathbf{k}}$ be the unit vector in $\mathbf{k}$ direction. By decomposing $\mathbf{x}$ into components parallel and perpendicular to $\mathbf{k}$, $\mathbf{x} = x^{\parallel}\,\hat{\mathbf{k}} + \mathbf{x}^{\perp}$, the phase becomes

$$\Phi = \omega\,t - k\,x^{\parallel}, \tag{B.72}$$

where $k = |\mathbf{k}|$ is the length of the vector $\mathbf{k}$. Phases which differ by multiples of 2π give the same values for the wave W. For example, when we take $V_0 = 0$, the real part of the wave becomes

$$W_x = U_0\,\cos(\omega\,t - k\,x^{\parallel})$$

and $\Phi = 0, n\,2\pi, n = \pm 1, \pm 2, \ldots$ describes the wave crests. From Eq. (B.72) it follows that the crests pass by our observer with speed $\mathbf{u} = u\,\hat{k}$, where $u = \frac{\omega}{k}$. Setting $\Phi = 0$, we get

$$x^{\parallel} = \frac{\omega}{k}\,t. \tag{B.73}$$

Let us in particular transform to the primed coordinate system in which the wave fronts are horizontal. This view of the wave is shown in Fig. B.6. The velocity of the primed coordinate system divided by the speed of light is $v/c = \lambda/cT$. This follows from the earlier conclusion that the slope of the wave fronts in the unprimed frame is cT/λ. In the primed frame of Fig. B.6, these wave fronts are horizontal, or parallel to the lines of simultaneity of the primed frame. Thus, they are parallel to the wave fronts in the unprimed frame as well, and therefore also have the slope cT/λ in this frame. Finally, we note that the slope of a line of simultaneity is equal to the velocity of the associated reference frame divided by c, from which Eq. (B.70) follows.

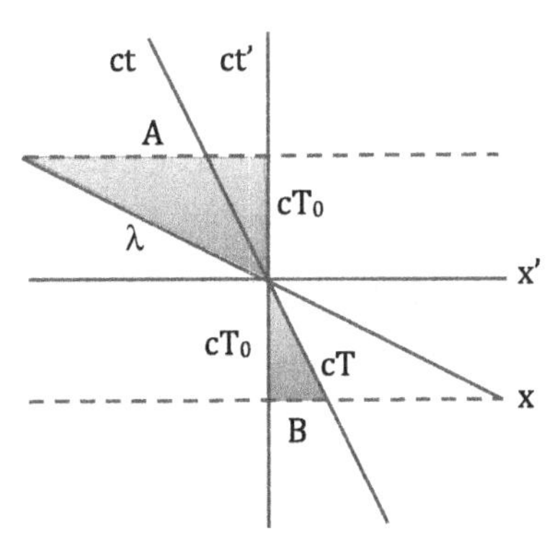

Figure B.6. View of the wave of Figure B.5 in a reference frame in which the wave fronts are horizontal.

If the period of the wave in the primed frame is T_0, we can relate the wavelength and period in the unprimed frame to T_0 and v using the *Pythagorean theorem of spacetime.* In particular, $A = cT_0/(v/c) = c^2T_0/v$ because in time cT_0 a wave with speed v/c propagates by one-wavelength (in the unprimed frame). We also obtain $B = cT_0(v/c) = vT_0$ because the slope of the line of simultaneity in unprimed frame is tilted with a slope v/c in the primed system. Then, since $\lambda^2 = A^2 - c^2T_0^2$ and $c^2T^2 = c^2T_0^2 - B^2$, we find that $\lambda = c^2T_0/v\gamma$ and $T = T_0/\gamma$, where $\gamma = (1 - v^2/c^2)^{-1/2}$.

If we express wave parameters in terms of angular velocity, $\omega = 2\pi/T$, and wave number, $k = 2\pi/\lambda$, and further define $\omega_0 = 2\pi/T_0$, then we easily show that

$$\omega = \omega_0\gamma, \quad k = \frac{\omega_0 v\gamma}{c^2}. \tag{B.74}$$

we further note that

$$\omega^2 - k^2c^2 = \omega_0^2\gamma^2 - \frac{\omega_0^2\gamma^2v^2}{c^2} = \omega_0^2. \tag{B.75}$$

This is the *dispersion relation* for relativistic waves. The phase and group velocities of a wave can be obtained from the dispersion relation. Since ω_0 is a constant, we immediately see that $\omega d\omega = c^2kdk$, which means that the *group velocity,* $u_g \equiv d\omega/dk$ is proportional to the inverse of the phase velocity: $u_\phi \equiv \omega/k = \lambda/T = c^2/v$. In particular, we find that

$$u_g = \frac{d\omega}{dk} = \frac{c^2k}{\omega} = v. \tag{B.76}$$

If the idea of a wave front is relativistically invariant, then the dispersion relation of the wave must take the form given by equation (B.75). Moreover, the group velocity resulting from this dispersion relation behaves sensibly when a Lorentz transformation is performed, in that it transforms like other velocities. Note in particular that the group velocity is zero when the wave fronts are horizontal in spacetime. This follows from equation (B.76) because $k = 0$ in this case. Furthermore, in a coordinate system moving with velocity $-v$ with respect to this special coordinate system, the group velocity is $+v$. This may seem like a trivial result, but in fact the *phase velocity* does *not* behave this way at all!

Planck and Einstein established that light consisted of photons with energy $E = \hbar\omega$. Louis de Broglie extended this idea to ordinary particles and further showed by arguments related to those given above that the particle momentum had to be related to the wave number by $p = \hbar k$.

The key in this argument is to note the relationship between the relativistic wave dispersion relation, $\omega^2 - k^2c^2 = \omega_0^2$, and the relativistic relationship between energy and momentum, $E^2 - p^2c^2 = m^2c^4$. In addition to the relationship between momentum and wave number, we see that $\omega_0 = mc^2$.

B.4.4. *Doppler effect*

If our observer counts the number of wave crests passing by, he can ask how the wave (B.67) is described in another inertial frame F'. An observer in F' who counts the number of wave crests, passing through the same space-time point at which our first observer already counts, must get the same number. When in frame F the wave takes its maximum at the space-time point (x^α) it must also be at its maximum in F' at the same space-time point in appropriately transformed coordinates (x'^α). More generally, this is holds for every value of the phase, because it is a scalar.

As (k^α) is a 4-vector the transformation law for angular frequency and wave vector is just a special case of Eqs. (B.62),

$$\begin{aligned} k'^0 &= k^0 \cosh(\xi) - k^{\|} \sinh(\xi) = \gamma(k^0 - \beta k^{\|}), \\ k'^{\|} &= -k^0 \sinh(\xi) + k^{\|} \cosh(\xi) = \gamma(k^{\|} - \beta k^0), \\ \mathbf{k}'^{\perp} &= \mathbf{k}^{\perp}, \end{aligned} \tag{B.77}$$

where the notation $k^{\|}$ and $k^{\perp}$ is with respect to the relative velocity of the two frames, $\mathbf{v}$.

These transformation equations for the frequency and the wave vector describe the relativistic Doppler effect. To illustrate their meaning, let us specialize to the case of a light source, which is emitted in F and the observer F' moves in wave vector direction away from the source, i.e., $\mathbf{v} \parallel \mathbf{k}$. The equation for the wave speed (B.73) implies

$$c = \frac{\omega}{k} \Rightarrow k = |\mathbf{k}| = \frac{\omega}{c} = k^0$$

and choosing directions so that $k'^{\|} = k$ holds, Eq. (B.77) becomes

$$k'^0 = \gamma\,(k^0 - \beta\,k) = \gamma\,(1-\beta)\,k^0 = k^0\sqrt{\frac{1-\beta}{1+\beta}}$$

or

$$\omega' = \frac{\nu'}{2\pi} = \omega\sqrt{\frac{1-\beta}{1+\beta}} = \frac{\nu}{2\pi}\sqrt{\frac{1-\beta}{1+\beta}}.$$

But, $c = \nu\lambda = \nu'\lambda'$, where λ is the wavelength in F and λ' the wavelength in F'. Consequently, we have

$$\lambda' = \lambda\sqrt{\frac{1+\beta}{1-\beta}}.$$

For a receding observer, or source receding from the observer, $\beta > 0$ in our conventions for F and F'. Then the wavelength λ' is larger than it is for a source at rest. This is an example of the red-shift, of major importance when one analyzes spectral lines in astrophysics. A single light signal suffices to obtain position and speed of a distant source. Note that for $\beta \ll 1$ one gets

$$\lambda' \simeq \lambda(1+\beta), \tag{B.78}$$

which is applicable to several important cases, including the pioneer redshift analysis of galaxies made by Hubble [Hu29].

B.5. Four-Momentum

B.5.1. *Energy–momentum of a particle*

Consider a point-like particle in its rest-frame and denote its mass there by m_0. In any other frame the rest-mass of the particle is still m_0, which is why it is a scalar. In the non-relativistic limit the momentum is defined by $\mathbf{p} = m_0\mathbf{u}$. Now, we want to define $\mathbf{p}$ as part of a relativistic 4-vector (p^α).

Consider a particle at rest in frame F, i.e., $\mathbf{p} = 0$. Assume that frame F' is moving with a small velocity $\mathbf{v}$ with respect to F. Then the non-relativistic limit is correct, and $\mathbf{p}' = -m_0\mathbf{v}$ has to hold approximately. On the other hand, the transformation laws (B.62), for vectors (note $\mathbf{p} \parallel \boldsymbol{\beta} = \mathbf{v}/c$) imply

$$\mathbf{p}' = \gamma\,(\mathbf{p} - \boldsymbol{\beta}\,p^0).$$

For $\mathbf{p} = 0$ we find $\mathbf{p}' = -\gamma\,\boldsymbol{\beta}\,p^0$. As in the non-relativstic limit $\gamma\,\beta \to \beta$, consistency requires $p^0 = c\,m_0$ in the rest frame, so that we get $\mathbf{p}' = -m_0\,\gamma\,v$. Consequently, for a particle moving with velocity $\mathbf{u}$ in frame F

$$\mathbf{p} = m_0\,\gamma\,\mathbf{u} \tag{B.79}$$

is the correct relation between relativistic momentum and velocity.

From the invariance of the scalar product, $p_\alpha p^\alpha = (p^0)^2 - \mathbf{p}^{\,2} = p'_\alpha p'^\alpha = m_0^2 c^2$, we get

$$p^{\,0} = +\sqrt{c^2\, m_0^2 + \mathbf{p}^{\,2}}, \tag{B.80}$$

which is of course consistent with calculating p^0 via the Lorentz transformation law (B.62). It should be noted that $c\,p^0$ has the dimension of an energy, i.e. the relativistic energy of a particle is

$$E = c\,p^0 = +\sqrt{c^4\, m_0^2 + c^2\,\mathbf{p}^{\,2}} = c^2\, m_0 + \frac{\mathbf{p}^{\,2}}{2m_0} + \,\ldots\;, \tag{B.81}$$

where the second term is just the non-relativistic kinetic energy $T = \mathbf{p}^{\,2}/(2m_0)$. The first term shows that (rest) mass and energy can be transformed into one another [Ein06].

Using the mass definition of mass, $m = c\,p^0$, together with (B.81) we obtain at this point the famous equation $E = m\,c^2 = \gamma m_0 c^2$. Avoiding this definition of m, because it is not the mass found in particle tables, where the mass of a particle is an invariant scalar, the essence of Einstein's equation is captured by

$$E_0 = m_0\, c^2,$$

where E_0 is the energy of a massive body (or particle) in its rest frame. Often, one does not use a subscript $_0$ and denote the rest mass simply by m.

The relativistic kinetic energy is the difference between the total energy and the rest energy, i.e.,

$$K = (\gamma - 1) m_0 c^2 \tag{B.82}$$

Figures B.7 and B.8 show a comparison between relativistic and non-relativistic momenta and energy.

Non-relativistic momentum conservation implies that, for two particles, $\mathbf{p}_1 + \mathbf{p}_2 = \mathbf{q}_1 + \mathbf{q}_2$, where $\mathbf{p}_i$, $(i = 1, 2)$ are the momenta of two incoming particles, and $\mathbf{q}_i$, $(i = 1, 2)$ are the momenta of two outgoing particles. This becomes a relativistic *energy–momentum conservation* law:

$$p_1^\alpha + p_2^\alpha = q_1^\alpha + q_2^\alpha. \tag{B.83}$$

One obtains the useful relations

$$\gamma = \frac{p_0}{m_0\, c} = \frac{E}{m_0\, c^2} \quad \text{and} \quad \beta = \frac{|\mathbf{p}|}{p^0}. \tag{B.84}$$

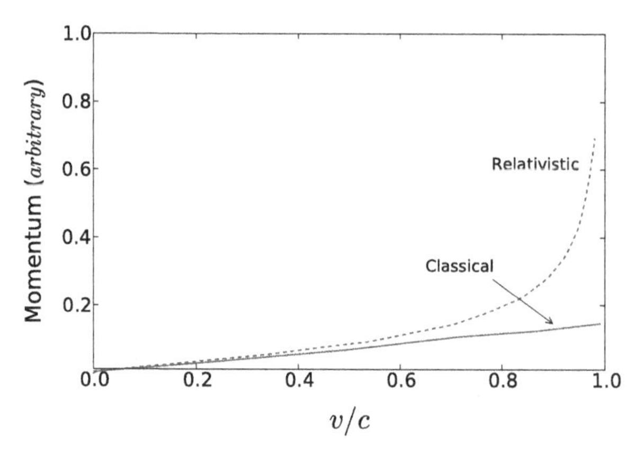

Figure B.7. Comparison between relativistic and non-relativistic momenta. Note that even an infinite amount of momenttum is not enough to achieve c.

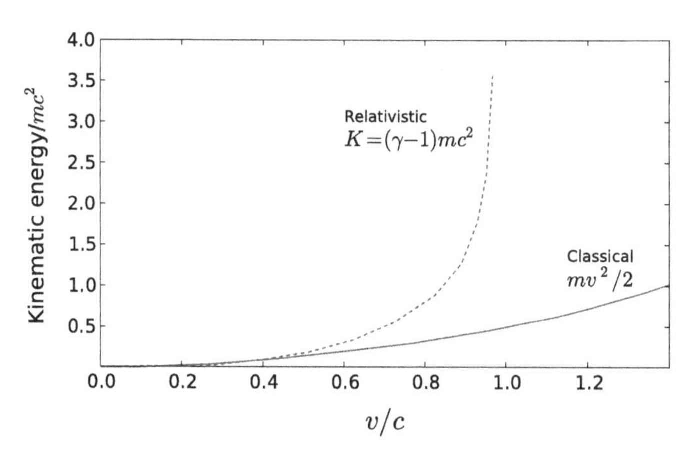

Figure B.8. Comparison between relativistic and non-relativistic energies. Note that even an infinite amount of energy is not enough to achieve c.

The contravariant generalization of the velocity vector is given by

$$U^\alpha = \frac{dx^\alpha}{d\tau} = \gamma u^\alpha \quad \text{with} \quad u^0 = c, \tag{B.85}$$

with the definition of the infinitesimal proper time (B.27). The relativistic generalization of the force is then the four-vector

$$f^\alpha = \frac{d\,p^\alpha}{d\,\tau} = m_0\,\frac{dU^\alpha}{d\tau}, \tag{B.86}$$

where the last equality can only be used for particles with non-zero rest mass.

B.5.2. *Particle collisions*

The energy available for particle production in collisions between particles depend on the total energy in the *center-of-mass* (CM) frame. We define the CM as the frame in which the total vector momentum of the collision is zero. By definition then, in the CM frame we have for two 4-vectors (p_1, p_2):

$$(p_1 + p_2) = (E_1 + E_2, \mathbf{p}_1 + \mathbf{p}_2) = (E_1 + E_2, 0). \tag{B.87}$$

If the masses of the two particles are equal as in the case of proton–antiproton collisions then the above reduces to

$$\begin{aligned}(p_1 + p_2) &= (E_1 + E_2, \mathbf{p}_1 + \mathbf{p}_2) = (E_1 + E_2, 0)\\ &= (2E, 0) \text{ (twice energy of either particle).}\end{aligned} \tag{B.88}$$

In high-energy physics, the square of the energy in the CM is often called s, so that $s =$ magnitude of $(p_1 + p_2)^2$.

In handling elementary particles, a proper unit for energy is needed, to avoid working with very large numbers. The work done in accelerating a charge through a potential difference is given by $W = qV$. For a proton, with the charge $e = 1.602 \times 10^{-19}\,\text{C}$ and a potential difference of $1\,\text{V}$, the work done is:

$$W = (1.602 \times 10^{-19}\ \text{C})(1\text{V}) = 1.602 \times 10^{-19}\ \text{J}. \tag{B.89}$$

The work done to accelerate the proton across a potential difference of $1\,\text{V}$ could also be written as: $\text{W} = (1\,\text{e})(1\,\text{V}) = 1\,\text{eV}$. Thus eV, pronounced "electron volt", is also a unit of energy. It's related to the SI (Système International) unit Joule by:

$$1\ \text{eV} = 1.602 \times 10^{-19}\ \text{J}. \tag{B.90}$$

Other associated units are the MeV ($10^6\,\text{eV}$), the GeV ($10^9\,\text{eV}$), etc. (See Appendix A).

Consider the squared center-of-mass energy $s = (p_a + p_b)^2 = (p_c + p_d)^2$ with four-momenta $p_i = (E_i, \mathbf{p}_i)$ in the $2 \to 2$ scattering $a + b \to c + d$,

$$s = (p_a + p_b)^2 = m_a^2 + m_b^2 + 2(E_a E_b - \mathbf{p}_a \cdot \mathbf{p}_b) = m_a^2 + m_b^2 + 2E_a E_b (1 - \beta_a \beta_b \cos\theta), \tag{B.91}$$

where $\beta_i = v_i = p_i/E_i$. We are interested often in the threshold energy $s_{\min}^{1/2}$ of a certain process. An example is the reaction of a proton and a photon yielding a proton and a pion, $p + \gamma \to p + \pi^0$. If in this process the proton is at rest,

$$s = m_p^2 + 2E_\gamma m_p \geq (m_p + m_\pi)^2 = m_p^2 + 2m_p m_\pi + m_\pi^2 \tag{B.92}$$

or

$$E_\gamma \geq m_\pi + \frac{m_\pi^2}{2m_p} \approx 145\,\text{MeV}. \tag{B.93}$$

Thus the photo-production of pions on protons at rest is only possible for $E_\gamma \geq 145\,\text{MeV}$. Photons with lower energy do not interact via this reaction, while more energetic ones do. That plays an important role in the transition from a transparent Universe at low energies to an opaque one at high energies.

The second important quantity characterizing a scattering process is the *four-momentum transfer* $t = (p_a - p_c)^2 = (p_b - p_d)^2$. As an example, we consider electron–proton scattering $e^- + p \to e^- + p$. Then

$$t = (p_e - p_e')^2 = 2m_e^2 - 2E_e E_e'(1 - \beta_e \beta_e' \cos\theta). \tag{B.94}$$

For high energies, $\beta_e, \beta_e' \to 1$, and

$$t \approx -2E_e E_e'(1 - \cos\theta) = -4E_e E_e' \sin^2\theta/2. \tag{B.95}$$

Because of four-momentum conservation, the variable t corresponds to the squared momentum Q of the exchanged virtual photon, $t = Q^2 = (p_e - p_e')^2 < 0$. Thus a *virtual particle* does not fulfill the relativistic energy–momentum relation. The quantum *energy–time uncertainty* (or Heisenbeg) relation $\Delta E \Delta t \gtrsim \hbar$ allows such a violation, if the virtual particle is exchanged only during a short enough time. Note also that the angular dependence of Rutherford scattering, $d\sigma/d\Omega \propto 1/\sin^4\theta/2$ is obtained if $d\sigma/d\Omega \propto 1/t^2$.

Appendix C

General Relativity

C.1. Christoffel Symbols

Consider a vector field $A_\mu(x^\nu)$ as a function of contravariant coordinates. Let us introduce a shorthand for the derivative as

$$A_{\mu,\nu} = \frac{\partial A_\mu}{\partial x^\nu}. \tag{C.1}$$

We want to know whether the derivative $A_{\mu,\nu}$ is a tensor. That is, does $A_{\mu,\nu}$ transform according to

$$A'_{\mu,\nu} = \frac{\partial x^\alpha}{\partial x'^\mu}\frac{\partial x^\beta}{\partial x'^\nu}A_{\alpha,\beta}?$$

To find out, we evaluate the derivative explicitly

$$A'_{\mu,\nu} = \frac{\partial A'_\mu}{\partial x'^\nu} = \frac{\partial}{\partial x'^\nu}\left(\frac{\partial x^\alpha}{\partial x'^\mu}A_\alpha\right) = \frac{\partial x^\alpha}{\partial x'^\mu}\frac{\partial A_\alpha}{\partial x'^\nu} + \frac{\partial^2 x^\alpha}{\partial x'^\nu \partial x'^\mu}A_\alpha. \tag{C.2}$$

But A_α is a function of x^ν not x'^ν, i.e. $A_\alpha = A_\alpha(x^\nu) \neq A_\alpha(x'^\nu)$. Therefore

$$\frac{\partial A_\alpha}{\partial x'^\nu} = \frac{\partial A_\alpha}{\partial x^\gamma}\frac{\partial x^\gamma}{\partial x'^\nu},$$

so that

$$\begin{aligned} A'_{\mu,\nu} &= \frac{\partial A'_\mu}{\partial x'^\nu} = \frac{\partial}{\partial x'^\nu}\left(\frac{\partial x^\alpha}{\partial x'^\mu}A_\alpha\right) \\ &= \frac{\partial x^\alpha}{\partial x'^\mu}\frac{\partial x^\gamma}{\partial x'^\nu}\frac{\partial A_\alpha}{\partial x^\gamma} + \frac{\partial^2 x^\alpha}{\partial x'^\nu \partial x'^\mu}A_\alpha \\ &= \frac{\partial x^\alpha}{\partial x'^\mu}\frac{\partial x^\gamma}{\partial x'^\nu}A_{\alpha,\gamma} + \frac{\partial^2 x^\alpha}{\partial x'^\nu \partial x'^\mu}A_\alpha. \end{aligned} \tag{C.3}$$

We see therefore that the tensor transformation law for $A_{\nu,\mu}$ is spoiled by the second term. Thus $A_{\nu,\mu}$ is not a tensor.

The definition of the derivative is,

$$A_{\mu,\nu} = \frac{\partial A_\mu}{\partial x'^\nu} = \lim_{dx \to 0} \frac{A_\mu(x + dx) - A_\mu(x)}{dx^\nu}. \tag{C.4}$$

The problem however is that the numerator is not a vector because $A_\mu(x+dx)$ and $A_\mu(x)$ are located at different points. The difference between two vectors is only a vector if they are located at the same point. The difference between two vectors located at separate points is not a vector because the transformations laws (B.15) depend on position. In the usual vector algebra when we represent two vectors A and B as little arrows, the difference $A - B$ is not even defined (i.e., is not a vector) if A and B are at different points. We need first to slide one of the vectors to the other one and only then can we visualize the difference between them. Thus to compare two vectors (i.e., compute $A - B$) we must first put them at the same spacetime point. This sliding is achieved by moving one of the vectors parallel to itself (called *parallel transport*), which is easy to do in flat space.

In order to calculate $A_\mu(x + dx) - A_\mu(dx)$ we must first define what is meant by parallel transport in a general curved space. When we parallel transport a vector in flat space its components do not change when we move it around, but they do change in curved space. Imagine an arrow standing upward on the curved surface of an sphere. If one moves the tail of the arrow along the surface keeping its tip pointing upward (in other words, transporting the vector parallel to itself), then an outside observer will notice that the arrow points in different directions as it moves and concludes that it is not the same vector. Thus, parallel transport produces a different vector. Vector A has changed into a different vector B. To fix this situation, the observer corrects the direction of the arrow at every little move of its tail so that it points in the same direction, i.e., making it the same vector at all times.

If denote δA_μ as the change produced in vector $A_\mu(x^\alpha)$ located at x^α by an infinitesimal parallel transport of a distance dx^ν, we expect A_μ to be directly proportional to dx^ν and to A_μ; the bigger our arrow, the more noticeable its change will be. Thus

$$A_\mu \propto A_\nu dx^\alpha. \tag{C.5}$$

The only sensible constant of proportionality will have to have covariant μ and α indices and a contravariant ν index as

$$\delta A_\mu = \Gamma^\nu_{\mu\alpha} A_\nu dx^\alpha. \tag{C.6}$$

where $\Gamma^\nu_{\mu\alpha}$ are called *Christoffel symbols.*

Equation (C.6) defines parallel transport. δA_μ is the change produced in vector A_μ by an infinitesimal transport of a distance dx^α to produce a new vector $C_\mu = A_\mu + \delta A_\mu$. To obtain parallel transport for a contravariant vector B^μ note that a scalar defined as $A_\mu B_\mu$ cannot change under parallel transport. Thus

$$\delta(A_\mu B_\mu) = 0 \tag{C.7}$$

from which it follows that

$$\delta A_\mu = -\Gamma^\mu_{\nu\alpha} A^\nu dx^\alpha. \tag{C.8}$$

We shall also assume symmetry under exchange of lower indices,

$$\Gamma^\alpha_{\mu\nu} = \Gamma^\alpha_{\nu\mu}. \tag{C.9}$$

If $A_\mu(x^\alpha)$ is parallel transported by an infinitesimal distance dx^α, the new vector C_μ will be

$$C_\mu = A_\mu + \delta A_\mu, \tag{C.10}$$

whereas the old vector $A_\mu(x^\alpha)$ at the new position $x^\alpha + dx^\alpha$ will be A_μ $(x^\alpha + dx^\alpha)$. The difference between them is

$$dA_\mu = A_\mu(x^\alpha + dx^\alpha) - [A_\mu(x^\alpha) + \delta A_\mu], \tag{C.11}$$

which by construction is a vector. Thus we are led to a new definition of derivative (which is a tensor — note the symbol ';' in the subscript)

$$A_{\mu;\nu} = \frac{dA_\mu}{dx^\nu} = \lim_{dx\to 0} \frac{A_\mu(x+dx) - [A_\mu(x) + \delta A_\mu]}{dx^\nu}. \tag{C.12}$$

Using (C.4) in (C.11) we have

$$dA_\mu = \frac{\partial A_\mu}{dx^\nu} dx^\nu - \delta A_\mu = \frac{\partial A_\mu}{dx^\nu} dx^\nu - \Gamma^\epsilon_{\mu\alpha} A_\epsilon dx^\alpha,$$

and (C.12) becomes

$$A_{\mu;\nu} = dA_\mu/dx^\nu = \partial A_\mu/\partial x^\nu - \Gamma^\epsilon_{\mu\nu} A_\epsilon,$$

(because $dx^\alpha/dx^\nu = \delta^\alpha_\nu$) which we shall from now on write as

$$A_{\mu;\nu} = A_{\mu,\nu} - \Gamma^\epsilon_{\mu\nu} A_\epsilon, \tag{C.13}$$

where $A_{\mu,\nu} = \partial A_\mu / \partial x^\nu$. The derivative $A_{\mu;\nu}$ is often called the covariant derivative (with the word covariant not meaning the same as before) and one can easily verify that $A_{\mu;\nu}$ is a second rank tensor. From (C.8),

$$A^\mu_{;\nu} = A^\mu_{,\nu} + \Gamma^\mu_{\nu\epsilon} A^\epsilon. \tag{C.14}$$

For tensors of higher rank the results are, for example,

$$A^{\mu\nu}_{;\lambda} = A^{\mu\nu}_{,\lambda} + \Gamma^\mu_{\lambda\epsilon} A^{\epsilon\nu} + \Gamma^\nu_{\lambda\epsilon} A^{\mu\epsilon}, \tag{C.15}$$

and

$$A_{\mu\nu;\lambda} = A_{\mu\nu,\lambda} - \Gamma^\epsilon_{\mu\lambda} A_{\epsilon\nu} - \Gamma^\epsilon_{\nu\lambda} A_{\mu\epsilon}, \tag{C.16}$$

and

$$A^\mu_{\nu;\lambda} = A^\mu_{\nu,\lambda} + \Gamma^\mu_{\lambda\epsilon} A^\epsilon_\nu - \Gamma^\epsilon_{\nu\lambda} A^\mu_\epsilon, \tag{C.17}$$

and

$$A^{\mu\nu}_{\alpha\beta;\lambda} = A^{\mu\nu}_{\alpha\beta,\lambda} + \Gamma^\mu_{\lambda\epsilon} A^{\epsilon\nu}_{\alpha\beta} + \Gamma^\nu_{\lambda\epsilon} A^{\mu\epsilon}_{\alpha\beta} - \Gamma^\epsilon_{\alpha\lambda} A^{\mu\nu}_{\epsilon\beta} - \Gamma^\epsilon_{\beta\lambda} A^{\mu\nu}_{\alpha\epsilon}. \tag{C.18}$$

Γ plays the same role for the gravitational field as the field strength tensor does for the electromagnetic field.

C.2. Metric Tensor

The process of covariant differentiation should never change the length of a vector. This means that the covariant derivative of the metric tensor should always be identically zero,

$$g_{\mu\nu;\lambda} = 0. \tag{C.19}$$

Using (C.15),

$$g_{\mu\nu;\lambda} = g_{\mu\nu,\lambda} - \Gamma^\epsilon_{\mu\lambda} g_{\epsilon\nu} - \Gamma^\epsilon_{\nu\lambda} g_{\mu\epsilon} = 0. \tag{C.20}$$

Thus

$$g_{\mu\nu,\lambda} = \Gamma^\epsilon_{\mu\lambda} g_{\epsilon\nu} + \Gamma^\epsilon_{\nu\lambda} g_{\mu\epsilon}, \tag{C.21}$$

and permuting the $\mu\nu\lambda$ indices cyclically gives

$$g_{\mu\nu,\lambda} = \Gamma^\epsilon_{\lambda\nu} g_{\epsilon\mu} + \Gamma^\epsilon_{\mu\nu} g_{\lambda\epsilon}, \tag{C.22}$$

and

$$g_{\nu\lambda,\mu} = \Gamma^{\epsilon}_{\nu\mu} g_{\epsilon\lambda} + \Gamma^{\epsilon}_{\lambda\mu} g_{\nu\epsilon}. \tag{C.23}$$

Now adding (C.22) and (C.23) and subtracting (C.21) gives

$$g_{\lambda\mu,\nu} + g_{\nu\lambda,\mu} - g_{\mu\nu,\lambda} = 2\Gamma^{\epsilon}_{\mu\nu} g_{\lambda\epsilon} \tag{C.24}$$

because of the symmetries of Eq. (C.9). Multiplying (C.24) by $g^{\lambda\alpha}$ and using Eq. (B.43) (to give $g_{\lambda\epsilon}g^{\lambda\alpha} = g_{\epsilon\lambda}g^{\lambda\alpha} = \delta^{\alpha}_{\epsilon}$) yields

$$\Gamma^{\alpha}_{\mu\nu} = \frac{1}{2} g^{\lambda\alpha}(g_{\lambda\mu,\nu} + g_{\nu\lambda,\mu} - g_{\mu\nu,\lambda}), \tag{C.25}$$

which is an important formula giving the Christoffel symbol in terms of the metric tensor and its derivatives.

Using $g^{\alpha\epsilon}g_{\epsilon\beta,\alpha} = g^{\alpha\epsilon}g_{\beta\alpha,\epsilon}$ (obtained using the symmetry of the metric tensor and swapping the names of indices) and contracting over $\alpha\nu$, Eq. (C.25) becomes (first and last terms cancel)

$$\Gamma^{\alpha}_{\beta,\alpha} = \frac{1}{2} g^{\alpha\epsilon}(g_{\epsilon\beta,\alpha} + g_{\epsilon\alpha,\beta} - g_{\beta\alpha,\epsilon}) = \frac{1}{2} g^{\alpha\epsilon} g_{\epsilon\alpha,\beta}. \tag{C.26}$$

Defining g as the determinant $|g_{\mu\nu}|$ and using (B.44) it follows that

$$\frac{\partial g}{\partial g_{\mu\nu}} = g g^{\mu\nu}, \tag{C.27}$$

a result which can be easily checked. Thus (C.26) becomes

$$\Gamma^{\alpha}_{\beta,\alpha} = \frac{1}{2g}\frac{\partial g}{\partial g_{\lambda\alpha}}\frac{\partial g_{\lambda\alpha}}{\partial x^{\beta}} = \frac{1}{2g}\frac{\partial g}{\partial x^{\beta}} = \frac{1}{2}\frac{\partial \ln(-g)}{\partial x^{\beta}}, \tag{C.28}$$

where we write $\ln(-g)$ instead of $\ln g$ because g is always negative.

The concept of *parallel transport* of a vector has important implications. We cannot define globally parallel vector fields. We can define local parallelism. In the Euclidean space, a straight line is the only curve that parallel transports its own tangent vector. In curved space, we can draw "nearly" straight lines by demanding parallel transport of the tangent vector. These "lines" are called *geodesics.* A geodesic is a curve of extremal length between any two points. The equation of a geodesic is

$$\frac{d^2x^{\alpha}}{d\lambda^2} + \Gamma^{\alpha}_{\mu\beta}\frac{dx^{\mu}}{d\lambda}\frac{dx^{\beta}}{d\lambda} = 0. \tag{C.29}$$

The parameter λ is called an *affine* parameter. A curve having the same path as a geodesic but parametrized by a non-affine parameter is not a geodesic curve. Geodesics in flat space maintain their separation; those in curved spaces do not.

C.3. Riemann Curvature Tensor

The Riemann curvature tensor is one of the most important tensors in general relativity. If it is zero then it means that the space is flat. If it is nonzero then we have a curved space. This tensor is most easily derived by considering the order of double differentiation on tensors. Firstly we write in general,

$$A^{\mu}_{,\alpha\beta} = \frac{\partial^2 A^{\mu}}{\partial x^{\alpha} \partial x^{\beta}}. \tag{C.30}$$

It is also common to write instead $A^{\mu}_{,\alpha\beta} = A^{\mu}_{,\alpha,\beta}$ or $A^{\mu}_{;\alpha\beta} = A^{\mu}_{;\alpha;\beta}$. We will use both notations.

In general, it turns out that even though $A^{\mu}_{,\alpha\beta} = A^{\mu}_{,\beta\alpha}$, in general it is true that $A^{\mu}_{;\alpha\beta} \neq A^{\mu}_{;\beta\alpha}$. To check this, first consider the second derivative of a scalar ϕ. A scalar does not change under parallel transport, therefore $\phi_{;\mu} = \phi_{,\mu}$. From Eq. (C.13) we have ($\phi_{;\mu}$ is a vector, not a scalar)

$$\phi_{;\mu;\nu} = \phi_{,\mu;\nu} = \phi_{,\mu,\nu} - \Gamma^{\epsilon}_{\mu\nu}\phi_{,\epsilon}, \tag{C.31}$$

but because $\Gamma^{\epsilon}_{\mu\nu} = \Gamma^{\epsilon}_{\nu\mu}$ it follows that $\phi_{;\mu\nu} = \phi_{;\nu\mu}$ meaning that the order of differentiation does not matter for a scalar. Consider now a vector. Let's differentiate Eq. (C.14). Note that $A^{\mu}_{;\nu}$ is a second rank tensor, so we use Eq. (C.17) as follows

$$\begin{aligned} A^{\mu}_{;\nu;\lambda} &= A^{\mu}_{;\nu,\lambda} + \Gamma^{\mu}_{\lambda\epsilon}A^{\epsilon}_{;\nu} - \Gamma^{\epsilon}_{\nu\lambda}A^{\mu}_{;\epsilon} = \frac{\partial}{\partial x^{\lambda}}(A^{\mu}_{;\nu}) + \Gamma^{\mu}_{\lambda\epsilon}A^{\epsilon}_{;\nu} - \Gamma^{\epsilon}_{\nu\lambda}A^{\mu}_{;\epsilon} \\ &= A^{\mu}_{,\nu,\lambda} + \Gamma^{\mu}_{\nu\epsilon,\lambda}A^{\epsilon} + \Gamma^{\mu}_{\nu\epsilon}A^{\epsilon}_{,\lambda} + \Gamma^{\mu}_{\lambda\epsilon}A^{\epsilon}_{;\nu} - \Gamma^{\epsilon}_{\nu\lambda}A^{\mu}_{;\epsilon}. \end{aligned} \tag{C.32}$$

If we now interchange the order of differentiation, i.e., swapping the ν and λ indices, we get

$$A^{\mu}_{;\lambda;\nu} = A^{\mu}_{,\lambda,\nu} + \Gamma^{\mu}_{\lambda\epsilon,\nu}A^{\epsilon} + \Gamma^{\mu}_{\lambda\epsilon}A^{\epsilon}_{,\nu} + \Gamma^{\mu}_{\nu\epsilon}A^{\epsilon}_{;\lambda} - \Gamma^{\epsilon}_{\lambda\nu}A^{\mu}_{;\epsilon}. \tag{C.33}$$

Subtracting the two last equations, we get

$$A^{\mu}_{;\nu;\lambda} - A^{\mu}_{;\lambda;\nu} = A^{\epsilon}(\Gamma^{\mu}_{\nu\epsilon,\lambda} - \Gamma^{\mu}_{\lambda\epsilon,\nu} + \Gamma^{\mu}_{\lambda\theta}\Gamma^{\theta}_{\nu\epsilon} - \Gamma^{\mu}_{\nu\theta}\Gamma^{\theta}_{\lambda\epsilon}) = A^{\epsilon}R^{\mu}_{\lambda\epsilon\nu}, \tag{C.34}$$

with the *Riemann curvature tensor* defined as

$$R^{\mu}_{\lambda\epsilon\nu} = -\Gamma^{\alpha}_{\beta\gamma,\delta} + \Gamma^{\alpha}_{\beta\delta,\gamma} + \Gamma^{\alpha}_{\epsilon\gamma}\Gamma^{\alpha}_{\beta\delta} - \Gamma^{\alpha}_{\epsilon\delta}\Gamma^{\epsilon}_{\epsilon\delta}. \tag{C.35}$$

The Riemann tensor tells us everything essential about the curvature of a space. For Cartesian space the Riemann tensor is zero.

The Riemann tensor has the following useful symmetry properties

$$R^{\alpha}_{\beta\gamma\delta} = -R^{\alpha}_{\beta\delta\gamma} \tag{C.36}$$

$$R^{\alpha}_{\beta\gamma\delta} + R^{\alpha}_{\gamma\delta\beta} + R^{\alpha}_{\delta\beta\gamma} = 0, \tag{C.37}$$

and

$$R_{\alpha\beta\gamma\delta} = -R_{\beta\alpha\gamma\delta}. \tag{C.38}$$

All other symmetry properties of the Riemann tensor may be obtained from these. For example

$$R_{\alpha\beta\gamma\delta} = -R_{\beta\alpha\gamma\delta}. \tag{C.39}$$

A few more definitions are used in general relativity. The *Ricci tensor* is obtained by contracting on a pair of indices

$$R_{\alpha\beta} = R^{\epsilon}_{\alpha\epsilon\beta}, \tag{C.40}$$

which has the property

$$R_{\alpha\beta} = R_{\beta\alpha}. \tag{C.41}$$

An important property of the Ricci tensor is that $R_{\alpha\beta} = 0$ for empty space.

The *Riemann scalar* (also called the *scalar curvature*) is defined by

$$R = R^{\alpha}_{\alpha} = g^{\alpha\beta}R_{\alpha\beta}. \tag{C.42}$$

Finally, the *Einstein tensor* is defined as

$$G_{\mu\nu} = R_{\mu\nu} - \frac{1}{2}Rg_{\mu\nu}. \tag{C.43}$$

C.4. The Energy–Momentum Tensor

The *stress–energy tensor* (also called *energy–momentum tensor*) is defined as the flux of the four-momentum across a surface of constant x^{β}. In component form, we have:

1. T^{00} = Energy density = ρ;
2. T^{0i} = Energy flux (energy may be transmitted by heat conduction);

3. T^{i0} = Momentum density (even if the particles do not carry momentum, if heat is being conducted, then the energy will carry momentum);
4. T^{ij} = Momentum flux (also called *stress*).

The energy–momentum tensor for a perfect fluid is

$$T^{\mu\nu} = (p + \rho)u^{\mu}u^{\nu} - p\eta^{\mu\nu}, \tag{C.44}$$

where ρ is the energy density and p is the pressure. Instead of deducing this from first principles, we shall now work this out for several specific cases.

First we consider a *motionless dust* representing a collection of particles at rest. In this case $u = (c, 0)$, so that $T_{00} = \rho$. The equation of state for dust according to Eq. (2.8) is $p = 0$ so that $T^{ii} = 0 = T^{0i} = T^{ij}$. Thus

$$T^{\mu\nu} = \begin{pmatrix} \rho & 0 & 0 & 0 \\ 0 & 0 & 0 & 0 \\ 0 & 0 & 0 & 0 \\ 0 & 0 & 0 & 0 \end{pmatrix}. \tag{C.45}$$

Second, we consider a *motionless fluid* represented by a collection of particles all moving randomly (such that they exert a pressure) but the whole collection is at rest, such as a gas of particles at non-zero temperature that is confined in a motionless container. In this case $u^{\mu} = (c, 0)$ again, but now $p \neq 0$. Thus again $T^{00} = \rho$. But now $T^{ii} = p$ and $T^{ij} = 0$, so that

$$T^{\mu\nu} = \begin{pmatrix} \rho & 0 & 0 & 0 \\ 0 & p & 0 & 0 \\ 0 & 0 & p & 0 \\ 0 & 0 & 0 & p \end{pmatrix}. \tag{C.46}$$

Finally, there is *motionless radiation*, which is characterized by the equation of state $p = \rho/3$, following Eq. (2.12). The radiation is confined to a container so that $u = (\gamma c, 0)$. Hence,

$$T^{\mu\nu} = \frac{4}{3}\rho u^{\mu}u^{\nu} - \frac{1}{3}\rho\eta^{\mu\nu} = \begin{pmatrix} \rho & 0 & 0 & 0 \\ 0 & \frac{1}{3}\rho & 0 & 0 \\ 0 & 0 & \frac{1}{3}\rho & 0 \\ 0 & 0 & 0 & \frac{1}{3}\rho \end{pmatrix}. \tag{C.47}$$

Thus the general case is the motionless fluid energy–momentum tensor in Eq. (C.46). The special cases of motionless dust or motionless radiation are obtained with the respective substitutions of $p = 0$ or $p = \frac{1}{3}\rho$ in Eq. (C.47).

In classical electrodynamics the four-current density is $j^{\mu} = (c\rho, \mathbf{j})$ and the covariant conservation law is $\partial_{\mu} j^{\mu} = 0$, which results in the continuity equation continuity $\partial_t \rho + \boldsymbol{\nabla} \cdot \mathbf{j} = 0$. The four Maxwell equations are entirely equivalent to only three Maxwell equations plus the continuity equation. In analogy with electrodynamics the conservation law for the energy–momentum tensor is

$$T^{\mu\nu}_{;\nu} = 0. \tag{C.48}$$

C.5. Einstein Field Equations

The curvature of space-time is necessary and sufficient to describe gravity. The latter can be shown by considering the Newtonian limit of the geodesic equation. We require that (a) the particles are moving slowly with respect to the speed of light, (b) the gravitational field is weak so that it may be considered as a perturbation of flat space, and, (c) the gravitational field is static.

In this limit, the geodesic equation changes to,

$$\frac{d^2 x^{\mu}}{d\tau^2} + \Gamma^{\mu}_{00} \left(\frac{dt}{d\tau} \right)^2 = 0. \tag{C.49}$$

The Christoffel symbol also simplifies to

$$\Gamma^{\mu}_{00} = -\frac{1}{2} g^{\mu\lambda} \partial_{\lambda} g_{00}. \tag{C.50}$$

In the weak gravitational field limit, we can lower or raise the indices of a tensor using the Minkowskian flat metric, e.g.,

$$\eta^{\mu\nu} h_{\mu\rho} = h^{\nu}_{\rho}. \tag{C.51}$$

Then, the Christoffel symbol is written as

$$\Gamma^{\mu}_{00} = -\frac{1}{2} \eta^{\mu\lambda} \partial_{\lambda} h_{00}. \tag{C.52}$$

The geodesic equation then reduces to

$$\frac{d^2 x^{\mu}}{d\tau^2} = \frac{1}{2} \eta^{\mu\lambda} \left(\frac{dt}{d\tau} \right)^2 \partial_{\lambda} h_{00}. \tag{C.53}$$

The space components of the above equation are

$$\frac{d^2x^i}{d\tau^2} = \frac{1}{2}\left(\frac{dt}{d\tau}\right)^2 \partial_i h_{00}, \tag{C.54}$$

or,

$$\frac{d^2x^i}{dt^2} = \frac{1}{2}\partial_i h_{00}. \tag{C.55}$$

The concept of an *inertial mass* arises from Newton's second law,

$$\mathbf{f} = m_i \mathbf{a}. \tag{C.56}$$

According to the law of gravitation, the gravitational force exerted on an object is proportional to the gradient of a scalar field Φ, known as the scalar gravitational potential. The constant of proportionality is the gravitational mass, m_g,

$$\mathbf{f}_g = -m_g \nabla\Phi. \tag{C.57}$$

According to the *equivalence principle*, the inertial and gravitational masses are the same,

$$m_i = m_g. \tag{C.58}$$

Hence,

$$\mathbf{a} = -\nabla\Phi. \tag{C.59}$$

Comparing Eqs. (C.55) and (C.59), we find that they are the same if we identify,

$$h_{00} = -2\Phi. \tag{C.60}$$

Thus,

$$g_{00} = -(1 + 2\Phi). \tag{C.61}$$

The curvature of space-time is sufficient to describe gravity in the Newtonian limit as long as the metric takes the form (C.61). All the basic laws of physics, beyond those governing freely-falling particles, adapt to the curvature of space-time (that is, to the presence of gravity) when we are working in Riemannian normal coordinates. The tensorial form of any law is coordinate-independent and hence, translating a law into the language of tensors (that is, replacing the partial derivatives by the covariant derivatives), we will have an universal law which holds in all coordinate systems. This procedure is sometimes called the *principle of equivalence.* For example, the conservation equation for the energy–momentum tensor $T^{\mu\nu}$ in flat space-time, viz.,

$$\partial_\mu T^{\mu\nu} = 0 \tag{C.62}$$

is immediately adapted to the curved space-time as,

$$T^{\mu\nu}_{;\mu} = 0. \tag{C.63}$$

This equation expresses the conservation of energy in the presence of a gravitational field.

We can now introduce Einstein's field equations which governs how the metric responds to energy and momentum. We would like to derive an equation which will supersede the Poisson equation for the Newtonian potential,

$$\nabla^2\Phi = -4\pi G\rho, \tag{C.64}$$

where $\nabla^2 = \delta^{ij}\partial_i\partial_j$ is the Laplacian in space and ρ is the mass density. A relativistic generalization of this equation must have a tensorial form so that the law is valid in all coordinate systems. The tensorial counterpart of the mass density is the energy–momentum tensor, $T^{\mu\nu}$. The gravitational potential should be replaced by the metric. Thus, we guess that our new equation will have $T^{\mu\nu}$ proportional to some tensor which is second-order in the derivatives of the metric,

$$T_{\mu\nu} = \kappa A_{\mu\nu}, \tag{C.65}$$

where $A_{\mu\nu}$ is the tensor to be found. The requirements on the equation above are: (a) By definition, the R.H.S must be a second-rank tensor; (b) It must contain terms linear in the second derivatives or quadratic in the first derivative of the metric; (c) The R.H.S must be symmetric in μ and ν as $T^{\mu\nu}$ is symmetric; (d) Since $T^{\mu\nu}$ is conserved, the R.H.S must also be conserved.

The first two conditions require the right-hand side to be of the form

$$\alpha R_{\mu\nu} + \beta R g_{\mu\nu} = T_{\mu\nu}, \tag{C.66}$$

where $R_{\mu\nu}$ is the Ricci tensor, R is the scalar curvature and α and β are constants. This choice is symmetric in μ and ν and hence satisfies the third condition. From the last condition, we obtain

$$g^{\nu\sigma}(\alpha R_{\mu\nu} + \beta R g_{\mu\nu})_{;\sigma} = 0. \tag{C.67}$$

This equation cannot be satisfied for arbitrary values of α and β. This equation holds only if α/β is fixed. As a consequence of the *Bianchi identity*, viz.,

$$R^{;\mu}_{\mu\nu} = \frac{1}{2}R_{;\nu}, \tag{C.68}$$

we choose

$$\beta = -\frac{1}{2}\alpha. \tag{C.69}$$

With this choice, the Equation (C.66) becomes

$$\alpha\left(R_{\mu\nu} - \frac{1}{2}Rg_{\mu\nu}\right) = T_{\mu\nu}. \tag{C.70}$$

In the weak field limit,

$$g_{00} \approx -2\Phi, \tag{C.71}$$

the 00-component of Eq. (C.70) becomes

$$-\alpha\nabla^2 g_{00} = T_{00} \Rightarrow 2\alpha\nabla^2\Phi = \rho. \tag{C.72}$$

Compare this result with Newton's Equation (C.64), we obtain,

$$2a = \frac{1}{4\pi G}. \tag{C.73}$$

Thus, we obtain the Einstein field equations in their final form as

$$R_{\mu\nu} - \frac{1}{2}Rg_{\mu\nu} = 8\pi G T_{\mu\nu}. \tag{C.74}$$

Introducing the cosmological constant, the *Einstein's field equations* are

$$G_{\mu\nu} = 8\pi G T_{\mu\nu} + \Lambda g_{\mu\nu}, \tag{C.75}$$

which are a set of 16 coupled equations which will give $g_{\mu\nu}$ (buried inside $G_{\mu\nu}$) for a given $T_{\mu\nu}$. Actually there are only 10 independent equations because of the symmetry $g_{\mu\nu} = g_{\nu\mu}$.

The solutions of the Einstein's field equations are "metrics" of spacetime. These metrics describe the structure of spacetime including the inertial motion of objects in spacetime. As the field equations are nonlinear, they cannot always be completely solved (i.e., without making approximations). For example, there is no known complete solution for a spacetime with two massive bodies in it (which is a theoretical model of a binary star system). Approximations are usually made in these cases. But there are numerous cases where the field equations have been solved completely. These solutions lead to the prediction of black holes and to different models of evolution of the Universe.

After obtaining $g_{\mu\nu}$ from Einstein's field equations, we can calculate the paths of light rays, the orbits of planets, etc. In practice the solution of the Einstein's field equations is exceedingly difficult and only a few exact solutions are known. In practice, the way one usually solves the Einstein's equations is to specify a metric in general terms which contains unknown coefficients. This metric is substituted into the Einstein's equations and one solves for the unknown coefficients. Thus we need to learn how to derive the metric for the spaces under consideration. We will do this for the Friedmann–Robertson–Walker (FRW) metric which is the metric appropriate to a homogeneous and isotropic Universe but where size can change with time. We will also discuss the Schwarzschild metric which describes spacetime in the vicinity of a non-rotating massive spherically-symmetric object. This will be relevant to discuss the structure of compact stars.

Appendix D

Riemann Geometry

Here we will extend the discussion of General Relativity in a cosmological context. The Universe is assumed to be homogeneous and isotropic (cosmological principle). On large scales, we need to find a spacetime manifold that looks the same in all directions and from all places. In two dimensions, the infinite plane looks the same everywhere on it. So does the sphere: there are no special points anywhere. Other, not so obvious, possibilities are the "hyperbolic paraboloid" or saddle shape, as shown in Figure D.1. We can understand curvature when a manifold is embedded in a higher-dimensional spacetime. For example, Figure D.1 shows the two-manifolds embedded in at three-dimensional space and then projected down to two-dimensional space.

If extend a straight line on the surface of a sphere, it would generate a great circle. Two parallel tracks eventually meet. A triangle made of segments of three such great circles does not have a total interior angle of 180° but a greater value. Similarly, on the saddle-shaped surface, parallel lines can converge and diverge, and a triangle always contains less than 180°. Since one of the results of GR is that light follows geodesic paths, we can use light-rays (astronomical observations) to probe the geometry of the Universe.

In $3 + 1$ dimensions we have the same set of homogeneous and isotropic manifolds, but promoted to higher dimensionality. The most general spatially homogeneous and isotropic manifolds are indeed flat space, the three-sphere, and the three-hyperboloid. The three-sphere is not a "ball" in three dimensions but a three-dimensional "surface" with no boundaries in four-dimensions. The construction of such surfaces is the subject of the following sections.

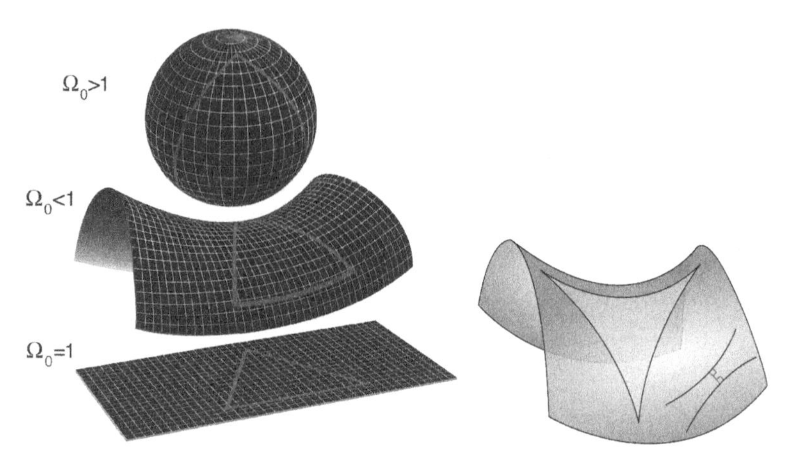

Figure D.1. Possible homogeneous and isotropic manifolds in two dimensions. Left: All three constant-curvature manifolds [Nas12]. Right: Hyperbolic paraboloid [Wik12].

D.1. Curved Spaces

D.1.1. *Polar coordinates*

A general N-dimensional Riemannian space, denoted by $\mathcal{R}_N$, is one in which the distance ds between two neighboring points can be written

$$ds^2 = g_{\mu\nu} dx^\mu dx^\nu. \tag{D.1}$$

If coordinates can be found such that

$$ds^2 = dx^\mu dx^\nu. \tag{D.2}$$

over the whole space then the space is said to be *Euclidean* and is denoted by $\mathcal{E}_N$. Clearly $\mathcal{E}_N$ is a special case of $\mathcal{R}_N$.

Next we shall restrict our discussion to the spatial part of the metric denoted as

$$d\ell^2 = h_{ij} dx^i dx^j. \tag{D.3}$$

The special relativity metrics can be written as

$$ds^2 = c^2 dt^2 - d\ell^2, \tag{D.4}$$

so that

$$h_{ij} = -g_{ij}. \tag{D.5}$$

Consider the two dimensional space where

$$d\ell^2 = dr^2 + r^2 d\theta^2, \tag{D.6}$$

because $h_{11} = 1$, $h_{12} = h_{21} = 0$, $h_{22} = r^2$. The space looks like $\mathcal{R}_2$ but actually it is $\mathcal{E}_2$ because we can find coordinates such that Eq. (D.2) is true. These coordinates are the two-dimensional plane polar coordinates

$$x = r\cos\theta \quad \text{and} \quad y = r\sin\theta, \tag{D.7}$$

in which

$$d\ell^2 = dx^2 + dy^2, \tag{D.8}$$

with $g_{11} = g_{22} = 1$ and $g_{12} = g_{21} = 0$ or $g_{\mu\nu} = \delta_{\mu\nu}$.

For a three dimensional space, the spherical polar coordinates are

$$x = r\sin\theta\cos\phi, \quad y = r\sin\theta\sin\phi, \quad \text{and} \quad z = r\cos\theta, \tag{D.9}$$

and the increments of length $d\ell_r$, $d\ell_\theta$, $d\ell_\phi$ in the ϵ_r, ϵ_θ and ϵ_ϕ directions respectively are

$$d\ell_r = dr, \quad d\ell_\theta = rd\theta, \quad \text{and} \quad r\sin\theta d\phi. \tag{D.10}$$

Thus the surfaces of a sphere is an example of a space which is $\mathcal{R}_2$ and cannot be reduced to $\mathcal{E}_2$. On the surface of the sphere the distance between two points is

$$d\ell^2 = d\ell_\theta^2 + d\ell_\phi^2, \tag{D.11}$$

where $h_{11} = r$, $h_{12} = h_{21} = 0$, $h_{22} = r^2\sin^2\theta$. For this surface it is not possible to find x, y such that $ds^2 = dx^2 + dy^2$ and therefore the surface of a sphere is not $\mathcal{E}_\in$ but rather a genuine $\mathcal{R}_\in$ space.

D.1.2. *Change of coordinates*

Infinitesimal length method

In this method one identifies the infinitesimal increments of length and simply multiplies them together to get the volume element. In Cartesian coordinates we have $d\ell_x = dx$, $d\ell_y = dy$, and $d\ell_z = dz$ to give

$$dV = d\ell_x d\ell_y d\ell_z = dxdydz. \tag{D.12}$$

In two dimensional plane polar coordinates $d\ell_r = dr$ and $d\ell_\theta = rd\theta$ to give

$$dV = d\ell_r d\ell_\theta = rdrd\theta dV = d\ell_x d\ell_y d\ell_z, \tag{D.13}$$

(actually this "volume" is an area). In 3-D spherical polar coordinates $d\ell_r = dr$, $d\ell_\theta = d\theta$ and $d\ell_\phi = r\sin d\phi$ to give

$$dV = d\ell_r d\ell_\theta d\ell_\phi = r^2 \sin\theta dr d\theta d\phi. \tag{D.14}$$

Jacobian method

Suppose $x = x(u,v)$ and $y = y(u,v)$ then

$$\iint f(x,y)dxdy = \iint f\left[x(u,v), y(u,v)\right] |J(u,v)|\, dudv, \tag{D.15}$$

where $|J(u,v)|$ is the modulus of the Jacobian defined as

$$J(u,v) = \begin{vmatrix} \partial x/\partial u & \partial x/\partial v \\ \partial y/\partial u & \partial y/\partial u \end{vmatrix}. \tag{D.16}$$

For three dimensions with $x = x(u,v,w)$, $y = y(u,v,w)$ and $z = z(u,v,w)$, we have

$$J(u,v) = \begin{vmatrix} \partial x/\partial u & \partial x/\partial v & \partial x/\partial w \\ \partial y/\partial u & \partial y/\partial u & \partial y/\partial w \\ \partial z/\partial u & \partial z/\partial u & \partial z/\partial w \end{vmatrix}. \tag{D.17}$$

For Cartesian coordinates obviously $J(x,y,z) = 1$. For plane polar coordinates, $x = r\cos\theta$, $y = r\sin\theta$, and

$$J(u,v) = \begin{vmatrix} \cos\theta & -r\sin\theta \\ \sin\theta & r\cos\theta \end{vmatrix} = r. \tag{D.18}$$

For spherical polar coordinates, $x = r\cos\theta$, $y = r\sin\theta\cos\phi$, $z = \cos\theta$,

$$J(u,v) = \begin{vmatrix} \sin\theta\cos\phi & r\cos\theta\cos\phi & -r\sin\theta\sin\phi \\ \sin\theta\sin\phi & r\cos\theta\sin\phi & r\sin\theta\cos\phi \\ \cos\theta & -r\sin\theta & 0 \end{vmatrix} = r^2\sin\theta. \tag{D.19}$$

The volume element in 2D is

$$dV = |J(u,v)|\, dudv, \tag{D.20}$$

and in 3D it is

$$dV = |J(u,v,w)|\, dudvdw, \tag{D.21}$$

which then reproduce Eqs. (D.12–D.13) and (D.14) for Cartesian, plane polar and spherical polar coordinates.

Metric tensor method

The volume element in any coordinate system reads

$$dV = \sqrt{h} du dv dw, \tag{D.22}$$

where h is the determinant of the spatial metric tensor. Thus

$$|J| = \sqrt{h}. \tag{D.23}$$

For a plane polar coordinates $ds^2 = dr^2 + r^2 d\theta^2 = g_{ij} dx^i dx^j$ so that

$$h = \begin{vmatrix} 1 & 0 \\ 0 & r^2 = \end{vmatrix} = r^2, \tag{D.24}$$

giving $\sqrt{h} = r$ so that $dV = \sqrt{h} dr d\theta$, in agreement with Eq. (D.13). For spherical polar coordinates $ds^2 = dr^2 + r^2 d\theta^2 + r^2 \sin^2\theta d\phi^2 = h_{ij} dx^i dx^j$, giving

$$h = \begin{vmatrix} 1 & 0 & 0 \\ 0 & r^2 & 0 \\ 0 & 0 & r^2 \sin^2\theta \end{vmatrix} = r^4 \sin^2\theta, \tag{D.25}$$

which yields $\sqrt{h} = r^2 \sin\theta$, so that $dV = \sqrt{h} dr d\theta d\phi = r^2 \sin\theta dr d\theta d\phi$, in agreement with Eq. (D.14).

Thus, one can just define

$$d^2x = du dv, \quad d^3x = du dv dw, \quad d^4x = du dv dw dt, \tag{D.26}$$

so that the measure is

$$dV = \sqrt{h} d^2x, \quad \text{or} \quad \sqrt{h} d^3x, \tag{D.27}$$

depending on the number of dimensions. It is important to remember that d^3x or d^4x is not $d\ell_1 d\ell_2 d\ell_3$ or $d\ell_1 d\ell_2 d\ell_3 d\ell_4$, but simply only the coordinates. For example in spherical polar coordinates

$$d^3x = dr d\theta d\phi, \tag{D.28}$$

or with time

$$d^4x = dr d\theta d\phi dt. \tag{D.29}$$

The measure is the volume is obtained with $dV = \sqrt{h} d^3x$ or $\sqrt{-g} d^4x$ (because for 4D we use $g_{\mu\nu}$ and $h_{ij} = -g_{ij}$ and $h = -g$). Thus in general

$$dV \neq d^3x, \quad \text{or} \quad dV \neq d^4x. \tag{D.30}$$

D.1.3. *Differential geometry*

The definition of a circle, $x^2 + y^2 = r^2$ can be written generally as $y = y(x)$. The same equation can be expressed parametrically in terms of the parameter θ as $x = r\cos\theta$ and $y = r\sin\theta$, or generally as $x = x(\theta)$ and $y = y(\theta)$. Another way to write our equation for the circle (radius = 1) is

$$\mathbf{x} = \cos\theta\hat{\mathbf{e}}_1 + \sin\theta\hat{\mathbf{e}}_2, \tag{D.31}$$

where $\hat{e}_1$ and $\hat{e}_2$ are basis vectors in $\mathcal{E}_2$.

For many curves the form $y = y(x)$ can be clumsy and nowadays mathematicians always prefer the parametric representation. Thus a general curve is expressed as

$$\mathbf{x} = \mathbf{x}(t), \tag{D.32}$$

where t is the parameter. If the basis is chosen to be $\mathcal{E}_2$ then $\mathbf{x} = \mathbf{x}(t)$ is equivalent to two scalar equations $x_1 = x_1(t)$ and $x_2 = x_2(t)$. Thus a curve can be specified in any number of dimensions. For our circle above we have $x_1 = x_1(\theta) = \cos\theta$ and $x_2 = x_2(\theta) = \sin\theta$. A general surface is expressed as

$$\mathbf{x} = \mathbf{x}(u, v). \tag{D.33}$$

If, for example, the basis in $\mathcal{E}_3$ then

$$x(u, v) = x_1(u, v)\hat{\mathbf{e}}_1 + x_2(u, v)\hat{\mathbf{e}}_2 + x_3(u, v)\hat{\mathbf{e}}_3. \tag{D.34}$$

D.1.4. *One dimensional curve*

Let us first consider the circle, often called the one sphere, denoted by S^1. Recall that for a circle of radius R, the proper way to express it is in terms of the one dimensional parameter θ as

$$x(\theta) = R(\cos\theta\hat{\mathbf{e}}_1 + \sin\theta\hat{\mathbf{e}}_2). \tag{D.35}$$

However it is common to introduce a fictitious extra dimension and embed the 1D curve in a 2D Euclidean space via

$$x^2 + y^2 = R^2. \tag{D.36}$$

which we recognize as the equation for a circle. Remember though this equation is really overkill. It is a 2D equation for a 1D curve! The 1-parameter equation (D.35) is much better. We can also write

$$x_1^2 + x_2^2 = R^2. \tag{D.37}$$

The element of length in the 2D Euclidean space is

$$d\ell^2 = dx_1^2 + dx_2^2. \tag{D.38}$$

In an ordinary 2D Euclidean space, x and y (or x_1 and x_2) are free to vary independently and this is how the whole 2D space gets covered. Equation (D.38) is true in general. However the reason that Eq. (D.36) or (D.37) describes a circle is because it constrains the value of y in terms of x. This constraint (D.36) picks out only those points in $\mathcal{E}_2$ which give the circle.

Equation (D.38) covers all of $\mathcal{E}_2$. We can constrain it for the circle by reducing the two parameters x_1 and x_2 to only one parameter. Thus we will have $d\ell$ for the circle. We do this using the 2D constraint (D.37) and writing $y = \sqrt{R^2 - x^2}$ and $dy = -x/\sqrt{R^2 - x^2}$ so that

$$dy^2 = \frac{x^2}{\sqrt{R^2 - x^2}} dx^2. \tag{D.39}$$

Note that $dy^2 = (dy)^2$ and $dy^2 \neq d(y^2)$. Thus Eq. (D.39) becomes

$$d\ell^2 = dx^2 + \frac{x^2}{\sqrt{R^2 - x^2}} dx^2 = \frac{R^2}{\sqrt{R^2 - x^2}}. \tag{D.40}$$

This can also be written in terms of the dimensionless coordinate $r = x/R$, to give

$$d\ell^2 = R^2 \frac{dr^2}{1 - r^2}, \tag{D.41}$$

where R is the radius of the space (the circle).

Another convenient coordinate system for the circle uses the angle θ from plane polar coordinates specified via

$$x = R\cos\theta, \quad y = R\sin\theta. \tag{D.42}$$

Identifying the increments of length $d\ell_R$ and $d\ell_\theta$ in the $\hat{e}_R$ and $\hat{e}_\theta$ directions as

$$d\ell_R = dR, \quad d\ell_\theta = Rd\theta, \tag{D.43}$$

then

$$d\ell^2 = d\ell_R^2 + d\ell_\theta^2 = dR^2 + R^2 d\theta^2, \tag{D.44}$$

which gives the distance $d\ell$ in the 2D space. To restrict ourselves to the rim of the circle (curved 1D space) we fix $d\ell_R = dR = 0$ and get

$$d\ell^2 = R^2 d\theta^2, \tag{D.45}$$

which makes it obvious that the space is the one sphere (circle) of radius R. Using simple trigonometry one can show that (D.45) is the same as (D.40).

Using $d\ell^2 = h_{ij}dx^i dx^j$, we evidently have

$$h_{ij} = (R^2), \tag{D.46}$$

which is a one dimensional "motion". The determinant is obviously $h = R^2$ giving $\sqrt{h} = R$. This allows us to calculate the volume (we are calling the length a geological volume) as

$$V = \int \sqrt{h} d^1 x = \int_0^{2\pi} R d\theta = 2\pi R. \tag{D.47}$$

The 1D curve that we described above is the circle or one sphere denoted S^1. However there are three 1D spaces which are homogeneous and isotropic. These are (i) the flat x line (R^1), (ii) the positively curved one sphere (S^1) derived above and (iii) the negatively curved hyperbolic curve (H^1).

The formulas for a space of constant negative curvature can be obtained with the replacement $R \to iR$, to yield

$$d\ell^2 = \frac{-R^2}{-R^2 - x^2} dx^2 = \frac{R^2}{R^2 + x^2} dx^2, \tag{D.48}$$

or using $r = x/R$,

$$d\ell^2 = R^2 \frac{dr^2}{1 + r^2} dx^2. \tag{D.49}$$

These results are also obtained by embedding in Minkowski space.

The line element for a space of zero curvature is obviously just

$$d\ell^2 = dx^2, \tag{D.50}$$

or using $r = x/R$,

$$d\ell^2 = R^2 dr^2. \tag{D.51}$$

These formulas are obtained from S^1 or H^1 by letting $R \to \infty$.

We can collect our results for R^1, S^1 and H^1 into a single formula

$$d\ell^2 = \frac{R^2}{R^2 - kx^2} dx^2, \tag{D.52}$$

or, using $r = x/R$,

$$d\ell^2 = R^2 \frac{dr^2}{1 - kR^2}, \tag{D.53}$$

where $k = 0, +1, -1$ for flat, closed and open curves, respectively (i.e., for R^1, S^1 and H^1, respectively).

D.1.5. *Two dimensional surface*

A surface is represented by two parameters u, v and expressed as $x = x(u, v)$, as mentioned previously. However we shall introduce a surface fictitious coordinate (three parameters) and embed the surface in $\mathcal{E}_3$. Thus with 3 parameters the equation for the two sphere is

$$x^2 + y^2 + z^2 = R^2, \tag{D.54}$$

which we recognize as the equation for a sphere. However this equation is overkill. It is a 3D equation for a 2D surface. We can also write

$$x_1^2 + x_2^2 + x_3^2 = R^2. \tag{D.55}$$

Any 3D Euclidean space $\mathcal{E}_3$ has length element

$$d\ell^2 = dx_1^2 + dx_2^2 + dx_3^2, \tag{D.56}$$

which under normal circumstances would map out the whole 3D volume. However Eq. (D.55) restricts x_3 according to

$$x_3^2 = R^2 - x_1^2 - x_2^2. \tag{D.57}$$

Writing

$$dx_3 = \frac{\partial x_3}{\partial x_1} dx_1 + \frac{\partial x_3}{\partial x_2} dx_2, \tag{D.58}$$

and with $\partial x_3/\partial x_1 = -x_1/\sqrt{R^2 - x_1^2 - x_2^2}$, we have

$$dx_3^2 = -\frac{x_1 dx_1 + x_2 dx_2}{\sqrt{R^2 - x_1^2 - x_2^2}}, \tag{D.59}$$

or

$$dx_3^2 = \frac{(x_1 dx_1 + x_2 dx_2)^2}{R^2 - x_1^2 - x_2^2}, \tag{D.60}$$

to give

$$d\ell^2 = dx_1^2 + dx_2^2 + \frac{(x_1 dx_1 + x_2 dx_2)^2}{R^2 - x_1^2 - x_2^2}, \tag{D.61}$$

which is rewritten as

$$d\ell^2 = \frac{1}{R^2 - x_1^2 - x_2^2}[(R^2 - x_2^2)dx_2^2 + x_1 x_2 dx_1 dx_2 + x_2 x_1 dx_2 dx_1]. \tag{D.62}$$

Let us introduce plane polar coordinates in the x_3 plane as

$$x_1 = r' \cos\theta, \quad \text{and} \quad x_2 = r' \sin\theta. \tag{D.63}$$

Thus, $\theta : 0 \to 2\pi$, and $r' : 0 \to R$. Using

$$dx_i = \frac{\partial x_i}{\partial x^1} dr' + \frac{\partial x_i}{\partial \theta} d\theta, \tag{D.64}$$

then Eqs. (D.61) and (D.62) become

$$d\ell^2 = \frac{R^2}{R^2 - r'^2} dr'^2 + r'^2 d\theta^2. \tag{D.65}$$

This can also be written in terms of the dimensionless coordinate $r \equiv r'/R$, to give

$$d\ell^2 = R^2 \left[\frac{dr^2}{1 - r^2} + r^2 d\theta^2 \right], \tag{D.66}$$

where $r : 0 \to 1$.

Another convenient coordinate system for the two sphere uses angles θ and ϕ from spherical polar coordinates specified via

$$x = R \sin\theta \cos\phi, \quad y = R \sin\theta \sin\phi, \quad \text{and} \quad z = R \cos\theta. \tag{D.67}$$

Substituting into Eq. (D.56) directly yields

$$d\ell^2 = R^2 (d\theta^2 + \sin^2\theta d\phi^2). \tag{D.68}$$

Equation (D.67) is alternatively obtained by identifying the increments of length $d\ell_R$, $d\ell_\theta$, $d\ell_\phi$ in the $\hat{\mathbf{e}}_R$, $\hat{\mathbf{e}}_\theta$ and $\hat{\mathbf{e}}_\phi$ directions as

$$d\ell_R = dR, \quad d\ell_\theta = R d\theta, \quad \text{and} \quad d\ell_\phi = R \sin\theta d\phi, \tag{D.69}$$

then

$$d\ell^2 = d\ell_R^2 + d\ell_\theta^2 + d\ell_\phi^2 = dR^2 + R^2(d\theta^2 + \sin^2\theta d\phi^2), \tag{D.70}$$

gives the distance ds in the 3D space. To restrict ourselves to the surface of the sphere (curved 2D space) we set $d\ell_R = dR = 0$ and get Eq. (D.68).

Using $d\ell^2 \equiv h_{ij}dx^i dx^j$ we evidently have for S^2

$$h_{ij} = \begin{pmatrix} R^2 & 0 \\ 0 & R^2 \sin^2 \theta \end{pmatrix}. \tag{D.71}$$

The determinant is obviously

$$h = R^4 \sin^2 \theta, \tag{D.72}$$

giving $\sqrt{h} = R^2 \sin \theta$. The volume (we are calling the surface area a generalized volume) is

$$V = \int \sqrt{h} d^2 x = \int \sqrt{h} d\theta d\phi + R^2 \int_0^\pi \sin \theta \int_0^{2\pi} d\phi = 4\pi R^2. \tag{D.73}$$

Actually there are three 2D spaces which are homogenous and isotropic. These are (i) the flat $x - y$ plane (R^2), (ii) the positively curved two sphere (S^2) and (iii) the negatively curved two hyperbola (H^2).

As before we can obtain the formula for H^2 with the replacement $R \to iR$, to yield

$$d\ell^2 = R^2 \left[\frac{dr^2}{1 + r^2} + r^2 d\theta^2 \right]. \tag{D.74}$$

This result is also obtained by embedding in Minkowski space.

We can collect our results for R^2, S^2 and H^2 into a single formula

$$d\ell^2 = R^2 \left[\frac{dr^2}{1 - kr^2} + r^2 d\theta^2 \right]. \tag{D.75}$$

where $k = 0, +1, -1$ for flat (R^2), closed (S^2) and open (H^2) surfaces, respectively.

The volume can also be calculated using $ds^2 = h_{ij}dx^i dx^j$ in (D.75). We have

$$h_{ij} = \begin{pmatrix} \frac{R^2}{1 - kr^2} & 0 \\ 0 & R^2 r^2 \end{pmatrix}, \tag{D.76}$$

giving the determinant

$$h = \frac{R^4 r^2}{1 - kr^2}, \tag{D.77}$$

or $\sqrt{h} = R^2r^2/\sqrt{1-kr^2}$. The volume is

$$V = \int \sqrt{h}d^2x = \int \sqrt{h} = R^2 \int_0^0 \frac{rdr}{\sqrt{1-kr^2}} \int_0^{2\pi} = 2\pi R^2 \int \frac{rdr}{\sqrt{1-kr^2}}. \tag{D.78}$$

What this really means is

$$\int_0^0 dr = 2\int_0^R dr, \tag{D.79}$$

where $r = 0$ at $\theta = 0$, $r = R$ at $\theta = \pi/2$ and $r = 0$ again at $\theta = \pi$.

The integral and its limits are more clearly done with the substitution $\sqrt{k}r \equiv \sin\chi$ where χ varies between 0 and π. Thus $\int_0^\infty d\chi$ is equivalent to $\int_0^0 dr = 2\int_0^R dr$ and the volume in Eq. (D.78) becomes

$$V = 2\pi R^2 \int_0^\pi \frac{1}{\sqrt{k}}\frac{1}{\sqrt{k}} \sin\chi d\chi, \tag{D.80}$$

giving

$$V = \frac{4\pi R^2}{k}. \tag{D.81}$$

Thus for $k = +1$ we have $V = 4\pi R^2$, as before. For $k = 0$ we have $V = \infty$ and for $k = -1$ we need to do the integral again. We find $V = \infty$ for $k = -1$, too.

D.1.6. *Three dimensional hypersurface*

Proceeding upwards in our number of dimensions we might inquire a curved volume. But the curvature can really only be imagined with respect to embedding in a 4D Euclidean space $\mathcal{E}_4$. We call the 4D space a *hypersurface.*

Our 4D Euclidean space (into which we will embed the hypersurface) has length element

$$d\ell^2 = dx_1^2 + dx_2^2 + dx_3^2 + dx_4^2 \tag{D.82}$$

(it is $d\ell^2$ and not ds^2).

A hypersurface is represented by three parameters u, v, w and is expressed as $\mathbf{x} = \mathbf{x}(u, v, w)$. We now introduce an extra fictitious coordinate for the three sphere S^3 as

$$x^2 + y^2 + z^2 + w^2 = R^2, \quad x_1^2 + x_2^2 + x_3^2 + x_4^2 = R^2, \tag{D.83}$$

which restricts x_4 as

$$x_4^2 = R^2 - x_1^2 - x_2^2 - x_3^2. \tag{D.84}$$

Writing

$$dx_4 = \frac{\partial x_4}{\partial x_1}dx_1 + \frac{\partial x_4}{\partial x_2}dx_2 + \frac{\partial x_4}{\partial x_3}dx_3, \tag{D.85}$$

and with $\partial x_4/\partial x_1 = -x_1/\sqrt{R^2 - x_1^2 - x_2^2 - x_3^2}$, etc., we have

$$dx_4 = -\frac{x_1 dx_1 + x_2 dx_2 + x_3 dx_3}{\sqrt{R^2 - x_1^2 - x_2^2 - x_3^2}}, \tag{D.86}$$

or

$$d\ell^2 = dx_1^2 + dx_2^2 + dx_3^2 + \frac{(x_1 dx_1 + x_2 dx_2 + x_3 dx_3)^2}{R^2 - x_1^2 - x_2^2 - x_3^2}. \tag{D.87}$$

Let us introduce spherical polar coordinates in the x_4 hyperplane as

$$x_1 = r' \sin\theta cos\phi, \quad x_2 = r' \sin\theta \sin\phi \quad \text{and} \quad x_3 = r' \cos\theta, \tag{D.88}$$

where

$$\theta : 0 \to \pi, \quad \phi : 0 \to 2\pi \quad \text{and} \quad r' : 0 \to R. \tag{D.89}$$

Using

$$dx_i = \frac{\partial x_i}{\partial r'}dr' + \frac{\partial x_i}{\partial \theta}d\theta + \frac{\partial x_i}{\partial \phi}d\phi, \tag{D.90}$$

then (D.87) becomes

$$d\ell^2 = \frac{R^2}{R^2 - r'^2}dr'^2 + r'^2 d\theta^2 + r'^2 \sin^2\theta d\phi^2. \tag{D.91}$$

Introducing the dimensionless coordinate $r = r'/R$, gives

$$d\ell^2 = R^2 \left[\frac{dr^2}{1 - r^2}dr^2 + r^2 d\theta^2 + r^2 \sin^2\theta d\phi^2\right], \tag{D.92}$$

where $r : 0 \to 1$.

Another convenient coordinate system for the three sphere uses angles χ, θ, ϕ from 4D hyperspherical polar coordinates specified via

$$\begin{aligned} x = R\sin\chi\sin\theta\cos\phi, \quad y = R\sin\chi\sin\theta\sin\phi, \\ z = R\sin\chi\cos\theta, \quad w = R\cos\chi. \end{aligned} \tag{D.93}$$

Substituting into Eq. (D.92) directly yields

$$d\ell^2 = R^2[d\chi^2 + \sin^2\chi(d\theta^2 + \sin^2\theta d\phi^2)]. \tag{D.94}$$

Using $d\ell^2 = h_{ij}dx^i dx^j$, we have

$$h_{ij} = \begin{pmatrix} R^2 & 0 & 0 \\ 0 & R^2\sin^2\chi & 0 \\ 0 & 0 & R^2\sin^2\chi\sin^2\theta \end{pmatrix}. \tag{D.95}$$

The determinant is $h = R^6\sin^4\chi\sin^2\theta$, giving $\sqrt{h} = R^3\sin^2\chi\sin\theta$. The volume is

$$V = \int\sqrt{h}d^3x = \int\sqrt{h}d\chi d\theta d\phi = R^3\int_0^\pi \sin^2\chi d\chi\int_0^\pi \sin\theta d\theta\int_0^{2\pi} d\phi, \tag{D.96}$$

where the limits $\int_0^\pi d\chi$ are the same as in the previous section. Thus

$$V = 4\pi R^3\int_0^\pi \sin^2\chi d\chi = 4\pi R^3\left[\frac{\chi}{2} - \frac{\sin 2\chi}{4}\right]_0^\pi, \tag{D.97}$$

giving

$$V = 2\pi^2 R^3 \tag{D.98}$$

for the volume of our hypersphere. Compare this to the volume of a Euclidean sphere $4\pi R^3$.

For a flat, open and closed hyperspheres the metric is

$$d\ell^2 = R^2\left[\frac{dr^2}{1-kr^2} + r^2d\theta^2 + r^2\sin^2\theta d\phi^2\right]. \tag{D.99}$$

The volume can be calculated with another method. Using $d\ell = h_{ij}dx_i dx_j$ in (D.99), we have

$$h_{ij} = \begin{pmatrix} \frac{R^2}{1-kr^2} & 0 & 0 \\ 0 & R^2r^2 & 0 \\ 0 & 0 & R^2r^2\sin^2\theta \end{pmatrix}, \tag{D.100}$$

giving the determinant $h = R^6 r^4 \sin^2\theta/(1-kr^2)$. The volume is

$$V = \int \sqrt{h} d^3x = \int \sqrt{h} dr d\theta d\phi = R^3 \int_0^0 \frac{r^2 dr}{\sqrt{1-kr^2}} \int_0^\pi \sin\theta d\theta \int_0^{2\pi} d\phi$$

$$= 4\pi R^3 \int_0^0 \frac{r^2 dr}{\sqrt{1-kr^2}}. \tag{D.101}$$

The limits of integration are the same as discussed in Equation (D.79).

Using the substitution $\sqrt{k}r = \sin\chi'$, with $\chi' : 0 \to \pi$, because $\chi : 0 \to \pi$. Thus the volume is

$$V = \frac{2\pi^2 R^3}{k^{3/2}}. \tag{D.102}$$

For $k = +1$ this agrees with our previous result.

D.2. The FRW Metric

The metric of special relativity is $ds^2 = c^2 dt^2 - (dx^2 + dy^2 + dz^2)$. Clearly the spatial part is a 3D Euclidean flat space. We have seen that the spatial metric for a homogeneous, isotropic curved space with a size $R(t)$ that can change in time is

$$ds^2 = R^2(t)\left[\frac{dr^2}{1-kr^2} + r^2(d\theta^2 + \sin^2\theta d\phi^2)\right]. \tag{D.103}$$

Replacing the spatial part of the special relativity metric with (D.103) we have the famous Friedmann–LeMaïtre–Robertson–Walker metric (FLRW), although sometimes LeMaïtre is left out and one refers to the *Friedmann–Robertson–Walker* (FRW) metric,

$$ds^2 = c^2 dt^2 - R^2(t)\left[\frac{dr^2}{1-kr^2} + r^2(d\theta^2 + \sin^2\theta d\phi^2)\right]. \tag{D.104}$$

where $R(t)$ is called the *scale factor* and the constant k can be 0 or ± 1, depending on the curvature.[1] The FRW metric is an exact solution of Einstein's field equations of general relativity; it describes a simply connected, homogeneous, isotropic expanding or contracting universe.

The coordinates r, θ, ϕ are such that the circumference of a circle corresponding to t, r, θ all being constant is given by $2\pi R(t) r$, the area

[1] Note that in the literature it is common to denote the scale factor by the letter a, as we did in Chapter 2 and the following Chapters.

of a sphere corresponding to t and r constant is given by $4\pi R^2(t)r^2$, but the physical radius of the circle and sphere is given by

$$R_U = R(t)\int_0^r \frac{dr'}{\sqrt{1-kr'^2}}. \tag{D.105}$$

For $k = +1$ the Universe is closed (but without boundaries), and $R(t)$ may be interpreted as the "radius" of the Universe at time t. If $k = 0, -1$, the Universe is flat/open and infinite in extent.

The time coordinate t appearing in the FRW metric is the so-called *cosmic time*. It is the time measured on the clock of an observer moving along with the expansion of the universe. The isotropy of the universe makes it possible to introduce such a global time coordinate. Observers at different points can exchange light signals and agree to set their clocks to a common time t when, e.g., their local matter density reaches a certain value. Because of the isotropy of the universe, this density will evolve in the same way at the different locations, and thus once the clocks are synchronized they will stay so.

Writing $ds^2 = g_{\mu\nu}dx^\mu dx^\nu$ and identifying $x = ct$, $x^1 = r$, $x^2 = \theta$, $x^3 = \phi$, we have

$$g_{00} = 1, \quad g_{11} = -\frac{R^2}{1-kr^2}, \quad g_{22} = -R^2r^2, \quad g_{33} = -R^2r^2\sin^2\theta. \tag{D.106}$$

Defining the determinant

$$g \equiv \det g_{\mu\nu} = g_{00}g_{11}g_{22}g_{33} = -\frac{R^6r^4\sin^2\theta}{1-kr^2}. \tag{D.107}$$

(Note that this is not $g = \det g^{\mu\nu}$.) Thus,

$$\sqrt{-g} = \frac{R^3r^2\sin\theta}{\sqrt{1-kr^2}}. \tag{D.108}$$

If $g_{\mu\nu}$ is represented by a matrix $[g_{\mu\nu}]$, then we found previously that $g^{\mu\nu}$ is just the inverse of this metric namely $[g_{\mu\nu}]^{-1}$. For a diagonal matrix (which we have for the FRW metric) each matrix element is simply given by $g^{\mu\nu} = 1/g_{\mu\nu}$. Thus, it is easy to get

$$g^{00} = 1, \quad g^{11} = -\frac{1-kr^2}{R^2}, \quad g^{22} = -\frac{1}{R^2r^2}, \quad g^{33} = -\frac{1}{R^2r^2\sin^2\theta}. \tag{D.109}$$

D.2.1. *Christoffel symbols*

We now calculate the Christoffel symbols using Equation (C.26). Using the symmetry $\Gamma^{\alpha}_{\beta\gamma} = \Gamma^{\alpha}_{\gamma\beta}$ we obtain

$$\begin{aligned}\Gamma^{\alpha}_{\beta\gamma} &= \frac{1}{2} g^{\alpha\epsilon}(g_{\epsilon\beta,\gamma} + g_{\alpha\gamma,\beta} - g_{\beta\gamma,\epsilon}) = \Gamma^{\alpha}_{\gamma\beta} \\ &= \frac{1}{2} g^{\alpha\alpha}(g_{\alpha\beta,\gamma} + g_{\alpha\gamma,\beta} - g_{\beta\gamma,\alpha}),\end{aligned} \tag{D.110}$$

which follows because $g^{\alpha\epsilon} = 0$, unless $\epsilon = \alpha$. ($g^{\mu\nu}$ is a diagonal matrix for the FRW metric.). The only non-zero Christoffel symbols are the following:

$$\Gamma^{1}_{11} = \frac{1}{2} g^{00}(g_{01,1} + g_{01,1} - g_{11,0}) = \frac{1}{2} g_{11,0}, \tag{D.111}$$

because $g_{01} = 0$ and $g_{00} = 1$. This becomes (here we use $c = 1$)

$$\begin{aligned}\Gamma^{0}_{11} &= -\frac{1}{2} g_{11,0} = -\frac{1}{2}\frac{\partial}{\partial t}\left(-\frac{R^2}{1 - kr^2}\right) \\ &= \frac{1}{2}\frac{1}{1 - kR^2}\frac{\partial r^2}{\partial t} = \frac{2R\dot{R}}{2(1 - kr^2)} = \frac{R\dot{R}}{1 - kr^2},\end{aligned} \tag{D.112}$$

because $r \neq r(t)$ and $R = R(t)$. Also,

$$\begin{gathered}\Gamma^{0}_{22} = -\frac{1}{2} g_{22,0} = -\frac{1}{2}\frac{\partial}{\partial t}(-R^2 r^2) = r^2 R\dot{R}, \quad \Gamma^{0}_{33} = r^2 \sin^2\theta R\dot{R}, \\ \Gamma^{1}_{11} = \frac{kr}{1 - kr^2}, \quad \Gamma^{1}_{22} = -r(1 - kr^2), \quad \Gamma^{1}_{33} = -r(1 - kr^2)\sin^2\theta, \\ \Gamma^{1}_{12} = \Gamma^{3}_{13} = \frac{1}{r}, \quad \Gamma^{2}_{33} = -\sin\theta\cos\theta, \quad \Gamma^{3}_{23} = \cot\theta, \quad \text{and} \\ \Gamma^{1}_{01} = \Gamma^{2}_{02} = \Gamma^{3}_{03} = \frac{\dot{R}}{R}.\end{gathered} \tag{D.113}$$

D.2.2. *Ricci tensor*

We can now calculate the Ricci tensor using Eq. (C.40). For the FRWmetric, $R_{\mu\nu} = 0$ for $\mu \neq \nu$, so that the non-zero components are R_{00}, R_{11}, R_{22}, R_{33}. Proceeding we have

$$R_{00} = \frac{1}{\sqrt{-g}}(\Gamma^{\epsilon}_{00}\sqrt{-g}), \quad -(\ln\sqrt{-g})_{,00} - \Gamma^{\epsilon}_{0\theta}\Gamma^{\theta}_{0\epsilon}. \tag{D.114}$$

But $\Gamma^{\epsilon}_{00} = 0$, giving

$$R_{00} = -(\ln\sqrt{-g}), 00 - \Gamma^{0}_{0\theta}\Gamma^{\theta}_{00} - \Gamma^{1}_{1\theta}\Gamma^{\theta}_{01} - \Gamma^{2}_{0\theta}\Gamma^{\theta}_{02} - \Gamma^{3}_{0\theta}\Gamma^{\theta}_{03}, \qquad \text{(D.115)}$$

where we have performed the sum over ϵ. The term $\Gamma^{\theta}_{0\theta} = 0$. In the last three terms we have $\Gamma^{\alpha}_{0\theta}$ where $\alpha = 1, 2, 3$. Now $\Gamma^{\alpha}_{0\theta} = 0$ for $\theta \neq \alpha$, so that we must have $\theta = 1, 2, 3$ in the third, fourth and fifth terms, respectively. Also, the second term contains $\Gamma^{0}_{0\theta}$ which is always 0. Thus,

$$\begin{aligned} R_{00} &= -(\ln\sqrt{-g})_{,00} - (\Gamma^{1}_{01})^2 - (\Gamma^{2}_{02})^2 - (\Gamma^{3}_{03})^2 \\ &= -(\ln\sqrt{-g})_{,00} - 3\left(\frac{\dot{R}}{R}\right)^2. \end{aligned} \qquad \text{(D.116)}$$

But

$$(\sqrt{-g})_{,0}\frac{\partial\sqrt{-g}}{\partial x'} = \frac{\partial\sqrt{-g}}{\partial t} = \frac{r^2\sin\theta}{\sqrt{1-kr^2}}\frac{\partial R^3}{\partial t} = \frac{r^2\sin\theta}{\sqrt{1-kr^2}}3R^2\dot{R}, \qquad \text{(D.117)}$$

so that

$$(\ln\sqrt{-g})_{,00} = 3\frac{\partial}{\partial t}\left(\frac{\dot{R}}{R}\right) = 3\frac{R\ddot{R} - \dot{R}^2}{R^2} = 3\frac{\ddot{R}}{R} - 3\left(\frac{\dot{R}}{R}\right)^2. \qquad \text{(D.118)}$$

We thus find

$$R_{00} = -3\frac{\dot{R}}{R}. \qquad \text{(D.119)}$$

Similarly, one can show that

$$\begin{aligned} R_{11} = \frac{R\ddot{R} - 2\dot{R}^2 + 2k}{1 - kr^2}, \quad & R_{22} = r^2(R\ddot{R} + 2\dot{R}^2 + 2k), \\ R_{33} = r^2\sin^2\theta(R\dot{R} &+ 2\dot{R}^2 + 2k). \end{aligned} \qquad \text{(D.120)}$$

D.2.3. *Ricci scalar and Einstein tensor*

We now calculate the Ricci scalar $\mathcal{R} \equiv R\mathcal{R}^{\alpha}_{\alpha} \equiv g^{\alpha\beta}\mathcal{R}_{\alpha\beta}$. The only non-zero contributions are

$$\begin{aligned} \mathcal{R} &= g^{00}R_{00} + g^{11}R_{11} + g^{22}R_{22} + g^{33}R_{33} \\ &= -6\left[\frac{\ddot{R}}{R} + \left(\frac{\dot{R}}{R}\right)^2 + \frac{k}{R^2}\right]. \end{aligned} \qquad \text{(D.121)}$$

Finally, we calculate the Einstein tensor $G_{\mu\nu} = R_{\mu\nu} - \mathcal{R}g_{\mu\nu}/2$. The only non-zero component is for $\mu = \nu$. We obtain

$$G_{00} = 3\left[\left(\frac{\dot{R}}{R}\right) + \frac{k}{R^2}\right], \quad G_{11} = -\frac{1}{1-kr^2}(2\ddot{R}R + \dot{R}^2 + k),$$

$$G_{22} = -r^2(2\ddot{R}R + \dot{R}^2 + k), \quad G_{33} = -r^2\sin^2\theta(2\ddot{R}R + \dot{R}^2 + k). \tag{D.122}$$

D.2.4. *Energy–momentum tensor*

For a perfect fluid the energy–momentum tensor is given by Eq. (C.44). The tensor for $T_{\mu\nu}$ is given in Eq. (2.82) for the metric of special relativity. For an arbitrary metric in general relativity we have

$$T_{\mu\nu} = (\rho + p)u_\mu u_\nu - pg_{\mu\nu}, \tag{D.123}$$

where we will use $g_{\mu\nu}$ from the FRW model. For a motionless fluid, we have $u^\mu = (c, \mathbf{0})$ or $u_\mu = (c, \mathbf{0}) = (c, \mathbf{0}) = (1, \mathbf{0})$ for $c \equiv 1$. Thus,

$$T_{00} = \rho + p - p = \rho, \tag{D.124}$$

and

$$T_{ii} = -pg_{ii}, \tag{D.125}$$

because $u_i = 0$. Upon substitution of the FRW metric given its equations (D.106), we have

$$T^{\mu\nu} = \begin{pmatrix} \rho & 0 & 0 & 0 \\ 0 & p\dfrac{R^2}{1-kr^2} & 0 & 0 \\ 0 & 0 & pR^2r^2 & 0 \\ 0 & 0 & 0 & pR^2r^2\sin^2\theta \end{pmatrix}. \tag{D.126}$$

D.2.5. *Friedmann equation, again*

Finally, we substitute our results into the Einstein field equations $G_{\mu\nu} = 8\pi G T_{\mu\nu} + \Lambda g_{\mu\nu}$. The $\mu\nu = 00$ component is

$$3\left[\left(\frac{\dot{R}}{R}\right)^2 + \frac{k}{R^2}\right] = 8\pi G\rho + \Lambda, \tag{D.127}$$

giving

$$H^2 = \left(\frac{\dot{R}}{R}\right)^2 = \frac{8\pi G}{3}\rho - \frac{k}{R^2} + \frac{\Lambda}{3}. \tag{D.128}$$

The $\mu\nu = 11$ component is

$$-\frac{1}{1-kr^2}(2\ddot{R}R + \dot{R}^2 + k) = 8\pi G p \frac{R^2}{1-kr^2} + \Lambda\left(-\frac{R^2}{1-kr^2}\right), \tag{D.129}$$

giving

$$2\frac{\ddot{R}}{R} + \left(\frac{\dot{R}}{R}\right)^2 + \frac{k}{R^2} = -8\pi G p + \Lambda. \tag{D.130}$$

Using (D.126) in the equation above we get the Friedmann equation

$$\frac{\ddot{R}}{R} = -\frac{4\pi G}{3}(\rho + p) + \frac{\Lambda}{3}. \tag{D.131}$$

The Friedmann equation are in fact two equations: Eqs. (D.128) and (D.131). They seem to involve four unknowns: the scale factor R, the spatial curvature parameter k, the matter/energy density ρ, and the pressure p. Since they are independent, we have only two equations for four unknowns. A little thinking shows, however, that the spatial curvature parameter is not a big problem. From Eq. (D.128) we can write (with $\Lambda = 0$)

$$k = \frac{8\pi G}{3}\rho(t)R^2(t) - R^2(t), \tag{D.132}$$

where the time argument is explicitly shown. In solving the differential equations we must supply boundary or initial conditions on the solutions. We are free to choose when to impose these boundary conditions, and the most convenient choice is to use the present time, which we will denote by t_0. The scale factor just measures the amount of expansion from a given reference time, and we are therefore free to choose $R(t_0) = R_0 = 1$. With this choice, the present value of the Hubble parameter is given by $H_0 = H(t_0) = \dot{R}(t_0)/R(t_0) = \dot{R}(t_0)$. If we furthermore denote $\rho(t_0) \equiv \rho_0$, we can write

$$k = \frac{8\pi G}{3}\rho_0 - H_0^2. \tag{D.133}$$

We thus see that if we specify initial conditions by choosing values for H_0 and ρ_0, e.g., by using measurements of them, then the spatial curvature is determined for all times.

There still remains three unknown functions $R(t)$, $\rho(t)$, and $p(t)$, and we have only two independent equations for them. Clearly, we need one more equation to close the system. The common way of doing this is by specifying an *equation of state* (EOS), that is, a relation between pressure p and matter/energy density ρ.

Appendix E

References

ABG48 R. A. Alpher, H. Bethe and G. Gamow, *Phys. Rev.* 73, 803 (1948).

Ade11 E. Adelberger, *et al.*, Solar fusion reactions, *Rev. Mod. Phys.* 83, 195 (2011).

AH48 R. A. Alpher and R. Herman, *Nature* 162, 774 (1948).

AI76 P. W. Anderson and N. Itoh, *Nature* 256, 25 (1975).

AL97 F. C. Adams and G. Laughlin, *Rev. Mod. Phys.* 69, 337 (1997).

Ald56 K. Alder, *et al.*, *Rev. Mod. Phys.* 28, 432 (1956).

Ali01 M. Aliotta, *et al.*, *Nucl. Phys.* A690 (2001) 790.

App88 J. H. Applegate, C.J. Hogan, and R.J. Scherrer, *Ap. J.* 329, 592 (1988).

Arn89 W. D. Arnett, J. N. Bahcall, R. P. Kirshner, and S. E. Woosley, *Ann. Rev. Astron. Astrop.* 27, 629 (1989).

As72 M. Abramowitz and I. A. Stegun (eds.), *Handbook of Mathematical Functions*, Dover, New York, 1972.

Ass87 H. J. Assenbaum, K. Langanke, and C. Rolfs, *Z. Phys.* A327 (1987) 461.

Ba10 H. Bateman, *Proc. Cambridge Phil. Soc.* 15 (1910) 423.

Ba96 G. Baur, *et al.*, *Phys. Lett. B* 368, 251 (1996).

Bah69 J. N. Bahcall and R. M. May, *Ap. J.* 155, 501 (1969).

Bah69b J. N. Bahcall and C. P. Moeller, *Ap. J.* 155, 511 (1969).

Bah89 J. N. Bahcall, *Neutrino Astrophysics* (Cambridge: Cambridge University Press), 1989.

Bah98 J. N. Bahcall, P. I. Krastev, and Yu. A. Smirnov, *Phys. Rev.* D58 (1998) 096016.

Bar94 F. C. Barker, *Nucl. Phys.* A575 (1994) 361.

Bay85 G. Baym, E. W. Kolb, L. McLerran, T. P. Walker, and R. L. Jaffe, *Phys. Lett.* B160, 181 (1985).

Ber07 C. A. Bertulani, *Nuclear Physics in a Nutshell*, Princeton Press, Princeton, 2007.

Ber10 G. Bertone (ed.), *Particle Dark Matter: Observations, Models and Searches*, Cambridge Univ. Press, (2010).

Bet37 H. A. Bethe, *Rev. Mod. Phys.* 9, 69 (1937).

Bet38 H. A. Bethe and C. L. Critchfield, *Phys. Rev.* 54, 248 (1938).

Bet39 H. A. Bethe, *Phys. Rev.* 55, 434 (1939).

Bet86 H. Bethe, *Phys. Rev. Lett.* 56, 1305 (1986).

BHK02 C. A. Bertulani, H.-W. Hammer and U. van Kolck, *Nucl. Phys. A* 712 (2002) 37.

Bir23 G. D. Birkhoff, *Relativity and Modern Physics*, Chapters 4, 5, 11, Harvard University Press, Cambridge, MA (1923).

BK02 P. Bedaque and U. van Kolck, *Annu. Rev. Nucl. Part. Sci.* 52 (2002) 339.

BM69 A. Bohr and B. Mottelson, *Nuclear Structure*, Vols. I and II (New York, Benjamin), 1969.

Bo36 N. Bohr, *Nature* 137, 344 (1936).

Boy08 Richard N. Boyd, *An Introduction to Nuclear Astrophysics*, University of Chicago Press (2008).

BP12 C. A. Bertulani and J. Piekarewicz, Editors, *Neutron Star Crust*, Nova Science Publishers, Hauppage, NY, 2012.

Brei59 G. Breit, *Theory of Resonances Reactions and Allied Topics*, Springer Verlag (1959).

BS93 D. M. Brink and G. R. Satchler, *Angular Momentum*, 3rd edition, Clarendon, Oxford, 1993.

Bu57 E. M. Burbidge, G. R. Burbidge, W. A. Fowler, and F. Hoyle, *Rev. Mod. Phys.* 29, 547 (1957).

But00 M. N. Butler and J. W. Chen, *Nucl. Phys.* A675 (2000) 575; M. N. Butler, J. W. Chen and X. Kong, *Phys. Rev. C* 63 (2001) 035501; M. Butler and J. W. Chen, *Phys. Lett.* B520 (2001) 87.

Cha89 S. Chandrasekhar, *Selected Papers on Stellar Structure and Stellar Atmospheres*, The University of Chicago Press (1989).

Cho01 M. Chown, *The Magic Furnace: The Search for the Origins of Atoms*, New York: Oxford University Press (2001).

Cho74 A. Chodos, R. L. Jaffe, K. Johnson, C. B. Thorn, and V. F. Weisskopf, *Phys. Rev.* D9, 3471 (1974).

Cla84 D. D. Clayton, *Principles of Stellar Evolution and Nucleosynthesis*, University of Chicago Press, Chicago, 1984.

Coo57 C. W. Cook, W. A. Fowler, C. C. Lauritsen, and T. Lauritsen, *Phys. Rev.* 107, 508 (1957).

Co06 National Institute of Standards and Technology (NIST), http://physics.nist.gov/

CRS99 J.-W. Chen, G. Rupak and M. J. Savage, *Nucl. Phys.* A653, 386 (1999).

DB10 P. Descouvemont and D. Baye, *Rep. Prog. Phys.* 73, 036301 (2010).

DLL02 P. Danielewicz, R. Lacey, and W. G. Lynch, *Science* 298 (2002) 1592.

Des04 P. Descouvemont, *Phys. Rev. C* 70, 065802 (2004).

Des04b P. Descouvement, A. Adahchour, C. Angulo, Alain Coc, and E. Vangioni-Flam, *At. Data and Nucl. Data Tables* 88 (2004) 203.

Ed60 A. R. Edmonds, *Angular Momentum in Quantum Mechanics*, Princeton University Press, Princeton, (1960).

EG88 J. Eisenberg and W. Greiner, *Excitation Mechanisms of the Nucleus*, Elsevier; 3rd edition (1988).

Ein05 A. Einstein, *Zur Elektrodynamik bewegter Körper*, Annalen der Physik 17 (1905) 891.

Ein06 A. Einstein, Ist die Trägheit eines Körpers von seinem Energieinhalt abhängig?, *Annalen der Physik* 18 (1906) 639.

Ein16 A. Einstein, *Annalen der Physik* 49 (1916) 769.

Fil83 B. W. Filippone *et al.*, *Phys. Rev. Lett.* 50, 412 (1983); *Phys. Rev. C* 28, 2222 (1983).

Fir96 D. J. Fixsen *et al.*, *Ap. J.* 473, 576 (1996).

Fu98 Y. Fukuda *et al.*, *Phys. Rev. Lett.* 81 (1998) 1158.

Fr90 M. W. Friedlander, *Cosmic Rays* (Harvard Univ. Press), 1990.

Fri22 A. Friedmann, *Zeitschrift fuer Physik*, 10, 377 (1922)

Fri24 A. Friedmann, *Zeitschrift fuer Physik*, 21, 326 (1924).

Fri51 E. Frieman and L. Motz, *Phys. Rev.* 89, 648 (1951).

Ga38 G. Gamow, *Phys. Rev.* 55 (1939) 718.

Ga39 G. Gamow, *Phys. Rev.* 55, 718 (1939).

Ga70 G. Gamov, *My World Line*, Viking Press (1970). ISBN 978-0670503766

Gin01 Owen Ginerich, *The Most Brilliant PhD Thesis Ever Written in Astronomy." In The Starry Universe: The Cecilia Payne-Gaposchkin Centenary.*, Schenectady, New York: L. Davis Press, 2001.

Gle90 N. K. Glendenning, *Mod. Phys. Lett.* A5, 2197 (1990).

Gut81 A. Guth, *Phys. Rev. D* 23, 347 (1981).

Gut97 A. H. Guth, *The Inflationary Universe*, Reading, MA: Perseus Books (1997).

Gut98 A. H. Guth, *The Inflationary Universe: Quest for a New Theory of Cosmic Origins*, Vintage Books (1998).

Hae07 P. Haensel, A. Y. Potekhin, and D. G. Yakovlev, *Neutron Stars 1*, Springer Verlag (2007).

Haw75 S. W. Hawking, *Commun. Math. Phys.* 43, 199 (1975).

Hei02 H. Heiselberg , *Neutron Star Masses, Radii and Equation of State*, arXiv:astro-ph/0201465.

HH52 W. Hauser and H. Feshbach, *Phys. Rev.* 87 (1952) 366.

Hoy53 F. Hoyle, D. N. F. Dunbar, W. A. Wenzel, and W. Whaling, *Phys. Rev.* 92, 1095 (1953).

Hoy54 F. Hoyle, *Ap. J. Sup.* 1, 121 (1954).

Hu29 E. P. Hubble, *Publ. Nat. Acad. Sci.* 15 (1929) 168.

Ili07 Christian Iliadis, *Nuclear Physics of Stars*, Wiley-VCH (2007).

IOT94 S. Ichimaru, S. Ogata , K. Tsuruta, *Phys. Rev.* E50, 2977 (1994).

Jac98 J. D. Jackson, *Classical Electrodynamics*, Third Edition, John Wiley & Sons, 1998.

Jar11 N. Jarosik, N. *et al.*, *Ap. J. Sup.* 192, 14 (2011).

JM53 H. L. Johnson and W. W. Morgan, *Astrophys. J.* 117, 313 (1953).

JRS06 P. Jaikumar, S. Reddy, and A. Steiner, *Phys. Rev. Lett.* 96, 041101 (2006).

Jun10 J. Huang, C. A. Bertulani, and V. Guimarães, *At. Data and Nucl. Data Tables*, 96, 824 (2010).

Kr66 J. D. Kraus, *Radio Astronomy*, McGraw-Hill, 1966.

KR99 X. Kong and F. Ravndal, *Nucl. Phys.* A656 (1999) 421; *Nucl. Phys.* A665 (2000) 137; *Phys. Lett.* B470 (1999) 1; *Phys. Rev. C* 64 (2001) 044002.

KS00 L. M. Krauss and G. D. Starkman, *Astrophys. J.* 531, 22 (2000).

KSW96 D. B. Kaplan, M. J. Savage and M. B. Wise, *Nucl. Phys.* B478 (1996) 629.

KT90 E. W. Kolb and M. Turner, *The Early Universe*, Addison-Wesley, Redwood City, CA, 1990.

La80 R. D. Lawson, *Theory of the Nuclear Shell Model*, Oxford: Clarendon Press, 1980.

Lea08 Henrietta S. Leavitt, *Annals of Harvard College Observatory.* LX(IV), 87 (1908).

LeC54 J. M. B. Lang and K. J. Le Coutuer, *Proc. Phys. Soc.* 67A 586, 1954.

Lem27 G. Lemaître, *Annales de la Socit Scientifique de Bruxelles* A47, 49 (1927).

Lem31 G. Lemaître, *Nature* 128, 699 (1931).

Lem31b G. Lemaître, *Monthly Notices of the Royal Astronomical Society* 9, 483 (1931).

Lem33 G. Lemaître, *Annales de la Socit Scientifique de Bruxelles* A53: 51 (1933).

Lep00 G. Peter Lepage, *Lattice QCD for Novices, Proceedings of HUGS 98*, edited by J. L. Goity, World Scientific (2000). arXiv:hep-lat/0506036

LK83 K. Langanke and S. Koonin, *Nucl. Phys.* A410, 334 (1983).

LP01 J. M. Lattimer and M. Prakash, *Astrophys. J.* 550, 426 (2001).

LT58 A. M. Lane and R. G. Thomas, *Rev. Mod. Phys.* 30, 257 (1958).

Mad01 P. Fisher, B. Kayser, K. S. McFarland, *Ann. Rev. Nucl. Part. Sci.* 49 (2001) 481.

Mes61 A. Messiah, *Quantum Mechanics*, John Wiley, 1961.

MJ55 M. G. Mayer and J. H. D. Jensen, *Elementary Theory of Nuclear Shell Structure*, Wiley, New York and Chapman Hall, London, 1955.

Mo95 P. Moller, J. R. Nix, W. D. Myers and W. J. Swiatecki, Atomic Data and Nuclear Data Tables, 59, 185 (1955).

Mor97 R. Morlock, R. Kunz, M. Jaeger, A. Müller, and J. W. Hammer, P. Mohr, H. Oberhummer, A. Mayer G. Staudt, and V. Kölle, *Phys. Rev. Lett.* 79, 3837 (1997).

MS85 S. P. Mikheyev and A. Smirnov, Sov. *J. Nucl. Phys.* 42, 913 (1985).

MT90 A. M. Mukhamedzhanov and N. K. Timofeyuk, *JETP Lett.* 51, 282 (1990).

Mv67 K. W. McVoy, *Ann. Phys.* 43 (1967) 91.

MW79 C. Mahaux and H. A. Weidenmüller, *Ann. Rev. Part. Nucl. Sci.* 29, 1 (1979).

Nav06 P. Navràtil, C. A. Bertulani, and E. Caurier, 2006a, *Physics Lett. B* 634, 191 (2006); *Phys. Rev. C* 73, 065801 (2006).

Nas12 Credit: NASA/WMAP Science Team (2012).

OC07 Dale A. Ostlie and Bradley W. Carrol, *An Introduction to Modern Stellar Astrophysics*, Addison-Wesley (2007).

Opi51 E. J. Öpik, *Proc. Roy. Irish Acad.* A54, 49 (1951).

Pac71 B. Paczynski, *Ann. Rev. Astronomy and Astrophysics* 9, 183 (1971).

Per99 S. Perlmutter *et al.* (The Supernova Cosmology Project), *Astrophys. J.* 517, 565 (1999).

Pod04 Ph. Podsiadlowski, *et al.*, *Astrophys. J.* 607, 17 (2004).

Pra96 M. Prakash, in *The Nuclear Equation of State*, ed. by A. Ausari and L. Satpathy, World Scientific Publishing Co., Singapore, 1996.

PW65 A. A. Penzias and R. W. Wilson, *Astrophys. J.* 142 (1965) 419.

Rap06 W. Rapp, J. Görres, M. Wiescher, H. Schatz, and F. Käppeler, *Astrophys. J.* 653, 474 (2006).

Ree59 H. Reeves and E. E. Salpeter, *Phys. Rev.* 116, 1505 (1959).

Rie98 Adam G. Riess *et al.* (Supernova Search Team) (1998), *Astronomical J.* 116, 1009 (1998).

Rind82 W. Rindler, *Introduction to Special Relativity*, Clarendon Press, Oxford 1982.

Rob35 H. P. Robertson, *Astrophys. J.* 82, 284 (1935).

Rob36 H. P. Robertson, *Astrophys. J.* 83, 187 (1936).

Rob36b H. P. Robertson, *Astrophys. J.* 83, 257 (1936).

RS95 C. Rolfs and E. Somorjai, *Nucl. Instrutn. Methods Phys. Res. Sect. B*, 99, 297 (1995).

RR88 C. Rolfs and W. S. Rodney, *Cauldrons in the Cosmos*, University of Chicago Press, Chicago, 1988.

RT01 T. Rauscher and F.-K. Thielemann, *At. Data Nucl. Data Tables* 79, 47 (2001).

Rup00 G. Rupak, *Nucl. Phys.* A678, 405 (2000).

Rus14 H. N. Russell, *Pop. Astr.* 22 (1914) 275.

SA09 B. D. Fields and S. Sarkar, *Phys. Lett. B* 667, 1 (2008, 2009). Also in http://pdg.lbl.gov (2011).

Sak67 A. D. Sakharov, *JETP Lett.* 5, 24 (1967).

Sal52 E. E. Salpeter, *Phys. Rev.* 88, 547 (1952); *Ap. J.* 115, 326 (1952).

Sal54 E. E. Salpeter, *Aust. J. Phys.* 7, 373 (1954).

Sal57 E. E. Salpeter, *Phys. Rev.* 107, 516 (1957).

Sch16 K. Schwarzschild, *Sitzungsberichte der Kniglich Preussischen Akademie der Wissenschaften* 7, 189 (1916).

Sh14 H. Shapley, *Astrophysical Journal*, 40, 448 (1914); H. Shapley and H. D. Curtis, *Bulletin of the National Research Council*, 2, 171 (1921).

SR04 R. R. Silbar and S. Reddy, *Am. J. Phys.* 72, 892 (2004).

Stei07 G. Steigman, *Annu. Rev. Nucl. Part. Sci.* 57, 463 (2007).

Str08 F. Strieder, *J. Phys. G* 35, 014009 (2008).

Sue56 H. E. Suess and H. C. Urey, *Rev. Mod. Phys.* 28, 53 (1953).

SvH69 E. E. Salpeter and H. M. van Horn, *Astrophys. J.* 155, 183 (1969).

TG04 N. D. Tyson and D. Goldsmith, *Origins: Fourteen Billion Years of Cosmic Evolution*, New York: Norton (2004).

Way02 Patrick A. Wayman, Cecilia Payne-Gaposchkin: Astronomer extraordinaire., *Astronomy and Geophysics* 43, 27 (2002).

WE47 E. P. Wigner and L. Eisenbud, *Phys. Rev.* 72, 29 (1947).

Wei35 C. F. von Weizsäcker, *Zeit. Phys. (Journal of Physics)* 96, 431 (1935).

Wein08 S. Weinberg, *Cosmology*, Oxford University Press, 2008.

Wik12 Credit: http://en.wikipedia.org/wiki/File:Hyperbolic_triangle.svg (2012).

Wil40 W. A. Wildhack, *Phys. Rev.* 57 (1940) 81.

Wit84 E. Witten, *Phys. Rev.* D30, 272 (1984).

Wol78 L. Wolfenstein, *Phys. Rev.* D17, 2369 (1978).

Won94 C. Y. Wong, *Introduction to High Energy Heavy Ion Collisions*, World Scientific, Singapore, 1994.

Woo86 S. E. Woosley and T. A. Weaver, *Ann. Rev. Astr. Astrophys.* 24, 205 (1986).

WW40 V. F. Weisskopf and D. H. Ewing, *Phys. Rev.* 57, 935 (1940).

Yao06 W.-M. Yao *et al.*, *J. Phys. G* 33 (2006) 1.

You92 Hugh Young, *University Physics*, 8th ed., Addison-Wesley, Reading MA, 1992, Sec. 19–5, *Speed of a Longitudinal Wave.*

Zel68 Ya. B. Zeldovich, *Sov. Phys. — Uspekhi* 11, 381 (1968).

Zw33 F. Zwicky, *Helv. Phys. Acta* 6, 110 (1933).

Index